Lecture Notes in Computer Science 11871

More information about this series at http://www.springer.com/series/7409

Hujun Yin · David Camacho ·
Peter Tino · Antonio J. Tallón-Ballesteros ·
Ronaldo Menezes · Richard Allmendinger (Eds.)

Intelligent Data Engineering and Automated Learning – IDEAL 2019

20th International Conference
Manchester, UK, November 14–16, 2019
Proceedings, Part I

 Springer

Editors
Hujun Yin (iD)
University of Manchester
Manchester, UK

David Camacho (iD)
Technical University of Madrid
Madrid, Spain

Peter Tino (iD)
University of Birmingham
Birmingham, UK

Antonio J. Tallón-Ballesteros (iD)
University of Huelva
Huelva, Spain

Ronaldo Menezes
University of Exeter
Exeter, UK

Richard Allmendinger (iD)
University of Manchester
Manchester, UK

ISSN 0302-9743   ISSN 1611-3349 (electronic)
Lecture Notes in Computer Science
ISBN 978-3-030-33606-6   ISBN 978-3-030-33607-3 (eBook)
https://doi.org/10.1007/978-3-030-33607-3

LNCS Sublibrary: SL3 – Information Systems and Applications, incl. Internet/Web, and HCI

This Springer imprint is published by the registered company Springer Nature Switzerland AG
The registered company address is: Gewerbestrasse 11, 6330 Cham, Switzerland

Preface

This year saw the 20th edition of the International Conference on Intelligent Data Engineering and Automated Learning (IDEAL 2019), held for the second time in Manchester, UK – the birthplace of one of the world's first electronic computers as well as artificial intelligence (AI) marked by Alan Turing's seminal and pioneering work. The IDEAL conference has been serving its unwavering role in data analytics and machine learning for the last 20 years. It strives to provide an *ideal* platform for the scientific communities and researchers from near and far to exchange latest findings, disseminate cutting-edge results, and to forge alliances on tackling many real-world challenging problems. The core themes of the IDEAL 2019 include big data challenges, machine learning, data mining, information retrieval and management, bio-/neuro-informatics, bio-inspired models (including neural networks, evolutionary computation, and swarm intelligence), agents and hybrid intelligent systems, real-world applications of intelligent techniques, and AI.

In total, 149 submissions were received and subsequently underwent rigorous peer reviews by the Program Committee members and experts. Only the papers judged to be of the highest quality and novelty were accepted and included in the proceedings. These volumes contain 94 papers (58 for the main track and 36 for special sessions) accepted and presented at IDEAL 2019, held during November 14–16, 2019, at the University of Manchester, Manchester, UK. These papers provided a timely snapshot of the latest topics and advances in data analytics and machine learning, from methodologies, frameworks, and algorithms to applications. IDEAL 2019 enjoyed outstanding keynotes from leaders in the field, Thomas Bäck of Leiden University and Damien Coyle of University of Ulster, and an inspiring tutorial from Peter Tino of University of Birmingham.

IDEAL 2019 was hosted by the University Manchester and was co-sponsored by the Alan Turing Institute and Manchester City Council. It was also technically co-sponsored by the IEEE Computational Intelligence Society UK and Ireland Chapter.

We would like to thank our sponsors for their financial and technical support. We would also like to thank all the people who devoted so much time and effort to the successful running of the conference, in particular the members of the Program Committee and reviewers, organizers of the special sessions, as well as the authors who contributed to the conference. We are also very grateful to the hard work by the local Organizing Committee at the University of Manchester, in particular, Yao Peng and

Jingwen Su for checking through all the camera-ready files. Continued support, sponsorship, and collaboration from Springer LNCS are also greatly appreciated.

September 2019
Hujun Yin
David Camacho
Peter Tino
Antonio J. Tallón-Ballesteros
Ronaldo Menezes
Richard Allmendinger

Organization

General Chairs

Hujun Yin	University of Manchester, UK
David Camacho	Universidad Politecnica de Madrid, Spain
Peter Tino	University of Birmingham, UK

Programme Co-chairs

Hujun Yin	University of Manchester, UK
David Camacho	Universidad Politecnica de Madrid, Spain
Antonio J. Tallón-Ballesteros	University of Huelva, Spain
Peter Tino	University of Birmingham, UK
Ronaldo Menezes	University of Exeter, UK
Richard Allmendinger	University of Manchester, UK

International Advisory Committee

Lei Xu	Chinese University of Hong Kong, Hong Kong, China, and Shanghai Jiaotong University, China
Yaser Abu-Mostafa	CALTECH, USA
Shun-ichi Amari	RIKEN, Japan
Michael Dempster	University of Cambridge, UK
Francisco Herrera	Autonomous University of Madrid, Spain
Nick Jennings	University of Southampton, UK
Soo-Young Lee	KAIST, South Korea
Erkki Oja	Helsinki University of Technology, Finland
Latit M. Patnaik	Indian Institute of Science, India
Burkhard Rost	Columbia University, USA
Xin Yao	Southern University of Science and Technology, China, and University of Birmingham, UK

Steering Committee

Hujun Yin (Chair)	University of Manchester, UK
Laiwan Chan (Chair)	Chinese University of Hong Kong, Hong Kong, China
Guilherme Barreto	Federal University of Ceará, Brazil
Yiu-ming Cheung	Hong Kong Baptist University, Hong Kong, China
Emilio Corchado	University of Salamanca, Spain
Jose A. Costa	Federal University of Rio Grande do Norte, Brazil
Marc van Hulle	KU Leuven, Belgium

Samuel Kaski   Aalto University, Finland
John Keane   University of Manchester, UK
Jimmy Lee   Chinese University of Hong Kong, Hong Kong, China
Malik Magdon-Ismail   Rensselaer Polytechnic Institute, USA
Peter Tino   University of Birmingham, UK
Zheng Rong Yang   University of Exeter, UK
Ning Zhong   Maebashi Institute of Technology, Japan

Publicity Co-chairs/Liaisons

Jose A. Costa   Federal University of Rio Grande do Norte, Brazil
Bin Li   University of Science and Technology of China, China
Yimin Wen   Guilin University of Electronic Technology, China

Local Organizing Committee

Hujun Yin   Richard Allmendinger
Richard Hankins   Yao Peng
Ananya Gupta   Jingwen Su
Mengyu Liu

Program Committee

Ajith Abraham   Josep Carmona
Jesus Alcala-Fdez   Mercedes Carnero
Richardo Aler   Carlos Carrascosa
Davide Anguita   Andre de Carvalho
Anastassia Angelopoulou   Joao Carvalho
Ángel Arcos-Vargas   Pedro Castillo
Romis Attux   Luís Cavique
Martin Atzmueller   Darryl Charles
Dariusz Barbucha   Richard Chbeir
Mahmoud Barhamgi   Songcan Chen
Bruno Baruque   Xiaohong Chen
Carmelo Bastos Filho   Sung-Bae Cho
Lordes Borrajo   Stelvio Cimato
Zoran Bosnic   Manuel Jesus Cobo Martin
Vicent Botti   Leandro Coelho
Edyta Brzychczy   Carlos Coello Coello
Fernando Buarque   Roberto Confalonieri
Andrea Burattin   Rafael Corchuelo
Robert Burduk   Francesco Corona
Aleksander Byrski   Nuno Correia
Heloisa Camargo   Luís Correia

Paulo Cortez
Jose Alfredo F. Costa
Carlos Cotta
Raúl Cruz-Barbosa
Alfredo Cuzzocrea
Bogusław Cyganek
Ireneusz Czarnowski
Ernesto Damiani
Amit Kumar Das
Bernard De Baets
Javier Del Ser
Boris Delibašić
Fernando Díaz
Juan Manuel Dodero
Jose Dorronsoro
Dinu Dragan
Gérard Dreyfus
Jochen Einbeck
Florentino Fdez-Riverola
Joaquim Filipe
Juan J. Flores
Simon James Fong
Pawel Forczmanski
Giancarlo Fortino
Felipe M. G. França
Dariusz Frejlichowski
Hamido Fujita
Marcus Gallagher
Yang Gao
Salvador Garcia
Pablo García Sánchez
Luis Javier Garcia Villalba
María José Ginzo Villamayor
Fernando Gomide
Antonio Gonzalez-Pardo
Anna Gorawska
Marcin Gorawski
Juan Manuel Górriz
Manuel Graña
Maciej Grzenda
Jerzy Grzymala-Busse
Pedro Antonio Gutierrez
Barbara Hammer
Julia Handl
Richard Hankins
Ioannis Hatzilygeroudis

Álvaro Herrero
J. Michael Herrmann
Ignacio Hidalgo
James Hogan
Jaakko Hollmén
Wei-Chiang Samuelson Hong
Vahid Jalali
Dariusz Jankowski
Piotr Jedrzejowicz
Vicente Julian
Rushed Kanawati
Mario Koeppen
Mirosław Kordos
Marcin Korzeń
Dariusz Krol
Pawel Ksieniewics
Raul Lara-Cabrera
Bin Li
Lei Liu
Wenjian Luo
José F. Martínez-Trinidad
Giancarlo Mauri
Cristian Mihaescu
Boris Mirkin
José M. Molina
João Mp Cardoso
Grzegorz J. Nalepa
Valery Naranjo
Susana Nascimento
Tim Nattkemper
Antonio Neme
Rui Neves Madeira
Ngoc-Thanh Nguyen
Paulo Novais
Fernando Nuñez
Ivan Olier-Caparroso
Eva Onaindia
Sandra Ortega-Martorell
Vasile Palade
Jose Palma
Juan Pavón
Yao Peng
Carlos Pereira
Barbara Pes
Marco Platzner
Paulo Quaresma

Cristian Ramírez-Atencia
Ajalmar Rêgo Da Rocha Neto
Izabela Rejer
Victor Rodriguez Fernandez
Matilde Santos
Pedro Santos
Jose Santos
Rafal Scherer
Ivan Silva
Leandro A. Silva
Dragan Simic
Anabela Simões
Marcin Szpyrka
Jesús Sánchez-Oro
Ying Tan
Qing Tian
Renato Tinós
Stefania Tomasiello

Pawel Trajdos
Carlos M. Travieso-González
Bogdan Trawinski
Milan Tuba
Turki Turki
Eiji Uchino
Carlos Usabiaga Ibáñez
José Valente de Oliveira
Alfredo Vellido
Juan G. Victores
José R. Villar
Lipo Wang
Tzai-Der Wang
Dongqing Wei
Raymond Kwok-Kay Wong
Michal Wozniak
Xin-She Yang
Huiyu Zhou

Additional Reviewers

Sabri Allani
Ray Baishkhi
Samik Banerjee
Avishek Bhattacharjee
Nurul E'zzati Binti Md Isa
Daniele Bortoluzzi
Anna Burduk
Jose Luis Calvo Rolle
Walmir Caminnhas
Meng Cao
Hugo Carneiro
Giovanna Castellano
Hubert Cecotti
Carl Chalmers
Lei Chen
Zonghai Chen
Antonio Maria Chiarelli
Bernat Coma-Puig
Gabriela Czanner
Mauro Da Lio
Sukhendu Das
Klaus Dietmayer
Paul Fergus
Róża Goścień
Ugur Halici

Hongmei He
Eloy Irigoyen
Suman Jana
Ian Jarman
Jörg Keller
Bartosz Krawczyk
Weikai Li
Mengyu Liu
Nicolás Marichal
Wojciech Mazurczyk
Philippa Grace McCabe
Eneko Osaba
Kexin Pei
Mark Pieroni
Alice Plebe
Girijesh Prasad
Sergey Prokudin
Yu Qiao
Juan Rada-Vilela
Dheeraj Rathee
Haider Raza
Yousef Rezaei Tabar
Patrick Riley
Gastone Pietro Rosati Papini
Salma Sassi

Ivo Siekmann
Mônica Ferreira Da Silva
Kacper Sumera
Yuchi Tian
Mariusz Topolski

Marley Vellasco
Yu Yu
Rita Zgheib
Shang-Ming Zhou

Special Session on Fuzzy Systems and Intelligent Data Analysis

Organizers

Susana Nascimento	NOVA University, Portugal
Antonio J. Tallón-Ballesteros	University of Huelva, Spain
José Valente de Oliveira	University of Algarve, Portugal
Boris Mirkin	National Research University Moscow, Russian

Special Session on Machine Learning towards Smarter Multimodal Systems

Organizers

Nuno Correia	NOVA University of Lisboa, Portugal
Rui Neves Madeira	NOVA University of Lisboa, Polytechnic Institute of Setúbal, Portugal
Susana Nascimento	NOVA University, Portugal

Special Session on Data Selection in Machine Learning

Organizers

Antonio J. Tallón-Ballesteros	University of Huelva, Spain
Raymond Kwok-Kay Wong	University of New South Wales, Australia
Ireneusz Czarnowski	University of Novi Sad, Serbia
Simon James Fong	University of Macau, Macau, China

Special Session on Machine Learning in Healthcare

Organizers

Ivan Olier	Liverpool John Moores University, UK
Sandra Ortega-Martorell	Liverpool John Moores University, UK
Paulo Lisboa	Liverpool John Moores University, UK
Alfredo Vellido	Universitat Politècnica de Catalunya, Spain

Special Session on Machine Learning in Automatic Control

Organizers

Matilde Santos	University Complutense of Madrid, Spain
Juan G. Victores	University Carlos III of Madrid, Spain

Special Session on Finance and Data Mining

Organizers

Fernando Núñez Hernández	University of Seville, Spain
Antonio J. Tallón-Ballesteros	University of Huelva, Spain
Paulo Vasconcelos	University of Porto, Portugal
Ángel Arcos-Vargas	University of Seville, Spain

Special Session on Knowledge Discovery from Data

Organizers

Barbara Pes	University of Cagliari, Italy
Antonio J. Tallón-Ballesteros	University of Huelva, Spain
Julia Handl	University of Manchester, UK

Special Session on Machine Learning Algorithms for Hard Problems

Organizers

Pawel Ksieniewicz	Wroclaw University of Science and Technology, Poland
Robert Burduk	Wroclaw University of Science and Technology, Poland

Contents – Part I

Contents – Part II

Special Session on Finance and Data Mining

Special Session on Knowledge Discovery from Data

Special Session on Machine Learning Algorithms for Hard Problems

Orchids Classification Using Spatial Transformer Network with Adaptive Scaling

Watcharin Sarachai$^{(\boxtimes)}$, Jakramate Bootkrajang$^{(\boxtimes)}$,
Jeerayut Chaijaruwanich$^{(\boxtimes)}$, and Samerkae Somhom$^{(\boxtimes)}$

Department of Computer Science, Chiang Mai University, Chiang Mai, Thailand
sarachaii@gmail.com, {jakramate.b,jeerayut.c,samerkae.s}@cmu.ac.th

Abstract. The orchids families are large, diverse flowering plants in the tropical areas. It is a challenging task to classify orchid species from images. In this paper, we proposed an adaptive classification model of the orchid images by using a Deep Convolutional Neural Network (D-CNN). The first part of the model improved the quality of input feature maps using an adaptive Spatial Transformer Network (STN) module by performing a spatial transformation to warp an input image which was split into different locations and scales. We applied D-CNN to extract the image features from the previous step and warp into four branches. Then, we concatenated the feature channels and reduced the dimension by an estimation block. Finally, the feature maps would be forwarded to the prediction network layers to predict the orchid species. We verified the efficiency of the proposed method by conducting experiments on our data set of 52 classes of orchid flowers, containing 3,559 samples. Our results achieved an average of 93.32% classification accuracy, which is higher than the existing D-CNN models.

Keywords: Orchids images · Classification · Deep learning

1 Introduction

Classification of orchid species is a challenging task because of its diversity and widespread families of flowering plants. The orchid species can be classified by its genes and visual characteristics such as shape, color, seed, and roots. The orchid specialists can classify the orchid species by their unique floral patterns. However, the flowers of several orchid species bare many similarities even though they are not the same species, so that the orchid specialists would require a substantial amount of time and efforts to classify them.

Hobbyist is using cameras, mobile phones, and other devices to capture a flower image. In a field of computer vision, we can attempt to classify the orchid species by using the captured flower images. There are various methods in the field of computer vision that could be used to classify the orchid species by using

© Springer Nature Switzerland AG 2019
H. Yin et al. (Eds.): IDEAL 2019, LNCS 11871, pp. 1–10, 2019.
https://doi.org/10.1007/978-3-030-33607-3_1

flower images. Several features, including color, texture, shape, and statistical information, are widely used to identify the difference among species of flowering plants. These features are processed by traditional hand-crafted techniques, which typically used in classification tasks such as a histogram of oriented gradients (HOGs), scale-invariant feature transform (SIFT), and speeded up robust features. The selected feature will be used as an input of a classifier such as the support vector machines (SVMs) [10,11,14,21,22], the neural network (NN) [4,17] and k-Nearest Neighbor (kNN) classifier [16]. To improve the accuracy of the learning algorithm, before extracting the features, an input image will be segmented a flower object from its background [7,13,14,16,17,21].

It is difficult to used traditional hand-crafted discriminative features to extract a feature of orchid images, due to the diversity and widespread families of orchid plants. Moreover, the specific hand-made features might not be generalized to every class of orchid species. Therefore, the deep convolutional neural network (D-CNN) technique have recently been used by many researchers because the accuracy is outperforming the traditional hand-crafted features. In this paper, we adopt the STN [8] to warp and scale the input image before sending to the D-CNN for handling the sophisticated features across overall of orchid classes. The STN is a differentiable module that warped the essential part of input feature maps. STN can be inserted to existing D-CNN architectures and simplified the complicated of orchid image input of the flower objects which vary in size and location.

The main contributions of this paper are the following: First, we proposed a methodology which tailors toward orchid image classification task. Second, we formulate the boundary lost function to constrain the value of location and scale within each boundary of the image region. Third, we purposed the training procedure to be used to train the orchids classification model to achieve higher accuracy. Lastly, we collected and curated a new corpus of orchid flower images in the northern part of Thailand.

The overview of this paper is as follows: We describe the background, and the related work of flowers classification approaches in Sect. 2. The orchid dataset is described in Sect. 3. Our methodology and learning algorithm are described in Sect. 4. Experimental results and performance evaluation are explained in Sect. 5.

2   Related Works

Since the literature on image classification is vast, in this section, we focused on natural flower and orchid flowers classification approaches. The traditional approaches extracted the features from an input image that comes from the spatial region of the image. These features are essential for distinctive a characteristic of each object class. The authors in [2] used color and segmentation-based fractal texture analysis (SFTA) to extract the features of Paphiopedilum orchid flowers and used as the input for a neural network. They achieved 97.64% of recognition accuracy. However, instead of extracted a feature by hands, the convolutional layers of D-CNN are modules for learning to find the most important feature in an input image. The first convolutional layers extracted the general feature, and the second layers processed a feature map abstractly. [20] used an inception-v3 [19] model that included the learn-able convolutional layers to extract the features and forward it to the fully-connected layer to classify the Oxford 17 [13] and Oxford 102 [14] flower dataset. [12] used prior knowledge that the flower has high brightness, combined with saliency and luminance map before forwarded to the D-CNN as inputs which can enhance performance on flower classifications.

Furthermore, several of researches did the classification by locating the saliency object in the background image and cropped them before being forward to the D-CNN, in [3], they used D-CNN with linear regression to predict the bounding box and confident score, which result in K-dimension output to find a location of a possible object in the input image. [9] utilized the STN to warped part of the input image before forwarding to the classification method. In our work, we utilized the STN as a differentiable attention mechanism to warp a part of the image, which contains scale and position. The action of STN conditioned on individual data samples, with the appropriate behavior learned during end-to-end training for the orchid class. The STN module learned to locate the required features in the input image and warped set of attention output images before forwarding to D-CNN to classify the orchid species.

3   The Orchid Dataset

We have created orchid flower dataset of 52 classes, the orchid flowers selected from the northern part of Thailand. The dataset contains Thai native orchid flowers, and each class contains at least 20 samples. For the Paphiopedilums which often called Slipper Orchids or Lady's Slippers, we merge our dataset with the dataset from [2]. The orchid dataset including of 52 species and the visual characteristics of the flower are varying in terms of shape, color, texture, flower size, and the other parts of the orchid plant like a leaf, inflorescences, roots, and surroundings. All images are taken from many devices such as a digital camera, a mobile phone, and other equipment. The orchids dataset contains 3,559 flower images from 52 categories. We divided all datasets into a training set and a test set with a ratio of 80% and 20% respectively. Our orchids classification dataset is publicly available upon request.

4   Orchids Classification

We apply an adaptive model of the spatial transformer network to classify orchid images [8], shown in Fig. 1(a). The first module is a spatial transformer network (STN). The STN warped part of an input image into a different location and scale, the dimension of each warped output images is $299 \times 299 \times 3$ and four different scales. Let $N = 4$; the STN produce N warped output images that will be forward to the D-CNN modules as the input, which is a learnable module to extract the feature map from an input image. All of D-CNN configuration is derived from inception-v3 models [19] of TensorFlow [1] platform exclude the final fully-connected layers. All of inception-v3 are initialized with the learned model [19], which is a pre-trained model on the ILSVRC-2012 [15] image classification dataset. All of the feature maps from each D-CNN branch will be concatenated and send to the estimation block (Fig. 2) which is an inception module from [19] reduced the number of feature map channels before forwarding to the prediction network.

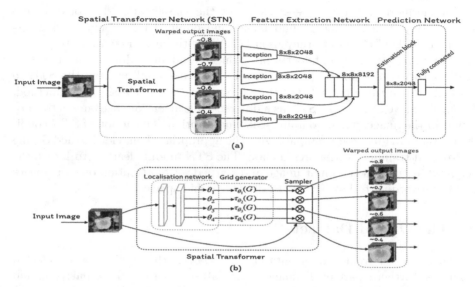

Fig. 1. (a) The orchids classification models. (b) The STN module warped the part of input image into difference location and scale.

4.1 Spatial Transformer Networks (STN)

The STN, shown in Fig. 1(b). The spatial information of the input images will be located and transformed a set of cropped output images which depend on each input image. The image has any channel C (red, green, blue) that will be warped by STN at the same location to every channel of an input image. The computation of STN begins with the localization network module that takes the input image, and through the D-CNN output parameters of spatial information that should be applied to the input image and produce warped output images, this gives a transformation conditional that depend on the input image features. The localization network produced set of N parameters, and the grid generator used it to create N set of sample grid which is a set of a point where an input image should be sampled to produce the transformed output. The sampler module takes all of the sample grids and input image U as input to produce warped output images.

The localization network $f_{loc}()$ takes the input image $U \in \mathbb{R}^{H \times W \times C}$ (width W, height H, channel C) produced the parameter θ, thus $\theta = f_{loc}(U)$. The transformation $\mathcal{T}_\theta$ takes the parameter θ as input and applied to the input image to produce a warped output through *grid generator* and *sampler* module. Each size of parameters θ is 3-dimension corresponding to the regression output of coordination x, y and scale s to generate the number of N sample grid. The output size of $f_{loc}(U)$ will be $4 \times \theta$ which the result is 12-dimension output say $\theta_j; j = 0, 1, 2, ..., 11$ (j denotes number of 12-dimension). Thus, the output of the localization network is $f_{loc}(U) = \sum_{n=0}^{N-1} \{\theta_{n*3+1}^x, \theta_{n*3+2}^y, \theta_{n*3+3}^s\}$.

The localization network is a D-CNN derive from the inception-v3 [19] model including five convolution layers and two max-pooling in the first part and the second part including of 11 inception blocks replace the final fully-connected layers with the two hidden layers: 128-dimension and 12-dimension fully-connected layer. The final 12-dimension output will be performed with L2-NORM and tanh activation. The L2-NORM improved the generalization of parameters θ and tanh activation constraint bounding box of each output value to $-1 \lneqq \theta \lneqq 1$.

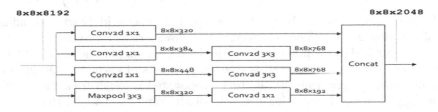

Fig. 2. Estimation block reduce dimension from $8 \times 8 \times 8192$ to $8 \times 8 \times 2048$.

The STN module of orchids classification allowing a network to be spatially invariant to the position of features, to improve the overall of model accuracy, we conducted series of experiments to find the best values of the translation-invariant scale and found that the best value is 80%, 70%, 60%, and 40%. Thus, we fixed the value to 0.8, 0.7, 0.6, and 0.4 to allows each warped window from STN varies at a different scale. Let $s_1 = 0.8$, $s_2 = 0.7$, $s_3 = 0.6$ and $s_4 = 0.4$, thus, the set of translation-invariant scale is $\mathbb{S} = \{s_1, s_2, s_3, s_4\}$ respectively. We adopt the parameterization of [9] to regress offset from a location and translation-invariant scale, which outputs the 12-dimension of localization network, as shown in Eq. 1.

$$f_{loc}(U)' = \sum_{n=0}^{N-1} \{\theta_{n*3+1}^{x'}, \theta_{n*3+2}^{y'}, \theta_{n*3+3}^{s'}\} \tag{1a}$$

$$f_{loc}(U)' = \sum_{n=0}^{N-1} \left\{ \frac{\theta_{n*3+1}^x}{e^{\theta_{n*3+3}^s} \times s_{n+1}}, \frac{\theta_{n*3+2}^y}{e^{\theta_{n*3+3}^s} \times s_{n+1}}, e^{\theta_{n*3+3}^s} \times s_{n+1} \right\} \tag{1b}$$

The input image U will be warped by the sampling grid centered at a particular location parameterize by parameters $\theta^{x'}$, $\theta^{y'}$ and $\theta^{s'}$.

In order to warp the input image at the location $(\theta^{x'}, \theta^{y'})$ and scale by $\theta^{s'}$, the general grid will be created, the size of the general grid corresponds to width w_{out} and high h_{out} of warped output image, where (x^t, y^t) is the target coordinate of general grid in the warped output, (x^{src}, y^{src}) are the source coordinate in the input image, such that the target and source are normalized coordinates: $-1 \leq (x^t, y^t) \leq 1$ and $-1 \leq (x^{src}, y^{src}) \leq 1$.

The transformation T_θ is a 2D affine transformation A_θ takes the output from $f_{loc}(U)'$ that constrained by location $\theta^{x'}$, $\theta^{y'}$ and scale $\theta^{s'}$, each transformation of warped output image I are:

$$A_\theta = \sum_{n=0}^{N-1} \left\{ \begin{bmatrix} \theta^{s'}_{n*3+3} & 0 & \theta^{x'}_{n*3+1} \\ 0 & \theta^{s'}_{n*3+3} & \theta^{y'}_{n*3+2} \end{bmatrix}_n \right\} \tag{2}$$

and the pointwise transformation is:

$$I = \sum_{n=0}^{N-1} \left\{ \begin{pmatrix} x^{src}_n \\ y^{src}_n \end{pmatrix} \right\} = \sum_{n=0}^{N-1} \{T_{\theta_n}(G)\} = \sum_{n=0}^{N-1} \left\{ A_{\theta_n} \begin{pmatrix} x^t \\ y^t \\ 1 \end{pmatrix} \right\} \tag{3}$$

Each sampling grid is the result of warping regular grid G with an affine transformation $T_{\theta_n}(G)$. The transformation allows cropping, translation, and scale to be applied to the input image varies by regression parameters $\theta^{x'}, \theta^{y'}, \theta^{s'}$ which each requires six parameters (the six elements of each A_θ). During the training time, the gradients will be back-propagated through from the sample point $T_{\theta_n}(G)$ to the localization network. In the sampler module, the warped output images will be performed by an apply set of sampling point $T_{\theta_n}(G)$ to an input image U. We followed this approach from [8].

4.2 Features Extraction Network and Prediction Network

Feature extraction network including of N-branches of D-CNN. In Fig. 1(a). Each branch derives a configuration from inception-v3 model [19] but excluded the final fully-connected layers. It takes a warped output image from the STN module produces the output of $8 \times 8 \times 2048$ dimension. Each output of N-branches of D-CNN are concatenated, which resulted in $8 \times 8 \times 8192$ dimension, and it would be reduced the channels from 8192 to 2048 by the estimation block. The prediction network downsamples the feature maps from $8 \times 8 \times 2048$ to $1 \times 1 \times 2048$ by global average pooling before forwarding to the fully-connected layer with a final 52-way softmax output to predict the orchid species.

4.3 Training Objective

The overall objective loss function is a softmax cross entropy loss (e) and the boundary loss (b): $L(x, c, \theta^{x'}, \theta^{y'}, \theta^{s'}) = L_e(x, c) + \alpha L_b(\theta^{x'}, \theta^{y'}, \theta^{s'})$.

The valid value of $\theta^{x'}$, $\theta^{y'}$ and $\theta^{s'}$ that constrained the bounding box within the boundary of the input image such that the valid values are $(|\theta^{x'}| + \theta^{s'}) \leq 1$

and $(|\theta^{y\prime}|+\theta^{s\prime}) \leqq 1$ and the weight term α is set to 1. Thus, we control the warped output image bounding box within the image input region by lost function:

$$L_b = \frac{1}{N}\sum_{n=0}^{N-1}\left(max(0,(|\theta^{x\prime}_{n*3+1}|+\theta^{s\prime}_{n*3+3})-1)\right)^2 + \frac{1}{N}\sum_{n=0}^{N-1}\left(max(0,(|\theta^{y\prime}_{n*3+2}|+\theta^{s\prime}_{n*3+3})-1)\right)^2 \quad (4)$$

4.4   Training Procedure

To train the orchids classification, first, we excluded the STN module by feeding the input images to N-branches of D-CNN with full size without warped and scaling. In this step, we trained the estimation block and 52-way fully-connected layers but kept the rest of the module variables unchangeable. All branches of D-CNN initialized with the pre-trained variable from models of [19] which trained from ILSVRC-2012 [15] image classification dataset to produce the feature maps output. The final 52-way prediction output will be trained first because of the variables of the STN module updated by back-propagation of the learning algorithm, which needs the final 52-way prediction output to be prepared.

Second, we adopt the localization network configuration with the inception-v3 model by replacing the final layers with the 12-dimension fully-connected layer. In this step, we integrated all of orchids classification modules and trained only 12-dimension fully-connected layer of the localization network module by keeping other variables unchangeable. The localization network initialized with pre-trained variables from [19] model. The updated gradient would flow back from 52-way through all networks to the localization network layers, but the variable update will have happened only in 12-dimension fully-connected layers which trained by stochastic gradient descent and updated the weight by momentum optimizer.

Finally, we would fit the overall of orchids classification by training all modules together. The networks are trained with back-propagation, using stochastic gradient descent, and updated variables by RMSProp optimizer.

5   Experiments

We conducted experiments on existing models and the orchids classification models. The architecture of orchids classification combined two existing methodology STN and D-CNN in a way that the STN warped an important part of input images before forwarding to the D-CNN as input. The orchids classification is an end-to-end training model by adjusting the parameters by back-propagation techniques. The STN module can be trained and used complicated features of orchid flower images. This module can produce significant parameter θ that would be used to warp important parts of orchid flower image and simplify the classification of the D-CNN network. We create three models of orchids classification to evaluate the performances on three datasets. The networks architecture and training parameters were initialized with the same settings except for the last fully-connected layer. The last layer has a final n-way softmax output, that depends on the number of classes of each dataset. The input image is an RGB image resized to 299 × 299 before feeding to the network. The dropout applied on the outputs of the estimation block before forwarding to the fully-connected layers. All our experiments are performed on a server with NVIDIA Tesla M10 GPU 8G memory.

Table 1. Accuracy of our approach compared to existing models on Oxford 17, the Oxford 102 and orchids dataset.

Method	Segmentation	Oxford 17	Oxford 102	Orchids
FCN-CNN w/augmentation [7]	Y	98.50%	97.10%	–
Nilsback and Zisserman [14]	Y	88.33%	72.80%	–
Inception-V1 (our) [18]	N	95.22%	95.72%	86.89%
Inception-V3 [20]	N	95.00%	94.00%	–
Inception-V3 (our) [19]	N	97.43%	96.91%	88.24%
ResNet-V1-50 (our) [5]	N	97.01%	96.02%	89.05%
ResNet-V2-50 (our) [6]	N	97.01%	95.48%	88.11%
Our model	N	98.23%	98.15%	**93.32%**

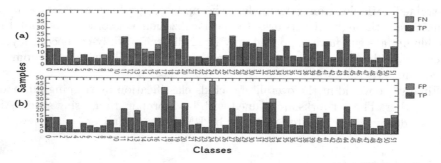

Fig. 3. (a) The recall of orchids classification models. (b) The precision of orchids classification models.

We compare the orchids classification with several existing models, as shown in Table 1. The accuracy of our model on Oxford 17 and Oxford 102 datasets are in the same order of magnitude but achieved a higher accuracy on orchids dataset, which has the best overall performance. Two aspects of a confusion matrix can determine the performance of our model for orchids on 52 classes. In Fig. 3(a), the highest false-negative value of recall has on class number 24 and in Fig. 3(b), the false-positive value of precision has on class number 18 and 24 are higher than other classes. Those two values mean that the features extracted from those classes are shared similar features with other classes.

6   Conclusion

In this paper, we used an adaptive STN module to extract a feature from orchid images to warp and scale the input image before forwarding to the D-CNN. The four branches of D-CNN models are designed to take the previous input feature maps, which each branch extracts the essential features from orchid images. The output of D-CNN would be concatenated and will be reduced by the estimation block. The estimation block decreases channels of feature maps by choosing the more useful features before sending the next layers. The final fully-connected layers predict the orchid class, which results achieved 93.32% average classification accuracy on orchids dataset, which was higher than the existing DCNN classification models. In the future, we will decrease the size of the overall model structure for better performance of the classification models.

References

1. Abadi, M., et al.: TensorFlow: a system for large-scale machine learning. In: OSDI, vol. 16, pp. 265–283 (2016)
2. Arwatchananukul, S., Charoenkwan, P., Xu, D.: POC: paphiopedilum orchid classifier. In: 2015 IEEE 14th International Conference on ICCI*CC, pp. 206–212. IEEE (2015)
3. Erhan, D., Szegedy, C., Toshev, A., Anguelov, D.: Scalable object detection using deep neural networks. In: Proceedings of the IEEE Conference on CVPR, pp. 2147–2154 (2014)
4. Guru, D., Kumar, Y.S., Manjunath, S.: Textural features in flower classification. Math. Comput. Model. **54**(3–4), 1030–1036 (2011)
5. He, K., Zhang, X., Ren, S., Sun, J.: Deep residual learning for image recognition. In: Proceedings of the IEEE Conference on CVPR, pp. 770–778 (2016)
6. He, K., Zhang, X., Ren, S., Sun, J.: Identity mappings in deep residual networks. In: Leibe, B., Matas, J., Sebe, N., Welling, M. (eds.) ECCV 2016. LNCS, vol. 9908, pp. 630–645. Springer, Cham (2016). https://doi.org/10.1007/978-3-319-46493-0_38
7. Hiary, H., Saadeh, H., Saadeh, M., Yaqub, M.: Flower classification using deep convolutional neural networks. IET Comput. Vision **12**(6), 855–862 (2018)
8. Jaderberg, M., Simonyan, K., Zisserman, A., et al.: Spatial transformer networks. In: Advances in Neural Information Processing Systems, pp. 2017–2025 (2015)
9. Johnson, J., Karpathy, A., Fei-Fei, L.: DenseCap: fully convolutional localization networks for dense captioning. In: Proceedings of the IEEE Conference on CVPR, pp. 4565–4574 (2016)
10. Li, L., Qiao, Y.: Flower image retrieval with category attributes. In: 2014 4th IEEE International Conference on ICIST, pp. 805–808. IEEE (2014)
11. Liu, W., Rao, Y., Fan, B., Song, J., Wang, Q.: Flower classification using fusion descriptor and SVM. In: 2017 International ISC2, pp. 1–4. IEEE (2017)
12. Liu, Y., Tang, F., Zhou, D., Meng, Y., Dong, W.: Flower classification via convolutional neural network. In: International Conference on FSPMA, pp. 110–116. IEEE (2016)
13. Nilsback, M.E., Zisserman, A.: A visual vocabulary for flower classification. In: 2006 IEEE Computer Society Conference on CVPR, vol. 2, pp. 1447–1454. IEEE (2006)

14. Nilsback, M.E., Zisserman, A.: Automated flower classification over a large number of classes. In: Sixth Indian Conference on Computer Vision, Graphics & Image Processing, 2008, ICVGIP 2008, pp. 722–729. IEEE (2008)
15. Russakovsky, O., et al.: ImageNet large scale visual recognition challenge. IJCV **115**(3), 211–252 (2015). https://doi.org/10.1007/s11263-015-0816-y
16. Sari, Y.A., Suciati, N.: Flower classification using combined a* b* color and fractal-based texture feature. Int. J. Hybrid Inf. Technol. **7**(2), 357–368 (2014)
17. Siraj, F., Ekhsan, H.M., Zulkifli, A.N.: Flower image classification modeling using neural network. In: 2014 International Conference on IC3INA, pp. 81–86. IEEE (2014)
18. Szegedy, C., et al.: Going deeper with convolutions. In: Proceedings of the IEEE Conference on CVPR, pp. 1–9 (2015)
19. Szegedy, C., Vanhoucke, V., Ioffe, S., Shlens, J., Wojna, Z.: Rethinking the inception architecture for computer vision. In: Proceedings of the IEEE Conference on CVPR, pp. 2818–2826 (2016)
20. Xia, X., Xu, C., Nan, B.: Inception-v3 for flower classification. In: 2017 2nd International Conference on ICIVC, pp. 783–787. IEEE (2017)
21. Zawbaa, H.M., Abbass, M., Basha, S.H., Hazman, M., Hassenian, A.E.: An automatic flower classification approach using machine learning algorithms. In: 2014 International Conference on ICACCI, pp. 895–901. IEEE (2014)
22. Zhang, C., Liang, C., Li, L., Liu, J., Huang, Q., Tian, Q.: Fine-grained image classification via low-rank sparse coding with general and class-specific codebooks. IEEE Trans. Neural Netw. Learn. Syst. **28**(7), 1550–1559 (2017)

Scalable Dictionary Classifiers for Time Series Classification

Matthew Middlehurst$^{(\boxtimes)}$, William Vickers, and Anthony Bagnall

School of Computing Sciences, University of East Anglia, Norwich, UK
M.Middlehurst@uea.ac.uk

Abstract. Dictionary based classifiers are a family of algorithms for time series classification (TSC) that focus on capturing the frequency of pattern occurrences in a time series. The ensemble based Bag of Symbolic Fourier Approximation Symbols (BOSS) was found to be a top performing TSC algorithm in a recent evaluation, as well as the best performing dictionary based classifier. However, BOSS does not scale well. We evaluate changes to the way BOSS chooses classifiers for its ensemble, replacing its parameter search with random selection. This change allows for the easy implementation of contracting (setting a build time limit for the classifier) and check-pointing (saving progress during the classifiers build). We achieve a significant reduction in build time without a significant change in accuracy on average when compared to BOSS by creating a fixed size weighted ensemble selecting the best performers from a randomly chosen parameter set. Our experiments are conducted on datasets from the recently expanded UCR time series archive. We demonstrate the usability improvements to randomised BOSS with a case study using a large whale acoustics dataset for which BOSS proved infeasible.

Keywords: Time series · Classification · Dictionary · Contracting

1 Introduction

Dictionary based learning is commonly employed in signal processing, computer vision and audio processing to capture recurring discriminatory features. The approach has been successfully applied to time series classification (TSC) in a variety of ways. An extensive experimental study [1] found that the best dictionary approach was the ensemble classifier the Bag of Symbolic Fourier Approximation Symbols (BOSS). It was shown that dictionary based classifiers detect a fundamentally different type of discriminatory features than other TSC approaches, and the addition of the BOSS ensemble to the meta ensemble the Hierarchical Vote Collective of Transformation-based Ensembles (HIVE-COTE) [12] leads to a significant improvement in accuracy. BOSS is an ensemble classifier that evaluates a range of parameter combinations over a grid, then retains all classifiers that are within 92% of the best combination, as measured by a leave-one-out cross-validation on the train data.

© Springer Nature Switzerland AG 2019
H. Yin et al. (Eds.): IDEAL 2019, LNCS 11871, pp. 11–19, 2019.
https://doi.org/10.1007/978-3-030-33607-3_2

The BOSS ensemble has some drawbacks. Firstly, the need to cross-validate each parameter combination means that it scales poorly. Secondly, the fact that it retains a variable number of base classifiers means that it is often very memory intensive. Thirdly, the histograms can be very large, so storing them all for each base classifier also requires a significant memory commitment for large problems. One proposed method to solve this, the BOSS vector space (BOSS-VS) classifier [15], has been shown to be significantly less accurate than the full BOSS [13,16]. We investigate whether we can mitigate against these problems without this significant loss in accuracy. Our primary contribution is to propose a new classifier, cBOSS, that uses an alternative ensemble mechanism to provide an order of magnitude speed up without loss of accuracy. Our secondary contributions include reproducing results from a related study [16] and making cBOSS both contractable (i.e. it is able to build the best possible classifier in a fixed amount of time) and check-pointable (i.e. the build classifier stage can be stopped and restarted). Our experiments are easily reproducible and we have released a Weka compatible version of BOSS, cBOSS and other tested classifiers that is integrated into the UEA Codebase[1]. There is also a Python version of the classifiers developed for the Alan Turing Institute sktime package[2].

The rest of this paper is structured as follows. Section 2 gives a brief description of dictionary based classification algorithms. Section 3 describes the alterations we make to the BOSS algorithm to make cBOSS. Section 4 presents the results of our experimental evaluation, and Sect. 5 concludes and offers some ideas for future work.

2   Time Series Dictionary Based Classifiers

Classifiers that use frequency of words as the basis for finding discriminatory features are often referred to as dictionary based classifiers [1]. They have close similarities to bag-of-words based approaches that are commonly used in computer vision. Informally, a dictionary based approach will be useful when the discriminatory features are repeating patterns that occur more frequently in one class then other classes.

Bag of SFA Symbols (BOSS)

A single BOSS [14] base classifier proceeds as follows. For each series, it extracts the windows sequentially, normalising the window if the parameter p is true. It then applies a Discrete Fourier Transform (DFT) to the resulting subseries, ignoring the first coefficient if p is true. The DFT coefficients are truncated to include only the first $l/2$ Fourier terms (both real and imaginary). The truncated samples are then discretised into α possible values using an algorithm called Multiple Coefficient Binning (MCB) (see [14]). MCB involves a preprocessing step to find the discretising break points by estimating the distribution of the Fourier coefficients. Consecutive windows producing the same word are only counted as

---

[1] https://github.com/TonyBagnall/uea-tsc.
[2] https://github.com/alan-turing-institute/sktime.

a single instance of the word. A bespoke BOSS distance function is used with a nearest neighbour classifier to classify new instances. The distance function is non-symmetrical, only including the distance for features that are non-zero in the first feature vector given. The BOSS base classifier has four parameters: window length w, word length l, whether to normalise each window p and alphabet size α. The BOSS ensemble (also referred to as just BOSS), evaluates all BOSS base classifiers in the range $w \subset \{10\ldots m\}$, $l \in \{16, 14, 12, 10, 8\}$ and $p \in \{true, false\}$. This parameter search is used to determine which base classifiers are used in the ensemble. Following [14], the alphabet size is fixed to 4 for all experiments. The number of window sizes is a function of the series length m. All BOSS base classifiers with a training accuracy within 92% of the best performing base classifier are kept for the ensemble. This dependency on series length and variability of ensemble size is a factor that can significantly impact on efficiency. Classification of new instances is then done using majority vote from the ensemble.

Word Extraction for Time Series Classification (WEASEL)
WEASEL [16] is a dictionary based classifier that is an extension of BOSS. WEASEL is a single classifier rather than an ensemble. WEASEL concatenates histograms for a range of parameter values of w and l, then performs a feature selection to reduce the feature space. Like BOSS, WEASEL performs a Fourier transform on each window. DFT coefficients are no longer truncated and instead the most discriminative real and imaginary features are retained, as determined by an ANOVA F-test. The retained values are then discretised into words using information gain binning, similar to the MCB step in BOSS. WEASEL does not remove adjacent duplicate words as BOSS does. The word and window size are used as keys to index the histogram. A further histogram is formed for bigrams. The number of features is reduced using a chi-squared test after the histograms for each instance are created, removing any words which score below a threshold. WEASEL uses a logistic regression classifier to make the predictions for new cases. WEASEL performs a parameter search for p and a reduced range of l and uses a 10-fold cross-validation to determine the performance of each set. The alphabet size α is fixed to 4 and the *chi* parameter is fixed to 2.

3   BOSS Enhancements (cBOSS)

Our changes to BOSS mainly focus on the ensemble technique of the classifier, which is computationally expensive and unpredictable. Ensembling has been shown to be an essential component of BOSS, resulting in significantly higher accuracy [8]. We assess whether we can replace the current ensemble mechanism with a more stable and efficient scheme without a significant reduction in accuracy. We found that complete randomisation, i.e. selecting random parameter combinations for a fixed number of base classifiers, worked reasonably well but on some data performed very badly. Hence, we retain an internal evaluation of each possible member through leave-one-out cross validation on the train data, then use this value to both select and weight classifier votes. The primary difference to BOSS is that we do not determine which members to retain after the complete search. Rather, we introduce a new parameter, k, the fixed ensemble size, and maintain a list of weights. The option to replace the ensemble size k with a time limit t through contracting is made available through this change in building process, when set the classifier will continue building until the time since building started is greater than t. Even with weighting, classifiers with poor parameters for a particular problem can degrade the overall classifier. Because of this we set a max ensemble size s to filter these out, with any classifier built past this value replacing the current lowest accuracy member of the ensemble if its accuracy is higher. To further diversify the ensemble and increase efficiency we take a randomly selected 70% subsample of the training data for each individual classifier. The parameter space we randomly sample for cBOSS is the same as that which BOSS searches exhaustively through. For classifying new instances, we adopt the exponential weighting scheme used in the Cross-validation Accuracy Weighted Probabilistic Ensemble (CAWPE) [9] to amplify small difference in weights and results in significantly improved performance. cBOSS is more formally described in Algorithm 1.

Algorithm 1. cBOSS_build(A list of n cases length m, $\mathbf{T} = (\mathbf{X}, \mathbf{y})$)

Parameters: the ensemble size k, the max ensemble size s
 1: Let w be window length, l be word length, p be normalise/not normalise and α be alphabet size.
 2: Let $\mathbf{C}$ be a list of s BOSS classifiers $(\mathbf{c}_1, \ldots, \mathbf{c}_s)$
 3: Let $\mathbf{E}$ be a list of s classifier weights $(\mathbf{e}_1, \ldots, \mathbf{e}_s)$
 4: Let $\mathbf{R}$ be a set of possible BOSS parameter combinations
 5: $i \leftarrow 0$
 6: $lowest\_acc \leftarrow \infty, lowest\_acc\_idx \leftarrow \infty$
 7: **while** $i < k$ AND $|\mathbf{R}| > 0$ **do**
 8:     $[l, a, w, p] \leftarrow random\_sample(\mathbf{R})$
 9:     $\mathbf{R} = \mathbf{R} \setminus \{[l, a, w, p]\}$
10:     $\mathbf{T}' \leftarrow$ subsample_data($\mathbf{T}$)
11:     $cls \leftarrow$ build_base_BOSS($\mathbf{T}', l, a, w, p$)
12:     $acc \leftarrow$ LOOCV(cls) { $train\ data\ accuracy$}
13:     **if** $i < s$ **then**
14:         **if** $acc < lowest\_acc$ **then**
15:             $lowest\_acc \leftarrow acc, lowest\_acc\_idx \leftarrow i$
16:         $c_i \leftarrow cls, e_i \leftarrow acc^4$
17:     **else if** $acc > lowest\_acc$ **then**
18:         $c_{lowest\_acc\_idx} \leftarrow cls, e_{lowest\_acc\_idx} \leftarrow acc^4$
19:         $[lowest\_acc, lowest\_acc\_idx] \leftarrow$ find_new_lowest_acc($\mathbf{C}$)
20:     $i \leftarrow i + 1$

---

4   Results

We compare cBOSS to the dictionary base classifiers BOSS [14] and WEASEL [16] in terms of accuracy and speed. Two known dictionary based classifiers not included in our comparison are Bag of Patterns (BOP) [10] and SAX-VSM [17]. Both of these classifiers were found to be significantly worse than the common time series classification benchmark, one nearest neighbour dynamic time warping [1]. Each of the classifiers are tested on the same 30 random resamples from the 114 datasets without missing values in the UCR repository. For these experiments we set the values for the cBOSS k and s values to 250 and 50 respectively. Figure 1 shows the critical difference diagram for these three classifiers. Solid bars indicate cliques where there is no significant difference between classifiers. Tests of difference are performed using pairwise Wilcoxon signed rank tests with the Holm correction. The results confirm that WEASEL is indeed significantly better than BOSS. They also demonstrate that there is no significant difference between BOSS and cBOSS.

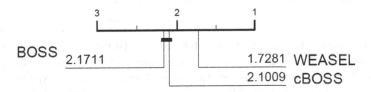

Fig. 1. Critical difference diagram showing accuracy ranks and cliques for the three dictionary based classifiers.

The aim of cBOSS is to be not significantly less accurate than BOSS in significantly less time than BOSS. Figure 2 shows the average build time plotted against average rank for the four classifiers. cBOSS is on average 12 times faster than BOSS, with no significant loss of accuracy.

To reduce the risk of bias or any suggestion of cherry picking, we have conducted experiments with all the datsets in the archive. However, many of these problems are small. To examine the effect on larger problems, we take a closer look at the results for 16 problems on which BOSS takes over an hour to build. Table 1 shows the pattern of results is the same on these 16 data. cBOSS is not significantly worse than BOSS, but an order of magnitude faster.

Table 1. Average large dataset performance by classifier using accuracy rank, Area Under the Receiver Operating Characteristic curve (AUROC) and Negative Log-Likelihood (NLL). Included is total build time over all datasets relative to BOSS.

Classifier	AccRank	AUROC	NLL	Build time
WEASEL	1.375	0.9529	0.9753	85.48%
cBOSS	2.25	0.9558	0.7951	7.7%
BOSS	2.375	0.953	0.7923	100%

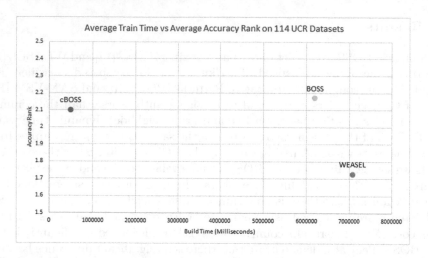

Fig. 2. Average build time against average accuracy rank for each dictionary classifier, with a lower value for both axis being a better performance.

To demonstrate the effectiveness of building cBOSS using a time contract Fig. 3 shows the change in accuracy over a range of t values on the 16 large problems. As shown an increase in time the classifier contracted for increases accuracy on average, though this increase slows as the parameter search progresses. The versatility of being able to build the best classifier within a train time limit is a large usability boost to the classifier, making this aspect much more predictable.

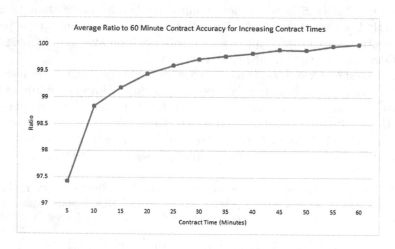

Fig. 3. Average accuracy value for a range of contract times on 16 large datasets. Times range from 5 to 60 min linearly spaced in increments of 5.

4.1 Whale Acoustics Use Case

Our interest in making BOSS scalable arose due to our desire to use it with an application in classifying animals based on acoustic samples. Recently, the protection of endangered whales has been a prominent global issue. Being able to accurately detect marine mammals is important to monitor populations and provide appropriate safeguarding for their conservation. North Atlantic right whales are one of the most endangered marine mammals with as few as 350 individual remaining in the wild [7]. We use a dataset from the Marinexplore and Cornell University Whale Detection Challenge[3] that features a set of right whale up-calls. Up-calls are the most commonly documented right whale vocalisation with an acoustic signature of approximately 60 Hz–250 Hz, typically lasting 1 s. Right whale calls can often be difficult to hear as the low frequency band can become congested with anthropogenic sounds such as ship noise, drilling, piling, or naval operations [3]. Each series is labelled as either containing a right whale or not with the aim to correctly identify the series that contain up-calls.

Previous work has been done in classifying the presence of whale species using acoustic data with whale vocalisations [6,18]. However, TSC approaches have not been applied. This problem is a good example of a large dataset for which it is infeasible to use BOSS. The dataset contains 10,934 train cases and 5885 test cases. Each case is a two second audio segment sampled at 2 kHz, giving a series length of 4000. We have done no preprocessing: the purpose of these experiments is to provide benchmark results for bespoke audio approaches and to test the scalability of cBOSS. The problem is large. It exceeds the largest train set from ElectricDevices of 8926 and series length from Rock of 2844 in the 128 UCR archive [4] (we will donate this data to the archive for the next release). For benchmarking alongside cBOSS on this dataset, we test the five classifiers that make up the HIVE-COTE ensemble [12], as well as two potential candidates for the ensemble WEASEL and Proximity Forest (PF) [13]. Only two of these classifiers, Time Series Forest (TSF) [5] and Random Interval Spectral Ensemble (RISE) [12], will complete within 28 days. The Shapelet Transform (ST) [2,11] is contractable, and we present results with a contracted time limit of five days. For fairness we ran PF on a single thread, but the capability to run using multiple threads is available and more likely to finish below the 28 day limit. BOSS did

Table 2. Accuracy and build time in hours for each of the potential HIVE-COTE components and cBOSS.

Classifier	Accuracy	Build time (hours)
cBOSS	0.8114	119.39
ST	0.818	120.05
RISE	0.7859	141.96
TSF	0.7712	16.23

---

[3] https://www.kaggle.com/c/whale-detection-challenge/data.

not complete, and also required huge amounts of memory (greater than 100 GB) to run at all. Attempts to build WEASEL ceased after a 300 GB memory limit was exceeded. cBOSS completed within 5 days, without a contract. Table 2 shows the accuracy and build time on the whales dataset each finished classifiers. These exploratory results suggest that both dictionary and shapelet approaches may be useful for this application.

5   Conclusion

We present cBOSS, a more scalable version of the BOSS classifier that uses a new ensemble mechanism. The replacement of the parameter search with randomly selected parameter sets provides a considerable speed up, and the introduction of subsampling for increased diversity and weighted voting means cBOSS is not significantly less accurate than BOSS. The inclusion of a fixed ensemble size, the ability to contract the build time and save progress with check-pointing make the classifier more robust and more predictable.

We have independently recreated the published results for the WEASEL classifier and verified the findings in [16]. WEASEL is significantly better than both BOSS and cBOSS. However, it also the slowest classifier to build on average, and has an equally large memory footprint as BOSS. This indicates this it is a suitable BOSS replacement on smaller datasets, it has the same scalability issues as BOSS. cBOSS is a viable alternative for large problems if a dictionary based classifier is required.

cBOSS scales well up to problems with tens of thousands of cases. However, building models with several hundred thousand instances may still cause an issue in requirements for space and time. Simple expedients such as subsampling can facilitate building models on large data, but doing this automatically whilst maintaining accuracy is challenging. Furthermore, cBOSS is still comparatively memory intensive, since it uses a nearest neighbour classifier. Our attempts to use alternative less memory intensive classifiers have been unsuccessful. Instead, we intend to introduce a contract for memory, setting the sampling size and ensemble size accordingly. Investigations into the feasibility of this and any affect on the classifiers performance could be an interesting future work in further improving dictionary based scalability.

Acknowledgements. This work is supported by the UK Engineering and Physical Sciences Research Council (EPSRC) iCASE award T206188 sponsored by British Telecom. The experiments were carried out on the High Performance Computing Cluster supported by the Research and Specialist Computing Support service at the University of East Anglia.

References

1. Bagnall, A., Lines, J., Bostrom, A., Large, J., Keogh, E.: The great time series classification bake off: a review and experimental evaluation of recent algorithmic advances. Data Min. Knowl. Disc. **31**(3), 606–660 (2017)

2. Bostrom, A., Bagnall, A.: Binary shapelet transform for multiclass time series classification. Trans. Large-Scale Data Knowl. Centered Syst. **32**, 24–46 (2017)
3. Cox, T.M., et al.: Understanding the impacts of anthropogenic sound on beaked whales. Technical report, Space and Naval Warfare Systems Centre, San Diego, CA, USA (2006)
4. Dau, H., et al.: The UCR time series archive. ArXiv e-prints arXiv:1810.07758 (2018)
5. Deng, H., Runger, G., Tuv, E., Vladimir, M.: A time series forest for classification and feature extraction. Inf. Sci. **239**, 142–153 (2013)
6. Dugan, P.J., Rice, A.N., Urazghildiiev, I.R., Clark, C.W.: North Atlantic right whale acoustic signal processing: part i. comparison of machine learning recognition algorithms. In: 2010 IEEE Long Island Systems, Applications and Technology Conference, pp. 1–6. IEEE (2010)
7. Kraus, S.D., et al.: North Atlantic right whales in crisis. Science **309**(5734), 561–562 (2005)
8. Large, J., Bagnall, A., Malinowski, S., Tavenard, R.: On time series classification with dictionary-based classifiers. Intell. Data Anal. (2019)
9. Large, J., Lines, J., Bagnall, A.: A probabilistic classifier ensemble weighting scheme based on cross validated accuracy estimates. Data Min. Knowl. Disc. (2019)
10. Lin, J., Khade, R., Li, Y.: Rotation-invariant similarity in time series using bag-of-patterns representation. J. Intell. Inf. Syst. **39**(2), 287–315 (2012)
11. Lines, J., Davis, L., Hills, J., Bagnall, A.: A shapelet transform for time series classification. In Proceedings the 18th ACM SIGKDD International Conference on Knowledge Discovery and Data Mining (2012)
12. Lines, J., Taylor, S., Bagnall, A.: Time series classification with HIVE-COTE: the hierarchical vote collective of transformation-based ensembles. ACM Trans. Knowl. Disc. Data **12**(5), 52:1–52:35 (2018)
13. Lucas, B., et al.: Proximity forest: an effective and scalable distance based classifier for time series. Data Min. Knowl. Disc. **33**(3), 607–635 (2019)
14. Schäfer, P.: The BOSS is concerned with time series classification in the presence of noise. Data Min. Knowl. Disc. **29**(6), 1505–1530 (2015)
15. Schäfer, P.: Scalable time series classification. Data Min. Knowl. Disc. **30**(5), 1273–1298 (2016)
16. Schäfer, P., Leser, U.: Fast and accurate time series classification with weasel. In Proceedings of the 2017 ACM on Conference on Information and Knowledge Management, pp. 637–646. ACM (2017)
17. Senin, P., Malinchik, S.: SAX-VSM: interpretable time series classification using sax and vector space model. In Proceedings 13th IEEE International Conference on Data Mining (ICDM) (2013)
18. Shamir, L., et al.: Classification of large acoustic datasets using machine learning and crowdsourcing: application to whale calls. J. Acoust. Soc. Am. **135**(2), 953–962 (2014)

Optimization of the Numeric and Categorical Attribute Weights in KAMILA Mixed Data Clustering Algorithm

Nádia Junqueira Martarelli[✉] and Marcelo Seido Nagano

Laboratory of Applied Operational Research, University of São Paulo, São Carlos,
São Paulo 13566-590, Brazil
{nadia.martarelli,drnagano}@usp.br

Abstract. The mixed data clustering algorithms have been timidly emerging since the end of the last century. One of the last algorithms proposed for this data-type has been KAMILA (KAy-means for MIxed LArge data) algorithm. While the KAMILA has outperformed the previous mixed data algorithms results, it has some gaps. Among them is the definition of numerical and categorical variable weights, which is a user-defined parameter or, by default, equal to one for all features. Hence, we propose an optimization algorithm called Biased Random-Key Genetic Algorithm for Features Weighting (BRKGAFW) to accomplish the weighting of the numerical and categorical variables in the KAMILA algorithm. The experiment relied on six real-world mixed data sets and two baseline algorithms to perform the comparison, which are the KAMILA with default weight definition, and the KAMILA with weight definition done by the traditional genetic algorithm. The results have revealed the proposed algorithm overperformed the baseline algorithms results in all data sets.

Keywords: Attributes weighting · Mixed data clustering · Biased Random-Key Genetic Algorithm · KAMILA algorithm

1 Introduction

Clustering algorithms aim to divide the data into groups of similar objects [2, 14]. Among pattern recognition techniques, data clustering task is considered the most challenging by the literature [17], because the pattern is obtained from unlabeled data [18].

This work was supported by the Conselho Nacional de Desenvolvimento Científico e Tecnológico (CNPq) - Brazil under grant number 306075/2017-2 and 430137/2018-4 and by Coordenação de Aperfeiçoamento de Pessoal de Nível Superior (CAPES) - Brazil.

H. Yin et al. (Eds.): IDEAL 2019, LNCS 11871, pp. 20–27, 2019.
https://doi.org/10.1007/978-3-030-33607-3_3

Many fields have used clustering algorithms for a variety of purposes, such as customer segmentation, stock trend analysis, taxonomy definition, gene and protein function identification, disease diagnosis, analysis of social networks and many others [21].

In most of the real cases, the data set is mixed, which means that some attributes are expressed by numeric data, and some others by categorical data. Although there are mixed data clustering algorithms proposed in the literature, this area is considered new when it is compared to the advancement of mono-type data clustering algorithms [1].

Although it is possible to transform a mixed data set into a mono-type data set, it is more interesting to work with a dissimilarity function that considers both types of data (numerical and categorical), so that there is no distortion in the scale and loss of readability [15,19].

In this vein, in 1988, Zhexue Huang [12,13] proposed the K-prototype algorithm, which becomes known as the seminal work in mixed data clustering. The k-prototype is a partitional algorithm, based on the k-means algorithm, which considers the simple matching dissimilarity for categorical data and the Euclidean distance for numerical ones.

Over the years, new algorithms based on k-means have been developed for mixed data clustering. One of them was proposed by Foss et al. [4], in 2016, called KAMILA (KAy-means for MIxed LArge data). The KAMILA aims to maximize the sum of two functions, the density function for numerical attributes, which considers the Euclidean distance, and probability mass function for categorical attributes.

Although the KAMILA has taken good results compared to other recent methods, it has some gaps. Among them is the attribute weights definition [3,4]. To cover this kind of problem, which is characterized as a continuous optimization, it is usual to use meta-heuristics, such as Genetic Algorithm (GA) and its variations. One of these GA versions is the Biased Random-Key Genetic Algorithm (BRKGA) [9], which improved the traditional approach proposing the maintenance of good individuals in the population over the generation, the creation of a proportional part of the population for mutant individuals and a crossover procedure that considers position by position.

These modifications generated excellent results for many problems that used BRKGA, such as in scheduling [8,11], orthogonal packing [7,10], and assembly line balancing [6], to mention a few.

Therefore, in this paper, we propose the modeling of BRKGA for the problem of attribute weighting in the KAMILA algorithm, naming it as Biased Random-Key Genetic Algorithm for Features Weighting (BRKGAFW).

This paper is organized as follows. In Sect. 2, we explain the objective function of KAMILA algorithm. In Sect. 3 we bring the modeling of BRKGA for this problem (BRKGAFW). In Sect. 4, we explain the design of experiments. Section 5 shows the results and the discussion. Finally, Sect. 6 presents the final remarks.

2 KAMILA Algorithm

The KAMILA algorithm [4] aims to maximize the following objective function, which will be named from now of Objective Function of the KAMILA Algorithm (OFKA)

$$OFKA = \sum_{i=1}^{N} max\{H_i^{(final)}(g)\} \tag{1}$$

where $H_i^{(final)}(g)$ is the best result of the assignment of the observation (i) for the gth group in each iteration t of the algorithm. Specifically, H is calculated on each iteration t as follows

$$H_i^{(t)} = log\left[\hat{f}_V^{(t)}\sqrt{\sum_{p=1}^{P}[\xi_p(\nu_{ip} - \hat{\mu}_{gp}^{(t)})]^2)}\right] + log\left[\sum_{q=1}^{Q}\xi_q log(m(w_{iq}; \hat{\theta}_{gq}^{(t)}))\right] \tag{2}$$

where $\hat{f}_V^{(t)}$ is a Kernel density estimator that contributes to the balance of the numeric attribute impact in the final equation; $\sqrt{\sum_{p=1}^{P}[\xi_p(\nu_{ip} - \hat{\mu}_{gp}^{(t)})]^2)}$ is the Euclidean distance between the observation i from the attribute p and the estimator for the centroid of group g and of the attribute p, $\hat{\mu}_{gp}^{(t)}$, where ξ_p is the user-defined weighting parameter of the attribute p; $\sum_{q=1}^{Q}\xi_q log(m(w_{iq}; \hat{\theta}_{gq}^{(t)}))$ represents the dissimilarity calculation for categorical data, where ξ_q is the user-defined weighting parameter of the attribute q.

This article aims to realize a better definition for ξ_p and ξ_q values using the proposed algorithm, BRKGAFW.

3 BRKGAFW

The BRKGAFW follows the same seven steps of the traditional GA (initial population, codification, evaluation, selection, reproduction and stopping rules). However, some procedures in specific steps are modified to improve the results. All the steps are explained ahead.

The initial population of the BRKGAFW is composed of n_p individuals randomly generated by the selection of continuous values from the interval $[0, 1]$. Each individual is represented by a string with *attr* positions, where *attr* is the total of attributes in the full data set.

The decoder algorithm is very specific to the problem. Indeed, it is the only link to the problem. Other genetic procedures are independent (see more in [9]). In the BRKGAFW, the decoder read all the positions as the weights of the attributes of the KAMILA algorithm.

As a fitness function, the BRKGAFW considers the OFKA. Hence, the bigger the better.

The selection step consists of classifying the individuals of the current population in descending order of the OFKA values. Thus, the population is divided into two groups, the first (elite), contains the $(p_e) * n_p$ best individuals, and the second group (non-elite), is composed of $(1 - p_e) * n_p$ individuals.

Three steps are followed to build the new population: (i) the elite group is fully copied to the new population, which is now called TOP; (ii) another group, called BOT, is created with $(p_m) * n_p$ mutant individuals; (iii) the rest of the individuals $((1 - p_e - p_m) * n_p)$ are created by a crossover procedure.

Mutant individuals are created by selecting a random individual from the entire population and changing p_{pm} percent of its positions by a value randomly generated from the $[0, 1]$ interval.

Individuals resulting from the crossover procedure are created as follows. Two individuals (parent 1 and parent 2) are selected from the population. One comes from TOP group (parent 1) and the other is selected from the entire population (parent 2). Both are selected randomly from their respective groups.

The new individuals (son) are built with information from parent 1 and parent 2. To support this process, an auxiliary string with the same size of an individual is generated. Each position of this string is filled with random numbers, belonging to $[0, 1]$ interval. This string does not represent a solution. It is created only to support the crossover process. Besides the auxiliary string, a predetermined parameter, called ρ_s, is also defined.

From here, each position of any participant individual is called allele. Therefore, the recombination process is accomplished comparing each position to the auxiliary string with the ρ_s value. The parent 1 allele is selected to compose the son string if the correspondent allele of the auxiliary string is smaller than ρ_s. Otherwise, the corresponding parent 2 allele feeds the son sequence.

The BRKGAFW stops when it achieves $maxIt$ iterations or when its improvement of quality is less than $minInc$ after $minIt$ successive generations.

4   Experiments

The computational experiments were run in RStudio Version 1.1.456, in a computer with Intel Core i7-4790K CPU @ 4 GHz along with 16 GB of RAM under Windows 10.

Three algorithms were considered, two baselines and our proposed algorithm, which are, respectively: (i) NO Features Weighting (NOFW) that considers the default condition of KAMILA algorithm (all weights equal to one); (ii) Genetic Algorithm for Features Weighting (GAFW); (iii) Biased Random-Key Genetic Algorithm for Features Weighting (BRKGAFW).

Each algorithm was run 100 times. The number of the groups was defined as the one that resulted in the best KAMILA objective function value in each data set, considering the interval $[2, 10]$. The KAMILA package was used for results generation [3].

In respect to data sets, we considered six real-world, which were extracted from the machine learning repository from the University of California Irvine[1] (UCI) [16]. For the best KAMILA performance, the numeric attributes were normalized by columns. Table 1 shows the data sets information.

Table 1. Data sets information.

| Data sets | Area | Instances | Number of attributes | | |
			Total	Numerical	Categorical
1 rContraceptive1473F9	Health	1473	9	7	10
2 rCredit1000F20	Finance	1000	20	13	21
3 rFlag194F29	Politics	194	29	19	30
4 rHeart270F13	Health	270	13	7	14
5 rSponge76F45	Biology	76	45	3	42
6 rTeaching151F5	Education	151	5	4	6

With respect to the values of the parameters of each algorithm, we considered the following values for all data sets: (i) GAFW, crossover rate $p_c = 0.8$, mutation rate $p_m = 0.1$, and percentage of the positions that receive mutation $p_{pm} = 0.1$; (ii) BRKGAFW, elite percentage $p_e = 0.3$, percentage of mutation group $p_m = 0.2$, percentage of the positions that receive mutation $p_{pm} = 0.2$, and $\rho = 0.7$.

For the GAFW's modeling, we used the roulette wheel to select individuals for the reproduction and two points method for the crossover.

Regarding general parameters, we used $n_p = 30$, $maxIt = 10$, $minIt = 5$, and $minInc = 0.01$.

To evaluate the methods, we executed the Wilcoxon rank sum test, a non-parametric statistical test [20], to identify if the algorithms results are statistically different. To do this, we utilized the wilcox.test function from the R software. The significance level adopted was 0.05.

5   Results

Table 2 presents the mean and standard deviation of the OFKA values resulted from 100 executions for all data sets and algorithms. As can be seen, the BRK-GAFW obtained the best OFKA values in all data sets as well as the smaller standard deviation.

In order to verify the real difference of the algorithms results, we applied the Wilcoxon rank sum test, a non-parametric statistical test. All p-values were zero. Therefore, the statistical test confirmed the results of the algorithms are all different. Consequently, the BRKGAFW is the best algorithm for all data set.

---

[1] https://archive.ics.uci.edu/ml/index.htm.

Table 2. Mean and the standard deviation (mean ± std) of the objective function value of the KAMILA algorithm (OFKA) of all methods, in all data sets. The best results are highlighted in bold font (the higher the better).

Data sets	BRKGAFW	GAFW	NOFW
1	**−2,972.3 ± 679.93**	−3,223.09 ± 968.42	−592,134.06 ± 102,820.2
2	**−3,017.36 ± 1,218.72**	−4,575.73 ± 1,424.20	−426,374.32 ± 30,822.46
3	**−1,337.23 ± 390.55**	−1,909.44 ± 575.75	−117,758.75 ± 67,061.39
4	**−610.00 ± 285.73**	−872.17 ± 519.94	−19,033.51 ± 203.49
5	**−363.60 ± 55.86**	−461.23 ± 87.95	−3,687.02 ± 0.00
6	**−86.26 ± 39.65**	−132.34 ± 49.08	−3,229.88 ± 2,372.39

With respect to the execution time, Table 3 shows the runtime values for all data sets and algorithms. As a result, it is possible to notice that, although the BRKGAFW overperformed the other algorithms, it spent more time than the NOFW in all data sets, analyzing the mean values.

Table 3. Mean and the standard deviation (mean ± std) of execution time (in seconds) of all methods, in all data sets. The best results are highlighted in bold font (the smaller the better).

Data sets	BRKGAFW	GAFW	NOFW
1	21.66 ± 1.23	20.43 ± 1.15	**0.26 ± 0.01**
2	38.65 ± 2.62	40.85 ± 2.10	**0.66 ± 0.03**
3	11.04 ± 0.84	11.23 + 0.77	**0.13 ± 0.02**
4	11.40 ± 1.62	10.04 ± 0.93	**0.13 ± 0.01**
5	10.47 ± 2.29	10.25 ± 0.87	**0.12 ± 0.01**
6	7.90 ± 0.51	7.15 ± 0.25	**0.07 ± 0.01**

In our point of view, the efficiency and effectiveness of the BRKGAFW should be analyzed together. One way to do this is to calculate the Average Percentage Relative Deviation (APRD) for the quality of the solution and for the runtime.

The APRD aims to highlight the best solution found (in this case, APRD is zero) and shows how the values distanced themselves from the best one (see in Framinan and Nagano [5]), following the equation

$$APRD = \frac{1}{N} \sum \left(\frac{f - f^*}{f^*} * 100 \right) \tag{3}$$

where f^* represents the best solution value found, f represents the solution value of the algorithm.

To measure the APRD for the quality of the solution, the best objective function value is the highest one. On the other hand, with respect to execution time, the best value is the smallest one.

After calculating the APRD, we found that in the first three data sets the APRD value for the quality of the solution from the NOFW to the BRKGAFW (the best value) was greater than the APRD value for the runtime from BRK-GAFW to the NOFW (the best value). This do not happened in the last three data sets.

Although the execution time of the BRKGAFW is high compared with the other approaches, it is an acceptable execution time for real-world scenarios. However, we believe that this time can be improved by decreasing population size by the deterministic generation of the initial population since if the algorithm already starts with good individuals its initial population can be smaller, decreasing thus the computational efforts, that is, execution time.

6   Final Remarks

This paper proposed an algorithm of continuous optimization to solve the problem of numerical and categorical attributes weighting in the KAMILA algorithm, which obtained excellent results. In all data sets, BRKGAFW obtained the best OFKA mean value, statistically proven.

The KAMILA algorithm, which was proposed by [4] aims to group mixed data (numerical and categorical) by non-transformation of data (mixed-type to mono-type). For this, the algorithm determines two ways of calculating the dissimilarity of both attributes separately. However, it is left to the user to define the weights of the numerical and categorical variables, otherwise, the algorithm defines, as standard, weight one for all the attributes. This gap was filled in this work by the proposed algorithm, BRKGAFW, using the same objective function of the KAMILA algorithm (OFKA) to optimize the choice of weights for both attributes.

References

1. Ahmad, A., Khan, S.S.: Survey of state-of-the-art mixed data clustering algorithms. IEEE Access **7**, 31883–31902 (2019). https://doi.org/10.1109/ACCESS.2019.2903568
2. Berkhin, P.: A survey of clustering data mining techniques. In: Kogan, J., Nicholas, C., Teboulle, M. (eds.) Grouping Multidimensional Data, pp. 25–71. Springer, Heidelberg (2006). https://doi.org/10.1007/3-540-28349-8_2
3. Foss, A., Markatou, M.: KAMILA: clustering mixed-type data in R and hadoop. J. Stat. Softw. **83**(13), 1–44 (2018). https://doi.org/10.18637/jss.v083.i13
4. Foss, A., Markatou, M., Ray, B., Heching, A.: A semiparametric method for clustering mixed data. Mach. Learn. **105**(3), 419–458 (2016). https://doi.org/10.1007/s10994-016-5575-7
5. Framinan, J.M., Nagano, M.S.: Evaluating the performance for makespan minimisation in no-wait flowshop sequencing. J. Mater. Process. Technol. **197**(1–3), 1–9 (2008). https://doi.org/10.1016/j.jmatprotec.2007.07.039
6. Gonçalves, J.A., Almeida, J.F., Raimundo, J.: A hybrid genetic algorithm for assembly line balancing. J. Heuristics **8**, 629–642 (2002). https://doi.org/10.1023/A:1020377910258

7. Gonçalves, J.F.: A hybrid genetic algorithm-heuristic for a two-dimensional orthogonal packing problem. Eur. J. Oper. Res. **183**, 1212–1229 (2007). https://doi.org/10.1016/j.ejor.2005.11.062
8. Gonçalves, J.F., Mendes, J.J.M., Resende, M.G.C.: A hybrid genetic algorithm for the job shop scheduling problem. Eur. J. Oper. Res. **167**, 77–95 (2005). https://doi.org/10.1016/j.ejor.2004.03.012
9. Gonçalves, J.F., Resende, M.G.C.: Biased random-key genetic algorithms for combinatorial optimization. J. Heuristics **17**, 487–525 (2011). https://doi.org/10.1007/s10732-010-9143-1
10. Gonçalves, J.F., Resende, M.G.C.: A parallel multi-population genetic algorithm for a constrained two-dimensional orthogonal packing problem. J. Comb. Optim. **22**, 180–201 (2011). https://doi.org/10.1007/s10878-009-9282-1
11. Gonçalves, J.F., Resende, M.G.C., Mendes, J.J.M.: A biased random-key genetic algorithm with forward-backward improvement for the resource constrained project scheduling problem. J. Heuristics **17**, 467–486 (2011). https://doi.org/10.1007/s10732-010-9142-2
12. Huang, Z.: Clustering large data sets with mixed numeric and categorical values. In: The First Pacific-Asia Conference on Knowledge Discovery and Data Mining, PAKDD 1997, Singapore, pp. 1–34 (1997)
13. Huang, Z.: Extensions to the k-means algorithm for clustering large data sets with categorical values. Data Min. Knowl. Disc. **2**(3), 283–304 (1998)
14. Jain, A.K., Dubes, R.C.: Algorithms for Clustering Data. Prentice-Hall, Upper Saddle River (1988)
15. Ji, J., Bai, T., Zhou, C., Ma, C., Wang, Z.: An improved k-prototypes clustering algorithm for mixed numeric and categorical data. Neurocomputing **120**(23), 590–596 (2013)
16. Lichman, M.: UCI machine learning repository (2013)
17. Saxena, A., et al.: A review of clustering techniques and developments. Neurocomputing **267**, 664–681 (2017). https://doi.org/10.1016/j.neucom.2017.06.053
18. Tan, P.N., Steinbach, M., Kumar, V.: Introduction to Data Mining. Addison-Wesley Longman Publishing Co., Inc., Boston (2005)
19. Wei, M., Chow, T.W.S., Chan, R.H.M.: Clustering heterogeneous data with k-means by mutual information-based unsupervised feature transformation. Entropy **17**(3), 1535–1548 (2015)
20. Wilcoxon, F.: Individual comparisons by ranking methods. Biometrics Bull. **1**(6), 80–83 (1945)
21. Xu, R., Wunsch, D.: Clustering. Wiley-IEEE Press, Hoboken, Piscataway (2009)

Meaningful Data Sampling for a Faithful Local Explanation Method

Peyman Rasouli[✉] and Ingrid Chieh Yu[✉]

Department of Informatics, University of Oslo, Oslo, Norway
{peymanra,ingridcy}@ifi.uio.no

Abstract. Data sampling has an important role in the majority of local explanation methods. Generating neighborhood samples using either the Gaussian distribution or the distribution of training data is a widely-used procedure in the tabular data case. Generally, this approach has several weaknesses: first, it produces a uniform data which may not represent the actual distribution of samples; second, disregarding the interaction between features tends to create unlikely samples; and third, it may fail to define a compact and diverse locality for the sample being explained. In this paper, we propose a sampling methodology based on observation-level feature importance to derive more meaningful perturbed samples. To evaluate the efficiency of the proposed approach we applied it to the LIME explanation method. The conducted experiments demonstrate considerable improvements in terms of fidelity and explainability.

1 Introduction

Explanation methods provide interpretable models for complicated, black box Machine Learning (ML) algorithms such as Random Forests (RF) and Deep Neural Networks [2,7]. These approaches simulate the behaviour of a black box using a simple, interpretable predictor (e.g., linear regression or decision tree) to provide explanations and justifications for the decision being made by the black box. Although black box machine learning models are in demand for automating safety-critical decision systems (e.g., self-driving cars), their applicability is disrupted due to lack of transparency and explainability. Furthermore, black box models are created based on human-made data which may contain bias and prejudices [5]. Thus, a resultant learned model may inherited such biases, unconsciously making unfair and incorrect decisions. Creating a faithful explanation method is considered as a first step to elucidate and address these concerns.

The main concern in creating an explanation method is to establish a trade-off between fidelity and interpretability. By fidelity we mean to which extent the interpretable model is able to accurately imitate a black-box prediction and often is measured in terms the F1-score and Mean Squared Error (MSE) [6]. A less-fidelity interpretable model may produce less accurate predictions compared to the black box model. To trust the explanations, the model must therefore be accurate and interpretable simultaneously, as it is unsound to trust explanations

© Springer Nature Switzerland AG 2019
H. Yin et al. (Eds.): IDEAL 2019, LNCS 11871, pp. 28–38, 2019.
https://doi.org/10.1007/978-3-030-33607-3_4

that are generated based on incorrect predictions. There are a wide range of research works which aim to address the described issue [10, 14, 15].

Deriving global while accurate explanations of a black box model has been seen as a challenging task [10]. In contrast, several local explanation methods have been developed by approximating an interpretable model based on the neighbourhood of a given sample to explain the corresponding prediction of the black box [4, 10–12, 16]. They assume that every complex model is linear on a local scale. Local Interpretable Model-agnostic Explanations (LIME), is a fully agnostic method for outcome explanation [10]. Given a sample x, it generates perturbed samples in the vicinity of x, which are weighted based on their proximity to the x. LIME maps the perturbed samples to a binary representation for the sake of interpretability. Using a sparse linear model trained on the perturbed samples, LIME explains the reasons behind an outcome in the form of feature importance. In [19], authors presented a Model Explanation System (MES) that explains the prediction of a black box model using a Monte Carlo algorithm. Based on formal requirements, this method constructs a scoring system for finding the best explanations in the form of simple logical rules. MES is both model-agnostic and data-agnostic. Local Rule-based Explanations (LORE) [4], is a model-agnostic approach that addresses black box outcome explanation problem by means of decision rules. It assumes a higher availability of clear and simple decision boundary in the neighbourhood of a data point rather than the whole data space. LORE provides decision rules and counter-factual rules to explain a given sample.

To increase the accountability of the generated explanations, our goal is to achieve an accurate interpretable model similar to the black box model. Specifying the neighbourhood of a given sample is the core task of several perturbation-based explanation methods [4, 10–12, 16]. The locality is defined based on the perturbed data, which later is used for creating an interpretable model. Deriving perturbed samples using either the Gaussian distribution or the distribution of training data is a widely-used procedure. Generally, this approach has several shortcomings; first, it creates a uniform data which may not represent the actual distribution of samples; second, it disregards the interaction between features which tends to generate unlikely data points; and third, it produces a synthetic data set similar to the training data which may fail to describe a compact locality for a given sample. Besides, relying solely on a distance function for weighting the neighborhood samples is not efficient enough [10]. Additive feature attribution methods convert feature values in the original representation to a binary representation for the sake of interpretability [8]. This conversion excludes the expressiveness of the data to a great extent and drops the amount of provided information drastically. This causes informative gaps between the original and interpretable representation of the data. The combination of ineffective sampling and information reduction may result in an unrepresentative data set, and accordingly more likely to create unfaithful interpretable models.

Our main contribution is a sampling methodology to produce representative data sets needed for local perturbation-based explanation methods. The proposed sampling strategy can be adopted by any model-agnostic explanation method applied on tabular data. Simply put, it exploits the feature importance and the distribution of training data to create a perturbed data set for describing the locality of a given sample. We employ observation-level (not model-level) feature importance which is unique regarding each sample and describes each feature's contribution in the prediction of the black box. The intuition behind the proposed approach is to create a neighborhood data by looking at the locality from both feature space and feature importance perspectives. As each view provides a different insight about the relationship of data points, we leverage them to generate a more expressive and informative data set. In addition we show how our sampling approach can be used in LIME to improve the locality definition in tabular data setting. The sampling in LIME varies in each execution, which means there are instabilities in the generated explanations [9]. Through exploiting feature importance, we are bounding the sampling w.r.t the black-box predictions that results in generating more stable explanations. We conducted several experiments on several data sets and compared the results to LIME. The evaluation shows a significant improvement in terms of fidelity and the quality of the generated explanations.

The rest of the paper is organized as follows. In Sect. 2 we introduce our sampling methodology. In Sect. 3 we describe the experiments and illustrate the results. Finally, we discuss future work and conclude the paper in Sect. 4.

2 Meaningful Data Sampling

Sampled data points from either the Gaussian distribution or the distribution of training data (e.g., in LIME) may not be representative enough to define a satisfactory locality for the sample being explained. In this paper, we remedy this problem by exploiting feature importance in the sampling process. To this end, we use the sampling function $\phi(.)$ defined in Eq. 1 that takes a sample x as input, the distribution of the training data $\mathcal{D}$, a black box function f, and number of perturbed samples N; it returns a set of neighborhood samples $\mathcal{Z}$ for creating a local interpretable model,

$$\mathcal{Z} = \phi(x, \mathcal{D}, f, N) \tag{1}$$

Our proposed sampling method utilizes observation-level feature importance for creating neighborhood data. To achieve the contribution of each feature in the prediction of a black box model, we exploit the random forests in combination with the *treeinterpreter* technique [13]. The goal is to use the locality information derived from both feature space and target space viewpoints to create a meaningful locality. The proposed method is described in Algorithm 1.

Random Data Generation. In this step, N perturbed samples, denoted by $\mathcal{Z}'$, are drawn from the given distribution $\mathcal{D}$. The proposed method starts with random data generation to achieve perturbed samples with a high degree of diversity and coverage of the feature space. We believe this is an appropriate initialization towards creating a meaningful data set.

Algorithm 1. Meaningful Data Sampling

> **procedure** $\phi(x, \mathcal{D}, f, N)$
>> $\mathcal{Z}' \leftarrow$ Random Data Generation$(\mathcal{D}, N)$
>> $T \leftarrow$ Random Forest Construction$\left(\mathcal{Z}', f(\mathcal{Z}')\right)$
>> $\Gamma \leftarrow$ Feature Importance Extraction$(T, \mathcal{Z}')$
>> $\mathcal{Z} \leftarrow$ Sample Manipulation$(x, \mathcal{Z}', \Gamma)$
>> **return** $\mathcal{Z}$

Random Forest Construction. Random Forests (RFs) are widely-used predictive models for both regression and classification tasks [1]. The strength of random forests originates from creating a forest of several random, uncorrelated decision trees and then merging them together in order to gain predictive results. Compared to a decision tree, a random forest adds randomness to the model by means of diverse subsets of samples and features while growing the trees which results in more accuracy and robustness against over-fitting.

Using $\left(\mathcal{Z}', f(\mathcal{Z}')\right)$ as training data, a random forest classifier, T, is created to mime the black box f. There are two important reasons for constructing the surrogate model T; first, observation-level features' importance of each perturbed sample $z' \in \mathcal{Z}'$ can be achieved via treeinterpreter; and second, it makes the sampling process to be independent from the type of the black box, therefore, it can be adopted by any local model-agnostic/model-specific explanation method that depends on neighborhood information for explaining an instance. Moreover, it is possible to train T with varied sizes of training data (generated using perturbed samples) and various hyper-parameter settings which allow us to create a robust surrogate model with similar performance to the black box f. In this paper, we use standard random forests with default hyper-parameters, leaving the exploration of different configurations to future work.

Feature Importance Extraction. Random forests are typically treated as black box which are infeasible to interpret. Nevertheless, it is possible to turn it into a "white box" using the treeinterpreter. The treeinterpreter is a Python library that uses the underlying trees in a random forest to explain how each feature contributes the end value. Each decision in a tree (or a forest) is followed by a path (or paths) from the root of the tree to the leaf, consisting of a series of decisions, guarded by a particular feature, each of which contribute to the final predictions. Since each decision is guarded by a feature, and the decision either adds or subtracts from the value given in the parent node, the prediction can be defined as the sum of the feature contributions plus the bias. A prediction in a random forest is the average of the prediction of each decision tree. As a result, each prediction is calculated as the average of the bias terms plus the average contribution of each feature.

Model-level feature importance is computed over the entire training data, while treeinterpreter acts on each sample at a time, explaining why a random forest makes a particular decision. For a given sample x and a set of target classes $\mathcal{C}$, feature contributions are represented by vectors $\{\gamma_1, .., \gamma_{|\mathcal{C}|}\}$ indicating the contribution of x to each class. The values of γ can be positive, negative, or zero and that all together (plus the bias value) constitute the final prediction. A positive value means that the feature pushes the decision towards the target class, a negative value means that the feature pushes the decision away from the target class, and zero value means that the feature is neutral, so it does not impact the predication.

Using treeinterpreter, feature contribution vectors for all samples in $\mathcal{Z}'$ are extracted, forming a feature contribution matrix Γ. To ease the further operations, Γ is discretized using a quantile-based discretization method introduced in [3]. The number of bins for each feature is obtained using Sturges's rule [17].

Sample Manipulation. The feature values in $\mathcal{Z}'$ and the feature contributions in Γ, each provides a different type of information about the data. Using importance of features accompanying their values helps to capture the relationship between samples more efficiently. Therefore, we would be able to generate neighborhood instances according to the locality defined by both feature values (explicitly) and feature contributions (implicitly) simultaneously.

In this phase, neighborhood samples are manipulated through comparing the contributions of the given sample, Γ_x, with the contributions of each perturbed sample, $\Gamma_{z'}, z' \in \mathcal{Z}'$ in the following manner: let $\mathcal{C}$ be the set of target classes; for each pair of samples predicted to the same class by T, i.e., $\{x_c, z'_c\}$, where $c \in \mathcal{C}$, their feature contribution vectors for that class are compared; while for a pair of samples predicted to different classes by T, i.e., $\{x_c, z'_{c'}\}$, where $c, c' \in \mathcal{C}$ and $c \neq c'$, their feature contribution vectors are compared mutually with respect to c and c'. Since feature importance vector is related to the prediction, we must consider the class of the samples during the comparison. In addition, the contribution of samples to the other classes will not impact this process, which helps to preserve the diversity of the data. It should be noted that we calculate the procedure only for the features of a perturbed sample that have different values than the features of the given sample. Thus, for a given perturbed sample $z' \in \mathcal{Z}'$, our goal is to identify features with similar contributions to the target class, while their different values has made the z' to be located in a distant neighborhood of the given sample x. Finally, the perturbed set $\mathcal{Z}$, initialized with $\mathcal{Z} = \mathcal{Z}'$, will contain the desired sampled data, as each sample $z \in \mathcal{Z}$ is derived via Eq. (2),

$$
z_j = \begin{cases} x_j, & \text{if } \{\Gamma_{x_j} = \Gamma_{z_j} \mid x_j \neq z_j\} \\ z_j, & \text{otherwise} \end{cases} \tag{2}
$$

where x is the given sample and $j = 1, .., d$ is the dimension of features. The reason behind this replacement is that we assume if a specific feature in a pair of samples has two different values, but similar contributions to a target class, then from the feature importance perspective both samples are similar with respect to that feature; therefore, setting same value for the feature in both samples may not change their predictions significantly, rather making them closer to each other. Doing the mentioned process for all samples in $\mathcal{Z}$ results in a compact and diverse neighborhood of the given sample x.

3   Experiments and Discussions

In this section, the proposed method is evaluated with respect to several classification data sets and black box models. To restate, the main goal of this work is to improve the fidelity, showing to which extent the interpretable model g is able to accurately imitate the black box model f. We adopted linear regressions and decision trees as interpretable models simultaneously to offer a better insight about the fidelity of the explanation method.

Experiments and Benchmarks. The proposed sampling method has been developed in Python programming language and experiments were run on a system with Intel Core i7-8650U processor and 32 GB of memory. We used scikit-learn libraries for implementing the machine learning and data mining algorithms. We applied our sampling method on LIME and benchmarked it against LIME with its original sampling strategy. As a convention, we use S-LIME to refer the LIME that is armed with our sampling technique. The evaluation results are reported in three parts: (i) fidelity comparison, (ii) neighborhood analysis, and (iii) explanation comparison.

Experimental Setup. In the experiments, four tabular classification data sets including *recidivism*[1], *adult*, *german credit*, and *red wine quality*[2] were used. Each data set was split to 85% *train set* and 15% *test set*. A multi-class AdaBoost classifier [20] (**AdB**) and a Random Forest classifier (**RF**) each with 300 estimators were employed as black box models. The 50% of *test set* were used to quantify the fidelity of the explanation methods. The number of neighborhood samples was set to $N = 1000$. Each sample was explained using $K = \{2, .., 5\}$ features and the results of K with the highest performance was considered.

The following properties are used for evaluating the performance of interpretable models, i.e., decision tree g^D and liner regression g^L, in miming the local behavior of the black box f. The notions y and Y are the ground-truth labels and $\hat{y}$ and $\hat{Y}$ are the predicted labels of a sample and the entire data set:

- $hit_{g^D}(y, \hat{y}) \in [0, 1]$. It compares the prediction of g^D and black box f on the sample of interest x using F1-score.
- $hit_{g^L}(y, \hat{y}) \in [0, 1]$. It compares the prediction of g^L and black box f on the sample of interest x using Mean Absolute Error (MAE).
- $fidelity_{g^D}(Y, \hat{Y}) \in [0, 1]$. It compares the predictions of g^D and black box f on the training samples $\mathcal{Z}$ using F1-score.
- $fidelity_{g^L}(Y, \hat{Y}) \in [0, 1]$. It compares the predictions of g^L and black box f on the training samples $\mathcal{Z}$ using R2-score.

---

[1] Data set is available at: https://www.kaggle.com/danofer/compass.
[2] Data sets are available at: https://archive.ics.uci.edu/ml/datasets/.

The $fidelity_{gD}$ and $fidelity_{gL}$ metrics describe how good are the interpretable models at imitating the black box model. The desired values for hit_{gD}, $fidelity_{gD}$, and $fidelity_{gL}$ are 1.0, while it is 0.0 for hit_{gL}. We executed 10 iterations of the experiment for each data set and black box model, hence the average and standard deviation of the results are reported in the following.

Table 1. LIME vs S-LIME: comparison of fidelity for **RF** black box model.

Data set	Method	hit_{gD}	hit_{gL}	$fidelity_{gD}$	$fidelity_{gL}$
adult	LIME	0.845 ± 0.052	0.126 ± 0.007	0.910 ± 0.008	0.426 ± 0.013
	S-LIME	$\mathbf{0.915 \pm 0.026}$	$\mathbf{0.105 \pm 0.011}$	$\mathbf{0.914 \pm 0.008}$	$\mathbf{0.463 \pm 0.013}$
german	LIME	0.864 ± 0.030	0.103 ± 0.006	0.822 ± 0.008	0.347 ± 0.020
	S-LIME	$\mathbf{0.886 \pm 0.044}$	$\mathbf{0.093 \pm 0.004}$	$\mathbf{0.834 \pm 0.007}$	$\mathbf{0.378 \pm 0.025}$
recidivism	LIME	1.000 ± 0.000	0.027 ± 0.002	0.931 ± 0.003	0.791 ± 0.010
	S-LIME	$\mathbf{1.000 \pm 0.000}$	$\mathbf{0.023 \pm 0.002}$	$\mathbf{0.948 \pm 0.003}$	$\mathbf{0.831 \pm 0.008}$
red wine	LIME	$\mathbf{0.854 \pm 0.023}$	0.202 ± 0.012	0.658 ± 0.005	0.216 ± 0.008
	S-LIME	0.845 ± 0.019	$\mathbf{0.184 \pm 0.009}$	$\mathbf{0.685 \pm 0.005}$	$\mathbf{0.262 \pm 0.009}$

Fidelity Comparison. Tables 1 and 2 report the statistical results for **RF** and **AdB** black box models, respectively. They exhibit overall improvement of S-LIME over LIME concerning all data sets and black box models. The individual sample evaluation metrics, i.e., hit_{gD} and hit_{gL}, are considerably important in measuring the fidelity of the intepretable models; according to the listed values for these properties, S-LIME classified a higher proportion of samples correctly compared to LIME. Especially, for *adult* and *red wine quality* data sets the improvements are more prominent. In addition, $fidelity_{gD}$ and $fidelity_{gL}$ show more accurate predictions of the neighborhood samples in S-LIME; they indicate that the explanations were derived from faithful local interpretable models.

Table 2. LIME vs S-LIME: comparison of fidelity for **AdB** black box model.

Data set	Method	hit_{gD}	hit_{gL}	$fidelity_{gD}$	$fidelity_{gL}$
adult	LIME	0.840 ± 0.055	0.008 ± 0.001	0.853 ± 0.020	0.238 ± 0.040
	S-LIME	$\mathbf{0.889 \pm 0.041}$	$\mathbf{0.006 \pm 0.001}$	$\mathbf{0.859 \pm 0.019}$	$\mathbf{0.259 \pm 0.041}$
german	LIME	0.816 ± 0.045	0.002 ± 0.000	0.730 ± 0.012	0.075 ± 0.027
	S-LIME	$\mathbf{0.854 \pm 0.042}$	$\mathbf{0.002 \pm 0.001}$	$\mathbf{0.749 \pm 0.010}$	$\mathbf{0.085 \pm 0.034}$
recidivism	LIME	0.996 ± 0.003	0.002 ± 0.000	0.927 ± 0.005	0.738 ± 0.010
	S-LIME	$\mathbf{0.996 \pm 0.003}$	$\mathbf{0.001 \pm 0.000}$	$\mathbf{0.945 \pm 0.003}$	$\mathbf{0.784 \pm 0.008}$
red wine	LIME	0.894 ± 0.060	0.020 ± 0.011	0.572 ± 0.083	0.047 ± 0.032
	S-LIME	$\mathbf{0.970 \pm 0.015}$	$\mathbf{0.018 \pm 0.011}$	$\mathbf{0.620 \pm 0.087}$	$\mathbf{0.052 \pm 0.036}$

Figure 1(a) illustrates the number of interpretable components/features represented to the user. It reveals the required number of features for explaining a sample, i.e., k, varies regarding the data sets and black box models; however, both methods have shown that samples can be explained with 3 or 4 features roughly. Referring to Fig. 1(b), the only downside of S-LIME is a negligible extra *time* required for explaining a sample that is not remarkable compared to the faithfulness of explanations. It is worth noting that choosing a proper neighborhood size N as well as finding suitable hyper-parameters for T can further improve the performance of the explanation method.

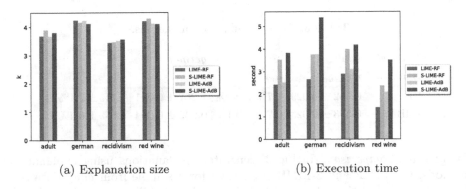

(a) Explanation size   (b) Execution time

Fig. 1. LIME vs S-LIME: comparison of explanation size and execution time.

Neighborhood Analysis. We analyzed the effect of the proposed sampling method in the formation of the locality for the sample being explained. As mentioned earlier, compactness and diversity are two important factors in generating neighborhood samples. According to our conducted experiments, we observed that S-LIME generates a higher number of data points in the proximity of the given sample. One example of this observation is illustrated in Fig. 2, which demonstrates the distribution of samples in the vicinity of a given sample from *recidivism* data set. For visualizing the data points in a 2-D space, Principal Component Analysis (PCA) [18] is used to reduce the feature dimension.

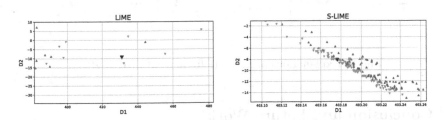

Fig. 2. Neighborhoods of LIME (left) and S-LIME (right). The '▼' denotes the sample being explained.

According to Tables 1 and 2, our devised sampling improves the fidelity of the explanation method significantly without creating exceedingly similar samples or an imbalanced data set. Based on analysis results listed in Table 3, the proposed method, on average, makes a minor number of features, i.e., $n_{features}$, in each neighborhood data point similar to the given sample. Moreover, the percentage of instances that are predicted differently to the opposite class, i.e., n_{preds}, is about 1%. As a result, the sampling procedure manipulates the features that are less prone to change the prediction of the sample. Therefore, by adopting the proposed sampling, we transform a randomly generated data to a meaningful one (with respect to the given sample) which is highly diverse and representative.

Table 3. Sample manipulation analysis.

Data set	adult		german	
Black box	$n_{features}$	n_{preds}	$n_{features}$	n_{preds}
RF	0.682 ± 0.029	$0.7\% \pm 0.1\%$	1.267 ± 0.091	$1.1\% \pm 0.1\%$
AdB	0.563 ± 0.023	$0.7\% \pm 0.1\%$	0.942 ± 0.054	$1.8\% \pm 0.4\%$

Explanation Comparison. In Fig. 3, generated explanations using explanation methods, i.e., LIME and S-LIME, are shown for a sample from *adult* data set. Above each explanation, the predicted class and probability by the black box f as well as the predicted class by decision tree g^D and the response value predicted by linear regression g^L are stated. An explanation is directly related to the prediction of the interpretable model, hence it is vital to establish the fidelity of its prediction before one count on the provided explanation. Fig. 3 reveals that S-LIME predicted the sample to the correct class with a high probability while LIME classifies the sample incorrectly. To make sure about the stability of predictions and corresponding explanations, we repeated the experiments multiple times; LIME experienced some fluctuation while S-LIME produced more uniform results in terms of prediction and explanation.

Fig. 3. Explanations of LIME (left) and S-LIME (right).

4   Conclusion and Future Work

In this paper, we introduced a meaningful data sampling technique applicable to local explanation methods. It exploits the similarity of feature values and feature importance viewpoints simultaneously to create a diverse and compact locality for the sample being explained. The methodology can be integrated into several local explanation methods for achieving meaningful and representative neighborhood samples. We demonstrated its applicability by applying it to the state-of-the-art explanation method, LIME. The obtained results are promising in terms of fidelity and explainability. In future work, we will study the effect of the proposed approach on rule-based explanation methods. Furthermore, its applicability on a real world use-case will be investigated.

References

1. Biau, G., Scornet, E.: A random forest guided tour. Test **25**(2), 197–227 (2016)
2. Breiman, L.: Random forests. Mach. Learn. **45**, 5–32 (2001)
3. Dougherty, J., Kohavi, R., Sahami, M.: Supervised and unsupervised discretization of continuous features. In: Proceedings Machine Learning, pp. 194–202 (1995)
4. Guidotti, R., Monreale, A., Ruggieri, S., Pedreschi, D., Turini, F., Giannotti, F.: Local rule-based explanations of black box decision systems. CoRR (2018)
5. Guidotti, R., Monreale, A., Ruggieri, S., Turini, F., Giannotti, F., Pedreschi, D.: A survey of methods for explaining black box models. ACM Comput. Surv. **51**(5), 93 (2018)
6. Han, J., Kamber, M., Pei, J.: Data Mining: Concepts and Techniques. Elsevier Science, Amsterdam (2011)
7. Krizhevsky, A., Sutskever, I., Hinton, G.E.: ImageNet classification with deep convolutional neural networks. In: Advances in Neural Information Processing Systems, pp. 1097–1105. Curran Associates Inc. (2012)
8. Lundberg, S.M., Lee, S.-I.: A unified approach to interpreting model predictions. In: Advances in Neural Information Processing Systems, pp. 4765–4774 (2017)
9. Molnar, C.: Interpretable Machine Learning: A Guide for Making Black Box Models Explainable (2019). https://christophm.github.io/interpretable-ml-book/
10. Ribeiro, M.T., Singh, S., Guestrin, C.: Why should I trust you? Explaining the predictions of any classifier. In: Proceedings of the 22nd ACM SIGKDD International Conference on Knowledge Discovery and Data Mining, pp. 1135–1144 (2016)
11. Robnik-Šikonja, M., Bohanec, M.: Perturbation-based explanations of prediction models. In: Zhou, J., Chen, F. (eds.) Human and Machine Learning. HIS, pp. 159–175. Springer, Cham (2018). https://doi.org/10.1007/978-3-319-90403-0_9
12. Robnik-Sikonja, M., Kononenko, I.: Explaining classifications for individual instances. IEEE TKDE **20**(5), 589–600 (2008)
13. Saabas, A.: TreeInterpreter Library (2019). https://github.com/andosa/treeinterpreter
14. Shrikumar, A., Greenside, P., Kundaje, A.: Learning important features through propagating activation differences. In: Proceedings of the 34th International Conference on Machine Learning, pp. 3145–3153 (2017)
15. Štrumbelj, E., Kononenko, I.: Explaining prediction models and individual predictions with feature contributions. Knowl. Inf. Syst. **41**, 647–665 (2014)
16. Štrumbelj, E., Kononenko, I., Robnik Šikonja, M.: Explaining instance classifications with interactions of subsets of feature values. Data Knowl. Eng. **68**(10), 886–904 (2009)

17. Sturges, H.A.: The choice of a class interval. J. Am. Stat. Assoc. **21**(153), 65–66 (1926)
18. Tipping, M.E., Bishop, C.M.: Probabilistic principal component analysis. J. R. Stat. Soc. Ser. B (Stat. Methodol.) **61**(3), 611–622 (1999)
19. Turner, R.: A model explanation system. In: IEEE 26th International Workshop on Machine Learning for Signal Processing, pp. 1–6 (2016)
20. Zhu, J., Zou, H., Rosset, S., Hastie, T.: Multi-class AdaBoost*. Stat. Interface **2**, 349–360 (2009)

Classifying Prostate Histological Images Using Deep Gaussian Processes on a New Optical Density Granulometry-Based Descriptor

Miguel López-Pérez[1]([✉])[iD], Adrián Colomer[2][iD], María A. Sales[3],
Rafael Molina[1][iD], and Valery Naranjo[2][iD]

[1] Departamento de Ciencias de la Computación e I.A., University of Granada,
Granada, Spain
mlopez@decsai.ugr.es
[2] Instituto de Investigación e Innovación en Bioingeniería,
I3B, Universitat Politècnica de València, Valencia, Spain
[3] Servicio de Anatomía Patológica, Hospital Clínico Universitario de Valencia,
Valencia, Spain

Abstract. The increasing use of whole slide digital scanners has led to an enormous interest in the application of machine learning techniques to detect prostate cancer using eosin and hematoxylin stained histopathological images. In this work the above problem is approached as follows: the optical density of each whole slide image is calculated and its eosin and hematoxylin concentration components estimated. Then, hand-crafted features, which are expected to capture the expertise of pathologists, are extracted from patches of these two concentration components. Finally, patches are classified using a Deep Gaussian Process on the extracted features. The new approach outperforms current state of the art shallow as well as deep classifiers like InceptionV3, Xception and VGG19 with an AUC value higher than 0.98.

Keywords: Prostate cancer · Optical Density · Texture features · Morphological features · Deep Gaussian Processes

1 Introduction

Prostate cancer is one of the most common causes of cancer death in men [8], being its early diagnosis essential. Nowadays, tissue histopathological slides can be digitized and stored thanks to the advent of digital scanners, which provide high-resolution images, called Whole-slide images (WSI). Access to this digital information and the need to reduce pathologists' workload, who have to daily deal with an increasing number of very high-resolution images, encourages the

This work was supported by the Spanish Ministry of Economy and Competitiveness through project DPI2016-77869.

© Springer Nature Switzerland AG 2019
H. Yin et al. (Eds.): IDEAL 2019, LNCS 11871, pp. 39–46, 2019.
https://doi.org/10.1007/978-3-030-33607-3_5

use of machine learning techniques on WSI for prostate cancer detection and grading, [3]. Two approaches can be found in the literature to tackle WSI classification problems. The first one is based on the use of deep learning methods [4], and will not be discussed here (a comparison will be provided in the experimental section). The second one is based on the extraction of hand-crafted features which are expected to capture the expertise of pathologists. From WSIs, their Hematoxylin and Eosin (H&E) components are extracted from either RGB or Optical Density (OD) space images [6]. Then, features such as LBP, HOG, and SIFT are used to encode image characteristics. Finally, classifications methods are applied. Recently, shallow Gaussian Processes (GP) have also been proposed with promising results [2].

In this work, in contrast to the approach massively utilized in the literature, we use the Optical Density (OD) space after performing color deconvolution as a more informative histological image space. In this space, texture and morphological hand-crafted features are extracted from H&E concentration images, and their interpretation provided. Finally, a Deep Gaussian Process classifier is used. We show that the followed approach outperforms classical as well as deep learning methods for the same task.

2   Color Deconvolution

For each WSI, the three-channel image information is the RGB intensity detected by a brightfield microscope observing a stained prostate histological slide. H&E are the stains usually used in pathology. Each $M \times N$ image is denoted by $\mathbf{I}$ with columns $\mathbf{i}_c = (i_{1c}, \ldots, i_{MN_c})^T$, $c \in \{R, G, B\}$.

According to the Lambert-Beer's law we can express the Optical Density (OD) for channel c of the slide as $\mathbf{y}_c = -\log(\mathbf{i}_c/i_c^0) \in \mathbb{R}^{MN \times 1}$, where $i_c^0 = 255$ is the incident light. Slides are stained using $\mathbf{n}_s = 3$ stains, $s \in \{H, E, Res\}$ (to obtain a unique stain decomposition we consider a third stain which represents the residual part) then the observed OD multichannel $\mathbf{Y} = [\mathbf{y}_R, \mathbf{y}_G, \mathbf{y}_B] \in \mathbb{R}^{MN \times 3}$ can be decomposed using $\mathbf{Y}^T = \mathbf{M}\mathbf{C}^T$, where $\mathbf{C} = [\mathbf{c}_1, \mathbf{c}_2, \mathbf{c}_3] \in \mathbb{R}^{MN \times 3}$ is the stain concentration matrix, with $\mathbf{c}_s$, the s-th column of $\mathbf{C}$, containing at each pixel position the concentration of stain color s and $\mathbf{M} \in \mathbb{R}^{3 \times 3}$ denoting the normalized stain matrix of the form specified in [6] where the s-th column of $\mathbf{M}$, $\mathbf{m}_s$, denotes the specific color of stain s.

The stain concentration matrix is recovered using $\mathbf{C}^T = \mathbf{M}^{-1}\mathbf{Y}^T$. Concentrations are transformed back to color (RGB) images using $\mathbf{y}_s^{sep} = \exp(-\mathbf{m}_s \mathbf{c}_s^T)$, $s \in \{H, E\}$. Features are usually extracted from the single channel images $\exp(-\mathbf{c}_s)$, $s \in \{H, E\}$ in the so called RGB space. In this work, we propose to perform this step in the OD, that is, directly on $\mathbf{c}_s, s \in \{H, E\}$. Figure 1 shows three different images from three different biopsies (and patients), one benign and two pathological, and their corresponding OD concentrations, Hematoxylin in the first row and Eosin in the second one. OD Hematoxylin captures nuclei information while OD Eosin contains information on stroma and cytoplasms.

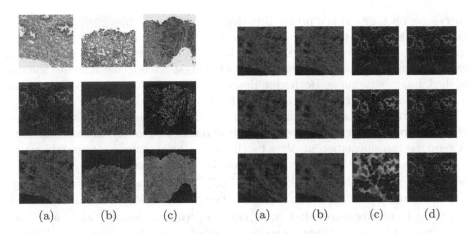

Fig. 1. Hematoxylin (second row) and Eosin (third row) in OD space for three samples: (a) Benign (b) and (c) malign.

Fig. 2. Granulometry profiles (steps $s = 1, 4, 16$) for image (a) in Fig. 1: (a) Π_φ; (b) Π_φ^r; (c) Π_γ; (d) Π_γ^r.

3 Feature Extraction

Let $\mathbf{z}$ be a gray-level image. We can define a morphological descriptor, using the opening operator $\gamma_i(\mathbf{z})$ applied to the image $\mathbf{z}$ with a SE (window) of size i. This opening operator can be expressed as the combination of an erosion ($\epsilon_i(\mathbf{z})$) followed by a dilation ($\delta_i(\mathbf{z})$), both with the SE of size i. When this opening is computed with a SE of increasing size (λ), we obtain a morphological opening pyramid (or granulometry profile) which can be formalized as:

$$\Pi_\gamma(\mathbf{z}) = \{\Pi_{\gamma\lambda} : \Pi_{\gamma\lambda} = \gamma_\lambda(\mathbf{z}), \forall \lambda \in [0, ..., n_{max}]\} \quad (1)$$

where n_{max} represents the maximum size of the structuring element, and the sizes increase in steps s.

By duality, a closing, $\varphi_i(\mathbf{z})$ is defined as the dilation of $\mathbf{z}$ followed by an erosion, both with a SE of size i. In the same way, a morphological closing pyramid is an anti-granulometry profile and can be computed on the image performing repeated closings with a SE of increasing size (λ) defined as:

$$\Pi_\varphi(\mathbf{z}) = \{\Pi_{\varphi\lambda} : \Pi_{\varphi\lambda} = \varphi_\lambda(\mathbf{z}), \forall \lambda \in [0, ..., n_{max}]\} \quad (2)$$

Making use of the opening pyramid (Π_γ), the granulometry curve or pattern spectrum of $\mathbf{z}$, $PS_\Gamma(\mathbf{z}, n)$, can be defined as:

$$PS_\Gamma(\mathbf{z}, n) = \frac{m(\Pi_{\gamma n}(\mathbf{z})) - m(\Pi_{\gamma n+1}(\mathbf{z}))}{m(\mathbf{z})}, \; n \geq 0 \quad (3)$$

where $m(\mathbf{z})$ is the Lebesgue measure of $\mathbf{z}$.

$PS_\Gamma(\mathbf{z}, n)$ (also called size density of $\mathbf{z}$) maps each size n to a measure of the bright image structures with this size: loss of bright image structures between two successive openings. It is a probability density function (a histogram) in which a large impulse in the pattern spectrum at a given scale indicates the presence of many image structures at that scale. By duality, the concept of pattern spectrum extends to anti-granulometry curve $PS_\Phi(\mathbf{z})$. This spectrum characterises the size of dark image structures. Both granulometry and anti-granulometry descriptors are concatenated to construct the final descriptor (*Gran*).

In this work, we also propose to use a variant of the granulometry, named geodesic granulometry (*GeoGran*), which is based on geodesic transformations. A geodesic transformation involves two images: a marker image (or patch) $\mathbf{y}$ and a reference image $\mathbf{z}$. The *geodesic dilation* is the iterative unitary dilation of $\mathbf{z}$ with respect to $\mathbf{y}$, that is, $\delta_{\mathbf{y}}^{(n)}(\mathbf{z}) = \delta_{\mathbf{y}}^{(1)}\delta_{\mathbf{y}}^{(n-1)}(\mathbf{z})$, being $\delta_{\mathbf{y}}^{(1)}(\mathbf{z}) = \delta_B(\mathbf{z}) \wedge \mathbf{y}$. The *reconstruction by dilation* is the successive geodesic dilation of $\mathbf{z}$ regarding $\mathbf{y}$ up to idempotence, that is, $R_{\mathbf{y}}^{\delta}(\mathbf{z}) = \delta_{\mathbf{y}}^{(i)}(\mathbf{z})$, so that $\delta_{\mathbf{y}}^{(i)}(\mathbf{z}) = \delta_{\mathbf{y}}^{(i+1)}(\mathbf{z})$. The *reconstruction by erosion* can be obtained as its dual operator, $R_{\mathbf{y}}^{\varepsilon}(\mathbf{z}) = [R_{\mathbf{y}^c}^{\delta}(\mathbf{z}^c)]^c$, being $\mathbf{z}^c$ the complement image (or patch). The reconstruction by dilation removes from the reference $\mathbf{z}$ the bright objects unconnected with the marker $\mathbf{y}$.

The underlying idea on which the new descriptor is based is to only consider in the granulometry spectrum the objects totally removed in each opening (closing) step. Using $\gamma(\mathbf{z})$ as indicated in Eq. (1) can lead to the inclusion in the pattern spectrum of fragments of objects partially removed in the process. To solve this shortcoming, we modify the granulometric profile (Eq. (1)) by using the geodesic opening given by $\gamma^r(\mathbf{z}) = R_{\gamma(\mathbf{z})}^{\delta}(\mathbf{z})$. By duality, the proposed geodesic closing, to be used in the computation of the anti-granulometric profile, (Eq. (2)) is $\varphi^r(\mathbf{z}) = R_{\varphi(\mathbf{z})}^{\varepsilon}(\mathbf{z})$. The new geodesic granulometric profiles will be denoted by Π_{γ}^r and Π_{φ}^r, and the descriptors $PS_\Gamma^r(\mathbf{z}, n)$ and $PS_\Phi^r(\mathbf{z}, -n)$, respectively.

The proposed framework, to discriminate between cancer and benign tissue in prostate, tries to mimic the way of analysis of a pathologist. Basically, the cancer destroys the tissue structure. A benign tissue is formed by glands, each of them with a lumen surrounded by cytoplasm and nuclei, distributed in a background of stroma (which also contains sparsely distributed nuclei) (Fig. 1-(a)). As cancer progresses, glands begin to proliferate and merge, destroying the structure of benign tissues. Cytoplasm and lumens disappear and stroma is invaded by nuclei. Figure 1, (first row), shows three different cancer stages ((a) benign, (b) grade 3, (c) grade 5). To capture in a descriptor the tissue structure, we propose to use PS_Φ with H as input image. This encodes the structure of the glands by recovering the structure of the nuclei which formed the gland frontiers (those that enclosed their lumen and cytoplasm). The granulometric profiles, Π_{φ}, for the three image examples are shown in Figs. 2-(c), 3-(c) and 4-(c). To capture stroma information, PS_Γ is applied on the E component. Figures 2-(a), 3-(a) and 4-(a) show the Π_{γ} profiles for the three examples. Figures 2, 3 and 4 also depict in columns (b) and (d) the geodesic profiles Π_{γ}^r and Π_{φ}^r, respectively. Note that Π_{φ}^r (columns (d)), for the three cases, shows that the results for different steps (different sizes of SEs) of the granulometric profile do not change. This suggests that stroma information more accurately extracted in

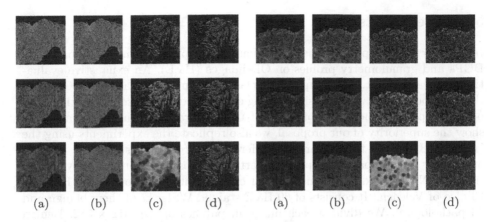

(a)   (b)   (c)   (d)   (a)   (b)   (c)   (d)

Fig. 3. Granulometry profiles (steps $s =$ 1, 4, 16) for image (b) in Fig. 1: (a) Π_φ; (b) Π_φ^r; (c) Π_γ; (d) Π_γ^r.

Fig. 4. Granulometry profiles (steps $s =$ 1, 4, 16) for image (c) in Fig. 1: (a) Π_φ; (b) Π_φ^r; (c) Π_γ; (d) Π_γ^r.

PS_Γ^r, is the most relevant information to discriminate between pathological and benign tissues (as results presented in the experimental section corroborate).

4 Deep Gaussian Process Classifier

In this section we provide a brief introduction to the use Gaussian Processes (GPs) in classification problems. Let us assume that we have n labelled training samples $\{(\mathbf{x}_i, y_i)\}_{i=1}^n$ where $\mathbf{x}_i \in \mathbb{R}^d$ is the feature vector and $y_i \in \{0, 1\}$ for a binary classification problem. Our observation model is $p(y_i|f_i) = \sigma^{y_i}(f_i)\sigma^{1-y_i}(f_i)$ where $\sigma(\cdot)$ denotes the sigmoid function. We use f_i instead of $f(\mathbf{x}_i)$ for simplicity.

In a GP based formulation of a supervised problem we assume that the distribution of $\mathbf{f} = (f_1, \ldots, f_n)^{\mathrm{T}}$ given $\mathbf{X}$ is a multivariate normal, where we assume zero mean for simplicity and a kernel function $k(\cdot, \cdot)$ defines our covariance matrix. In this paper we use the squared exponential kernel (SE) defined as $k(\mathbf{x}, \mathbf{x}') = C \exp(-\gamma\|\mathbf{x} - \mathbf{x}'\|^2)$, where the parameters C and γ will be estimated from the observations (the learning task).

Now we have all the ingredients we need to model our supervised learning problem using GPs. We first learn the model parameters (C, γ and for a regression problem ρ^2 as well) by maximizing on them the marginal log-likelihood $\log p(\mathbf{y})$ which will allow us to calculate $p(\mathbf{f}|\mathbf{y})$ and finally perform inference: given a new feature vector $\mathbf{x}^*$, we calculate $p(f_*|\mathbf{y}, \mathbf{x}_*, \mathbf{X})$ which will allow us to predict $\mathbf{y}_{\mathbf{x}^*}$.

In standard (single-layer) GPs, the output of the GP is directly used to model the observed response $\mathbf{y}$. However, this output could be used to define the input locations of another GP. If this is repeated L times, we obtain a hierarchy of GPs that is known as a Deep Gaussian Process (DGP) with $L + 1$ layers [7].

5   Experiments

In this section we present the validation of the proposed method, based on DGPs and granulometry profiles on OD images (H&E). An exhaustive evaluation was carried out. We compare the classification performance of DGP with the most popular shallow classifiers using classical texture descriptors, the granulometry profiles and a combination of them extracted from OD images. To show the superiority of our proposal, we also replicate the experiments using the H&E RGB images, to compare the different preprocessing strategies. Finally, our results are compared with state-of-art strategies based on a variety of pre-trained CNNs. The database used in our experiments was provided by Hospital Clinico of Valencia. It consists of 60 H&E stained WSIs, 17 of them benign and 43 pathological. We divided each image in patches of size 512×512. Benign patches are extracted from benign WSIs only. A patch is declared cancerous when it overlaps at least 20% of a region marked as tumoral by the pathologist. The final number of patches is 7867 (6725 benign and 1142 malign). Note that a 5-fold cross-validation strategy is carried out ensuring that patches from the same image are in the same fold. This process results in four partitions used to train the predictive models and a remaining fold employed to validate their performance.

As feature descriptors we computed the morphological descriptors PS_Φ and PS_Γ on H and E, respectively, and their geodesic versions PS_Φ^r and PS_Γ^r. PS_Φ and PS_Φ^r with SE of increasing size in steps of $s = 2$ from 0 to $n_{max} = 24$, and in steps of $s = 4$ for PS_Γ and PS_Γ^r from 0 to $n_{max} = 48$. Note that we use *Gran* and *GeoGran* labels to denote PS and PS^r descriptors, respectively. Besides that, to capture the texture information we use the uniform and rotationally invariant Local Binary Patterns (*LBP*) [5] as baseline descriptor (with neighbourhood of $R = 1$ and $P = 8$) and the combination of it with a contrast measure, according to the work of Guo et al. [1], obtaining an additional Local Binary Pattern Variance (*LBPV*) descriptor. The different combinations of descriptors have been labelled as *GranLBP*, *GranLBPV*, *GeoGranLBP* and *GeoGranLBPV*.

The shallow classifiers used in the experiments were Random Forest (RF), Extreme Gradient Boosting (xgboost) and Gaussian Processes (GP). We also included in the comparison a 3-layer DGP. This model is also compared with deterministic deep learning-based models. We use the area under the curve ROC (AUC) as the evaluation metric for this binary classification problem.

Table 1 shows the classification results obtained in RGB space while Table 2 shows AUC values in OD space. Analysing both tables, we observe how morphological features encode better the key tumoral information than *LBP(V)* in both spaces. More in depth, *LBPV* outperforms the basic *LBP* in RGB and OD while *GeoGran* performs clearly better than *Gran* in OD although both feature vectors show a similar behaviour in RGB. Regarding the classification technique, GPs and DGPs discriminate in a more successful way than the rest in both spaces. For every descriptor and classifier, the results obtained in the OD space are superior to those achieved in the RGB space, in which the majority of state-of-the-art works based their approaches. From Tables 1 and 2, we can observe

Table 1. Performance of descriptors and classifiers in RGB space

AUC	RF	XgBoost	GP	DGP
LBP	0.6663 ± 0.1400	0.6728 ± 0.1279	0.7003 ± 0.1190	0.6824 ± 0.1230
LBPV	0.7695 ± 0.0565	0.7912 ± 0.0674	0.8243 ± 0.0891	0.8281 ± 0.0917
Gran	0.8549 ± 0.0856	0.8778 ± 0.0735	0.8984 ± 0.0641	0.9004 ± 0.0619
GeoGran	0.9089 ± 0.0494	0.9095 ± 0.0454	0.8910 ± 0.0599	0.9001 ± 0.0547
GranLBP	0.8331 ± 0.0949	0.8551 ± 0.0842	0.9111 ± 0.0492	0.8979 ± 0.0570
GranLBPV	0.8758 ± 0.0611	0.8908 ± 0.0509	0.9280 ± 0.0349	$\mathbf{0.9458 \pm 0.0396}$
GeoGranLBP	0.8958 ± 0.0566	0.9048 ± 0.0469	0.9014 ± 0.0507	0.9019 ± 0.0473
GeoGranLBPV	$\mathbf{0.9174 \pm 0.0351}$	$\mathbf{0.9273 \pm 0.0329}$	$\mathbf{0.9307 \pm 0.0307}$	0.9333 ± 0.0309

how in the OD space, the relevant features of histopathological images related to both texture and morphology information are better captured. In addition, as we can see in Table 2, the best result is achieved in OD space using *GeoGranLBPV* together with DGPs (AUC = 0.9829). This fact suggests that texture and morphology features provide complementary information to characterize prostatic tumoral tissues.

Finally, with the aim to evaluate the discriminatory ability of our best predictive model (i.e. *GeoGranLBPV* and DGPs), a comparison with models based on convolutional neural networks (CNN) is established. CNNs automatically extract feature maps containing the relevant information of the input images. Due to the reduced number of samples of our data set to train from scratch a CNN, in this experiment, we fine-tuned three of the most relevant state-of-the-art architectures. In particular, Inceptionv3, Xception and VGG19 neural networks are initialized with the weights achieved in the ImageNet challenge and re-trained using our RGB histological images as input. Note that the re-training process is performed from different layers depending on the architecture: Inceptionv3 from 'mixed 7' layer, Xception from 'add 10' layer and VGG19 from 'block3_conv1' layer. This procedure is carried out during 100 epochs in which the binary cross-entropy loss function is minimized using Stochastic Gradient Descent optimizer with a learning rate of $lr = 1 \cdot 10^{-4}$. Data augmentation technique is used to synthetically increase the number of instances and a batch size of 16 samples is established due to the available memory of the NVIDIA Titan V GPU utilized in this work.

After training the predictive models based on CNNs along the five partitions, the same validation process as described before was carried out. InceptionV3 (AUC = 0.9196 ± 0.0302) and Xception (AUC = 0.9210 ± 0.0260) present comparable results to those obtained using GPs and morphological features in the RGB space. Notice that VGG19 (AUC = 0.9813 ± 0.0068) is clearly better than DGPs in this space. In the OD space, simple textural features as *LBP* (with independence of the classifier used) are competitive to InceptionV3 and Xception CNN models. In addition, the best predictive model in the OD space, i.e. *GeoGranLBPV* and DGPs with AUC = 0.9829 (see Table 2), outperforms the best result obtained by a CNN (i.e. VGG19 with AUC = 0.9813).

Table 2. Performance of descriptors and classifiers in Optical Density space

AUC	RF	XgBoost	GP	DGP
LBP	0.9300 ± 0.0603	0.9262 ± 0.0615	0.9253 ± 0.0635	0.9261 ± 0.0618
LBPV	0.9351 ± 0.0373	0.9421 ± 0.0243	0.9443 ± 0.0314	0.9521 ± 0.0259
Gran	0.9323 ± 0.0453	0.9461 ± 0.0322	0.9516 ± 0.0346	0.9551 ± 0.0339
GeoGran	0.9690 ± 0.0303	0.9688 ± 0.0249	0.9636 ± 0.0242	0.9666 ± 0.0216
GranLBP	0.9436 ± 0.0640	0.9541 ± 0.0524	0.9581 ± 0.0422	0.9649 ± 0.0379
GranLBPV	0.9370 ± 0.0340	0.9573 ± 0.0206	0.9696 ± 0.0175	0.9725 ± 0.0151
GeoGranLBP	0.9666 ± 0.0408	0.9700 ± 0.0304	0.9669 ± 0.0283	0.9715 ± 0.0244
GeoGranLBPV	**0.9692 ± 0.0241**	**0.9747 ± 0.0170**	**0.9807 ± 0.0097**	**0.9829 ± 0.0092**

6   Conclusions and Future Work

In this work we have introduced a new OD granulometry-based descriptor for prostate cancer classification in histopathological images. The proposed descriptor extracts the morphological information in the OD space as a more informative space. For classification we use an state of the art DGP model. The obtained results show the better performance of both texture and morphological descriptors on H&E OD images. Moreover, the OD geodesic granulometry descriptor reveals the importance of the stroma identifying cancer. With the combination of OD features and DGP we can discriminate almost perfectly whether a patch is cancerous or not. Besides, the proposed model outperforms current state-of-art CNNs.

References

1. Guo, Z., Zhang, L., Zhang, D.: Rotation invariant texture classification using LBP variance (LBPV) with global matching. Pattern Recogn. **43**(3), 706–719 (2010)
2. Kandemir, M., Zhang, C., Hamprecht, F.A.: Empowering multiple instance histopathology cancer diagnosis by cell graphs. In: Golland, P., Hata, N., Barillot, C., Hornegger, J., Howe, R. (eds.) MICCAI 2014. LNCS, vol. 8674, pp. 228–235. Springer, Cham (2014). https://doi.org/10.1007/978-3-319-10470-6_29
3. Komura, D., Ishikawa, S.: Machine learning methods for histopathological image analysis. Comput. Struct. Biotechnol. J. **16**, 34–42 (2018)
4. Litjens, G., et al.: Deep learning as a tool for increased accuracy and efficiency of histopathological diagnosis. Sci. Rep. **6**, 26286 (2016)
5. Pietikäinen, M., Hadid, A., Zhao, G., Ahonen, T.: Computer Vision Using Local Binary Patterns. Springer, Heidelberg (2011). https://doi.org/10.1007/978-0-85729-748-8
6. Ruifrok, A.C., Johnston, D.A.: Quantification of histochemical staining by color deconvolution. Anal. Quant. Cytol. Histol. **23**(4), 291–299 (2001)
7. Salimbeni, H., Deisenroth, M.: Doubly stochastic variational inference for deep Gaussian processes. In: NIPS, pp. 4591–4602 (2017)
8. Siegel, R.L., Miller, K.D., Jemal, A.: Cancer statistics. CA Cancer J. Clin. **68**(1), 7–30 (2018)

Adaptive Orthogonal Characteristics of Bio-inspired Neural Networks

Naohiro Ishii[1(✉)], Toshinori Deguchi[2], Masashi Kawaguchi[3], Hiroshi Sasaki[4], and Tokuro Matsuo[1]

[1] Advanced Institute of Industrial Technology, Tokyo, Japan
nishii@acm.org, matsuo@aiit.ac.jp
[2] Gifu College, National Institute of Technology, Gifu, Japan
deguchi@gifunct.ac.jp
[3] Suzuka College, National Institute of Technology, Mie, Japan
masashi@elec.suzukact.ac.jp
[4] Fukui University of Technology, Fukui, Japan
hsasaki@fukui-ut.ac.jp

Abstract. Neural networks researches are developed for the machine learning and the recent deep learning. Adaptive mechanisms of neural networks are prominent in the visual pathway. In this paper, adaptive orthogonal properties are studied, First, this paper proposes a model of the bio-inspired asymmetric neural networks. The prominent features are the nonlinear characteristics as the squaring and rectification functions, which are observed in the retinal and visual cortex networks. In this paper, the proposed asymmetric network with Gabor filters and the conventional energy model are analyzed from the orthogonality characteristics. Second, it is shown that the biological asymmetric network is effective for generating the orthogonality function using the network correlation computations. Finally, the asymmetric networks with nonlinear characteristics are able to generate independent subspaces, which will be useful for the creation of features spaces and efficient computations in the learning.

Keywords: Asymmetric neural network · Gabor filter · Correlation and orthogonality analysis · Energy model · Linear and nonlinear pathways

1 Introduction

Neural networks currently plays an important role in the processing complex tasks for the visual perception and the deep learning. To estimate the visual motion and adaptation, biological sensory information models have been studied [1–3]. For the deep learning efficiently, an independent projection from the inputs are studied in the convolutional neural networks [4]. Widrow et al. [5] showed neurons nonlinear characteristics generate independent outputs for network learning. For the efficient deep learning, the orthogonalization in the weight matrix of neural networks are studied using optimization methods [6]. Further, the feature vectors from different classes are expected to be as orthogonal as possible for efficient learning [7]. It is important to make clear the network structures how to generate the independence and orthogonality

© Springer Nature Switzerland AG 2019
H. Yin et al. (Eds.): IDEAL 2019, LNCS 11871, pp. 47–59, 2019.
https://doi.org/10.1007/978-3-030-33607-3_6

relations, which will generate feature spaces, effectively. By using Gabor filters, a symmetric network model was developed for the motion detection, which is called energy model [3, 10]. This paper develops an asymmetrical networks, which is based on the catfish retina [14, 15]. In this paper, adaptive orthogonal properties are shown in the asymmetrical neural networks. Then, the orthogonality characteristics between the asymmetric networks and the conventional energy model are compared. In the biological visual systems, the nonlinear characteristics are observed as the squaring function and rectification function [8, 9]. It is shown that the asymmetric network with Gabor filters has orthogonal properties strongly under stimulus conditions, while the conventional energy symmetric network is weak in the orthogonality. Since the visual cortex is derived as the extended asymmetric networks, it is shown that their nonlinearities generate independent subspaces.

2 Asymmetric Neural Networks

2.1 Background of Asymmetric Neural Networks

In the biological neural networks, the structure of the network, is closely related to the functions of the network. Naka et al. [13] presented a simplified, but essential networks of catfish inner retina as shown in Fig. 1. Visual perception is carried out firstly in the retinal neural network as the special processing between neurons.

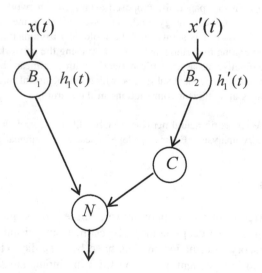

Fig. 1. Asymmetric network with linear and squaring nonlinear pathways

Visual perception is carried out firstly in the retinal neural network as the special processing between neurons. The following asymmetric neural network is extracted from the catfish retinal network [13]. The asymmetric structure network with a quadratic nonlinearity is shown in Fig. 1, which composes of the pathway from the bipolar cell B to the amacrine cell N and that from the bipolar cell B, via the amacrine cell C to the N [12, 13]. Figure 1 shows a network which plays an important role in the movement perception as the fundamental network. It is shown that N cell response is realized by a linear filter, which is composed of a differentiation filter followed by a low-pass filter. Thus, the asymmetric network in Fig. 1 is composed of a linear pathway and a nonlinear pathway with the cell C, which works as a squaring function.

3   Orthogonal Characteristics of Asymmetric Networks

3.1   Orthogonality of Asymmetric Network Under the Stimulus Condition

The inner orthogonality under the constant value stimulus is computed in the asymmetric networks as shown in Fig. 2.

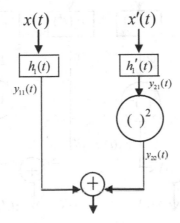

Fig. 2. Asymmetric network unit with Gabor filters

The variable t in the Gabor filters is changed to t', where by setting $\xi \triangleq 2\pi\omega$ in the Eq. (1), $t' = 2\pi\omega t = \xi t$ and $dt = dt/\xi$ hold. Then, Gabor filters are shown as

$$G_s(t') = \frac{1}{\sqrt{2\pi}\sigma}e^{-\frac{t'^2}{2\sigma^2\xi^2}}sin(t') \quad \text{and} \quad G_c(t') = \frac{1}{\sqrt{2\pi}\sigma}e^{-\frac{t'^2}{2\sigma^2\xi^2}}cos(t') \tag{1}$$

The impulse response functions $h_1(t)$ and $h_1'(t)$ are replaced by $G_s(t')$ and $G_c(t')$ or vice versa. The outputs of these linear filters are given as follows,

$$y_{11}(t) = \int_0^\infty h_1(t')x(t-t')dt' \tag{2}$$

$$y_{21}(t) = \int_0^\infty h_1'(t')x(t-t')dt' \tag{3}$$

Adelson and Bergen [3] proposed an energy model with Gabor filters and their squaring function as a functional unit. To verify the orthogonality among their units, a parallel units are located as shown in Fig. 3, in which the asymmetric networks units are proposed here. In Fig. 3, inputs and output of the left asymmetrical network unit are shown in $x(t), x'(t)$ and $y(t)$, while those of the right asymmetrical unit are shown in $v(t'), v'(t')$ and $z(t')$. Static conditions imply the brightness of input images do not change to be constant to time, while the dynamic conditions show they change to time. In Fig. 4, the first row (a) shows the same stimuli is inputted in $x(t), x'(t), v(t')$ and $v'(t')$, which are shown in white circle. In the second row (b) in Fig. 4, the rightmost stimulus $v'(t')$ is changed to new different one. The $(c), (d)$ and (e) in Fig. 4 show the stimulus changes are moved.

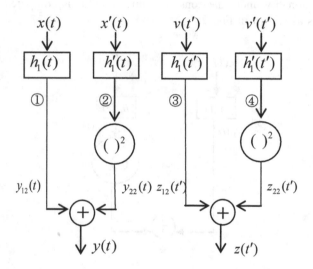

Fig. 3. Orthogonality computations between asymmetric networks units

	$x(t)$	$x'(t)$	$v(t')$	$v'(t')$
(a)	○	○	○	○
(b)	○	○	○	●
(c)	○	○	●	●
(d)	○	●	●	●
(e)	●	●	●	●

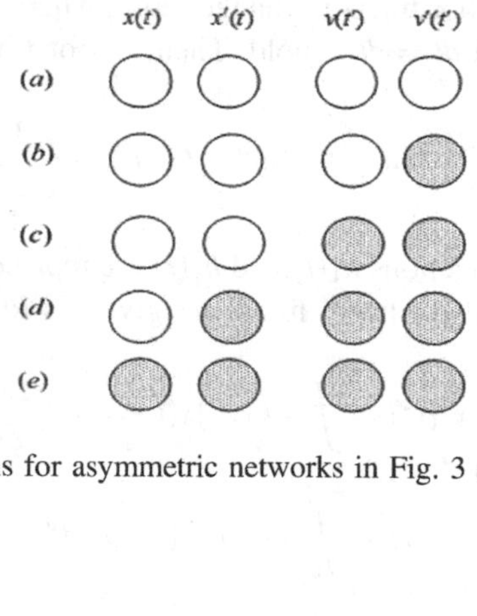

Fig. 4. Stimulus conditions for asymmetric networks in Fig. 3 and energy model in Fig. 5

3.2 Orthogonality Under Static Conditions

Static condition means the input brightness is constant, which does not change to time.. In the first row (a) in Fig. 4, the all the inputs are assumed to be the constant value B, while the rightmost input in (b) is assumed to be constant value. C. The orthogonality computation is performed in the following pairs pathways ①③, ①④, ②③, and ②④ in Fig. 3. Under the condition (a), the non-orthogonality between ① and ③ is computed as

$$\int_{-\infty}^{\infty} y_{12}(t)z_{12}(t)dt = \frac{B^2}{2\pi\xi\sigma^2} \int_{-\infty}^{\infty} e^{-\frac{t'^2}{\sigma^2\xi^2}} \sin(t')\sin(t')dt'$$

$$= \frac{B^2}{2\pi\xi\sigma^2} \int_{-\infty}^{\infty} e^{-\frac{t'^2}{\sigma^2\xi^2}} \sin^2(t')dt' > 0 \qquad (4)$$

Second, the orthogonality between ① and ④ is computed as

$$\int_{-\infty}^{\infty} y_{12}(t)z_{22}(t)dt = \frac{1}{\sqrt{2\pi}\sigma} \cdot \left(\frac{B^2}{2\pi\xi\sigma^2}\right) \int_{-\infty}^{\infty} e^{-\frac{3t'^2}{2\sigma^2\xi^2}} \cos^2(t')\sin(t')dt' = 0 \qquad (5)$$

Third, the orthogonality between ② and ③ holds also as the same Eq. (5). Finally, the non-orthogonality between ② and ④ are computed as follows,

$$\int_{-\infty}^{\infty} y_{22}(t)z_{22}(t)dt = \left(\frac{1}{2\pi\sigma^2}\right)^2 \cdot \frac{B^2}{2\xi} \int_{-\infty}^{\infty} e^{-\frac{t'^2}{2\sigma^2\xi^2}} dt' - \left(\frac{1}{2\pi\sigma^2}\right)^2 \cdot \frac{B^2}{2\xi} \int_{-\infty}^{\infty} e^{-\frac{t'^2}{2\sigma^2\xi^2}} \cos(2t')dt'$$

$$-\left(\frac{1}{2\pi\sigma^2}\right)^2 \cdot \frac{B^2}{2\xi} \int_{-\infty}^{\infty} e^{-\frac{t'^2}{2\sigma^2\xi^2}} \sin^4(t')dt' = \frac{B^2}{16\sqrt{2\pi}\sigma^2}(1 - e^{-2\xi^2\sigma^2}) > 0$$

$$(6)$$

Thus, the orthogonality ratio among pathways is defined to be 50% in case of (a). Under the condition (b), similar equations hold as Eqs. (4), (5) and (6) except only B^2 is changed to BC in Eq. (5) for ①④ and (6) for ②④. Thus, the orthogonality ratio under the condition (b) is 50%. Similarly, conditions $(c), (d)$ and (e) under the static conditions shows the orthogonality ratio to be 50%.

3.3 Orthogonality Under Dynamic Conditions

To verify the orthogonality properties between the asymmetric networks units, the white noise stimuli [11, 12] are schematically shown in Fig. 4. In the first row in Fig. 5, one white noise is a low pass filtered one with zero mean and its power p, which is shown in the circles only under the input variables $x(t), x'(t), v(t)$ and $v'(t')$. Similarly, in the second row in Fig. 4, the other white noise is a high pass filtered one with zero mean and its power p', which is shown in the grayed circles under input variable $v'(t')$. The impulse response functions $h_1(t)$ and $h'_1(t)$ are replaced by the Gabor filters, $G_s(t')$ and $G_c(t')$ as shown in the Eq. (1). The stimulus with the high pass filtered noise is moved from the right to the left according to (a), (b), (c), (d) and (e) in front of the visual space. Under the stimulus condition (a) in Fig. 4, the correlation between outputs $y(t)$ and $z(t)$ between the asymmetrical networks with Gabor filters in Fig. 3, is computed as follows,

$$\int_{-\infty}^{\infty} y(t)z(t)\,dt = \int_{-\infty}^{\infty} dt \{ \int_{0}^{\infty} h_1(\tau)x(t-\tau)d\tau$$

$$+\int_{0}^{\infty}\int_{0}^{\infty} h'_1(\tau_1)h'_1(\tau_2)x'(t-\tau_1)x'(t-\tau_2)d\tau_1 d\tau_2 \} \times$$

$$\{\int_{0}^{\infty} h_1(\sigma)v(t-\sigma)d\sigma + \int_{0}^{\infty}\int_{0}^{\infty} h'_1(\sigma_1)h'_1(\sigma_2)v'(t-\sigma_1)v'(t-\sigma_2)d\sigma_1 d\sigma_2\}$$

$$=\int_{0}^{\infty}\int_{0}^{\infty} h_1(\tau)h'_1(\sigma)d\tau d\sigma E[x(t-\tau)v(t-\sigma)]$$

$$+\int_{0}^{\infty}\int_{0}^{\infty}\int_{0}^{\infty} h_1(\tau)h'_1(\sigma_1)h'_1(\sigma_2)E[x(t-\tau)v'(t-\sigma_1)v'(t-\sigma_2)]d\tau d\sigma_1 d\sigma_2$$

$$+\int_{0}^{\infty}\int_{0}^{\infty}\int_{0}^{\infty} h_1(\sigma)h'_1(\tau_1)h'_1(\tau_2)E[v(t-\sigma)x'(t-\tau_1)x'(t-\tau_2)]d\sigma d\tau_1 d\tau_2$$

$$+\int_{0}^{\infty}\int_{0}^{\infty}\int_{0}^{\infty}\int_{0}^{\infty} h'_1(\tau_1)h'_1(\tau_2)h'_1(\sigma_1)h'_1(\sigma_2)E[x'(t-\tau_1)x'(t-\tau_2)v'(t-\sigma_1)v'(t-\sigma_2)]d\tau_1 d\tau_2 d\sigma_1 d\sigma_2$$

$$=\{\int_{0}^{\infty} h_1(\tau)d\tau \int_{0}^{\infty} h'_1(\sigma)d\sigma\}\cdot p + 0 + 0 + 3p^2\{\int_{0}^{\infty} h'_1(\tau)d\tau\}^4$$

$$\quad ①③ \qquad\qquad ①④\quad ②③ \qquad ②④$$

$$(7)$$

where the first term of (5) of the path ways ① and ③, shown in ①, and the second and third terms are to be 0 by ①④ and ② ③. The fourth term is by ②④.

The terms ①③ and ②④ are not zero, because the following equations hold,

$$\int_{0}^{\infty} h_1(\tau)d\tau = \frac{1}{\sqrt{2\pi}\sigma}\int_{0}^{\infty} e^{-\frac{\tau^2}{2\sigma^2\xi^2}}\sin(\tau)d\tau = \frac{\xi}{\sqrt{\pi}}e^{-\frac{1}{2}\sigma^2\xi^2}\int_{0}^{\frac{1}{\sqrt{2}}\sigma\xi} e^{\tau^2}d\tau > 0 \quad (8)$$

and

$$\int_{0}^{\infty} h'_1(\tau)d\tau = \frac{1}{\sqrt{2\pi}\sigma}\int_{0}^{\infty} e^{-\frac{\tau^2}{2\sigma^2\xi^2}}\cos(\tau)d\tau = \frac{\xi}{2}e^{-\frac{1}{2}\sigma^2\xi^2} > 0, \quad (9)$$

where ξ is the center frequency of the Gabor filter.

Thus, since two pathways are zero in the correlation, while other two pathways (25% and 25%) are non-zero, the orthogonality becomes 50% for the stimuli (a) in Fig. 4.

Under the stimulus condition (b) in Fig. 4, the correlation between outputs $y(t)$ and $z(t)$ between the asymmetrical networks in Fig. 3, is computed as (10),

$$\int_{-\infty}^{\infty} y(t)z(t)dt = p\{\int_{0}^{\infty} h_1(\tau)d\tau \int_{0}^{\infty} h_1'(\sigma)d\sigma\} + 0 + 0 + 0 \tag{10}$$

$$\textcircled{1}\textcircled{3}\quad \textcircled{1}\textcircled{4}\quad \textcircled{2}\textcircled{3}\quad \textcircled{2}\textcircled{4}$$

Since three pathways are zero in the correlation, while the first pathway is non-zero, (25%), the orthogonality becomes 75% for the stimuli (b) in Fig. 4. Similarly, under the stimulus condition (c) in Fig. 4, the orthogonality becomes 92%. Under the stimulus condition (d) in Fig. 4, the orthogonality becomes 75%. Under the stimulus condition (e) in Fig. 4, the orthogonality becomes 75%.

4　Orthogonal Properties of Conventional Energy Model

A conventional energy model [3] is extensively applied to the motion detection and physiological models as the fundamental model of the neural networks. In Fig. 5, two energy models are shown in parallel.

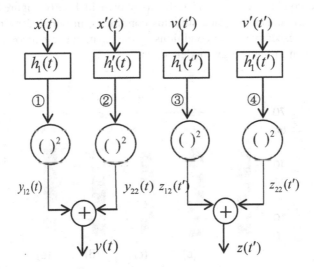

Fig. 5. Orthogonality computations between energy model units

4.1　Orthogonality Under Static Conditions

Under the condition (a) in Fig. 4, the non-orthogonality between $\textcircled{1}$ and $\textcircled{3}$ becomes

$$\int_{-\infty}^{\infty} y_{12}(t)z_{12}(t)dt = (\frac{B^2}{2\pi\sigma^2})^2 \int_{-\infty}^{\infty} e^{-\frac{2t'^2}{\sigma^2\xi^2}} sin^2(t')sin^2(t')dt'$$

$$= (\frac{B^2}{2\pi\xi\sigma^2}) \int_{-\infty}^{\infty} e^{-\frac{2t'^2}{\sigma^2\xi^2}} sin^4(t')dt' > 0 \tag{11}$$

Second, the non-orthogonality between ① and ④ is computed as

$$\int_{-\infty}^{\infty} y_{12}(t)z_{22}(t)dt = (\frac{B^2}{2\pi\sigma^2})^2 \int_{-\infty}^{\infty} e^{-\frac{2t^2}{\sigma^2\xi^2}}sin^2(t')cos^2(t')dt' > 0 \qquad (12)$$

Third, the non-orthogonality between ② and ③ is computed as the same Eq. (12). Fourth, the non-orthogonality between ② and ④

$$\int_{-\infty}^{\infty} y_{22}(t)z_{22}(t)dt = (\frac{B^2}{2\pi\sigma^2})^2 \int_{-\infty}^{\infty} e^{-\frac{2t^2}{\sigma^2\xi^2}}cos^2(t')cos^2(t')dt' > 0 \qquad (13)$$

Thus, under the condition (a), the orthogonality ratio becomes 0% in the energy model units in Fig. 6. Under the static conditions (b), (c), (d) and (e), the same equations hold as Eqs. (11), (12), (13). Thus, the orthogonality ratio 0% holds in the energy model units under all the conditions in Fig. 4. We can summarize the orthogonality ratio of the asymmetric units in the Sect. 3.2 and the energy model ones under the static conditions in the Sect. 4.3 as shown in Fig. 6. Figure 6 shows the orthogonality ration under the static stimulus conditions, in which 50% orthogonality ratio is shown in respective static conditions in the asymmetric networks units, while 0% ration is shown in the energy models units.

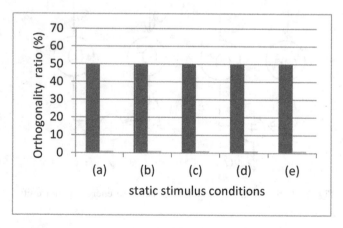

Fig. 6. Orthogonality ratio under static stimulus conditions. (a)–(e) show the stimulus conditions in Fig. 4.

4.2 Orthogonality Under Dynamic Conditions

Under the stimulus condition (a) in Fig. 4, the correlation between outputs $y(t)$ and $z(t)$ between the energy model units in Fig. 5, is computed as follows,

$$\int_{-\infty}^{\infty} y(t)z(t)\,dt = \int_{-\infty}^{\infty} dt \{ \int_0^{\infty}\int_0^{\infty} h_1(\tau_1)h_1(\tau_2)x(t-\tau_1)x(t-\tau_2)d\tau_1 d\tau_2 +$$

$$\int_0^{\infty}\int_0^{\infty} h_1'(\tau_1')h_1'(\tau_2')x'(t-\tau_1')x'(t-\tau_2')d\tau_1' d\tau_2' \}$$

$$\cdot \{ \int_0^{\infty}\int_0^{\infty} h_1(\sigma_1)h_1(\sigma_2)v(t-\sigma_1)v(t-\sigma_2)d\sigma_1 d\sigma_2 +$$

$$\int_0^{\infty}\int_0^{\infty} h_1'(\sigma_1')h_1'(\sigma_2')v'(t-\sigma_1')v'(t-\sigma_2')d\sigma_1' d\sigma_2' \}$$

(14)

The Eq. (14) consists of 4 cross- terms (4cross- pathways, ①③, ①④, ②③, ②④ in Fig. 5) computations. The solution of the Eq. (14) becomes

$$3p^2(\int_0^{\infty} h_1(\tau)d\tau)^4 + 3p^2(\int_0^{\infty} h_1(\tau)d\tau)^2 \cdot (\int_0^{\infty} h_1'(\sigma)d\sigma)^2 +$$

$$3p^2(\int_0^{\infty} h_1'(\tau)d\tau)^2 \cdot (\int_0^{\infty} h_1(\sigma)d\sigma)^2 + 3p^2(\int_0^{\infty} h_1'(\tau)d\tau)^4$$

(15)

The solution (15) shows these four terms are not zero, since the integral of the respective impulse functions of Gabor filters, are proved to be not zero by the Eqs. (8) and (9). Thus, the orthogonality ratio here is 0%. Under the stimulus condition (b) in Fig. 4, the orthogonality ratio is $(1/12) \times 4 = (1/3) \simeq 33\%$. Under the stimulus condition (c) in Fig. 4, the orthogonality ration is $(1/12) \times 6 = (1/2) = 50\%$. Under the stimulus condition (d) in Fig. 4, the orthogonality ratio is $(1/12) \times 4 = (1/3) \simeq 33\%$. Under the stimulus condition (e) in Fig. 4, the orthogonality ratio here is 0%.

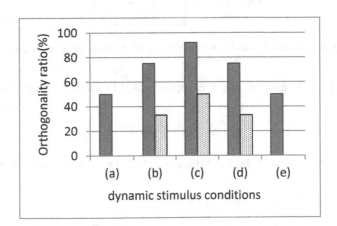

Fig. 7. Comparison of orthogonality under dynamic stimulus conditions. (a)–(e) show the stimulus conditions in Fig. 4.

4.3 Comparison of Orthogonal Properties Between the Asymmetric Network and the Energy Model

Orthogonal ratios between the asymmetric network and the conventional energy model are compared as shown in Fig. 7, in which the filled bar shows the orthogonality ratio by the asymmetric networks, while the dotted bar by the energy model.

5 Application of Asymmetric Networks to Bio-inspired Neural Networks

Here, we present a functional diagram of the neural network with half-wave rectification nonlinearity in the neural network of the brain cortex in Fig. 8 [8, 9].

5.1 Nonlinearity of Half-Wave Rectification to Generate Independent Subspaces

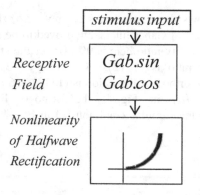

Fig. 8. Model of neural network V1 area of the biological brain cortex

The half-wave rectification in Fig. 8 is approximated in the following equation.

$$f(x) = \frac{1}{1 + e^{-\eta(x-\theta)}} \tag{16}$$

By Taylor expansion of the Eq. (16) at $x = \theta$, the Eq. (17) is derived as follows,

$$\begin{aligned}
f(x)_{x=\theta} &= f(\theta) + f'(\theta)(x - \theta) + \frac{1}{2!}f''(\theta)(x - \theta)^2 + \dots \\
&= \frac{1}{2} + \frac{\eta}{4}(x - \theta) + \frac{1}{2!}(-\frac{\eta^2}{4} + \frac{\eta^2 e^{-\eta\theta}}{2})(x - \theta)^2 + \dots
\end{aligned} \tag{17}$$

In Fig. 9, the nonlinear terms, $x^2, x^3, x^4, \dots$ are generated in the Eq. (17). Thus, the combination of Gabor function pairs ($G_{ab \sin}, G_{ab \cos}$) are generated in Fig. 9, in which the transformed network consists of the extended asymmetrical network in Fig. 2. Then, the characteristics of the extended asymmetric network have pathways with higher order nonlinearities.

5.2 Generation of Independent Sub-spaces

Combinations of Gabor function pairs $(G_{ab\ sin}, G_{ab\ cos})$ are generated according to the Eq. (18) as shown in Fig. 10. The characteristics of the extended asymmetric network have pathways with higher order nonlinearities. In Fig. 8, linear notation L shows term, $Ae^{-\frac{t'^2}{2\sigma^2\xi^2}}sin(t')$ or $Ae^{-\frac{t'^2}{2\sigma^2\xi^2}}cos(t')$, while S shows their doubles. Selective independent sub-spaces are generated in the layers in Fig. 10. The combination pairs of the Gabor filters are shown in (18) by approximated computations. Under the same stimulus condition in Fig. 4, the orthogonality is computed between items connecting solid lines in Fig. 10. Here, the connecting line shows over 50% orthogonality ratio in Fig. 10.

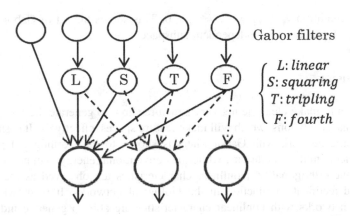

Fig. 9. A transformed network model for the layered network for one pathway in Fig. 6

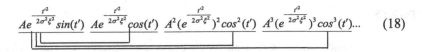

$$Ae^{-\frac{t'^2}{2\sigma^2\xi^2}}sin(t')\quad Ae^{-\frac{t'^2}{2\sigma^2\xi^2}}cos(t')\quad A^2(e^{-\frac{t'^2}{2\sigma^2\xi^2}})^2cos^2(t')\quad A^3(e^{-\frac{t'^2}{2\sigma^2\xi^2}})^3cos^3(t')... \quad (18)$$

Fig. 10. Orthogonal combination pairs among Gabor sine and cosines functions

Since $Ae^{-\frac{t'^2}{2\sigma^2\xi^2}}sin(t')$ and $Ae^{-\frac{t'^2}{2\sigma^2\xi^2}}cos(t')$ are orthogonal, thus they have independent relations which are shown in the solid line in Fig. 10. Similarly, $Ae^{-\frac{t'^2}{2\sigma^2\xi^2}}sin(t')$ and $A^2(e^{-\frac{t'^2}{2\sigma^2\xi^2}})^2cos^2(t')$ are orthogonal, thus they have independent relation. Then, similar independent relations hold between the left $cos(t')$ and the right $sin(t')$, $sin^3(t')$.. terms, respectively in the Eq. (19) in Fig. 11, which are shown in the solid line.

Similar independent relations hold between $sin^3(t')$ and $cos(t'), cos^2(t'), cos^3(t')$...
terms, respectively. Thus, selective independent subspaces are generated by the combination pairs of sine and cosine terms in Fig. 11.

$$Ae^{-\frac{t'^2}{2\sigma^2\xi^2}}cos(t') \quad Ae^{-\frac{t'^2}{2\sigma^2\xi^2}}sin(t') \quad A^3(e^{-\frac{t'^2}{2\sigma^2\xi^2}})^3 sin^3(t') \quad A^5(e^{-\frac{t'^2}{2\sigma^2\xi^2}})^5 sin^5(t')... \quad (19)$$

Fig. 11. Orthogonal combination pairs among Gabor cosines functions

Under the condition of every pairs of Gabor elements with the orthogonality relation, $\{x_i, x_j\}, i,j = 1 \sim m$, where x_i and x_j are Gabor elements, the following equation

$$\gamma_1 x_1 + \gamma_2 x_2 + \ldots + \gamma_m x_m = 0$$

holds at the condition $\gamma_1 = \gamma_2 = \ldots = \gamma_m = 0$. This shows the set of Gabor elements $\{x_1, x_2, \ldots, x_m\}$ generates an independent subspace.

6 Conclusion

It is important to make clear the network structures how to generate the independence and orthogonality relations, which will make feature spaces, effectively. It is shown that the asymmetric network with Gabor filters is shown to have orthogonal properties strongly under stimulus conditions, while the conventional energy symmetric network is weak in the orthogonality. Nonlinear characteristics are observed as the squaring function and rectification function in the biological networks. It is shown that the asymmetric networks with nonlinear characteristics are able to generate independent subspaces, which will be useful for the creation of features spaces.

References

1. Reichard, W.: Autocorrelation, a Principle for the Evaluation of Sensory Information by the Central Nervous System. Sensory Communication. MIT Press, Cambridge (2012)
2. Barlow, H.B.: Possible Principles Underlying the Transformations of Sensory Messages. In: Rosenblith, W.A. (ed.) Sensory Communication. MIT Press, Cambridge (2012)
3. Adelson, E.H., Bergen, J.R.: Spatiotemporal energy models for the perception of motion. J. Opt. Soc. Am. A **2**(2), 284–298 (1985)
4. Pan, H., Jiang, H.: Learning convolutional neural networks using hybrid orthogonal projection and estimation. In: Proceedings Machine Learning Research, ACML 2017, vol. 77, pp. 1–16 (2017)
5. Widrow, B., Greenblatt, A., Kim, Y., Park, D.: The *No-Prop* algorithm: a new learning algorithm for multilayer neural networks. Neural Netw. **37**, 182–188 (2013)

6. Huang, l., Liu, X., Lang, B., Yu, A.W., Wang, Y., Li, B.: Orthogonal weight normalization: solution to optimization over multiple dependent stiefel manifolds in deep neural networks. In: 32nd AAAI Conference on AI, AAAI-18, pp. 3271–3278 (2018)
7. Shi, W., Gong, Y., Cheng, D., Tao, X., Zheng, N.: Entropy orthogonality based deep discriminative feature learning for object recognition. Pattern Recogn. **81**, 71–80 (2018)
8. Simonceli, E.P., Heeger, D.J.: A model of neuronal responses in visual area MT. Vision. Res. **38**, 743–761 (1996)
9. Heeger D.J.: Models of motion perception, University of Pennsylvania, Department of Computer and Information Science. Technical rep No. MS-CIS-87-91, September 1987
10. Grzywacz, N.M., Yuille, A.L.: A model for the estimate of local image velocity by cells in the visual cortex. Proc. R. Soc. Lond. vol. B **239**, 129–161 (1990)
11. Marmarelis, P.Z., Marmarelis, V.Z.: Analysis of Physiological Systems – The White Noise Approach. Plenum Press, New York (1978)
12. Sakuranaga, M., Naka, K.-I.: Signal transmission in the Catfish Retina.III. Transmission to Type-C Cell. J. Neurophysiol. **53**(2), 411–428 (1985)
13. Naka, K.-I., Sakai, H.M., Ishii, N.: Generation of transformation of second order nonlinearity in catfish retina. Ann. Biomed. Eng. **16**, 53–64 (1988)
14. Ishii, N., Deguchi, T., Kawaguchi, M., Sasaki, H.: Motion Detection in Asymmetric Neural Networks. In: Cheng, L., Liu, Q., Ronzhin, A. (eds.) ISNN 2016. LNCS, vol. 9719, pp. 409–417. Springer, Cham (2016). https://doi.org/10.1007/978-3-319-40663-3_47
15. Ishii, N., Deguchi, T., Kawaguchi, M., Sasaki, H.: Distinctive Features of asymmetric neural networks with gabor filters. In: de Cos Juez, F., et al. (eds.) Intelligent Hybrid Artificial Systems, HAIS 2018. LNCS, vol. 10870, pp. 185–196. Springer, Cham (2018). https://doi.org/10.1007/978-3-319-92639-1_16

Using Deep Learning for Ordinal Classification of Mobile Marketing User Conversion

Luís Miguel Matos[1], Paulo Cortez[1(✉)], Rui Castro Mendes[1], and Antoine Moreau[2]

[1] ALGORITMI Centre, University of Minho, 4804-533 Guimarães, Portugal
id6929@alunos.uminho.pt, pcortez@dsi.uminho.pt, azuki@di.uminho.pt
[2] OLAmobile, Spinpark, 4805-017 Guimarães, Portugal
antoine.moreau@olamobile.com

Abstract. In this paper, we explore Deep Multilayer Perceptrons (MLP) to perform an ordinal classification of mobile marketing conversion rate (CVR), allowing to measure the value of product sales when an user clicks an ad. As a case study, we consider big data provided by a global mobile marketing company. Several experiments were held, considering a rolling window validation, different datasets, learning methods and performance measures. Overall, competitive results were achieved by an online deep learning model, which is capable of producing real-time predictions.

Keywords: Mobile performance marketing · Multilayer perceptron · Ordinal classification

1 Introduction

The massive adoption of smartphones has increased of value of mobile performance marketing. These markets are implemented using a Demand Side Platform (DSP), which matches users to ads and is used by publishers and advertisers [9, 11]. A publisher is a web content owner (e.g., games) that attracts users. The web content is funded by DSP dynamic ads, which when clicked redirects the user to an advertiser site. When there is a conversion, the DSP returns a portion of the sale value to the publishers. Under this market, a vital element is the prediction of the user conversion rate (CVR), i.e., if there will be a conversion when a user sees an ad [4]. Typically, CVR prediction is modeled as a binary classification task ("no sale", "sale") by using offline learning Machine Learning methods, namely Logistic Regression (LR) [4], Gradient Boosting Decision Trees (GBDT) [8], Random Forests (RF) [4] and Deep Learning Multilayer Perceptrons (MLP) [9].

In contrast with the binary CVR studies (e.g., [4,8]), in this paper we propose a novel ordinal classification of mobile CVR, which assumes five classes: "no sale", "very low", "low", "medium" and "high". This approach provides a

© Springer Nature Switzerland AG 2019
H. Yin et al. (Eds.): IDEAL 2019, LNCS 11871, pp. 60–67, 2019.
https://doi.org/10.1007/978-3-030-33607-3_7

good proxy to the client Lifetime Value (LTV) [11]. Thus, using such ordinal classifier, a DSP can better select the best ad campaign for a particular user by maximizing the expected conversion value. Following a recently proposed binary deep learning approach [9], we explore three main MLP strategies to handle ordinal classification: pure classification, regression and ordinal classification. These approaches are tested using two learning models, offline and online, using a realistic rolling window validation and real-world big data from a global DSP. Also, the deep learning models are compared with a LR method.

2   Materials and Methods

2.1   Data

We collected the data from a worldwide marketing company (OLAmobile). The DSP generates two main events: redirects – the user ad clicks; and sales – when there is a conversion. All redirects and sales are stored at the DSP data center, being associated with a timestamp when they arrive. The DSP managed by the company generates millions of redirects and thousands of sales per hour.

We had access to a secure web service that allowed us to retrieve NR redirects and NS sales from the data center. Our computing server, an Intel Xeon 1.70 GHz wth 56 cores and 2 TB of disk space, is limited when compared with the data center and thus we work with sampled data. The data was collected during a two week period, starting at 30th May of 2019, via a stream engine that uses K computing cores to continuously retrieve redirects and sales, "sleeping" every SR and SS seconds [9] (Table 1). The analyzed DSP contains two traffic modes: TEST – used to measure the performance of new campaigns; and BEST – with 80% of the traffic and including only the best TEST performing ads. The Y_{no} and Y_{yes} columns denote the number of collect "no sale" and "sale" events. Also, the R_Y column represents the sales ratio $R_Y = Y_{yes}/(Y_{no} + Y_{yes})$. Since we worked with sample data, the collected data ratio is higher than the expected real DSP one. To get a more realistic dataset, we randomly undersample [1] the number of sales (Y'_{yes}) such that a more realistic ratio ($R_{Target'}$) is obtained, which in this work was fixed to 5% for both BEST and TEST traffic. Thus, for each traffic mode there are two datasets: collected (C) and realistic (R). Table 2 presents a summary of the collected attributes. Most attributes are categorical and some present a high cardinality (e.g., city).

Table 1. Summary of the collected DSP datasets

Traffic	NR	NS	SR	SS	K	Y_{no}	Y_{yes}	Y'_{yes}	R_Y	$R_{Y'}$
TEST	100	100	300	10	2	290,279	29,596	15,393	9.3%	5.04%
BEST	100	100	300	10	2	328,028	156,637	17,389	32.2%	5.03%

Table 2. Summary of the DSP data attributes

Context	Attribute	Description (a – TEST traffic, b – BEST traffic)
User	Country	User country: 198^b or 225^a levels (e.g., Russia, Spain, Brazil)
	City	User city: 10690^a or 13423^b levels (e.g., Lisbon, Paris)
	Region	Region of the country: 23^{ab} levels (e.g., Asia, Europe)
	Browser	Browser name: 14^{ab} levels (e.g., Chrome, Safari)
	Operator	Mobile carrier or WiFi: 404^b or 448^a levels (e.g., Vodafone)
Advertiser	Vertical	Ad type: 4^a or 5^b levels (e.g., video, mainstream, dating)
	Campaign	Ad product identification: 1389^b or 1741^a categorical levels
	Special	Smart link or special offer: 1018^b or 1101^b levels
Publisher	Account	Publisher type: 8^b or 9^a levels (e.g, app developer, webmaster)
	Manager	Publisher account manager: 19^b or 34^a categorical levels
Sale	Value	Value of the conversion in EUR: numeric (e.g., 0.00, 0.01, 69.34)

After consulting the DSP experts, we created the ordinal Y target by grouping the value into $k = 5$ classes (Fig. 1). The grouping was achieved by using rounded EUR values after using quantiles over the collected sales values (when > 0): "none" – equal to 0; "very low" –>0 and <0.03; "low" –>= 0.03 and <0.10; "medium" –>= 0.10 and <0.30; "high" –>= 0.30. As shown in Fig. 1, when there is a sale, the ordinal classes are relatively balanced. Also, TEST traffic presents a lower number of sales when compared with BEST traffic.

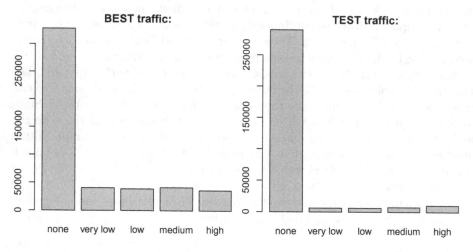

Fig. 1. Histograms for the ordinal sale classes for collected BEST and TEST datasets

2.2 Data Preprocessing and Ordinal Approaches

Since several attributes are sparse and present a high cardinality (Table 2), we applied the Percentage Categorical Pruning (PCP) transformation to all input attributes [9]. The transform works by merging the least 10% frequent levels in the training data into a "others" category. Then, the resulting values are preprocessed using the known one-hot coding, which assigns one binary input per level. In [9], the PCP transform allowed to substantially reduce the input memory (e.g., reduction of 94% for the city attribute) and processing requirements.

For the mobile marketing data, the number of conversions (sales) is typically much lower than the number of ad clicks (redirects). While the target CVR data is unbalanced, in [9] we found that the binary deep learning MLP classifier was capable of high quality predictions even without any training data balancing method. However, when approaching the ordinal task and in particular for the realistic datasets (which present the lowest 5% conversion rate), the training data becomes extremely unbalanced. Thus, we opted to balance the realistic training data by using the SMOTE method [1], which creates new synthetic examples for the minority classes. This balancing method was only applied to training data and thus the test sets were kept with the original target values.

In this paper, we apply three approaches for the ordinal classification: Multi-class Classification (MC), regression and the k-1 ordinal approach (OA). The first approach discards the ordering and performs a simpler 5-class classification task. The second approach transforms each ordinal class into a numeric score y, where $y \in [0, 1, 2, 3, 4]$ (R1) or $y \in [0, 1, 2, 4, 8]$ (R2). The first regression scale (R1) uses equal spaced values, while the second one (R2) assigns larger distances to the highest sales ("medium" and "high"), in an attempt to favor such classes due to the minimization of squared errors. In both scales, the ordinal class is associated with the nearest scale value to the prediction (e.g., in R2 a prediction of Sect. 4 is assumed as the "medium" class). The third approach transforms the ordinal target, with the classes $V1 < V2 < ... < V_k$, into k-1 binary tasks [5]. Each classifier learns the probability of $P(Y > V_k)$, $k \in \{1, ..., k-1\}$. Then:

$$P(V_1) = 1 - P(Y > V_1) \tag{1}$$

$$P(V_i) = \max(0, P(Y > V_{i-1}) - P(Y > V_i)), 1 < i < k \tag{2}$$

$$P(V_k) = P(Y > V_{k-1}) \tag{3}$$

The classifiers are independent and in a few cases we experimentally found it could occur that $P(Y > V_i) > P(Y > V_{i-1})$. Thus, we added the $\max(0, ...)$ function in Eq. (2) to avoid the computation of negative probabilities.

2.3   Deep Learning Methods

All learning methods are coded in Python using the Keras library [2]. In previous work, a very competitive deep learning MLP model was proposed for binary CVR prediction, outperforming a convolution neural network and a logistic regression [9]. We adapt the same MLP, also known as Deep Feedforward Neural Network (DFFN) [6]. Let $(L_0, L_1, ..., L_H, L_O)$ denote a vector with the layer sizes with $m \in \{1, ..., H\}$ hidden layers, where $L_0 = I$ is the input layer size (I is the total number of input levels after the PCP transform) and L_O is number of output nodes. The adopted model consists of a trapezoidal shaped MLP, with $H = 8$ hidden layers of decreasing size: $(I, 1024, 512, 256, 128, 64, 32, 16, 8, L_O)$. In all hidden layers ($\{1, ..., 8\}$) we used the popular ReLU activation function, due to its fast training and good convergence properties. For multi-class classification, the output layer contains $L_O = 5$ nodes and the softmax function is used to output class probabilities $P(V_k) \in [0, 1]$. For the regression models, only $L_O = 1$ linear output node is used. Finally, for the $k - 1$ ordinal classification, one logistic output node ($L_O = 1$) is used, trained such that $P(Y > V_k) \in [0, 1]$.

During the training phase, we used the AdaDelta gradient function [6], which is based on a stochastic gradient decent method. Following our previous work [9], we used two approaches to avoid overfitting: dropout, which randomly ignores neural weights (dropout values of 0.5 and 0.2 for hidden layers $m = 4$ and $m = 6$); and early stopping, which stops the training when the validation error does not improve 1% within 3 epoch runs of after a maximum of 100 epochs.

Since we work with stream data, the learning models should be dynamic, assuming a continuous learning through time. We compare two MLP learning modes (proposed in [9]) for ordinal classification: reset – offline mode, when new training data is available (new rolling window iteration, Sect. 2.4), the whole neural weights are randomly set; and reuse – online learning, any new training starts with the previous fitted MLP weights (from previous rolling window training) and only the new input node (due to appearance of new input levels) to first hidden layer connections are randomly set.

2.4 Evaluation

The learning models are evaluated using robust rolling window validation [10], which simulates a classifier usage through time, with multiple training and test updates. In the first iteration ($u = 1$), the model is adjusted to a training window with the W oldest examples and then predicts T test predictions. In the next iteration ($u = 2$), the training set is updated by discarding the oldest T records and adding T more recent ones. A new model is fit, producing T new predictions, and so on. In total, this produces $U = D - (W + T)$ model updates (training and test iterations), where D is the data length (number of examples). After consulting OLAmobile experts, we opted to use the realistic values of $W = 100,000$ and $H = 5,000$, which results in the model updates: $U = 43$ – collected TEST traffic; $U = 76$ – collected BEST traffic; $U = 41$ realistic TEST traffic; and $U = 49$ realistic BEST traffic.

For each rolling window iteration, we collect the test data measures and the computational effort (in seconds). We compute the F1-score, $F1_k$ for each class V_k, which considers both precision and recall [13]. The global measure is obtained by using the Macro-averaging F1-score (MF1), which weights equally the F1-score for each class. We note that the ordinal classes are unbalanced, and thus other F-score averaging measures, such as micro or weight averaging, would favor mostly models that classify well the no conversion "none" class. As a secondary global measure, we adopt the Mean Absolute Error for Ordinal Classification (MAEO) [12], which computes how far (using absolute errors) are the predictions from the target (e.g., the error is 2 if the prediction is "low" and the target is "high"). This measure is often used in ordinal classification but, similarly to the micro and weight averaging F1-score measures, it tends to produce low values when the classifier is more biased to correctly predict the "none" class. The rolling window results are averaged over all U iterations and the Wilcoxon test is used to check if paired MF1 differences are significant [7].

Table 3. Classification results (best values in **bold**; best models in *italic*).

Data	Traffic	Meth.	Model	Global		F1-score per class					Effort(s)
				MAEO	MF1	$F1_1$	$F1_2$	$F1_3$	$F1_4$	$F1_5$	
C	BEST ($U = 76$)	MC	MLP reset	0.48	0.57	0.87	0.49	0.35	0.54	0.58	61
			MLP reuse	**0.45**	**0.64**[a]	**0.88**	**0.56**	**0.48**	**0.61**	0.66	42
			LR	0.51	0.57	0.86	0.51	0.38	0.51	0.59	**35**
		R1	MLP reset	0.63	0.22	0.85	0.22	0.03	0.01	0.00	46
			MLP reuse	0.61	0.24	0.84	0.30	0.04	0.00	0.00	44
		R2	MLP reset	0.77	0.28	0.83	0.00	0.00	0.31	0.24	49
			MLP reuse	1.02	0.28	0.77	0.00	0.00	0.26	0.39	47
		OA	MLP reset	0.47	0.59	0.87	0.51	0.43	0.54	0.61	117
			MLP reuse	**0.45**	**0.64**[b]	**0.88**	0.55	**0.48**	**0.61**	**0.67**	94
			LR	0.51	0.54	0.86	0.47	0.35	0.49	0.54	39
	TEST ($U = 43$)	MC	MLP reset	0.25	0.19	0.95	0.00	0.00	0.00	0.00	46
			MLP reuse	0.19	0.51	**0.96**	**0.36**	0.22	0.45	0.55	45
			LR	0.21	0.43	0.96	0.26	0.14	0.36	0.45	**31**
		R1	MLP reset	0.22	0.22	**0.96**	0.08	0.02	0.01	0.01	51
			MLP reuse	0.22	0.23	**0.96**	0.15	0.03	0.00	0.00	45
		R2	MLP reset	0.28	0.30	0.95	0.00	0.00	0.24	0.33	50
			MLP reuse	0.27	0.31	0.95	0.00	0.00	0.24	0.34	47
		OA	MLP reset	0.21	0.31	**0.96**	0.10	0.09	0.20	0.19	118
			MLP reuse	**0.18**	**0.54**[b]	**0.96**	0.34	**0.31**	**0.50**	**0.57**	109
			LR	0.21	0.42	**0.96**	0.24	0.13	0.34	0.41	39
R	BEST ($U = 49$)	MC	MLP reset	0.37	0.37	0.91	0.20	**0.19**	0.24	0.29	100
			MLP reuse	0.31	**0.38**[c]	**0.93**	0.21	**0.19**	**0.26**	0.30	**84**
			LR	0.93	0.26	0.76	0.13	0.12	0.13	0.15	120
		R1	MLP reset	0.32	0.21	0.88	0.10	0.07	0.01	0.00	106
			MLP reuse	**0.30**	0.21	0.89	0.10	0.07	0.00	0.00	87
		R2	MLP reset	0.68	0.25	0.87	0.00	0.00	0.08	0.28	104
			MLP reuse	0.61	0.26	0.89	0.00	0.00	0.08	**0.33**	88
		OA	MLP reset	0.34	0.37	0.92	0.19	**0.19**	0.24	0.30	278
			MLP reuse	0.37	0.36	0.92	0.20	**0.19**	0.22	0.25	219
			LR	0.89	0.25	0.73	0.10	0.10	0.14	0.19	109
	TEST ($U = 41$)	MC	MLP reset	0.23	0.44	**0.96**	0.27	**0.26**	0.33	0.40	130
			MLP reuse	0.22	**0.45**[c]	**0.96**	0.27	**0.26**	**0.35**	0.41	99
			LR	0.69	0.30	0.83	0.13	0.11	0.17	0.23	156
		R1	MLP reset	0.19	0.25	0.95	0.18	0.10	0.03	0.00	132
			MLP reuse	**0.18**	0.23	0.95	0.17	0.04	0.00	0.00	103
		R2	MLP reset	0.30	0.28	0.95	0.00	0.00	0.14	0.33	130
			MLP reuse	0.29	0.31	0.95	0.00	0.00	0.15	**0.46**	**97**
		OA	MLP reset	0.21	0.44	**0.96**	0.24	0.24	0.34	0.42	303
			MLP reuse	0.21	0.44	**0.96**	0.25	**0.26**	0.33	0.42	240
			LR	0.65	0.29	0.81	0.09	0.11	0.18	0.28	123

a – Statistically significant when compared with MC reset and LR.
b – Statistically significant when compared with OC reset and LR.
c – Statistically significant when compared with MC LR.

3   Results

We tested two types of datasets (C and R), two types of traffic (BEST and TEST), three ordinal methods (MC, R1/R2, OA), two MLP learning modes (reset, reuse) and a LR baseline model. Table 3 shows the obtained average rolling window classification results. Overall, the best learning algorithm is MLP reuse, which tends to produce the highest MF1 values, often with statistical significance when compared with MLP reset or LR model. When compared with LR, MLP reuse presents an MF1 improvement that ranges from 7 to 15% points. The reuse learning always requires less computational effort when compared with the reset mode. As for the ordinal methods, the multi-class (MC) and $k-1$ OA achieve the best overall MLP reuse results, with slight F1-score differences. In general, the regression ordinal scales (R1 and R2) produce worst F1-score results. Only in two cases (R BEST and TEST), R2 obtained the best F1-score for the "high" conversion class (V_5). The $k-1$ OA method requires more computation than the MC approach. Yet, we note that in this work we used one processing core for each model, thus the OA effort could by substantially reduced if $k-1$ cores were used to fit each of its individual binary models.

4   Conclusions

In this paper, we used big data from a mobile marketing company. The goal was to predict the type of conversion rate (CVR) when an user clicks an ad, set in terms of five ordinal classes. Using a realistic rolling window validation, we compared three main ordinal methods (multi-class – MC, regression – R1/R2 and $k-1$ ordinal approach – OA) using two deep learning approaches (offline – MLP reset; and online – MLP reuse) and a logistic regression (LR) model.

The best results were achieved by the MLP reuse model and the MC and OA approaches. Such model is capable of real-time predictions. For instance, the 5,000 predictions for the C BEST MC setup require 42 s, which results in an average 8 ms per prediction. Interesting results were achieved for the collected (C) datasets, with most F1-scores above 50%, macro F1-scores of 64% and 54%, as well as a low MAEO error (lower than 0.5). As for the realistic (R) datasets (with lower amount of conversion cases), while low MAOE errors where obtained (e.g., lower than 0.30), the individual F1-scores are lower when compared with the collected datasets, resulting in an average macro F1-score of 38% and 45%.

Considering all obtained results, we recommend the MC MLP reuse model, which requires the training of just one classifier and is capable of real-time predictions. This model is potentially capable of providing value to the analyzed marketing company, since it currently does not have any method to estimate the value level of a conversion. In particular, for TEST traffic the company uses a random selection of ads, which produces much lower macro F1-scores. For instance, for the realistic TEST dataset, the random class assignment results in an average macro F1-score that is around 10%. In future work, we intend to improve the realistic ordinal results by adopting a multi-objective (F1-score for each class) evolutionary learning to train the MLP model [3].

Acknowledgments. This article is a result of the project NORTE-01-0247-FEDER-017497, supported by Norte Portugal Regional Operational Programme (NORTE 2020), under the PORTUGAL 2020 Partnership Agreement, through the European Regional Development Fund (ERDF). This work was also supported by Fundação para a Ciência e Tecnologia (FCT) within the Project Scope: UID/CEC/00319/2019.

References

1. Batista, G.E., Prati, R.C., Monard, M.C.: A study of the behavior of several methods for balancing machine learning training data. ACM SIGKDD Explor. Newsl. **6**(1), 20–29 (2004)
2. Chollet, F., et al.: Keras (2015). https://keras.io
3. Cortez, P.: Modern Optimization with R. UR. Springer, Cham (2014). https://doi.org/10.1007/978-3 319-08263-9
4. Du, M., State, R., Brorsson, M., Avanesov, T.: Behavior profiling for mobile advertising. In: Anjum, A., Zhao, X. (eds.) Proceedings of the 3rd BDCAT 2016, Shanghai, China, 6–9 December, pp. 302–307. ACM (2016)
5. Frank, E., Hall, M.: A simple approach to ordinal classification. In: De Raedt, L., Flach, P. (eds.) ECML 2001. LNCS (LNAI), vol. 2167, pp. 145–156. Springer, Heidelberg (2001). https://doi.org/10.1007/3-540-44795-4_13
6. Goodfellow, I., Bengio, Y., Courville, A.: Deep Learning. MIT Press, Cambridge (2016)
7. Hollander, M., Wolfe, D.A., Chicken, E.: Nonparametric Statistical Methods. Wiley, New York (2013)
8. Lu, Q., Pan, S., Wang, L., Pan, J., Wan, F., Yang, H.: A practical framework of conversion rate prediction for online display advertising. In: Proceedings of the ACM ADKDD, pp. 9:1–9:9. Halifax, August 2017
9. Matos, L., Cortez, P., Mendes, R., Moreau, A.: Using deep learning for mobile marketing user conversion prediction. In: Proceedings of the IEEE International Joint Conference on Neural Networks (IJCNN 2019). IEEE, Budapest, July 2019
10. Oliveira, N., Cortez, P., Areal, N.: The impact of microblogging data for stock market prediction: using twitter to predict returns, volatility, trading volume and survey sentiment indices. Expert Syst. Appl. **73**, 125–144 (2017)
11. Silva, S., Cortez, P., Mendes, R., Pereira, P.J., Matos, L.M., Garcia, L.: A categorical clustering of publishers for mobile performance marketing. In: Graña, M., et al. (eds.) SOCO 2018-CISIS 2018-ICEUTE 2018. AISC, vol. 771, pp. 145–154. Springer, Cham (2019). https://doi.org/10.1007/978-3-319-94120-2_14
12. Sousa, R.G., Yevseyeva, I., da Costa, J.F.P., Cardoso, J.S.: Multicriteria models for learning ordinal data: a literature review. In: Yang, X.S. (ed.) Artificial Intelligence, Evolutionary Computing and Metaheuristics - In the Footsteps of Alan Turing, pp. 109–138. Springer, Heidelberg (2013). https://doi.org/10.1007/978-3-642-29694-9_6
13. Witten, I.H., Frank, E., Hall, M.A.: Data mining: practical machine learning tools and techniques, 3rd edn. Morgan Kaufmann, San Francisco (2011)

Modeling Data Driven Interactions on Property Graph

Worapol Alex Pongpech[(⊠)]

Business Analytic and Data Science Graduate Program,
Faculty of Applied Statistics, NIDA, Bangkok 10240, Thailand
worapol@as.nida.ac.th

Abstract. Data Driven Transformation is a process where an organization transforms its infrastructure, strategies, operational methods, technologies, or organizational culture to facilitate and encourage *data driven decision-making* behaviors. We defined a data driven network as group/groups of people in an organization network working on common projects where making data driven decisions is necessary in the project. One important problem in data driven transformation is how to understand data driven behaviors in the network. Ability to model and compute on a DDIG is crucial for understanding data driven network behaviors. To this end, we modeled a data driven network as a property graph, in which nodes represent users, projects, or data sources and edges represent interactions between them. We termed this Data Driven Interactions Graph and modeled on the Neo4j graph platform. A graph model framework to represent data driven interactions such that a number of network properties can be computed from the proposed graph is introduced. Finally, we discussed data utilization and data collaboration behaviors of the social working network.

Keywords: Data Driven Transformation · Collaboration ·
Utilization · Property graph · Social network analysis · Clustering ·
Communities · Neo4j

1 Introduction

In the past couple of years, business organizations sought advantages over competitors by utilizing data driven decision-making on every projects in their organizations. This is commonly known as organization's data driven transformation. In this research, data driven transformation is defined as a process where organization transforms its infrastructure, strategies, operational methods, technologies, or organizational behaviors to facilitate and encourage *data driven decision-making* interactions. The aim of data driven transformation is to minimize no-data decisions-making and to transform legacy information technology toward modern data analytic enabled system. This process is a highly complex and challenging process involving not only technology but also human behaviors within the process.

© Springer Nature Switzerland AG 2019
H. Yin et al. (Eds.): IDEAL 2019, LNCS 11871, pp. 68–75, 2019.
https://doi.org/10.1007/978-3-030-33607-3_8

Decisions-making driven by personal experiences are not necessary incorrect, however it is often difficult to understand how the decisions had been made without any explanation from the decision-makers. In a number of cases, reason for decision-making is often resorted to *gut feeling*. Understanding how incorrect decisions had been made could prevent organizations from repeating the same mistake as it is often stated that those who do not understand the history are deemed to make the same mistake. Companies such as Google and Amazon also found that byproduct of data driven transformation is companies' ability to learn and understand from the behaviors patterns in the decisions-making.

The extant literature shown insufficient guidance to compute and evaluate data driven behaviors such as collaboration and utilization or to model data driven interactions in a network. Motivated by this research gap, we developed a graph based computational framework coupled with *social network theory* to facilitate model and compute data driven network properties of a given network. We termed this data driven interactions Graph (DDIG) and modeled on Neo4j platform. Ability to compute and model these activities allow organizations to extrapolate data driven behaviors from its data driven network. In this paper, we gave formal expressions for describing interactions between each node on the graph. To identify clustering and communities of the data driven interactions graph, we implemented triangle counting and clustering coefficient, betweenness centrality, and label propagation algorithms using Neo4j graph algorithms. Collaboration and data utilization behaviors of the graph are then extrapolated from these computed properties.

The rest of the paper is organized as follows. Related works is given in Sect. 2. Data driven interactivity graph is presented in Sect. 3. Future works, and research summary is discussed in the last section.

2   Related Works

Modeling human behaviors is not a simple consideration, and it is notoriously difficult and more when the need to predict human behaviors is necessary [1]. We have found a number of techniques that have been utilized in modeling human behaviors [2] and many of these works are in the area of the probabilistic models both relational and propositional types of models.

While there are a few works focusing on human work collaborations, we have yet to find any in the area of data collaboration and data utilization. Wood [3] provided a comprehensive discussion on theory of collaboration. Milojević [4] focused on investigating human work collaboration on researchers specifically on how these researchers citing behaviors. Hsieh [5] focused on investigating structural relation in to capture collaborative performance in a Public Service Agency.

There are a number of theory in social network analysis such as collaboration graph [6] which is a graph modeling some social network where the vertices represent participants of that network (usually individual people) and where two distinct participants are joined by an edge whenever there is a collaborative relationship between them of a particular kind.

Collaboration graphs are used to measure the closeness of collaborative relationships between the participants of the network. Some of the most well-studied collaboration graphs are the Erdos collaboration graph [7], the Hollywood graph or co-stardom network [8], and the Co-authorship graphs [9]. The Erdos collaboration graph is a collaboration graph of mathematicians joined by an edge whenever they co-authored a paper together (with possibly other co-authors present). These works have clearly demonstrated the potential of modeling data driven interactions using graph theory.

Graph model coupled with social network analysis allow a more complete data and users interactions investigation. One popular topic in social network analysis is community detection. Communities, or clusters, are usually groups of vertices having higher probability of being connected to each other than to members of other groups, though other patterns are possible. There are a number of works in community detection. Duch [10] focused is community detection in complex networks using extremal optimization. It is possible that there can be overlap community in a given graph, Yang's work is in detection of overlap community [11]. Another interesting work is presented by Leung on real-time community detection in large networks [12].

3   Data Driven Interaction Framework

In this paper, we defined data driven interaction as an occasion when two or more people or things on the same network communicate with or react to each other while they are working on a common project. We interested in model these interactions such that data collaboration and utilization of the network can be evaluated. When view under this light, data driven interactions can be investigated through social network analysis and graph theory to enable data collaboration and utilization evaluation of the network.

A framework is a real or conceptual structure intended to serve as a support or guide for the building of something that expands the structure into something useful. The data driven interactions framework defines entities in the data driven decision-making process, and defines interactions between entities.

The framework specifies three types of entities in data driven decision-making process. First entity is the project, second is the users who are working on the given project, and the third is the data sources utilized for the project. The framework also specifies two types of data driven interactions between users and users on the data. We give a data driven interactions framework as follows.

1. Interacting on Data (IoD): is the acts that users utilized the given sets of data toward making informed decision such as exploring the data, extracting the data, processing the data, and storing the data.
2. Interaction through Data (ItD): is the acts between two or more users collaborating through the given sets of data toward making informed decision such as transfer of data between users, discussion on data with the others, or request of data from the others.

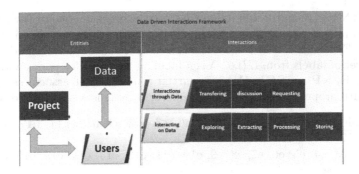

Fig. 1. xx

The given framework put the scope of investigating data driven transformation directly on to the path of social network analysis where each user can be represented as a node on a graph and interactions between users can be represented as edges on a graph. Edge attributes can be used to express these interactions (Fig. 1).

3.1   The Data Driven Interactions Graph Model

Data driven interaction is modeled as a labeled-property graph [13]. The term property graph has come to denote an attributed, multi-relational graph. That is, a graph where the edges are labeled and both vertices and edges can have any number of key/value properties associated with them. A labeled-property graph model is represented by a set of nodes, relationships, properties, and labels. Both nodes of data and their relationships are named and can store properties represented by key/value pairs. Nodes can be labeled to be grouped. The edges representing the relationships have two qualities: they always have a start node and an end node, and are directed; making the graph a directed graph. Relationships can also have properties. This is useful in providing additional metadata and semantics to relationships of the nodes. Direct storage of relationships allows a constant-time traversal.

Assume that $\mathbf{L}$ is an infinite set of labels (for nodes and edges), $\mathbf{P}$ is an infinite set of property names, $\mathbf{V}$ is an infinite set of atomic values, and $\mathbf{T}$ is a finite set of datatypes (e.g., integer). Given a set $\mathbf{X}$, we assume that $\mathrm{SET+}(\mathbf{X})$ is the set of all finite subsets of $\mathbf{X}$, excluding the empty set. Given a value v $\in \mathbf{V}$, the function type(v) returns the data type of v. The values in $\mathbf{V}$ will be distinguished as quoted strings.

Definition 1. A Data driven Interactions graph is a tuple $G = (N, E, \rho, \lambda, \sigma)$ where:

1. N is a finite set of nodes (also called vertices);
2. E is a finite set of edges such that E has no elements in common with N;

3. ρ: E $\rightarrow$ (N $\times$ N) is a total function that associates each edge in E with a pair of nodes in N (i.e., ρ is the usual incidence function in graph theory);
4. λ: (N $\cup$ E) $\rightarrow$ $SET^+(\mathbf{L})$ is a partial function that associates a node or edge with a set of labels from L (i.e., λ is a labeling function for nodes and edges);
5. σ: (N $\cup$ E) $\times \mathbf{P}$ $\rightarrow$ $SET^+(\mathbf{V})$ is a partial function that associates nodes or edges with properties, and for each property it assigns a set of values from $\mathbf{V}$.

Example N = {n1, n2, n3, n4, n5, n6, n7, n8, n9, n10, n11, n12, n13, n14, n15, n16, n17, n18}
E = {e1, e2, e3, e4, e5, e6, e7, e8, e9, e10, e11, e12, e13, e14, e15, e15, e17, e18, e19, e20}
(n1) = User, (n1, fname) = "Paul", (n1, lname) = "Tread"
(n2) = {$User$}, (n2, fname) = "Jean", (n2, lname) = "Scot"
(n3) = {$User$}, (n3, fname) = "Dan", (n3, lname) = "Zach"
(n11) = {$Project$}, (n11, title) = "customer-engagement"
(n12) = {$Project$}, (n12, title) = "customer-churn"
(n13) = {$Project$}, (n13, title) = "customer-care"
(n14) = {$Project$}, (n14, title) = "new-product"
(n15) = {$Project$}, (n15, title) = "new-campaign"
(n16) = {$Data$}, (n16, title) = "source1", (n16, keyword) = "CRM"
(n17) = {$Data$}, (n17, title) = "source2", (n17, keyword) = "Operation"
(n18) = {$Data$}, (n18, title) = "source3", (n18, keyword) = "Transaction"
(e1) = (n1, n12), (e1) = work-on
(e2) = (n1, n13), (e2) = {work-on"}
(e3) = (n1, n14), (e3) = {work-on"}
(e21) = (n1, n2), (e1) = ITD, (e21, interacts) = "taking"
(e22) = (n1, n3), (e2) = ITD, (e22, interacts) = "sharing"
(e23) = (n4, n12), (e3) = ITDn, (e23, interacts) = "giving"

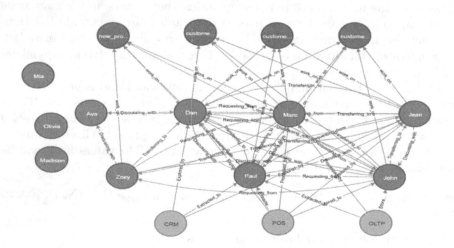

Fig. 2. Projects and employees collaboration graph

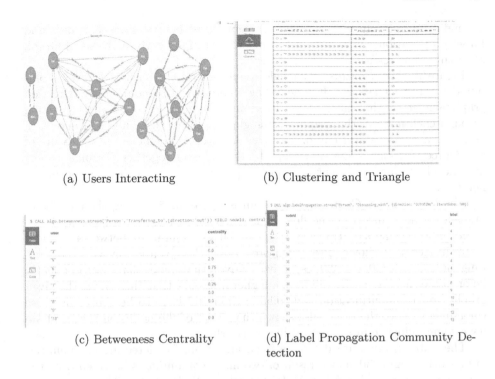

(a) Users Interacting (b) Clustering and Triangle

(c) Betweeness Centrality (d) Label Propagation Community Detection

Fig. 3. Data driven interacts graph

Follow the given formal definition, a property data driven interactions graph is illustrated in the Fig. 2 implemented through cypher, Neo4j Desktop. While some of the expressions are omitted in the example, it should be enough to demonstrated how to construct a data driven interactions graph through property graph model. In the next section, some of network properties that can be utilized to explain data driven behaviors is given.

3.2 Data Collaboration and Data Utilization

Ability to compute, observe, and model DDIs graph enable organization to better understand how decisions are made, and allow change managers to manage data driven transformation in an organization effectively. For example, an increasing or decreasing data driven interactions reflects and reveals state of data interactions in the organization. In this research, data driven interactions graph is modeled using Neo4j graph platform, and a number of data driven network properties were computed using graph algorithms in Neo4j.

While a simple query can provide a quick glance at the state of data driven interactions network as illustrated in Fig. 3a, a number of network properties can also be computed through graph algorithms. It is not our intention to illustrate all possible graph algorithms, we implemented a few algorithms in Neo4j [14] that can be utilized to give us the standing of our data driven network.

Triangle counting/clustering coefficient algorithms, betweenness centrality, and Label propagation community detection were computed through Neo4j Graph algorithm. The results are shown in Fig. 3.

Figure 3b shown results of triangle counting and clustering coefficient computed from the DDIs graph. The triangle counting is a community detection graph algorithm that is used to determine the number of triangles passing through each node in the graph. A triangle is a set of three nodes, where each node has a relationship to all other nodes. Clustering coefficient is a measure of the degree to which nodes in a graph tend to cluster together. Triangle counting and clustering coefficient are directly proportion with collaboration in the network.

The Betweenness Centrality algorithm is a way of detecting the amount of influence a node has over the flow of information in a graph. Figure 3c shown result of betweenness centrality of the given DDIs graph using betweenness algorithm in Neo4j. It shown three main brokers on the given graph with the score ranging from 6.5, 6.0. and 2.0 respectively. Given that nodes that most frequently lie on these shortest paths will have a higher influence that the others, the given graph shown that utilization and collaboration will have to be broker by these three nodes. Also, the result also shown that most of the nodes on the graph do have some influence in the data driven network.

The Label community detection is an algorithm for detecting communities in networks. Figure 3d shown result of two main communities detected from the given DDIs graph using label Algorithm in Neo4j. It is also shown a number of single node communities, which implied that number of nodes were not well collaborated with the others on the data driven network. Table 1 is given to illustrate possible results that can be extracted from these graph algorithms. For the purpose of illustration, collaboration and utilization scores are scaled from 1 to 5 and others networks properties are given nominal scale of high, medium, and low.

Table 1. Data driven collaboration and utilization evaluation.

DDIG	C. Coeff	Triangle	Betweenness	Label	Collaboration	Utilization
Graph-A	High	High	Medium	Medium	1	3
Graph-B	Low	Low	Low	Low	1	1
Graph-C	High	High	High	High	5	5
Graph-D	Medium	Low	High	Low	3	3

4   Summary

A large number of triangle counting and a high clustering coefficient imply a dense data driven network. A dense data driven network is result of a good collaboration, which a good data driven transformation should be aiming to achieve. To understand how collaboration in the data network manifested, it is important to identify influential node/nodes in the network. Betweenness centrality algorithm computes centrality weight that used to rank node's influence on the data driven network. Digging deeper into a dense data driven network is the number of communities where the higher number of communities implies possible more data utilization than the lower number of the communities. In this, the Label propagation was used to compute the number of communities on the data driven.

To this end, we concluded that data driven interactions can be modeled using a graph modeling where its network properties can be computed using graph algorithms. While the data used to model data driven interactions graph composed of only a few nodes and edges and they are generated synthetically. The results shown that the DDIs graph can be used to monitor collaboration and utilization status of a given data driven network.

References

1. Pentland, A., Liu, A.: Modeling and prediction of human behavior. Neural Comput. **11**(1), 229–242 (1999)
2. Sadilek, A.: Modeling human behavior at a large scale. Ph.D. thesis, Rochester, NY, USA (2012). AAI3543317
3. Wood, D.J., Gray, B.: Toward a comprehensive theory of collaboration. J. Appl. Behav. Sci. **27**(2), 139–162 (1991)
4. Milojević, S.: How are academic age, productivity and collaboration related to citing behavior of researchers? PLoS ONE **7**(11), 1 13 (2012)
5. Huich, J.Y., Liou, K.T.: Collaborative leadership and organizational performance: assessing the structural relation in a public service agency. Rev. Public Pers. Adm. **38**(1), 83–109 (2018)
6. Odda, T.: On properties of a well-known graph or what is your ramsey number? Ann. N. Y. Acad. Sci. **328**(1), 166–172 (1979)
7. Batagelj, V., Mrvar, A.: Some analyses of Erdős collaboration graph. Soc. Netw. **22**, 173–186 (2000)
8. Chung, F.R.K., Lu, L.: Complex graphs and networks. In: NSF-CBMS Regional Research Conference on the Combinatorics of Large Sparse Graphs, held at California State University San Marcos, 7–12 June 2004 (2006)
9. Malbas, V.S.R.: Mapping the collaboration networks of biomedical research in Southeast Asia. PeerJ PrePrints **3**, e936v1 (2015)
10. Duch, J., Arenas, A.: Community detection in complex networks using extremal optimization. Phys. Rev. E **72**, 027104 (2005)
11. Yang, J., Leskovec, J.: Community-affiliation graph model for overlapping network community detection. In: 2012 IEEE 12th International Conference on Data Mining (ICDM), pp. 1170–1175, December 2012
12. Leung, I.X.Y., Hui, P., Liò, P., Crowcroft, J.: Towards real-time community detection in large networks. Phys. Rev. E **79**, 066107 (2009)
13. Angles, R.: The property graph database model. In: AMW (2018)
14. Neo4j: The neo4j graph algorithms user guide v3.5

Adaptive Dimensionality Adjustment for Online "Principal Component Analysis"

Nico Migenda[1](✉), Ralf Möller[2], and Wolfram Schenck[1]

[1] Center for Applied Data Science Gütersloh, Faculty of Engineering and
Mathematics, Bielefeld University of Applied Sciences, Bielefeld, Germany
nico.migenda@fh-bielefeld.de

[2] Computer Engineering Group, Faculty of Technology, Bielefeld University,
Bielefeld, Germany

Abstract. Many applications in the Industrial Internet of Things and
Industry 4.0 rely on large amounts of data which are continuously generated. The exponential growth in available data and the resulting storage
requirements are often underestimated bottlenecks. Therefore, efficient
dimensionality reduction gets more attention and becomes more relevant.
One of the most widely used techniques for dimensionality reduction is
"Principal Component Analysis" (PCA). A novel algorithm to determine
the optimal number of meaningful principal components on a data stream
is proposed. The basic idea of the proposed algorithm is to optimize the
dimensionality adjustment process by taking advantage of several "natural" PCA features. In contrast to the commonly used approach to start
with a maximal set of principal components and apply some sort of stopping rule, the proposed algorithm starts with a minimal set of principal
components and uses a linear regression model in the natural logarithmic
scale to approximate the remaining components. An experimental study
is presented to demonstrate the successful application of the algorithm
to noisy synthetic and real world data sets.

Keywords: Principal Component Analysis · Adaptive dimensionality
adjustment · Stopping rule · Big data · Industry 4.0 · Industrial IoT

1 Introduction

Current developments in communication technology, data management and analysis, measurement technology and sensors as well as automation technology create previously inconceivable possibilities to collect, process, analyze, and evaluate
large amounts of heterogeneous data. This represents a great opportunity and
a great challenge for the players in industry and science. These possibilities are
correlated with an increasing demand in computational efficiency and storage
[7]. Additionally, many industrial processes run continuously, requiring the use
of efficient algorithms for the online processing of data streams.

H. Yin et al. (Eds.): IDEAL 2019, LNCS 11871, pp. 76–84, 2019.
https://doi.org/10.1007/978-3-030-33607-3_9

In the field of data analysis with big data [13], dimensionality reduction methods are typically applied to reduce the set of attributes without loosing important information. One of the most used approaches for dimensionality reduction is "Principal Component Analysis" (PCA) [6,12]. PCA applies an orthonormal transformation to transform a possibly correlated set of data into a set of linearly independent variables. High-dimensional patterns with n dimensions are approximated by a lower-dimensional subspace of m dimensions $\mathbb{R}^n \rightarrow \mathbb{R}^m$. A PCA model describes the subspace with m principal components (with $m \leq n$). In the following, the $n \times m$ matrix $\mathbf{W}$ denotes the estimated normalized eigenvectors $\mathbf{w}_i, i = 1, \ldots, m$ of the data covariance matrix, with one vector per column. These eigenvectors are identical to the principal components. The variance of the projection of the data distribution on the ith principal component $\mathbf{w}_i$ is equal to the eigenvalue λ_i. All eigenvalues λ_i are stored in a diagonal matrix $\mathbf{\Lambda}$ with a size of $m \times m$. An estimation of the remaining eigenvalues in the $n-m$ minor eigendirections is represented by the residual variance σ^2 ([11] p. 93). Additionally, a center vector $\mathbf{c} \in \mathbb{R}^n$ is required to center the PCA. This allows to represent multivariate data only with the matrices $\mathbf{W}, \mathbf{\Lambda}$, the vector $\mathbf{c}$ and σ^2.

When reducing the dimensionality of a data set, one has to define a stopping rule to stop adding more dimensions to the PCA space than necessary. In the following, state-of-the-art stopping rule approaches are reviewed, and it is shown that they are not applicable to online PCA methods [2] operating on continuous data streams. As a consequence, a novel approach to resolve the problem is presented. Furthermore, the results from an experimental study are presented, in which the proposed method is applied to noisy synthetic and real world data sets.

2   Review of Stopping Rule Methods

Many researchers have proposed methods for choosing the optimal number of meaningful principal components [4,9]. Some methods are heuristic, others statistical and different techniques often achieve different results. In the following, some of the most commonly used stopping rules are reviewed and it is shown why they are not applicable to online PCA. The set $V = \{\lambda_1, \lambda_2, \ldots, \lambda_m\}$ is used for further calculations. One criterion used to find the number of meaningful components is the eigenvalue-one criterion [4]. With this approach, every eigenvalue $\lambda > 1$ is kept, while all eigenvalues below that threshold are discarded. The output dimensionality

$$m = |\{b \in V \mid b > 1\}| \tag{1}$$

depends on the number of eigenvalues greater than one. Even though there are a number of problems associated with this criterion [10], it is still a commonly used method in data analysis due to its simplicity. Another approach follows the idea to keep all eigenvalues λ above the average $\frac{1}{n}\sum_{i=1}^{n}\lambda_i$. The output dimensionality

$$m = |\{b \in V \mid b > \frac{1}{n}\sum_{i=1}^{n}\lambda_i\}| \tag{2}$$

is the number of eigenvalues λ that are greater than the average. A similar approach is to checked if an eigenvalue λ represents more variance than a predefined precision $\theta \in [0,1]$ of the total variance. The resulting dimensionality

$$m = |\{b \in V \mid b > \theta\sum_{i=1}^{n}\lambda_i\}| \tag{3}$$

counts every eigenvalue fulfilling the inequality. An alternative criterion is to keep the minimal set of components that describe a certain amount of the total variance. The optimal dimensionality

$$m = \arg\min_{z}\left\{z \in \{1,\ldots,n\} \mid \sum_{i=1}^{z}\lambda_i \geq \theta\sum_{i=1}^{n}\lambda_i\right\} \tag{4}$$

is the smallest z that fulfills the above inequality. Please note that the eigenvalues λ_i have to be sorted by size in descending order for this criterion.

All reviewed and many other stopping rule algorithms [9] require the training of all n principal components before deciding how many are important for further usage. Providing the full set continuously makes the presented stopping rules unusable in online PCA. In the following, a novel approach is presented that solves this problem for the stopping rules reviewed above.

3   Adaptive Dimensionality Adjustment

The proposed algorithm for adaptive dimensionality adjustment is intended to enable the use of stopping rules in hierarchical online PCA. In order to avoid the training of all n principal components, the approach takes advantage of the natural PCA behavior. One characteristic of online hierarchical PCA models is that the eigenvalues λ_i are naturally sorted in a descending order. Additionally, a characteristic of real world data is that the variance is not evenly distributed, so that only some attributes yield important variance and a major part is negligible. The first characteristic is exploited by initializing the PCA output dimensionality with $m_0 = 2$ and only train the two most important principal components. In this way, the initial matrix size reduces to $n \times 2$ for $\mathbf{W}$ and 2×2 for $\mathbf{\Lambda}$. The two principal components are trained with an online hierarchical PCA approach [8]. In the following, an estimation of the remaining $n - m$ eigenvalues has to be obtained. For this, it is taken advantage of the second PCA characteristic. With the assumption of a rapid decline in the eigenvalues λ, a transformation into the logarithmic scale enables the use of a linear regression model [5].

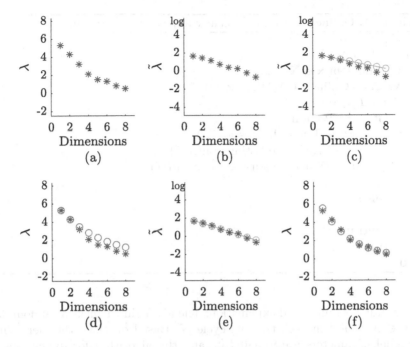

Fig. 1. Application of the algorithm to an 8-dim artificial data set. The real eigenvalues are represented by a star and the estimations by circles: (a–d) 1st step with $m = 2$; (e–f) 2nd step with $m = 4$.

In order to estimate the remaining $n - m$ eigenvalues, the trained eigenvalues λ_i ($i \in \{1, \ldots, m\}$) are first converted into log-eigenvalues $\widetilde{\lambda}_i = \log(\lambda_i)$ with $\widetilde{V}$ being the set of log-eigenvalues (the tilde denotes logarithmic values). Based on the log-eigenvalues $\widetilde{\lambda}_i$, the slope α and the offset β of a linear regression model in the logarithmic scale are calculated. The aim is to estimate the values of the remaining $n - m$ log-eigenvalues $\widetilde{\lambda}_i^* = \alpha i + \beta$ ($i \in \{m+1, \ldots, n\}$), where the star denotes estimated values. The estimated log-eigenvalues $\widetilde{\lambda}_i^*$ are used to extend the set $\widetilde{V}$: $\widetilde{U} = \widetilde{V} \cup \{\widetilde{\lambda}_{m+1}^*, \ldots, \widetilde{\lambda}_n^*\}$. The set $\widetilde{U}$ is converted back to the non-log domain by applying the exponential function, yielding the set U.

The process is illustrated in Fig. 1 for an eight-dimensional synthetic data set. The sorted eigenvalues in normal and logarithmic scale are shown in Fig. 1a–b. In the initial step with $m_0 = 2$, the line of best fit is a simple line through the two initial eigenvalues (Fig. 1c) which are then transformed back into the normal scale (Fig. 1d). Based on these reconstructions, the dimensionality m is adjusted. The newly added eigenvalues (two in this example, thus $m = 4$) are initialized with the linear regression model estimations and the eigenvectors with a random orthonormal system. If the contribution of a principal components is not needed to stay above the threshold, the dimension is discarded. The regression parameters are updated based on the extended set after a specific training period (Fig. 1e–f).

Algorithm 1. Online PCA dimensionality adjustment procedure

1: $m \leftarrow 2, \gamma \leftarrow 0$
2: **for** $i = 1$ to N **do**
3:     $\mathbf{x} \leftarrow$ random $\mathbf{x} \in \mathbf{X} \subset \mathbb{R}^n$
4:     $\mathbf{W}, \mathbf{\Lambda} \leftarrow$ Online PCA$(\mathbf{W}, \mathbf{\Lambda}, \mathbf{x}, m)$ ▷ [8]
5:     **if** $i > \Gamma$ **then**
6:         **if** $\gamma == 0$ **then**
7:             $\alpha, \beta \leftarrow$ Linear Regression$(\widetilde{V})$
8:             $U \leftarrow$ Eigenvalue Estimation(α, β)
9:             $m \leftarrow$ Dimensionality Adjustment(U)
10:             $\gamma \leftarrow \gamma_0$
11:         **else**
12:             $\gamma \leftarrow \gamma - 1$
13:         **end if**
14:     **end if**
15: **end for**

To provide a summarized explanation: The algorithm 1 picks a random data point $\mathbf{x} \in \mathbf{X} \subset \mathbb{R}^n$ in each training cycle N. Based on that, the hierarchical PCA model parameters are updated [8] and the algorithm for dimensionality adjustment is applied. However, in the beginning the first two principal components are trained for Γ training cycles, before dimensionality adjustment is activated. Additionally, the dimensionality adjustment is stopped for γ_0 training cycles to train newly added components. With that, the presented stopping rule approaches in Sect. 2 can be rewritten, depending on the already trained m eigenvalues and the $n - m$ reconstructions according to the line of best fit. In case of the eigenvalue-one criterion, (1) can be rewritten

$$m = |\{b \in U \mid b > 1\}| \tag{5}$$

as an equation based on the extended set U. The same process is applied to the approach of keeping all eigenvalues greater than the average (2). The total variance $\lambda_{total} = \sum_{i=1}^{n} \lambda_i \approx \sum_{i=1}^{m} \lambda_i + \sigma^2$ can be approximated by the already trained m eigenvalues and the residual variance σ^2 [11]. This yields the output dimensionality

$$m = |\{b \in U \mid b > \frac{1}{n}\lambda_{total}\}| \tag{6}$$

depending on the already trained m eigenvalues and the reconstructions on the line of best fit. For the approach of keeping all eigenvalues greater than a proportion θ of the total variance, the output dimensionality becomes

$$m = |\{b \in U \mid b > \theta\lambda_{total}\}|. \tag{7}$$

In the last case, the dimensionality

$$m = \arg\min_z \left\{ z \mid \sum_{i=1}^{m} \lambda_i + \exp(\beta) \sum_{i=m+1}^{z} \exp(\alpha i) \geq \theta \lambda_{total} \right\} \qquad (8)$$

can be rewritten as an inequality depending on the already trained m eigenvalues λ_i and the associated reconstructions expressed by the linear regression model. The approach can represent all possibly occurring component distributions for hierarchical online PCA. One extreme would be a fully symmetric data distribution with all dimensions yielding the same variance. A linear regression model with a slope α of zero perfectly estimates the remaining principal components. On the other hand, a distribution with all variance located in one dimension is correctly estimated with a large negative slope α. In this way, a simple line in a logarithmic scale can cover all possibly occurring eigenvalue distributions in the normal scale. In the following, the approach for adaptive dimensionality determination is applied to noisy synthetic and real world data.

Table 1. Dimensionality adjustment process over 100 repetitions with a varying θ for two different data sets.

Data set	θ	Training cycle mean $\pm s$	Dimensionality mean $\pm s$	Optimal dimensionality	Maximum overshoot
Robot [1]	0.9	2317 ± 1637	3 ± 0	3	4
	0.95	654 ± 489	4 ± 0	4	4
	0.99	207 ± 51	5 ± 0	5	3
Waveform [3]	0.9	1278 ± 553	33.2 ± 1.2	33	6
	0.95	1254 ± 900	36.9 ± 0.2	36	3
	0.99	535 ± 68	39.9 ± 0.3	40	1

4   Experimental Results and Analysis

This experiment aims at effectively adjusting the dimensionality for two medium-dimensional data sets, based on the approach presented in Sect. 3. The first data set contains ten-dimensional data $n_1 = 10$ about orientation, velocity and acceleration of a robot driving over a surface [1]. The second data set contains waveform data augmented by 19 dimensions which contain only gaussian noise with uniform variance (overall: $n_2 = 40$ [3]). Both data sets are trained with the same set of parameters[1]. As stopping rule criterion, (8) is chosen, due to its ability of describing all kinds of eigenvalue distributions. The aim is to test the algorithm with different accuracies $\theta = \{0.9, 0.95, 0.99\}$. The dimensionality adjustment process is repeated 100 times, so that meaningful results are obtained.

---

[1] Parameter set: $m_0 = 2$, $\Gamma = 100$, $\gamma_0 = 10$, $N = 5000$.

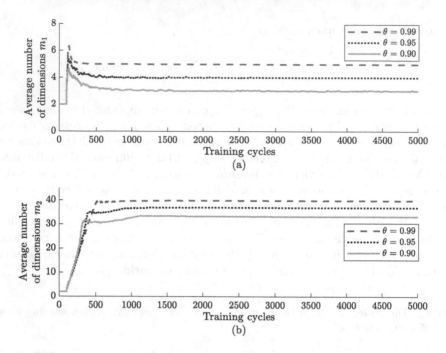

Fig. 2. Dimensionality adjustment process averaged over 100 repetition: (a) shows the average dimensionality adjustment m_1 for a real world data set [1]; (b) visualization of the average dim. adjustment m_2 for waveform data [3].

The results are presented in Table 1. In case of the robot data set [1], the algorithm was in all 100 repetitions, for all θ able to estimate the correct final-state dimensionality. The number of training cycles needed to estimate the final-state dimensionality varied over the accuracy factor θ. With a higher θ the mean was smaller. For the noisy waveform data set [3], the correct dimensionality was achieved in most but not all repetitions. The number of training cycles in a repetition needed to reach the final-state dimensionality is comparable with the results achieved for the robot data set. In Fig. 2, the averaged dimensionality adjustment process is visualized for both data sets. It is shown (Fig. 2a–b) that the dimensionality immediately increases towards the desired dimensionality when the algorithm is activated after Γ training cycles. It is noticeable that the algorithm adds several dimensions at once, speeding up the adjustment process and providing an advantage over the naive approach that simply adds or removes a single dimension. The results prove the robust dimensionality adjustment process with the proposed approach.

5 Conclusion

Driven by the technological developments in the Industrial IoT and in Industry 4.0, the further advancement of dimensionality reduction methods becomes strongly relevant. The method suggested in this work makes a step into this direction by enabling the use of stopping rules in hierarchical online PCA.

The algorithm exploits two "natural" PCA features, by starting with a minimal set of principal components which is extended based on estimations made by a linear regression model on log-eigenvalues. The proposed algorithm was tested in an experimental study with noisy synthetic and real world data sets, proving its robustness in dealing with noise. Furthermore, it is capable of adding many principal components at once, with properly initialized eigenvalues, which gives the algorithm a clear advantage compared to any naive dimensionality adjustment approach which just adds and removes single dimensions. For the same reason, the proposed algorithm is also compatible with various stopping rules.

In conclusion, the results achieved in this work suggest that the proposed extension to stopping rule approaches enables the more efficient use of dimensionality reduction for applications relying on high-dimensional data streams like in the Industrial Internet of Things and related areas.

Acknowledgements. This work was supported by the EFRE-NRW funding programme "Forschungsinfrastrukturen" (grant no. 34.EFRE-0300119).

References

1. Careercon 2019 - help navigate robots. https://www.kaggle.com/c/career-con-2019/data. Accessed 20 May 2019
2. Cardot, H., Degras, D.: Online principal component analysis in high dimension: which algorithm to choose? Int. Stat. Rev. **86**, 29–50 (2015)
3. Gordon, A.D., Breiman, L., Friedman, J.H., Olshen, R.A., Stone, C.J.: Classification and regression trees. Biometrics **40**(3), 874 (1984)
4. Guttman, L.: Some necessary conditions for common-factor analysis. Psychometrika **19**(2), 149–161 (1954)
5. Hancock, P., Baddeley, R., Smith, L.: The principal components of natural images. Netw. Comput. Neural Syst. **3**, 61–70 (1970)
6. Jolliffe, I.T., Cadima, J.: Principal component analysis: a review and recent developments. Philos. Trans. R. Soc. A Math. Phys. Eng. Sci. **374**(2065), 20150202 (2016)
7. Katal, A., Wazid, M., Goudar, R.H.: Big data: issues, challenges, tools and good practices. In: 2013 Sixth International Conference on Contemporary Computing (IC3), pp. 404–409 (2013)
8. Moeller, R.: Interlocking of learning and orthonormalization in RRLSA. Neurocomputing **49**(1–4), 429–433 (2002)
9. Peres-Neto, P.R., Jackson, D.A., Somers, K.M.: How many principal components? Stopping rules for determining the number of non-trivial axes revisited. Comput. Stat. Data Anal. **49**(4), 974–997 (2005)
10. Preacher, K.J., MacCallum, R.C.: Repairing tom swifts electric factor analysis machine. Underst. Stat. **2**(1), 13–43 (2003)

11. Schenck, W.: Adaptive Internal Models for Motor Control and Visual Prediction. MPI Series in Biological Cybernetics, 1st edn. Logos Verlag Berlin, Berlin (2008)
12. Tharwat, A.: Principal component analysis - a tutorial. Int. J. Appl. Pattern Recogn. **3**, 197–238 (2016)
13. Zhang, T., Yang, B.: Big data dimension reduction using PCA. In: 2016 IEEE International Conference on Smart Cloud (SmartCloud), pp. 152–157 (2016)

Relevance Metric for Counterfactuals Selection in Decision Trees

Rubén R. Fernández[✉], Isaac Martín de Diego, Víctor Aceña,
Javier M. Moguerza, and Alberto Fernández-Isabel

Data Science Lab, Rey Juan Carlos University, Móstoles, Spain
{ruben.rodriguez,isaac.martin,victor.acena,javier.moguerza,
alberto.fenandez.isabel}@urjc.es
http://www.datasciencelab.es

Abstract. Explainable Machine Learning is an emerging field in the Machine Learning domain. It addresses the explicability of Machine Learning models and the inherent rationale behind model predictions. In the particular case of example-based explanation methods, they are focused on using particular instances, previously defined or created, to explain the behaviour of models or predictions. Counterfactual-based explanation is one of these methods. A counterfactual is an hypothetical instance similar to an example whose explanation is of interest but with different predicted class. This paper presents a relevance metric for counterfactual selection called *sGower* designed to induce sparsity in Decision Trees models. It works with categorical and continuous features, while considering number of feature changes and distance between the counterfactual and the example. The proposed metric is evaluated against previous relevance metrics on several sets of categorical and continuous data, obtaining on average better results than previous approaches.

Keywords: Explainable Machine Learning · Example-based · Counterfactuals · Decision Trees

1 Introduction

In recent years, black-box Machine Learning (ML) models are quickly gaining popularity because of their ability to model complex data. This usually comes at expense of interpretability. However, some application domains face a trade-off between model performance and interpretability [5]. Model interpretability limits Artificial Intelligence applications because of ethical concerns, such as bias against minorities or high risk environments, and lack of trust due to users' inability to understand model decisions [9].

Explanaible ML is a emerging field of ML which aims to make models interpretable for humans. This would enable users to audit models, decreasing bias in models and allow its usage in high risk environments. Example-based methods have been intensively studied in the literature, and they are regarded as an effective method to provide explanations [1,10]. Examples are explained using other

© Springer Nature Switzerland AG 2019
H. Yin et al. (Eds.): IDEAL 2019, LNCS 11871, pp. 85–93, 2019.
https://doi.org/10.1007/978-3-030-33607-3_10

instances, by leveraging in the human capability to compare. In counterfactual-based methods, an example is explained with the feature changes needed to flip the outcome of the prediction.

In this paper, a novel relevance metric for selecting counterfactuals is proposed in the context of Decision Trees models. The relevance metric, called *sGower*, is a modified version of the Gower distance which induces sparsity. The metric differs from previous approaches as it allows categorical and continuous features while considering distance to the example to be explained.

The paper is structured as follows. Section 2 briefly introduces the Explainable ML field and explores related work. Section 3 addresses how counterfactuals can be calculated in Decision Trees and introduces *sGower*, a relevance metric for counterfactual selection. In Sect. 4, *sGower* is evaluated and compared with other relevance metrics on several data sets. Finally, Sect. 5 summarises the article and the obtained insights, and provides further research directions.

2   Related Work

ML explainability can be approached in two different levels, namely transparent models and post-hoc interpretability. Figure 1 presents a taxonomy of explainable ML methods based on [10]. Transparent models, which are the opposite of black-box models, refer to models that are considered interpretable due to their simple structure such as Linear Regression, Decision trees or Rule-based systems. However, transparent models become black-boxes as their number of parameters grows [1,10]. On the other hand, post-hoc techniques try to explain trained ML models. These include approximating a model using a surrogate model, calculating statistics using model internals, or explanations based on examples.

Explainable ML methods can be further classified based on its scope into local and global interpretability. Global interpretability means that a model can be fully understood at once [9]. This is often achieved by using transparent models, or approximating the model globally by a simpler model, or through example-based methods. On the other hand, local interpretability focuses on a region of the input space (usually an example). By restricting the input space, the decision surface of the model gets smoother. This allows providing approximate explanations to predictions, even in black-box models. Local interpretability is usually approached using local example-based methods or local surrogates, which approximate a small region around an example [1,4,11].

The paper scope is limited to local example-based explanation, which uses examples, previously defined or created, to explain a prediction. This explanations work well when the example can be represented in a meaningful way. For instance, images, text, and tabular data that can be summarised in few values [10]. Examples to explain instances can be calculated in different ways, and they might have different meanings. Influential examples measure which are the instances that affect the most to the prediction of a given example [7]. Prototypes and criticisms compare the example to individuals which are representative in

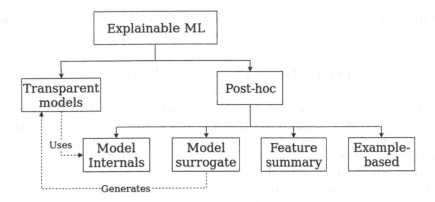

Fig. 1. A taxonomy of Explainable Machine Learning based on [10].

the data and individuals that are not well explained [6]. Counterfactual explanations focus on describing what changes would an observation requires to change the outcome to the expected outcome, often referred as *foil class*.

Counterfactual explanations are intuitive for people, as they resemble human-thinking. They usually are stated as follows "If X had not happened, then Y would not have occur". Furthermore, when the required feature changes are few, it is easier to understand than a full causal attribution [9]. Counterfactual explanations have been used to explain local transparent models [1,4,12], which can be surrogates, or black-box models [8]. They can also be classified as generating exact counterfactuals [4,12] or approximate [1,8].

Counterfactual extraction methods usually generate more than one counterfactual. However, users expect only one explanation, and iterate through them if they do not meet their expectations. Therefore, it is necessary to sort counterfactuals according to some relevance metric. Counterfactuals should involve as few feature changes as possible and be close to the original example [10]. Furthermore, the changes proposed should be feasible and coherent, for example, a counterfactual for a diagnose system should not propose to change the age, or to change the address in a credit-score system. In practise, these limitations can not be addressed in a general framework and they are application and user specific, as the features are handled without domain knowledge. Consequently, counterfactual relevance is usually measured using number of feature changes [4], distance between the example and the counterfactual, or both [8].

This paper focuses on counterfactual explanation for Decision Trees. Relevance metrics for counterfactual trees have been previously proposed focused on the number of feature changes [4]. However, the distance to the example to be explained is not considered. Another relevance metric is based on L1 distance [12], but this method has not been fully disclosed. Alternative metrics have been also proposed (see, for instance, [8]), but they do not work with categorical features (which is one of the strongest points of Decision Tree models). To address these issues it is proposed a novel metric for computing counterfactual relevance able to handle numeric and categorical features, inducing sparsity, and considering distance to the example to be explained.

3   Counterfactual Generation and Selection

Let t be a decision tree for a problem $\mathcal{X} \in \{\mathbf{x}_1, \mathbf{x}_2, \ldots, \mathbf{x}_n\}$ where $\mathbf{x}$ is a vector of dimension d and $y \in \{y_0, \cdots, y_n\}$ its associated label. Let $\mathbf{z}$ be the example to be explained, $\hat{y}$ the predicted class and $foil\_class$ the expected class such that $\hat{y} \neq foil\_class$. A decision tree is represented by a recursive node-based structure. Let a node be defined by a 4-uple (l, r, t, f), which represent the left and right children, the split value and the split column. A child can be either a node or a leaf. Let a leaf be characterised as the class it represents. Let $leaves$ be the set of all leaves in t. Let $\mathbf{c}$ be the vector of changes defined by a vector of dimension d, where c_i contains the operation to update z_i. Let z' be the result of applying the vector of changes $\mathbf{c}$ to $\mathbf{z}$. Finally, let C be the set of counterfactuals for the example $\mathbf{z}$ in the decision tree t.

Counterfactuals can easily be extracted from decision trees. Notice that, given an example $\mathbf{z}$ with predicted class $\hat{y}$, then all the tree paths which end in a leaf with class distinct to $\hat{y}$ are counterfactuals. In this context, a counterfactual path, ending in leaf l, can be represented as the feature changes $\mathbf{c}$ which make $\mathbf{z}$ fall in l. The changes in $\mathbf{c}$ depend on the feature type. In categorical features, the change is either asserting a value (e.g., $job = gardener$) or not belonging to a set of values (e.g., $transportation \notin \{car, plane\}$). On the other hand, in a numeric feature c_i represents a numerical value, which represents the difference with respect to z_i. If $c_i > 0$ then the feature z_i should be greater than $c_i + z_i$, otherwise $c_i < 0$ and the feature z_i should be less than or equals to $c_i + z_i$.

The calculation of the vector of changes $\mathbf{c}$ is described in Algorithm 1. It starts at the root node, with the change vector $\mathbf{c} = I^d$, where I represents the identity operation, thus no change is performed. Given a node e, if e is a leaf ($e \in leaves$), then $\mathbf{c}$ is added to the set of resulting counterfactuals providing that the class of e is equals to $foil\_class$. Otherwise, the node conditional is checked with $check\_split$, which checks the conditional considering the changes $\mathbf{c}$. If the condition is meet, then the algorithm is called recursively with the node e_l, and with e_r after having updated $\mathbf{c}$ to not meet the condition. On the other hand, if the condition is not meet, the roles of e_l and e_r are inverted, the method is first called recursively with e_r, and then e_l after updating $\mathbf{c}$.

A novel metric for computing the relevance of the counterfactuals is proposed. The metric is derived from Gower distance [3] (see Eq. 1), useful for categorical and real features. Gower distance works by computing a distance for each feature independently and then average the result over the number of features. The distance used in each feature depends on the type of data, having as only requirement a range between 0 and 1. In this proposal, Gower is modified to induce sparsity, thus penalising the number of changes in the counterfactual. As a result, few large changes are preferred over many small changes, which make a counterfactual explanation harder to understand. Given the set of feature changes

---

Algorithm 1. node_counterfactuals

Require: $(node, foil\_class, z, c, leaves)$
1: $C \leftarrow \emptyset$
2: **if** $node \in leaves$ **then**
3:     **if** $node_{class} = foil\_class$ **then**
4:         $C \leftarrow c$
5:     **end if**
6: **else**
7:     $col = node_f$
8:     **if** $check\_split(col, node_t, z, c)$ **then**
9:         $C = C \cup node\_counterfactuals(node_l, foil\_class, z, c, leaves)$
10:         $c = negate\_split(col, node_t, z, c)$
11:         $C = C \cup node\_counterfactuals(node_r, foil\_class, z, c, leaves)$
12:     **else**
13:         $C = C \cup node\_counterfactuals(node_r, foil\_class, z, c, leaves)$
14:         $c = negate\_split(col, node_t, z, c)$
15:         $C = C \cup node\_counterfactuals(node_l, foil\_class, z, c, leaves)$
16:     **end if**
17: **end if**
18: **return** C

$\mathcal{F} = \{1, \ldots, d | z_i \neq z_i'\}$ and being $\#\mathcal{F}$ the number of feature changes, $sGower$ is defined as follows:

$$Gower(z, z') = \frac{\sum_{i \in \mathcal{F}} dist(z_i, z_i')}{d} \tag{1}$$

$$sGower(z, z') = \frac{\sum_{i \in \mathcal{F}} dist(z_i, z_i')}{1 + d - \#\mathcal{F}} \tag{2}$$

$sGower$ lower bound is 0, when there are no changes the numerator is zero and the denominator is greater than zero. The upper bound is d, when all features are changed and its pairwise distances are equal to 1.

4   Experiments

The performance of the proposed metric is evaluated in two groups of experiments. The first group of experiments focus on continuous features data sets, whereas the second group focus on data sets containing both categorical and continuous features. The data sets were randomly split into train (70%) and test (30%), being the training set used to fit the decision tree and the test to generate counterfactuals. Continuous features were standardised in both experiments, so as to provide meaningful distance comparisons between counterfactuals and examples to be explained.

Given an example to be explained, all possible counterfactuals are generated using Algorithm 1. Then, the counterfactual which minimises the relevance metric with respect to the example is chosen. If more than one counterfactual get

Table 1. Continuous features data sets results. Lower values are better. Best result (excluding reference value) is in bold. m.c and m.d refers to minimum changes and minimum distance respectively.

Data set	Number of changes					Distance				
	Gower	GS	m.c.	sGower	m.d.	Gower	GS	m.c.	sGower	m.d.
Breast cancer	1.40	**1.38**	1.00	1.39	1.44	**1.39**	1.32	1.85	**1.39**	0.68
Glass	1.14	1.14	1.00	**1.12**	1.14	**0.26**	0.42	0.80	**0.26**	0.19
Ionosphere	1.39	1.41	1.18	**1.38**	1.39	**0.53**	0.59	1.36	**0.53**	0.53
Iris	1.16	1.19	1.00	**1.15**	1.21	0.53	**0.48**	0.81	0.54	0.47
Transfusion	1.04	1.06	1.00	**1.03**	1.08	0.41	**0.39**	0.53	0.41	0.07
Average	1.26	1.23	1.03	**1.21**	1.25	**0.62**	0.64	1.07	0.62	0.38

the best relevance, the counterfactual generated by the leaf closest to the leaf of the example is selected. The distance between the leafs is measured as the number of edges between them. The presented results are the average over all the counterfactuals generated for each relevance metric.

In the continuous features experiment, five data sets from *UCI* repository were used [2]. These data sets contain between 4 and 34 features, and the number of observations range from 150 to 569. The counterfactuals are evaluated according to the number of feature changes and L1 distance with respect to the example. This evaluation is performed separately. The goal is to select the counterfactuals with fewer feature changes and as close as possible to the example.

The relevance metrics evaluated are *sGower* and Gower, using the Euclidean distance for continuous features (normalised to be from 0 to 1), the absolute pairwise distance scaled by the inverse of the median absolute deviation as proposed by [8] (GS), minimum changes which selects the counterfactual with fewer feature changes [4], and minimum distance which selects the counterfactual which is closer according to L1 distance. The performance values for the methods minimum changes and minimum distance are the best possible for number of changes and L1 distance, respectively. Results are reported in Table 1.

sGower outperforms GS on average in both number of changes (1.21 vs. 1.23) and minimum distance (0.62 vs. 0.64), and performs better than Gower in number of changes (1.26) while obtaining similar results in minimum distance (0.62). Furthermore, *sGower* perform better than minimum number of changes and minimum distance in the complementary task. As an example to illustrate the proposed method, a counterfactual for the transfusion data set is obtained [13]. The data set contains 4 variables: months since last donation (Recency), total number of donation (Frequency), total blood donated in cc. (Monetary), months since first donation (Time) and the target variable that indicates whether the person donated (or not) blood in March 2007. The example to be explained was classified as donation in March 2007 with attributes: Recency 2 months, Frequency 50, Monetary 12500 and Time 98. In this case, the counterfactual explanation will focus on the changes required to classify the individual as no donation. *SGower* and GS counterfactuals change suggest to reduce Monetary to less than 1125, minimum number of changes to reduce Frequency to less than 25 and minimum distance to increase Recency to more than 6.5.

In the categorical and continuous feature experiment, the ability of *sGower* to handle categorical data is tested. The counterfactual will be evaluated using number of changes and Gower distance will be used to measure the distance between the counterfactual and the example. Gower will be parametrized similar to *sGower*, simple matching for categorical features (0 if the values are equals, and 1 otherwise) and the Euclidean distance for continuous features (normalised to be from 0 to 1). This experiment will use 3 data sets from *UCI* repository. The results are reported in the Table 2.

Table 2. Categorical and continuous features data sets results. Lower values are better. Best result (excluding reference value) is in bold. m.c and m.d refers to minimum changes and minimum distance (Gower) respectively.

Data set	Number of changes			Distance		
	m.d.	m.c.	sGower	m.d.	m.c.	sGower
Credit	1.88	1.10	**1.33**	0.86	0.88	**0.87**
Credit_aus	2.96	1.01	**1.14**	0.81	0.82	**0.81**
Thyroid-disease	1.80	1.00	**1.76**	0.10	0.14	**0.10**
Average	2.21	1.03	**1.41**	0.59	0.61	**0.60**

sGower outperforms alternative methods excluding the reference score. In addition, the comparison between Gower and *sGower* shows that *sGower* is able to introduce sparsity without heavily penalising the distance to the example.

5 Conclusions

In this paper, *sGower*, a relevance metric for counterfactual selection has been proposed. It considers the distance between counterfactual and original example and number of changes, while being able to handle categorical and continuous features. The *sGower* metric was evaluated against previous metrics in data sets with continuous features, and continuous and categorical features. The evaluation, averaged over all test instances of each data set, included distance between the counterfactual and the example, and number of changes.

The proposed *sGower* metric performs better on the continuous, and continuous and categorical experiments compared to previous relevance metrics. However, the relevance might not correlate with the ability of the users to understand an explanation, as they only maximise its desirable properties. In addition, the ability to understand explanations strongly depends on users' domain, skills and background. For this reason, general relevance metrics are only useful as prototypes which can be adapted to the domain and public.

sGower relevance metric has been tested only with the desired properties of a counterfactual explanation. Future work will include a user-oriented study to asses the quality of the explanations. This study should measure how counterfactual explanations among the different relevance metrics support users in decision making and transparency [1]. Further, future enhancements will consider restrictions and weight support in counterfactual changes. Restrictions in changes will prevent generating counterfactual changes that does not make sense in the domain. For example, a diagnose system that suggests a gender change to reduce the likelihood of having a disease. In such case, the gender might be a risk factor, but not the cause. On the other hand, weight support will enable to weight more those changes that are more expensive according to the domain.

Acknowledgements. Research supported by grant from the Spanish Ministry of Economy and Competitiveness: SABERMED (Ref: RTC-2017-6253-1); Retos-Investigación program: MODAS-IN (Ref: RTI2018-094269-B-I00); and NVIDIA Corporation.

References

1. Adhikari, A., Tax, D., Satta, R., Fath, M.: Example and feature importance-based explanations for black-box machine learning models. arXiv preprint arXiv:1812.09044 (2018)
2. Dua, D., Graff, C.: UCI machine learning repository (2017). http://archive.ics.uci.edu/ml
3. Gower, J.C.: A general coefficient of similarity and some of its properties. Biometrics **24**, 857–871 (1971)
4. Guidotti, R., Monreale, A., Ruggieri, S., Pedreschi, D., Turini, F., Giannotti, F.: Local rule-based explanations of black box decision systems. arXiv preprint arXiv:1805.10820 (2018)
5. Hall, P., Gill, N.: Introduction to Machine Learning Interpretability. O'Reilly Media, Sebastopol (2018)
6. Kim, B., Khanna, R., Koyejo, O.O.: Examples are not enough, learn to criticize! criticism for interpretability. In: Advances in Neural Information Processing Systems, pp. 2280–2288 (2016)
7. Koh, P.W., Liang, P.: Understanding black-box predictions via influence functions. In: Proceedings of the 34th International Conference on Machine Learning, vol. 70, pp. 1885–1894. JMLR. org (2017)
8. Laugel, T., Lesot, M.J., Marsala, C., Renard, X., Detyniecki, M.: Inverse classification for comparison-based interpretability in machine learning. arXiv preprint arXiv:1712.08443 (2017)
9. Miller, T.: Explanation in artificial intelligence: insights from the social sciences. Artif. Intell. **267**, 1–38 (2018)
10. Molnar, C.: Interpretable machine learning (2018)
11. Ribeiro, M.T., Singh, S., Guestrin, C.: Why should i trust you?: explaining the predictions of any classifier. In: Proceedings of the 22nd ACM SIGKDD International Conference on Knowledge Discovery and Data Mining, pp. 1135–1144. ACM (2016)

12. Sokol, K., Flach, P.A.: Glass-box: explaining AI decisions with counterfactual statements through conversation with a voice-enabled virtual assistant. In: IJCAI, pp. 5868–5870 (2018)
13. Yeh, I.C., Yang, K.J., Ting, T.M.: Knowledge discovery on RFM model using Bernoulli sequence. Expert Syst. Appl. **36**(3), 5866–5871 (2009)

Weighted Nearest Centroid Neighbourhood

Víctor Aceña[✉], Javier M. Moguerza, Isaac Martín de Diego,
and Rubén R. Fernández

Rey Juan Carlos University, c/ Tulipán s/n, 28933 Móstoles, Spain
{victor.acena,javier.moguerza,isaac.martin,ruben.rodriguez}@urjc.es
http://www.datasciencelab.es

Abstract. A novel binary classifier based on nearest centroid neighbours is presented. The proposed method uses the well known idea behind the classic k-Nearest Neighbours (k-NN) algorithm: one point is similar to others that are close to it. The new proposal relies on an alternative way of computing neighbourhoods that is better suited to the distribution of data by considering that a more distant neighbour must have less influence than a closer one. The relative importance of any neighbour in a neighbourhood is estimated using the SoftMax function on the implicit distance. Experiments are carried out on both simulated and real data sets. The proposed method outperforms alternatives, providing a promising new research line.

Keywords: Nearest Neighbours · Classification · Nearest Centroid Neighbourhood · Parameter selection · Similarity measure

1 Introduction

The k-Nearest Neighbours (k-NN) algorithm is a basic method that assigns the most frequent class label among the k training points closest to the target point. The main idea of assigning a value based on the most similar points fits into the human thought of constant search for similarities between objects to compare them. Because of this and its computational simplicity, it is widely used in Machine Learning, Pattern Recognition or Data Mining algorithms.

The k-NN setting parameters are the number of neighbours and the distance function. However, some other factors should be considered in order to evaluate the performance of the method: the size and shape of the neighbourhood, determined by the way it is computed; the number of neighbours, caused by the choice of k; the similarity among neighbours, determined by the chosen distance; and finally, the relative importance of a neighbour, class, or feature.

In this paper, it is presented a classifier built by using the information from all these factors. The method is named wk-NCN, weighted k-Nearest Centroid Neighbourhood. A novel way of calculating neighbourhoods based on centroids is used. Thus, a method for calculating an adaptive neighbours to the data

© Springer Nature Switzerland AG 2019
H. Yin et al. (Eds.): IDEAL 2019, LNCS 11871, pp. 94–101, 2019.
https://doi.org/10.1007/978-3-030-33607-3_11

distribution is provided. Furthermore, the well-known SoftMax function, on the implicit distance, is used to estimate the proper weight of each neighbour.

The rest of the paper is structured as follows. Some of the related k-NN improvements are presented in Sect. 2. Section 3 addresses the fundamentals of the proposed wk-NCN classifier. The experiments and results for simulated and real data sets are detailed in Sect. 4. Section 5 concludes and provides future guidelines.

2 Related Work

The analysis and improvements of the k-NN method has been object of study since it was firstly proposed by Cover and Hart in 1967 [3]. In this section, a brief review on some of the variations of the original algorithm is presented, specially focused on the relevant factors regarding the performance of the algorithm: the size and shape of the neighbourhood, the distance metric, and the importance of a neighbour, class or feature.

One of the main issues of k-NN is the selection of the k parameter. That is, how to define the best number of neighbours for classification tasks. Many authors have proposed methods for finding the optimal k in different contexts. One of the most relevant and used contribution is the optimal k by Silverman [11]: $k = n^{4/(d+4)}$, where n is the sample size and d is the number of features in the data set. In the same way, Ghosh [6] uses Bayesian methods to estimate k as opposed to the classical cross-validation and likelihood cross-validation. Jaiswal, Bhadouria and Sahoo [9] present an automated parameter selector guided by Cuckoo search, in which k and the distance metric among four usual metrics, are optimised together by a meta-heuristic method.

Beyond a search for common distance metrics, many other flexible or adaptable metrics can be designed, such as the proposed by Hastie's and Tibshirani [7]. In order to create searching methods for adaptive parameters, Hulett, Hall and Qu [8] build an automatic selection of neighbouring instances as defined by a dynamic local region. Zhang et al. [12] propose the S-k-NN algorithm, which uses the reconstruction of the correlation between test samples and training samples in order to automatically calculate the optimal k for any test sample.

Adaptive methods based on building neighbourhoods dependent on the data distribution have been developed. For instance, Chaudhuri [2] proposes a new definition of neighbourhood to capture the idea that the neighbours should be as near to the target point and as symmetrically placed around it as possible. This method has been called Nearest Centroid Neighbourhood (NCN). Under this definition of neighbourhood García et al. [5] developed the kNCN model for regression tasks.

Finally, a remote neighbour should have less influence than a nearby neighbour on the decision for a test point. This idea has been theoretically proven by Samworth, Richard et al. [10] and developed by Biswas et al. [1].

The proposal presented in this paper is based on these two ideas: to build a data distributed based neighbourhood, and to differently weight the neighbours according to their distance to the point of interest. It is possible to obtain a method that adapts to the data distribution by using the neighbourhood definition proposed in [2]. In addition, the effect of the nearest neighbours can be weighted in an effective way to create the weighted Nearest Neighbourhood method by using the SoftMax function.

3 Method

Let $X = \{(x_1, x_2, \ldots, x_n) \mid x_i \in \mathbb{R}^d\}$ be the set of training data points and $y = \{(y_1, y_2, \ldots, y_n) \mid y_i \in \{0,1\}\}$ the set of its corresponding class labels. Let $N_k(S, p)$ be the k-neighbourhood of a point p in a data set S, that is, the k points in S nearest to p.

3.1   The k-NN and k-NCN classifiers

The main idea behind the k-NN classifier is quite simple: to estimate the probability of a class as the proportion of points in the neighbourhood of a point that belong to that class. It is formalised as follows. Given a training data set X, a positive integer k and a new point to be classified p, the k-NN classifier identifies the k neighbours in the training data set that are nearest to p: $N_k(X, p)$. Then, the conditional probability of class $j \in \{0, 1\}$ can be estimated using the following equation:

$$P(y = j | X = p) = \frac{1}{k} \sum_{i \in N_k(X,p)} I(y_i = j) \tag{1}$$

The k-NCN method (first presented in [2]) is also based on the neighbourhood of a point, dependent on a parameter k. However, the neighbourhood is based on centroids calculation. Thus, the set of k neighbours whose centroid is closest to the point p is selected. Following [2], the class label will be estimated as indicated in Algorithm 1. The input of Algorithm 1 are the training data, $X = \{x_1, x_2, \ldots, x_n\}$, the number of centroid neighbours k, the distance metric ($Dist$) to compute the similarity between points, and the target point p whose class will be estimated. The output is the radius that defines the neighbourhood around the target point p.

The intuition behind the calculation of k-NCN is to establish a symmetry within the neighbourhood with respect to a point p and to preserve the closeness of the points contained in such neighbourhood. This intuition is confirmed during the second iteration of the *while* loop in Algorithm 1, since the second neighbour is not the second nearest neighbour. Instead, the algorithm selects a point in a direction diametrically opposite to the first neighbour with respect to p. Thus, advantages over the k-NN method are achieved.

In low-density areas of the data set, the neighbourhood are quite bigger than the obtained using the classic nearest neighbourhood. In high-density areas the neighbourhood radius is roughly the same in both methods. These issues are shown in Fig. 1.

Once the neighbourhoods are computed, it is possible to estimate the conditional probability of class in the same way as in the k-NN method (Eq. 1).

Algorithm 1. Nearest centroid neighbours

Input: $(X, k, p, Dist)$
Output: r
1:   $Q \leftarrow \emptyset$
2:   $q_1 \leftarrow findNN(X, p)$ ▷ q_1 is the nearest neighbour
3:   $Q \leftarrow q_1$
4:   $X_{aux} \leftarrow X - \{q_1\}$
5:   $r \leftarrow Dist(q_1, p)$
6:   $j \leftarrow 1$
7:   **while** $j \leq k$ **do**
8:      $j \leftarrow j + 1$
9:      $dist\_min \leftarrow \infty$
10:     **for all** $x_i \in X_{aux}$ **do**
11:        $M \leftarrow computeCentroid(Q \cup \{x_i\})$
12:        **if** $Dist(M, p) \leq dist\_min$ **then**
13:          $q_j \leftarrow x_i$
14:          $dist\_min \leftarrow Dist(M, p)$ ▷ update minimum distance
15:     $Q \leftarrow Q \cup \{q_j\}$
16:     $r \leftarrow max(r, Dist(q_j, p))$ ▷ update the radius when needed
17:     $X_{aux} \leftarrow X_{aux} - \{q_j\}$

3.2 *wk*-NCN classifier

The advantages of the k-NCN method over k-NN method could be a drawback in some situations. Notice that, when the neighbourhood is very wide, there could be points far from the target that contribute too much to the estimation of the class probability. One way to reduce this effect is to weight the contribution of each neighbour depending on the distance to the target. In this paper, it is proposed to incorporate this weighting by using the SoftMax function, defined as follows:

$$\sigma : \mathbb{R}^T \to [0, 1]^T$$

$$\sigma(\mathbf{z})_j = \frac{e^{z_j}}{\sum_{t=1}^{T} e^{z_t}} \quad j = 1, 2, \ldots, T$$

Thus, the estimation of the conditioned probability of a point is:

$$P(y = j | X = p) = \sum_{i \in N_k(X, p)} \frac{e^{-||x_i - p||_2}}{\sum_{i \in N_k(X, p)} e^{-||x_i - p||_2}} \cdot I(y_i = j) \tag{2}$$

where $|| \cdot ||_2$ defines the $L^2 - norm$. The *wk*-NCN method computes the nearest centroid neighbours as indicated in Algorithm 1, and estimates the probability of class using Eq. 2. Thus, the neighbourhood depends on the distribution of the data set and the neighbours are weighted based on their distance to the point of interest. The weights decrease exponentially with respect to the distance between each neighbour and the target point.

4   Experiments and Results

The performance of the proposed method is first evaluated on simulated data sets. Then, its relative performance regarding alternative methods is estimated on a battery of real data sets.

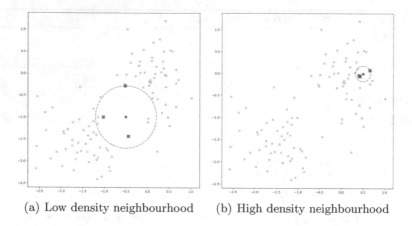

(a) Low density neighbourhood   (b) High density neighbourhood

Fig. 1. Nearest Centroid neighbours examples. Comparison between high and low density areas. The figure shows the computed neighbourhoods by 3-NCN method of two example points. The two colours in the plot represents the category labels. Within each neighbourhood, the points marked are the 3 Nearest Centroid Neighbours, that define the neighbourhood.

4.1 Simulated Data

Three different scenarios are considered: separated binary classes, overlapped binary classes, and highly overlapped binary classes. Thus, three binary classification data sets are generated using two bivariate normal distributions. In the first scenario, two separated classes are considered: $\mathcal{N}(\mu_i = 0, \sigma_i = 0.5)$ and $\mathcal{N}(\mu_i = -2, \sigma_i = 0.5)$, respectively. In the second scenario, two overlapped classes are considered: $\mathcal{N}(\mu_i = 0, \sigma_i = 0.5)$ and $\mathcal{N}(\mu_i = -1, \sigma_i = 0.5)$, respectively. In the third scenario, two highly overlapped classes are considered: $\mathcal{N}(\mu_i = 0, \sigma_i = 0.5)$ and $\mathcal{N}(\mu_i = -0.5, \sigma_i = 0.5)$, respectively.

With the settings, 150 samples of each class are generated for each example. In addition, 100 simulations are performed on each scenario. The methods wk-NCN, k-NCN and the classical k-NN are tested. In order to measure the error the leave one out evaluation method was considered. The average rate of misclassification over all simulations was calculated for each $k \in \{1, \ldots, 20\}$. As expected, in the scenario of separate classes the errors of the three methods are zero. But in the other two scenarios the methods wk-NCN and k-NCN outperform the k-NN for the same k-value as shown in Fig. 2. Given the normality of the generated data, there are no significant differences between the wk-NCN and k-NCN methods.

4.2 Real Data

In this experiment, the performance of the methods presented in the previous section is evaluated on real data sets obtained from the UCI repository [4]. A description of these data sets is detailed in Table 1. Each data set has been

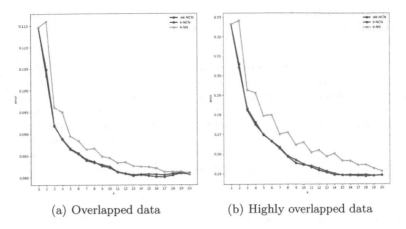

(a) Overlapped data   (b) Highly overlapped data

Fig. 2. Performance (proportion of misclassified instances) of the wk-NCN and alternative methods for simulated data sets.

divided into two sets: training set (70%) and test set (30%). Training sets are used to fit the models, and test sets are be used to evaluate the performance of the models.

The proposal, wk-NCN with k parameter selected by leave one out, is compared to six alternative methods: k-NCN and k-NN with k parameters selected by leave one out; k-NN where the parameter k is the optimal value proposed by Silverman, k-NN with k equals 1, that is 1-NN; Decision Tree and Random Forest models. Table 2 details the classification errors on the test sets for all the trained models.

Notice that the proposed wk-NCN method obtained better or equal results than the k-NCN method for all the considered data sets. Additionally, a model raking is built in order to achieve global summary of the results. For each data set, each model is scored according to its error in a bottom-up way. That is, the model with the smallest error has 1 point, the next one has 2 points, etc. Therefore, the lowest score is the best model. Figure 3 shows the average and

Table 1. Data sets summary

Data set	Samples	Features
Ionosphere	351	34
Mammographic	961	6
Congressional voting records	435	16
Breast cancer wisconsin (Diagnostic)	569	32
Banknote authentication	1372	5
Blood transfusion service center	748	5
Connectionist (Sonar, Mines vs. Rocks)	208	60

Table 2. Real data sets test errors (proportion of misclassified instances). Best results in bold for each data set.

Method	Test errors						
	Mammographic	Ionosphere	Congressional	Breast	Banknote	Blood	Connectionist
wk-NCN	**0.217**	0.151	0.100	0.041	**0.000**	0.213	**0.206**
k-NCN	0.229	0.151	0.100	0.053	**0.000**	0.227	**0.206**
k-NN (k-loo)	0.229	0.151	0.100	**0.029**	0.002	0.253	**0.206**
k-NN (k-Silverman)	0.225	0.151	0.086	0.035	0.015	**0.209**	**0.206**
1-NN	0.245	0.151	0.100	0.053	0.002	0.316	**0.206**
Decision tree	0.257	0.104	**0.014**	0.089	0.034	0.262	0.397
Random forest	0.233	**0.085**	**0.014**	0.029	0.005	0.284	0.333

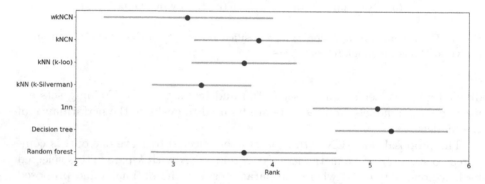

Fig. 3. Average rank and confidence interval for the proposed and alternative methods.

standard deviation for the rank of each method when all the data sets where considered. It is observed that the wk-NCN method obtains the best overall score, followed by the k-NN with k parameter selected by Silverman method. A hypothesis test to evaluate the differences among the methods is performed. The null hypothesis is that the alternative methods obtain lower error than the proposed one. The wk-NCN method statistically outperforms k-NCN, k-NN (k-loo), 1-NN (p-value < 0.01) and Decision Tree methods (p-value < 0.1).

5   Conclusions

This paper presents the weighted Nearest Centroid Neighbourhood method for binary classification. It is based on an alternative version of neighbourhood, that establish a symmetry within the neighbourhood with respect to the target point. Furthermore, the number of neighbours varies according to the distribution around the target point. The SoftMax function is used to weight the effect of each neighbour according to the implicit distance to the target.

The experimental results show that the wk-NCN outperforms some of the alternative methods. In particular, the proposed method equals or improves all

k-NCN experiments. This suggests that weighting the distances by the SoftMax function is effective. This opens the research line to look for other metrics to weight the effect of neighbours according to their distance.

Although wk-NCN is adaptable to the distribution of the data, this could be improved by considering two factors. First, a constant radius is generated for each neighbourhood. Therefore, an adaptive version of the neighbourhood should be explored without losing the useful symmetry structure provided by the method. Furthermore, the number of neighbours in the neighbourhood is inherently variable in this method. Therefore, when creating neighbourhoods, an adaptive k will be a natural variation.

Acknowledgements. Research supported by grant from the Spanish Ministry of Economy and Competitiveness, under the Retos-Colaboración program: SABERMED (Ref: RTC-2017-6253-1); Retos-Investigación program: MODAS-IN (Ref: RTI2018-094269-B-I00); and the support of NVIDIA Corporation with the donation of the Titan V GPU.

References

1. Biswas, N., Chakraborty, S., Mullick, S.S., Das, S.: A parameter independent fuzzy weighted k-nearest neighbor classifier. Pattern Recogn. Lett. **101**, 80–87 (2018)
2. Chaudhuri, B.: A new definition of neighborhood of a point in multi-dimensional space. Pattern Recogn. Lett. **17**(1), 11–17 (1996)
3. Cover, T.M., Hart, P.E., et al.: Nearest neighbor pattern classification. IEEE Trans. Inf. Theory **13**(1), 21–27 (1967)
4. Dua, D., Graff, C.: UCI Machine Learning Repository (2017). http://archive.ics.uci.edu/ml
5. García, V., Sánchez, J., Marqués, A., Martínez-Peláez, R.: A regression model based on the nearest centroid neighborhood. Pattern Anal. Appl. **21**(4), 941–951 (2018)
6. Ghosh, A.K.: On optimum choice of k in nearest neighbor classification. Comput. Stat. Data Anal. **50**(11), 3113–3123 (2006)
7. Hastie, T., Tibshirani, R.: Discriminant adaptive nearest neighbor classification and regression. In: Advances in Neural Information Processing Systems, pp. 409–415 (1996)
8. Hulett, C., Hall, A., Qu, G.: Dynamic selection of k nearest neighbors in instance-based learning. In: 2012 IEEE 13th International Conference on Information Reuse & Integration (IRI), pp. 85–92. IEEE (2012)
9. Jaiswal, S., Bhadouria, S., Sahoo, A.: KNN model selection using modified cuckoo search algorithm. In: 2015 International Conference on Cognitive Computing and Information Processing (CCIP), pp. 1–5. IEEE (2015)
10. Samworth, R.J., et al.: Optimal weighted nearest neighbour classifiers. Ann. Stat. **40**(5), 2733–2763 (2012)
11. Silverman, B.W.: Density Estimation for Statistics and Data Analysis. Routledge, New York (2018)
12. Zhang, S., Cheng, D., Deng, Z., Zong, M., Deng, X.: A novel KNN algorithm with data-driven k parameter computation. Pattern Recogn. Lett. **109**, 44–54 (2018)

The Prevalence of Errors in Machine Learning Experiments

Martin Shepperd[1(✉)], Yuchen Guo[2], Ning Li[3], Mahir Arzoky[1],
Andrea Capiluppi[1], Steve Counsell[1], Giuseppe Destefanis[1], Stephen Swift[1],
Allan Tucker[1], and Leila Yousefi[1]

[1] Brunel University London, London, UK
`martin.shepperd@brunel.ac.uk`
[2] Xi'an Jiaotong University, Xi'an, China
[3] Northwestern Polytechnical University, Xi'an, China

Abstract. *Context*: Conducting experiments is central to research machine learning research to benchmark, evaluate and compare learning algorithms. Consequently it is important we conduct reliable, trustworthy experiments.

Objective: We investigate the incidence of errors in a sample of machine learning experiments in the domain of software defect prediction. Our focus is simple arithmetical and statistical errors.

Method: We analyse 49 papers describing 2456 individual experimental results from a previously undertaken systematic review comparing supervised and unsupervised defect prediction classifiers. We extract the confusion matrices and test for relevant constraints, e.g., the marginal probabilities must sum to one. We also check for multiple statistical significance testing errors.

Results: We find that a total of 22 out of 49 papers contain demonstrable errors. Of these 7 were statistical and 16 related to confusion matrix inconsistency (one paper contained both classes of error).

Conclusions: Whilst some errors may be of a relatively trivial nature, e.g., transcription errors their presence does not engender confidence. We strongly urge researchers to follow open science principles so errors can be more easily be detected and corrected, thus as a community reduce this worryingly high error rate with our computational experiments.

Keywords: Classifier · Computational experiment · Reliability · Error

1 Introduction

In recent years there has been a proliferation in machine learning research and its deployment in a wide range of application domains. The primary vehicle for evaluation has been empirical via experiments. Typically an experiment seeks to assess the behaviour of learning algorithms over one or more data sets by varying the treatment and examining the response variables, e.g., predictive performance and execution time. Unfortunately, a challenge has been to construct a consistent

© Springer Nature Switzerland AG 2019
H. Yin et al. (Eds.): IDEAL 2019, LNCS 11871, pp. 102–109, 2019.
https://doi.org/10.1007/978-3-030-33607-3_12

or even coherent picture from the many experimental results. For instance, in software defect prediction, a major meta-analysis of results found that the single largest determinant of predictive performance was not the choice of algorithm but which research group undertook the work [17].

As a side effect of a recent systematic review of supervised and unsupervised classifiers [13] conducted by three of the authors (NL, MS and YG), we observed quality problems with a surprising proportion of studies when trying to reconstruct confusion matrices in order to obtain comparable classification performance statistics. These involved violations of simple integrity constraints regarding the confusion matrix such as marginal probabilities not summing to one [4]. One likely driver for these kinds of errors is that computational experiments can be extremely complex, involving data pre-processing, feature subset selection, imbalanced learning, complex cross-validation design and tuning of hyper-parameters over many, often large, data sets.

In this study we explore the phenomenon of simple arithmetical and statistical errors in machine learning experiments and investigate one particular domain of classifying software components as defect or not defect-prone. This is an active and economically important area. For overviews see [6,10]. Note we do not consider the more complex challenges of appropriate experimental design and analysis [11] nor the ongoing debate concerning the validity of null hypothesis significance testing (NHST) [7].

The remainder of this paper is organised as follows. The next section provides some background on error checking in experiments. We then summarise how our data were extracted from a systematic review comparing unsupervised and supervised learners in Sect. 3. Next, in Sect. 4 we explain how we checked for errors and the outcome of this analysis. We conclude with a discussion of the significance of our findings, possible steps the research community might adopt and suggestions for further work.

2   Background

For some time, researchers have expressed concerns about the reliability of individual experiments [9] from a range of causes including simple errors. Brown and Heathers [5] analysed a series of empirical psychology studies for simple arithmetic errors e.g., if there are 10 participants, a proportion could not take on the value of 17%. They entitle their method granularity-related inconsistency of means (GRIM) and found that of 71 testable articles around half (36/71) appeared to contain at least one inconsistent mean.

An alternative approach is presented by Nuijten et al. [15] who provide an R package to assist in the checking of inferential statistics such as χ^2, ANOVA and t-tests. However, their automated procedure requires the reporting of inferential statistics using the APA format which is not commonplace in computer science. Nevertheless it is sobering to note that in their analysis of 250,000 p-values from psychology experiments, half of all published papers contained at least one p-value that was inconsistent with its test statistic and degrees of freedom. In

12% of papers the error was sufficient to potentially impact the statistical conclusion.

More specific to experiments based on learning classifiers is the work by Bowes et al. [4] to reverse engineer the confusion matrix[1] from partial information, and check the satisfaction of various integrity constraints. This has been extended and applied by Li et al. [13].

This paper integrates the Li et al. [13] analysis with an additional category of error relating to performing multiple NHSTs without correction of the acceptance threshold, usually denoted α and conventionally set to 0.05. However, NHST with multiple tests becomes problematic [1]. When many tests are made, the probability of making at least one Type I error amongst the comparisons grows linearly with the number of tests. Since some experiments make many tens or even hundreds of comparisons and rely on NHST as a means of inferencing this is a very real threat to experimental validity. Therefore a correction should be made to the α acceptance threshold. The best known, though conservative, method is the Bonferroni correction which is $\alpha' = \alpha/n$ where n is the number of tests or comparisons. More modern approaches include Benjamini-Hochberg which controls for the false discovery rate [3] and the Nemenyi post hoc procedure [8].

3   Systematic Review

In a systematic review of studies comparing the performance of unsupervised and supervised learners, we (LN, MS and YC) identified 49 relevant studies that satisfy the inclusion criteria given in Table 1. An extended description can be found in [13]. The conduct of the review was guided by the method and principles set out by Kitchenham et al. [12]. We were then able to use these 49 primary studies as a convenience sample to assess the error-proneness of computational experiments in machine learning.

There is a one-to-many mapping from paper to result with the papers containing between 1 and 751 (median = 12) results apiece. The papers cover 14 different unsupervised prediction techniques (e.g., Fuzzy CMeans and Fuzzy Self Organising Maps) coupled with six different cluster labelling techniques (e.g., distribution-based and majority voting). The full list of papers and raw data may be found online[2].

4   Analysis

We examine three questions. First the prevalence of inconsistency errors relating to the confusion matrix. Here we investigate result by result. Second, we look

---

[1] A confusion matrix is a 2×2 contingency table where the cells represent true positives (TP), false negatives (FN), false positives (FP) and true negatives (TN) respectively. Most classification performance statistics, e.g. precision, recall and the Matthews correlation coefficient (MCC), can be defined from this matrix.

[2] Our data may be retrieved from Figshare http://tiny.cc/vvvqbz.

Table 1. Systematic review inclusion criteria

Criterion	Description
Language	Written in English
Topic	Applies at least one unsupervised learning method for predicting defect-prone software modules
Availability	Full content must be available
Date	January 2000 – March 2018
Reviewed	All papers that indicated some minimal peer review process, however, we observed some outlets appear on Geoff Beall's controversial 'predatory' publisher list [16]
Duplicates	Includes new (only the most recent version used when multiple reports of a single experiment) software defect prediction experiments. We do not count re-analysis of previously published experiments
Software system	Uses real data (not simulations)
Reporting	Sufficient detail to enable meta-analysis

at the occurrence of a particular statistical error relating to a failure to adjust the α threshold when using NHST form of statistical inference. Here the unit is paper since the error relates to making *multiple* tests. Third, we examine the extent to which these problems co-occur and therefore the proportion of papers implicated.

4.1   Inconsistency Errors in the Confusion Matrix

Our approach was to use the DConfusion tool [4] to reconstruct the confusion matrix of classification performance for each result. This can often, but not always, be accomplished from a partial set of reported results. For instance, if precision, recall and false positive rate are reported one can reconstruct the complete confusion matrix. Of course, some cases (for us ~33% or 823/2456) failed to report sufficient information so we cannot undertake consistency checking. Incomplete reporting also hinders meta-analysis.

The next step is to test for six integrity constraints (expressed as rules that should be false). If one or more rules are true then we know that there is some issue with the results as reported. The cause could be as simple as a wrongly transcribed value to some deeper error. However, from the perspective of our analysis all we can say is there is a problem. Note that the DConfusion tool also handles rounding errors which could lead to small differences in results and consequently the appearance of an inconsistency problem. In total, 262 out of 2456 experimental results were inconsistent (see Tables 2 and 3).

Table 2. Distribution of confusion matrix errors

Rule	Count
1: Performance metric out of range e.g., FMeasure $\notin [0,1]$ or MCC $\notin [-1,1]$	171
2: Recomputed defect density d is zero	7
3: Recomputed d differs from original reported defect percentage by more than 0.1 (i.e., we allow for small rounding errors)	60
4: Recomputed performance metrics differ from known original ones by more than 0.05 (i.e., we allow for small rounding errors). NB, the rounding error ranges are computed by adding ± 0.05 to the original data unlike the more conservative range 0.01 used in [4]	3
5: Internal consistency of the re-computed confusion matrix	19
6: Other obvious reporting errors within paper e.g., the confusion matrix is inconsistent with their dataset or dataset summary statistics	2
Total errors	262
Checkable and consistent	1371

For a full explanation for all consistency rules refer to the figshare project http://tiny.cc/vvvqbz

Table 3. Proportions (with rounding) of inconsistently reported experimental results

Result	Count	% of total
Inconsistent results	262	10.7%
All other results	2194	89.3%
(Other) Cannot check	823	33.5%
(Other) Can check - ok	1371	55.8%
Total	2456	100%

4.2 Failure to Adjust Acceptance Threshold for NHST Errors

Irrespective of one's views regarding the validity of null hypothesis significance testing (NHST) [7] it is demonstrably an error to set α at a particular level[3] and then undertake multiple tests without making some adjustment to this threshold [2]. A range of adjustment methods have been proposed subsequent to the classic Bonferroni method. In our sample, we noted researchers used either Benjamini-Hochberg [3] or the Nemenyi procedure [8] procedure. In terms of assessing the experiment we are agnostic as to which is the 'correct' adjustment.

The number of significance tests ranges from 1 to 2000 with the median $= 100$. Naturally the experiment that only undertakes a single NHST does not require to correct α, however the remaining 12 experiments do. Table 4 summarises the results. Note the experiment that makes partial corrections uses the Nemenyi post hoc test procedure for some analyses but not for the remaining 84 tests.

---

[3] Of 13 papers using NHST, 12 have $\alpha = 0.05$ and, unusually, one study interprets $0.05 < p < 0.1$ with $p = 0.077$ as being 'significant'.

We thus still consider this as an error. So more than half $((6+1)/13)$ of the experiments that make use of NHST-based analysis are in error. Since this is part of researchers' inferencing procedures e.g., to determine if classifier X is to be preferred to classifier Y, this is worrisome.

Table 4. Failure to adjust the acceptance threshold when performing *multiple* NHSTs

Adjust?	Count
No	6
Partial	1
Yes	5
Total	12

4.3 Do Different Types of Error Co-occur?

Finally, we ask the question: is a paper that commits one class of error more likely to commit other types of error? Table 5 gives the contingency table of the two types of error. There does not seem to be much evidence that experimenters who make errors with confusion matrices are more likely to incorrectly deploy NHST. Consequently a highly disturbing 22 (1+15+3+3) out of 49 papers contain demonstrable errors. The situation may be worse, since 11 papers (without NHST errors) provide insufficient information for us to check the consistency confusion matrices.

Table 5. Co-occurrence of different classes of error by paper

	NHST Error	No NHST Error
Inconsistent confusion matrix	1	15
Consistent confusion matrix	3	16
Incomplete reporting	3	11

5 Discussion and Conclusions

In this study we have audited 49 papers describing experiments based on supervised learners for software defect prediction. They were identified by a systematic review of research in this area. We then checked for arithmetic errors and inconsistencies related to confusion matrices. These are important since they form the basis for calculating most classification performance statistics; errors could therefore lead to wrong conclusions. In contrast, we also checked that experiments that made use of NHST type inferencing adjusted the significance threshold α when undertaking multiple tests to prevent inflation of the false discovery rate. This type of statistical error can also lead to wrong conclusions for researchers who wish to use p-values as a means of determining whether a result is significant or not.

Obviously there are other classes of error one could check for in experiments. We chose confusion matrix errors and failure to adjust significance testing for pragmatic reasons: they are objective and can be undertaken without access to the original data. Nevertheless one must have concern that the true picture is likely to be worse than we have uncovered. Moreover, ~33%% of experimental studies fail to report sufficient information for us to be able to check for consistency. Thus an overall (knowable) error rate of ~45% (22/49) of papers across all publication venues (which is likely to be an underestimate) does not inspire confidence in the quality of our machine learning experiments or at least our attention to detail.

In summary:

1. We have identified a number of inconsistencies or errors in a surprisingly high proportion of published machine learning experiments. These may or may not be consequential but do raise some concerns about the reliability of analyses.
2. There are also a proportion of studies that have not published sufficient information for checking and there are of course other errors that are difficult to detect using the procedures at our disposal.
3. Our sample of experiments is a convenience sample and so is not necessarily representative of other areas of machine learning.
4. We strongly recommend that researchers adopt the principles of Open Science [14] so that data, experimental results and code are available for scrutiny.
5. It is our intention to communicate with the affected authors to highlight data analysis issues that *seem* to require correction. However, we recognise that mistakes can be made by all of us, so error checking is a process that needs to be undertaken with civility and professionalism.

Our analysis of errors in a sample of 49 machine learning experiments has uncovered some worrying findings. Errors, both arithmetic and statistical are surprisingly commonplace with discoverable problems in almost 45% of papers. This appears broadly in line with similar analyses in experimental psychology [5,15]. Nor do error rates appear to be much improved in the more obviously peer-reviewed literature. We suggest three future lines of enquiry. First, our sample is relatively small and non-random. It would be interesting to see how other areas of machine learning research compare. Second, a wider range of errors—particularly statistical ones—might usefully be explored. Third, dialogue with authors might help us better understand the nature of errors and their significance. As we have stated our analysis of confusion matrices identifies inconsistencies but not the underlying causes. Then we will be in a better position to answer the question: do the errors have material impact upon experimental conclusions?

Finally, we strongly believe these findings should give additional impetus to the move to open science and publication of all research data, code and results. When conducting complex computational experiments, errors may be hard to completely avoid; openness better helps us to detect and fix them.

References

1. Benavoli, A., Corani, G., Demšar, J., Zaffalon, M.: Time for a change: a tutorial for comparing multiple classifiers through Bayesian analysis. J. Mach. Learn. Res. **18**(1), 2653–2688 (2017)
2. Bender, R., Lange, S.: Adjusting for multiple testing - when and how? J. Clin. Epidemiol. **54**(4), 343–349 (2001)
3. Benjamini, Y., Hochberg, Y.: Controlling the false discovery rate: a practical and powerful approach to multiple testing. J. Royal Stat. Soc.: Ser. B (Methodol.) **57**(1), 289–300 (1995)
4. Bowes, D., Hall, T., Gray, D.: DConfusion: a technique to allow cross study performance evaluation of fault prediction studies. Autom. Softw. Eng. **21**(2), 287–313 (2014)
5. Brown, N., Heathers, J.: The GRIM test: a simple technique detects numerous anomalies in the reporting of results in psychology. Soc. Psychol. Pers. Sci. **8**(4), 363–369 (2017)
6. Catal, C., Diri, B.: A systematic review of software fault prediction studies. Expert Syst. Appl. **36**(4), 7346–7354 (2009)
7. Colquhoun, D.: An investigation of the false discovery rate and the misinterpretation of p-values. Royal Soc. Open Sci. **1**, 140216 (2014)
8. Demšar, J.: Statistical comparisons of classifiers over multiple data sets. J. Mach. Learn. Res. **7**, 1–30 (2006)
9. Earp, B., Trafimow, D.: Replication, falsification, and the crisis of confidence in social psychology. Front. Psychol. **6**, 621 (2015)
10. Hall, T., Beecham, S., Bowes, D., Gray, D., Counsell, S.: A systematic literature review on fault prediction performance in software engineering. IEEE Trans. Softw. Eng. **38**(6), 1276–1304 (2012)
11. Ioannidis, J.: Why most published research findings are false. PLoS Med. **2**(8), e124 (2005)
12. Kitchenham, B., Budgen, D., Brereton, P.: Evidence-Based Software Engineering and Systematic Reviews. CRC Press, Boca Raton (2015)
13. Li, N., Shepperd, M., Guo, Y.: A systematic review of unsupervised learning techniques for software defect prediction. Inf. Softw. Technol. (2019, under review)
14. Munafò, M., et al.: A manifesto for reproducible science. Nat. Hum. Behav. **1**(1), 0021 (2017)
15. Nuijten, M., Hartgerink, C., van Assen, M., Epskamp, S., Wicherts, J.: The prevalence of statistical reporting errors in psychology (1985–2013). Behav. Res. Methods **48**(4), 1205–1226 (2016)
16. Perlin, M., Imasato, T., Borenstein, D.: Is predatory publishing a real threat? Evidence from a large database study. Scientometrics (2018, online). https://doi.org/10.1007/s11192-018-2750-6
17. Shepperd, M., Bowes, D., Hall, T.: Researcher bias: the use of machine learning in software defect prediction. IEEE Trans. Softw. Eng. **40**(6), 603–616 (2014)

A Hybrid Model for Fraud Detection on Purchase Orders

William Ferreira Moreno Oliverio[(⊠)], Allan Barcelos Silva,
Sandro José Rigo, and Rodolpho Lopes Bezerra da Costa

Applied Computing, Universidade do Vale do Rio dos Sinos – UNISINOS,
São Leopoldo, Brazil
william.moreno@gmail.com, albarsil@gmail.com,
rodolpho.lopes@gmail.com, rigo@unisinos.br

Abstract. Frauds on the purchasing area impacts companies all around the globe. One of the possibilities to tackle this issue is through the usage of audits, however, due to the massive volume of the data available today, it is becoming impossible to manually check all the transactions of a company, hence only a small sample of the data is verified. This work presents a new approach through the usage of signature detection with clustering techniques to increase the probability of inclusion of fraud-related documents in sample sets of transactions to be audited. Due to a non-existence of a public database related to the purchase area of companies for fraud detection, this work uses real procurement data to compare the probability of selecting a fraudulent document into a data sample via random sampling versus the proposed model as well as exploring what would be the best clustering algorithm for this specific problem. The proposed model improves the current state-of-the-art since it does not require pre-classified datasets to work, is capable of operating with a very high number of data records and does not need manual intervention. Preliminary results show that the probability of including a fraudulent document on the sample via the proposed model is approximately seven times higher than random sampling.

Keywords: Fraud detection · Clustering · Procurement · ERP

1 Introduction

Enterprise Resource Planning (ERP) are systems that provide complete automation for most business processes. While the automation increases the efficiency of the company, it opens possibilities for internal fraud if the controls available on the system are not robust enough to prevent it. Frauds represent, on average, 5% of the company revenue [1, 2] and combined with the fact that the procurement departments manage more than 60% of company expenditure [10], this context represents a relevant research topic.

This work focus on detecting frauds on the procurement area, which has the highest financial impact on organizations. One of the possibilities to tackle this issue is through the execution of audits; however, due to the massive volume of the data available today, it is becoming impossible to manually examine all the transactions of a company. Hence only a small sample of the data is verified during an audit. Due to the

H. Yin et al. (Eds.): IDEAL 2019, LNCS 11871, pp. 110–120, 2019.
https://doi.org/10.1007/978-3-030-33607-3_13

small number of frauds compared to the typical transactions, frequently, these fraudulent transactions may not be included in the sample and hence are not verified during the audit.

This paper presents a new approach using the techniques of signature detection associated with clustering to increase the probability of inclusion of fraud-related documents in the sample data to be audited.

Due to non-existence of a public database related to the purchase area of companies for fraud detection, this work uses real procurement data to compare the probability of selecting a fraudulent document into a data sample via random sampling versus the proposed model as well as exploring what would be the best clustering algorithm for this specific problem.

The proposed approach improves the current state-of-the-art since it does not require pre-classified datasets to work, is capable of operating with a very high number of data records and does not need manual intervention.

2 Background

In this section, we cover the basic concepts of fraud in the procurement area, together with their main symptoms, the concept of signature matching, and the main concepts of clustering as well.

Several authors have already studied fraudulent behaviour, which can be classified into different types. In the works of [3, 4] the main types of frauds are described and grouped into 8 categories, but they can be summarized into bid-rigging, double payment, kickback fraud, non-accomplice vendor, personal purchases, redirect payment fraud, shell company and pass through. Each of these frauds has different symptoms associated. One example on the redirect payment fraud would be the changes of banking details of a vendor on the corporate system before payment. This work focus on the occupational fraud described as: "the use of one's occupation for personal enrichment through the deliberate misuse or misapplication of the employing organization's resources or assets" [1].

The concept of signature matching was already used on papers related to several different areas, including software engineering [11], biology [5], and network security [7]. On this paper, the signature matching will be used to identify symptoms related to fraud based on the analysis of the records on the ERP system of a company. An example of a possible signature would be to identify which purchase orders (POs) had an invoice amount higher than the request made to the supplier.

Clustering, according to [6] is defined as "an automatic learning technique which aims at grouping a set of objects into clusters so that objects in the same clusters should be similar as possible, where objects in one cluster should be as dissimilar as possible from objects in other clusters". Besides, clustering can be classified into supervised and non-supervised approaches [9].

To the specific problem of fraud detection in procurement, a non-supervised approach will be applied due to the lack of available datasets with examples of fraudulent and non-fraudulent documents which are required by the supervised clustering algorithms. Among the several available clustering algorithms, the following were used in the proposed model: K-Means, DBSCAN, BIRCH, HDBSCAN, CURE, CLIQUE, ROCK, and Spectral Clustering.

3 Related Works

There are several papers related to fraud detection on the procurement area and nine papers were selected for a comparison covering the following points: scalability, the capability to process data in real-time, the requirement of domain knowledge by the operator, the adaptability to new types of fraud, how automated the solution is and the availability of metrics to compare the performance of the solution. The result of the comparison can be found in Fig. 1.

Paper	Scalable	Real time capable	Domain knowledge required	Adaptable	Automated	Metrics available
The Effective Use of Benford's Law to Assist in Detecting Fraud in Accounting Data	Yes	Partial	Yes	Yes	No	No
A business process mining application for internal transaction fraud mitigation	No	No	Yes	Yes	No	Yes
Fraud detection in ERP systems using Scenario matching	Partial	Yes	No	No	Yes	Yes
Reducing false positives in fraud detection: Combining the red flag approach with process mining	Partial	Partial	Partial	No	No	Yes
Fuzzy C-Means for Fraud Detection in Large transaction data sets	Partial	Partial	Partial	Yes	Partial	No
Screening for bid rigging – does it work?	No	No	Yes	Yes	No	Yes
Prediction of Enterprise Purchases using Markov models in Procurement Analytics Applications	Yes	Partial	No	Yes	Yes	Yes
K-Means: Fraud Detection Based on Signaling Data	Yes	No	Yes	Yes	No	Yes

Fig. 1. The overall analysis of the selected papers

After reviewing the work performed on this area, we identified that there is not a clear trend in the state of the art and each of the papers selected presented positive points and limitations as well while exploring different approaches. Out of all the papers reviewed, two were considered as the main inspiration for this work: [15] which provided a deep explanation about the different fraud types in procurement and [16] for detailing the concept of signature matching on the fraud detection field.

4 Proposed Model

In this section, we describe de approach proposal for this work. After an overview with the main motives for the proposal, the components are detailed and commented.

The model has three main modules, where the following activities take place:

(a) Data extraction and cleaning: The data is extracted from the ERP environment, filtered, cleaned, and standardized.
(b) Signature matching: The data previously extracted is verified on a series of checks to identify symptoms which could be an indication of fraud.
(c) Clustering: The data generated on the signature matching phase is used as input to a clustering algorithm.

The novelty of the proposed model is the possibility of automatic selection of data to be used in an audit sample without human interaction by selecting the clusters with the highest averages for the signatures.

4.1 Data Extraction and Cleaning

The proposed model starts with the data extraction from the ERP system which holds the corporate data related to the purchases performed by the organization.

The data extracted and used on the model are the following:

(a) Purchase requisition: The original request for purchase raised by the employee
(b) Purchase orders: The purchase order after the approval and any adjustments performed during the approval phase
(c) Goods receipt: The records of the reception of the physical goods in stock or confirmation of services provided for each purchase order
(d) Invoice receipt: The invoice received for the supplier
(e) Payments: The payments performed to the supplier

Accounting documents:

(f) Customer, supplier, plant, storage location and company codes master data: The details for each of the object (e.g.: creation date, banking account, address, etc.)
(g) System logs: The details for the changes performed on all the items above (e.g.: changes on supplier banking accounts, changes performed on the unit price after a purchase order is approved, etc.).

Since the ERP system holds data for more than 180 different countries, the methodology chosen was to extract the purchase orders which were created over a period of 1 year for 9 different countries.

The countries were chosen based on the Corruption Perception Index [8], which classify all the countries across the globe according to the perceived corruption level. Three countries were selected from among the most corrupted countries, three from the cleanest countries and three from the countries with an average score. The selected countries can be found in Fig. 2.

After the data was extracted from the ERP system, it was standardized and cleaned. Some of the activities performed on the phase where the conversion of the amounts paid to a single currency using the average exchange rate for the period taken into analysis. Another activity was removing the data related to intra-company transactions, which are the purchases performed across different companies of the same group. This occurs when a company in one country purchases raw materials from another country.

Position	Country	Corruption Prevention Index	Classification
3	Switzerland	85	Very Clean
11	Germany	81	Very Clean
13	Australia	77	Very Clean
60	Croatia	49	Average Results
61	Romania	48	Average Results
61	Malaysia	47	Average Results
149	Bangladesh	28	Highly Corrupt
149	Kenya	28	Highly Corrupt
168	Venezuela	18	Highly Corrupt

Fig. 2. Selected countries

4.2 Signature Matching

On the signature matching phase, we use SQL scripts to compare the data extracted previously in order to identify symptoms which could be related to fraud. A total of 17 signatures were created:

- Goods not delivered for a paid invoice
- Invoice amount higher than the order
- Sequential invoices number from a supplier
- Sleeping vendor
- PO performed without approval on the workflow
- Retrospective PO (PO created after the invoice date)
- Vendors without address
- Difference between supplier creation and first purchase
- A vendor with a bank account used by another vendor
- The average difference between purchase dates
- Purchase price increase after PO creation
- Split PO red flag (PO created by the same person, same vendor, same amount and date)
- Alternate payee (changes on bank master data related to the PO)
- Block/unblock (when a PO is blocked and unblocked)
- Supplier marked for deletion
- Changes in payment terms for a vendor
- Payment block

The output of the signature matching is calculated and stored on a new dimension in the dataset. The format of the data is always numeric so it can be used as input on the clustering module. Finally, the output can be either binary returning 1 if the purchase order is relevant to the signature or 0 if it is not relevant or a floating number.

The signature for goods not delivered for a paid invoice returns a binary value of whether or not the invoice was paid without a goods receipt while the signature for sequential invoices number from a supplier would return a score based on the formula below, where MaxINV is the highest invoice number received in the 1 year of data analysed, MinINV is the minimum invoice number received on the same period and NumINV is the number of invoices received.

$$Score = (MaxINV - MinINV)/NumINV$$

The last item generated on this module is the calculation of the average unit values for the quantity and prices of the purchased goods. This information is summarized at three levels, which are the supplier, the material and finally the supplier combined with the material. Based on the average values, new scores are created to identify when specific purchase orders have values significantly higher than the average quantity or price purchased in the period.

Finally, the PO creation date and the PO value are included as features for the clustering module, resulting in a total of 25 features. A number of 17 were generated via signature matching, being 6 based on averages of prices and quantity for materials and suppliers and 2 features generated directly from raw data (PO date and PO value).

4.3   Clustering and Evaluation

In the clustering approach, the values of the signatures are used as input for the clustering algorithm. However, some pre-processing activities are performed before the execution of the clustering algorithm. We choose to scale the data before the processing since the results of the signatures can have values with a very high variation. Some scores have a result of 1 (binary output) while others can have a substantially high number (e.g.: the difference of the price of a specific good price in one PO versus the historical average price).

The data were scaled using Scikit learn [12], through the usage of MinMaxScaling which scales and translates each feature individually, so the data is in the range of the data set. The result would be between 0 and 1.

Due to the very high number of dimensions and based on [13], we decided to reduce the dimensionality of the dataset from 25 dimensions to 6 and 7 dimensions trough application of Principal Component Analysis (PCA) in order to reduce the processing time and increase the cluster accuracy.

The following clustering algorithms were selected to be used on this experiment based on the review of clustering algorithms performed by [14]: K-Means, DBSCAN, BIRCH, HDBSCAN, CURE, OPTICS, Spectral Clustering, CLIQUE, ROCK.

For each clustering algorithm, we selected one parameter per clustering algorithm which varies until the best result is achieved according to the internal validation indexes Davies-Bouldin index, Calinski-Harabasz and Silhouette index as shown in Table 1.

The proposed model implements a loop which executes the clustering algorithm 100 times with different values for the parameters and compares the results of the internal validation index for each execution in order to select the best clustering

Table 1. Parameters selected for each clustering algorithm

Algorithm	Parameter	Description
K-Means	K	Number of clusters
DBSCAN	eps	The maximum distance between two samples
BIRCH	n_clusters	Number of clusters
HDBSCAN	min_clusters	Minimum number of elements to form a cluster
CURE	number_clusters	Number of clusters
OPTICS	min_cluster_size	Minimum cluster size
Spectral Clustering	n_clusters	Number of clusters
CLIQUE	intervals	Number of cells in the grid in each dimension
ROCK	number_clusters	Number of clusters

algorithm and the best value for each parameter provided the best results for the data being processed.

Finally, the cluster with the highest average values for the signatures are automatically selected as the cluster with the highest risk, however, the analysis can be performed manually as exemplified in the next session.

5 Results

For this experiment, all the POs and their related information (according to the details on the data extraction chapter) were extracted from the ERP system and copied to a separate database based on Microsoft SQL Server 2017. The data then was filtered by eliminating intra-company transactions (when a company buys a good or service from another company of the same group) and repetitive procurement, reducing the number of POs to 147,898 which became the scope of this exercise.

Next step performed was to normalize all the amounts available on the data to a single currency. After the data was filtered and normalized, it was processed using a SQL Script to generate the scores for the signatures. The score for each signature was saved on a column and then exported to a text file.

After that, we executed the PCA on the dataset available on SciKitLearn library. Results are available in Fig. 3, which shows that keeping 6 dimensions would maintain 97.9% of original data variance and 7 dimensions would maintain 98.6%.

Based on the results from PCA, the data was executed on the clustering algorithms in three different formats: (a) No PCA executed; (b) PCA executed and keeping the top 6 principal components; (c) PCA executed and keeping the top 7 principal components. Figure 4 shows the run-time of each clustering algorithm can be found as well as the number of executions of each clustering algorithm.

BIRCH clustering algorithm was executed 6 times since the clustering algorithm took more than 12 h to finish the processing of the data when the parameter Threshold was raised above 0.11. The algorithms identified with "Lack of memory" failed to start the data processing right at the start due to lack of system memory. The computer used to run the experiment was equipped with 32 GB of memory.

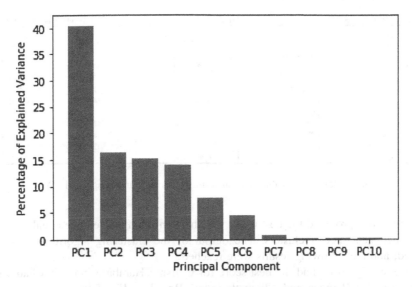

Fig. 3. Data variance by principal component

Algortihm	Average run-time in seconds					
	Source data		PCA - 7 Dimensions		PCA - 6 Dimensions	
	Number of executions	Average run time	Number of executions	Average run time	Number of executions	Average run time
K-Means	100	245.70	100	173.41	100	236.58
DBSCAN	100	725.14	100	446.92	100	526.66
BIRCH	6	10.83	6	6.56	6	5.89
HDBSCAN	241	210.18	241	13.29	241	7.56
CURE	-	Lack of memory		Timeout - 12 hours	-	Timeout - 12 hours
OPTICS	-	Lack of memory	-	Timeout - 12 hours	-	Timeout - 12 hours
Spectral Clustering	-	Lack of memory	-	Lack of memory	-	Lack of memory
CLIQUE	-	Lack of memory	-	Lack of memory	-	Lack of memory
ROCK	-	Lack of memory	-	Lack of memory	-	Lack of memory

Fig. 4. Performance of clustering algorithms

Algorithms marked with "Timeout – 12 h" started the data processing but didn't finish the processing after 12 h of execution and were cancelled. Applying PCA on the dataset reduced the processing time of all the algorithms, with HDBSCAN being the clustering algorithm which had the best positive impact on applying PCA to the dataset.

Algorithm HDBSCAN was executed 241 times in order to achieve the best results based on internal validation indexes. The other clustering algorithms achieved the best score in 100 executions.

The next step performed was to choose among all the executions which clustering algorithm and which parameter generated the best clusters based on internal cluster validation. Figure 5 shows the best result for each of the three indexes by clustering algorithm and utilization of PCA on the data.

K-Means algorithm had the best score for Calisnki Harabasz index but had worst results on both DB index and Silhouette index. Based on Fig. 5, the clustering algorithm chosen was HDBSCAN without application of PCA which have the best score for Silhouette index and DB index. The parameter used on HDBSCAN clustering

PCA Performed / Dimensions	Algorithm	Max of CH Index	Min of DB Index	Max of Silhouette Index
YES-6	BIRCH	7,184.1	1.18336	0.31691
YES-6	DBSCAN	31,300.0	0.78809	0.49017
YES-6	HDBSCAN	12,179.7	0.75381	0.49686
YES-6	K-Means	99,112.2	0.84192	0.48378
YES-7	BIRCH	9,036.4	1.09877	0.32323
YES-7	DBSCAN	31,300.0	0.79820	0.49207
YES-7	HDBSCAN	56,817.8	0.63224	0.53223
YES-7	K-Means	99,112.2	0.84155	0.48378
No PCA applied	BIRCH	15,582.1	0.68663	0.34748
No PCA applied	DBSCAN	15,621.0	0.68818	0.48885
No PCA applied	HDBSCAN	87,980.2	0.53937	0.55652
No PCA applied	K-Means	99,112.2	0.84163	0.48378

Fig. 5. Best index values by clustering algorithm and usage of PCA

algorithm which provided the best result was the minimum cluster size equals to 3400 POs, which generated 4 clusters. The output of the clustering algorithm was then analyzed, and the following results were reached.

K-Means algorithm had the best score for Calisnki Harabasz index but had worst results on both DB index and Silhouette index. Based on Fig. 5, the clustering algorithm chosen was HDBSCAN without application of PCA which have the best score for Silhouette index and DB index. The parameter used on HDBSCAN clustering algorithm which provided the best result was the minimum cluster size equals to 3400 POs, which generated 4 clusters. The output of the clustering algorithm was then analyzed, and the following results were reached.

Cluster 0: This cluster has the highest average values for retrospective PO for goods receipts and the second high for invoice receipts as well as all the vendors without phone numbers in the master data. The retroactive POs, even not following the default policy of the company, can not be an indication of fraud by themselves even when combined with the fact of missing. Based on this, we can not see any pieces of evidence of possible fraud on this cluster and hence classify it as low risk. This cluster contains 8836 POs.

Cluster 1: The only signature with high values on this cluster is the number of suppliers sharing a bank account. Even not being a common item, there are times that multiple vendors can share the same bank account, for example for payment of taxes, when a vendor is created for each tax but all of them should be paid to the government on the same account. Since there are no further signatures associated, we can classify this cluster as low risk. This cluster contains 74257 POs.

Cluster 2: There are no indications in any signatures on this cluster, hence it is classified as low risk. It contains 4333 POs.

Cluster 3: This cluster contains the highest signatures for the sequential invoices, which means that the suppliers involved send most of their invoices to the company object of this study coupled with the fact that all the POs were approved outside the default approval workflow. The only case foreseen for this are the invoices submitted by contractors directly hired by the company, hence most of their invoices (if not all of them) are sent to the same company. This behaviour by itself is not worrying and the cluster would be considered as low risk, however as a control to ensure this is really the case the suppliers which are part of this group could be verified to be really contractors in order to completely eliminate this risk.

In summary, this cluster is classified as low-risk and contains 3952 POs.

Cluster 4: Cluster 4 is very similar to cluster 3, with the only signature active is that all the POs were approved outside the default approval workflow. It contains 35068 documents and it is classified as low risk.

Outliers: The outlier is the cluster which should be included as the scope for an audit in this area since it has the highest score for 16 out of the 22 signatures available and hence there is a risk for several types of frauds. It contains 21452 POs and is classified as the only high-risk cluster.

The average results for the signatures used to perform the manual classification can be found in Fig. 6 with significant values highlighted in green and signatures not considered in amber.

Signatures	Outliers	Clusters generated				
		0	1	2	3	4
Invoice for undelivered goods/services	0.00000	0.00000	0.00000	0.00000	0.00000	0.00000
Invoice amount is higher than order	12.97786	9.86555	9.03913	3.11562	3.98305	7.66559
Sequential invoice numbers	4.86654	0.51494	0.03721	8.84537	9.08629	0.06773
Sleeping vendor	0.02694	0.00000	0.00000	0.00000	0.00000	0.00000
Po performed outside standard workflow	0.57230	0.00000	0.00000	0.00000	1.00000	1.00000
Retrospective po (goods receipt)	0.25396	1.75928	0.18738	0.04154	0.34514	0.02310
Retrospective po (invoice receipt)	5.27009	5.17587	0.79992	1.21348	0.24848	1.05569
Vendors without phone number	0.17775	1.00000	0.00000	0.00000	0.00000	0.00000
Vendors without address	0.00685	0.00000	0.00000	0.00000	0.00000	0.00000
Difference between supplier creation and first sale	977.30146	673.27230	1134.05867	965.35680	1339.25607	1363.75918
Vendor with the same bank account as another ven	1.05836	1.20032	1.23555	0.99792	0.98634	1.12513
Average difference between sales dates	42.45432	37.17021	28.67281	14.82022	10.71027	25.74612
Price increased after po creation	4.18561	0.71885	0.85943	0.31109	0.03264	0.05322
Price above average for supplier and material	1.12240	0.01829	0.03571	0.00000	0.00000	0.01178
Quantity above average for supplier and material	0.06075	0.02842	0.03704	0.00449	0.00026	0.00467
Price above average for supplier	10.87754	0.70644	1.21253	4.77470	0.80708	0.68366
Quantity above average for supplier	5.82948	0.33985	0.39617	0.22728	0.10648	0.27615
Price above average for material	12.99958	0.01829	0.03579	0.00237	1.42592	0.19439
Quantity above average for material	0.10398	0.06614	0.03742	0.00176	0.04113	0.01031
Multiple po for same creator, supplier, amount and	0.18050	0.00000	0.00000	0.00000	0.00000	0.00000
Bank data changed for supplier	0.00009	0.00000	0.00000	0.00000	0.00000	0.00000
Supplier blocked before a purchase	0.10656	0.00283	0.01212	0.00092	0.00177	0.03214
Supplier marked for deletion after a PO creation	0.10544	0.00283	0.01236	0.00092	0.00177	0.03291
Changes performed on payment terms before a pur	0.58559	0.13886	0.11474	0.07339	0.01645	0.12065
Purchase order blocked	0.01380	0.00136	0.00059	0.00000	0.00000	0.00137
Number of PO per cluster	21452	8836	74257	4333	3952	35068

Fig. 6. Average of the signature values per cluster generated

6 Conclusion

The utilization of signature matching coupled with clustering algorithms reduced the scope of data considered to be analyzed to just 14.5% of the original scope, from 147,898 documents to 21,452 documents by removing from the scope the low-risk documents.

The best clustering algorithm among all the ones compared for this specific problem was HDBSCAN which generated the best results according to both Silhouette index and Davies-Bouldin index, however, K-Means had the best results according to Calinski Harabasz index. Finally, it was identified that reducing the dimensionality of the data trough PCA improved the performance of the clustering algorithm but did not increase the results of the indexes for the clusters generated.

References

1. Association of Certified Fraud Examiners (ACFE). Report to the nations. Austin, USA (2016). https://www.acfe.com/rttn2016/docs/2016-report-to-the-nations.pdf. Accessed 3 Mar 2019
2. Association of Certified Fraud Examiners (ACFE). Report to the nations. Austin, USA (2018). https://s3-us-west-2.amazonaws.com/acfepublic/2018-report-to-the-nations.pdf. Accessed 3 Mar 2019
3. Huber, M., Imhof, D.: Machine Learning with screens for detecting Bid-Rigging Cartels. Working Papers SES. [s. l.], n. 494, pp. 1–28 (2018)
4. Imhof, D., Karagök, Y., Rutz, S.: Screening for bid rigging-does it work? J. Compet. Law Econ. **14**(2), 235–261 (2018)
5. Jonasson, J., Oloifsson, M., Monstein, H.J.: Classification, identification and subtyping of bacteria based on pyrosequencing and signature matching of 16S rDNA fragments. Apmis. **110**(3), 263–272 (2002)
6. Popat, S., Emmanuel, M.: Review and comparative study of clustering techniques. Int. J. Comput. Sci. Inf. Technol. **5**(1), 805–812 (2014)
7. Smith, R., et al.: Evaluating GPUs for network packet signature matching. In: Proceeding of the 2009 IEEE International Symposium on Performance Analysis of Systems and Software, Boston, MA, USA, pp. 175–184. IEEE (2009)
8. Transparency International. Corruption Perception Index (2017). https://www.transparency.org/news/feature/corruption_perceptions_index_2017#table. Accessed 3 Mar 2019
9. Xu, R., Wunsch II., D.: Survey of clustering algorithms. IEEE Trans. Neural Netw. **16**(3), 645–678 (2005)
10. Westerski, A., et al.: Prediction of enterprise purchases using Markov models in procurement analytics applications. Procedia Comput. Sci. **60**, 1357–1366 (2015)
11. Zaremski, A., Wing, J.: Signature matching: a key to reuse. In: Proceedings of the 1st ACM SIGSOFT Symposium on Foundations of Software Engineering, Los Angeles, California, USA, pp. 182–190. ACM (1993)
12. Pedregosa, et al.: Scikit-learn: machine learning in python. JMLR **12**, 2825–2830 (2011)
13. Ding, C., He, X.: K-means clustering via principal component analysis, vol. 29 (2004). https://doi.org/10.1145/1015330.1015408
14. Xu, D., Tian, Y.: A comprehensive survey of clustering algorithms. Ann. Data Sci. **2**(2), 165–193 (2015)
15. Baader, G., Krcmar, H.: Reducing false positives in fraud detection: Combining the red flag approach with process mining. Int. J. Account. Inf. Syst. **31**, 1–16 (2018)
16. Islam, A.K., et al.: Fraud detection in ERP systems using scenario matching. In: Rannenberg, K., Varadharajan, V., Weber, C. (eds.) SEC 2010. IFIP AICT, vol. 330, pp. 112–123. Springer, Heidelberg (2010). https://doi.org/10.1007/978-3-642-15257-3_11

Users Intention Based on Twitter Features Using Text Analytics

Qadri Mishael$^{(\boxtimes)}$ (ID), Aladdin Ayesh (ID), and Iryna Yevseyeva (ID)

School of Computer Science and Informatics, De Montfort University, Leicester, UK
{qadri.mishael,aayesh,iryna}@dmu.ac.uk

Abstract. Online Social networks are widely used in current times. In this paper, we investigate twitter posts to identify features that feed in intention mining calculation. The posts features are divided into three sets: tweets textual features, users features, and network contextual features. In this paper, our focus is on tweets analysing textual features. As a result of this paper, we were able to create intentions profiles for 2960 users based on textual features. The prediction accuracy of three classifiers was compared for the data set, using ten intention categories to test the features. The best accuracy was achieved for SVM classifier. In the future, we plan to include user and network contextual features aiming at improving the prediction accuracy.

Keywords: Intention mining · Online Social Network · Feature selection · Machine learning

1 Introduction

Online Social networks (OSN) are an essential part of people's life nowadays. OSN users like to share information by publishing posts about daily activities, feelings, opinions, interests or goals. The posts may include text, images, video clips, or Uniform Resource Locators (URLs). Many researchers attempted to extract and mine the future intentions from the users previous physical actions and verbal instructions or promises. The goal behind Intention Mining is to enhance the services that a system provides to users by building intention-based user profiles. However, this is a challenging process.

Researchers suggested various definitions for intention of an individual, and proposed a number of models to present the intention, based on data sources and target systems. For our research, we adopt the following definition of intention which was given by Purohit et al. [8]. In a given system, user's intention is represented in the form of the user's goal of performing an action or a set of actions. In this paper, we study OSN to identify features that feed in intention mining calculation. We used Twitter as an example for our work.

We extracted the features from Twitter meta-data of posts. The meta-data of the tweets present different features, such as tweets' topics or sentiments. In the following section, we review the literature on intention mining for users of OSN

© Springer Nature Switzerland AG 2019
H. Yin et al. (Eds.): IDEAL 2019, LNCS 11871, pp. 121–128, 2019.
https://doi.org/10.1007/978-3-030-33607-3_14

in research, followed by section on preliminaries and notation. We describe the proposed framework and methodology. The following section includes description on suggested features that we consider in our work. Next section describes the experiments and results. Finally, in the last section we conclude and give suggestion for future work.

2   Related Work

Many researchers worked on building online profiles for users intention [3,6,10]. Chen et al. in [3], worked on identifying users' intents over forums posts. Ding et al. [4] used word embedding features to identify the consumption intent of users over microblogging in China. Their model of consumption intention was based on a binary classification of the posts to decide if the sentence contains consumption OSN users. They used textual features from the users shared information over OSN using word embedding techniques. As seen in the literature, the use of textual features was the main focus of the researchers. In this paper, we include only textual based features of users profiles from OSN.

3   Problem Definition

We suppose that we have an n set of Twitter posts for each user u (tweets) $T : \{t_1, t_2, ..., t_n\}$ where t_{ij} represents a post i for a user u_j, with m users $U : \{u_1, u_2, ..., u_m\}$. For each tweet t_i in T, there exist a set of k features $F_{ti} : \{f_{ti1}, f_{ti2}, ..., f_{tik}\}$ and for each user u_j in U there is a set of features $F_{uj} : \{f_{uj_1}, f_{uj_2}, ..., f_{uj_k}\}$ that can be extracted from twitter stream.

$$\forall t \in T \exists F_t : \{F_{t1}, F_{t2}, ..., F_{tk}\} \tag{1}$$

$$\forall u \in U \exists F_u : \{F_{u1}, F_{u2}, ..., F_{uh}\} \tag{2}$$

For both F_t and F_u, we have $\hat{F}_t$ and $\hat{F}_u$ as subsets of the original sets that hold the minimum features of both the user and the posts, which is used to build intention users' profiles posts.

$$\exists \hat{F}_t \subset F_t , \exists \hat{F}_u \subset F_u \tag{3}$$

4   Framework and Methodology

We utilize number of techniques in our study which are based on linguistic approaches. The techniques are textual analysis techniques, semantic linguistic models [2], and sentiment analysis techniques [1]. We combine these techniques in our proposed framework that is divided into three stages, see Fig. 1. The first stage is to retrieve Twitter data using Application Programming Interface (API) with certain conditions or queries. The second stage is to preprocess information and build the database, this is carried out as follows:

- **Extract tweet data**: The tweets come with different attributes in standard format which need to be preprocessed and filtered.
- **Extract user Profiles**: Every user profile u on Twitter is retrieved based on their tweets to analyse their behaviour. The behaviour of each user is considered based on the features that described in more details in Sect. 5.
- **Extract Network data**: All the users profiles in the data set are examined and analysed to identify the users' followers and followees.
- **Build Database**: The three previous components are used to build a database which rows represent the users and the columns are the features.

In the third stage, the feature set is reduced by using feature selection techniques.

- **Hybrid features selection**: Set of feature selection techniques are used to have the minimum set of features that estimate the user intention.
- **Intention Classification**: Different classifiers such as Decision Tree, Naive Based, Support Vector Machine are used to validate the features predicate if the users data holds an intention.

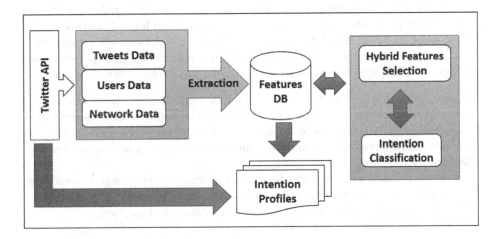

Fig. 1. User intention profile system

5   Sets of Textual Features

We divide the intention predictive features into three sets: tweets textual features (F_t), users features (F_u), and network features (F_n) that we extract from the other two sets. We first will focus on the tweets textual features set only and the role of these features in estimating the user intention.

Tweets Text (t_{text}): Different parts of tweet's text hold different textual features besides the message in the tweet, such as hashtags, mentions, and URL links. *Hashtags* in twitter are usually used to demonstrate the user's interests on a certain case or a topic.

We consider here the average number of hashtages ($avg.no.hash$) included in a tweet and the list of the most frequently used hashtags by user ($hash_{top}$). We assume that the user's attitude and interest are indicated by this feature. *Mentions* in tweets, if present, represent a way of communication between users, since it will capture the mentioned user attention and make him/her read the tweet. Every mention starts with "@" character. Analysing mentions provides information about a connection or a relation between two or more users, or the user's interest in a certain brand. We calculate the number of the mentions ($no.mention$) for each user in all the posts, also the average number of the mentions ($avg.no.mention$) in a user post. In addition, the type of the mentions as a person or a brand ($mention.type$) is computed. Analysing the type of URL links ($link.type$) in a tweet provides understanding about user intentions if it is related to a certain need. The type of the link is labelled into either shopping, travel, or job, manually.

In addition to the above textual features, we consider **text syntax** similarity of certain tweets to other tweets based on certain natural lingual pattern.

Moreover, the context of tweets is very important indicator for the user's intention. Hence, same set of words in different context can produce different meaning [1]. Therefore, we use the tweets' context to extract features such as:

– Topics ($topic$): When we take a good look at twitter users' posts, we notice users always try to post, retweet, and follow the topics that concern them more frequently than any other topics. Therefore, tweets' topics are important for estimating users' intention. The topic feature can be summarized by two features, which are:
  - Topic frequency ($tfreq$) can be extracted from the posts a user publishes regarding a certain topic over the number of total posts. This feature is used to extract a knowledge about user interest in that topic.
  - Top topic ($ttop$) tweeted by the user which gives indication of the user interest and needs.
– Sentiment ($senti$): Tweets are described to have positive or negative sentiments, and usually used prior to any future actions. Thus, analysing the sentiment of a tweet has a great role to understand the behaviour and the needs of the users [1]. We extracted and used the following sentiment:
  - Sentiment strength of tweets (sst) as positive or negative. The value of the sentiment strength is measured for the user's positive or negative tweets regarding a certain topic, see details in Sect. 6.3.
  - User average sentiment score ($sscore$) over period of time to measure the user interest about that topic.

6   Experiment and Result

6.1   Data Set

For our work we used crowd-sourced data set[1] that was collected from twitter for sentiment analysis of users' NewYearResolution hashtag of the year 2015. The tweets are dated between Dec 2014 and Feb 2015. The data set contains demographic and geographical data of users, and resolution categorizations. For our study, we assumed that the intention for each user is stated in his/her tweets. The tweets intentions are categorized based on the resolutions in ten unique topics listed in Table 1. Each category is labelled manually and combined with topic of the tweet, for example, fitness and health resolution which is mentioned in the tweet topic, such as eat healthy food, go to gym, quit smoking.

Before using the data set, we start by cleaning the data set from unavailable users and keep only the publicly existed users and posts. After the cleaning step, we ended up with 2965 users with single post each out of 4518 users. Following that step, we retrieve users time line over 60 days of posts, the period length is selected based on our assumption that the user will post tweets related to his intention in the near future. The total number of posts retrieved is 6931899 posts. For our experiment, we take a sample of 53 posts from each category in the data set to make a balanced sample. We display the distribution of the ten categories in Table 1. As Table 1 shows, the posts topics are distributed over several categories of intentions.

Table 1. The distribution of labelled topics over each extracted topic in percentage for 53 posts sample

Topic	T0	T1	T2	T3	T4	T5	T6	T7	T8	T9
Personal growth	11%	15%	19%	9%	6%	6%	6%	15%	9%	4%
Humor	19%	6%	13%	11%	9%	22%	4%	6%	6%	6%
Health/Fitness	8%	15%	4%	2%	15%	9%	21%	2%	13%	11%
Recreation/Leisure	19%	4%	9%	11%	6%	11%	11%	8%	13%	8%
Career	15%	13%	15%	15%	9%	13%	4%	4%	8%	4%
Education/Training	32%	6%	13%	6%	9%	2%	6%	13%	6%	8%
Finance	10%	10%	2%	6%	8%	12%	10%	10%	4%	31%
Family/Friends relationships	11%	6%	11%	21%	2%	2%	6%	19%	17%	6%
Time management/Organization	13%	8%	15%	11%	6%	8%	8%	13%	13%	6%
Philanthropic	21%	4%	8%	9%	6%	6%	6%	13%	4%	25%

6.2   Preprocessing

We start to extract the tweets textual features set as described in Sect. 5. We preprocess the post's text using tokenization, stemming, lemmatization and cleaning

---

[1] https://www.figure-eight.com/data-for-everyone/.

from stop words and symbols. The resulting tokens are used to build a words' dictionary which is used to create bag of word (BOW) corpus. The corpus is used to train a topic model to extract the topics as a textual features of each post. Other post parts such as hashtags, mentions and URL links are extracted separately.

6.3 Extracting Topic and Sentiment Features

Two main part of the tweets features set are the topic and the sentiments. The posts number per topics are not evenly distributed. Therefore, we used the generative probabilistic model Latent Dirichlet Allocation (LDA) [2] to create ten topics and build a model based on words w per topic t and topic t per document d which cover all the posts. We selected ten topics since it is the number of the predefined topics. Each post is assigned to a topic with highest confidence score. In Table 2, applying LDA gives different topics allocation, with more even distributions but different keywords set for each topic. Furthermore, we used each user's posts within 60 days window interval to test the frequency of the specified topic using the created LDA model. Sentiment and sentiment strength of each post are extracted using SentiSterength tool [9][2]. The tool calculates positive, negative and neutral sentiments and gives code in the range of $[-1, 0, 1]$, where 1 represents positive, -1 is negative, and zero is neutral. The sentiment strength is calculated by summing the positive and negative sentiment strength. Moreover, we calculate sentiment average score for each user by revisiting his/her 60 days time line posts.

Table 2. The distribution of the topics with top 10 keywords based on LDA using Bag of Words (bow) model

Topic	Keywords	#Posts
T0	start, new, learn, watch, tri, need, buy, go, help, goal	380
T1	gym, work, god, week, time, day, gain, stop, chang, start	302
T2	year, like, goal, let, ve, make, go, hope, happi, think	339
T3	peopl, stop, make, ralli, think, start, time, good, care, hour	271
T4	f***, twitter, stop, get, seve, break, word, past, lose, thing	252
T5	day, make, go, stop, tweet, tri, play, time, focus, selfi	262
T6	stop, drink, smoke, right, cigarett, chrisbrown, club, na, plan, supper	288
T7	better, friend, make, time, continu, love, shit, awesom, use, spend	279
T8	eat, happi, amp, new, healthi, game, year, cariloha, need, food	308
T9	make, money, possible, thank, live, life, tapp, run, happy, like	284

---

[2] http://sentiStrength.wlv.ac.uk/.

6.4   Results and Evaluation

For our experiment, we took a sample of 53 posts from each category in the data set to make a balanced sample, i.e. ten sets of 53 posts. We display the distribution of the categories in Table 1. As Table 1 shows, each post is classified in a different category. Therefore, we train three well-known classification models, which are Decision Tree Classifier (DT), Support Vector Machine (SVM), and Naive Bayes (NB) for the 53 sets of labelled tweets using One-vs-All classifier strategy [8]. We run experiments using different combination of sets of features and then we evaluate the performance improvement we achieve. These feature combinations are considering the textual set of features: (t_{text}), $(avg.no.hash)$, $(hash_{top})$, $(no.mention)$, $(avg.no.mention)$, $(mention.type)$, then considering the Topic and Sentiment set of features: $(tfreq)$, $(ttop)$, (sst), $(sscore)$. Finally, we use all these features sets together to perform prediction of intention.

In order to evaluate the performance of the classifiers, we use the measures of accuracy, and F-measure [5]. The F-measure is the mean value of precision and recall.

Larger values of Accuracy and F-measure give indication of better classifier performance. We validate the performance by applying a 10-fold cross validation process to grantee no over fitting [5]. We compare the performance of the classifiers with baseline model Information Gain (IG) [8]. IG depends only on text mining technique for feature selection. It uses the entropy to measure the uncertainty between text and target class with and without the features. This means the most important features to classify the tweets are used [7].

From the results shown in the Table 3, applying the SVM classifier on multi-classification in multi-intentions problem presents the best result with respect to accuracy when compared to DT and NB. In addition, predicting intentions, such as Recreation & Leisure and Education & Training both get the highest accuracy in all the classifiers, see Table 3, which indicates that features that describe the performance of those two categories are most often represented in the text.

Table 3. Results for using features to build three classification models SVM, DT, and NB with multi-intentions for each post

Category	SVM		DT		NB	
	Accuracy	F-measure	Accuracy	F-measure	Accuracy	F-measure
Personal growth	0.78	0.48	0.71	0.00	0.74	0.12
Humor	0.82	0.52	0.76	0.34	0.74	0.08
Health/Fitness	0.80	0.36	0.76	0.05	0.76	0.05
Recreation/Leisure	0.83	0.55	0.79	0.36	0.78	0.05
Career	0.77	0.62	0.70	0.30	0.75	0.44
Education/Training	0.90	0.54	0.87	0.47	0.85	0.00
Finance	0.81	0.40	0.81	0.37	0.78	0.1
Family/Friends/Relationships	0.77	0.45	0.72	0.20	0.73	0.14
Time management/Organization	0.72	0.76	0.57	0.70	0.65	0.74
Philanthropic	0.81	0.56	0.77	0.28	0.75	0.15

7 Conclusion

This paper study the problem of intention mining of users in twitter. We propose three set of features: tweet features, user features, and network features. We focus on the textual features of tweets. We used a framework to extract and analyse the tweets features for intention mining. Those features are used to train well-known machine learning classifiers to classify posts into intention classes. As a result of our work, we were able to create intentions profiles for 2960 users based on textual features. The prediction accuracy of three classifiers were compared for the data set, using predefined intention categories to test the extracted features. The best accuracy was achieved for SVM classifier. Our future research will focus on the user's profile-based features, such as the demographic information about users and network-based features that describe the user posting, socializing, and networking behaviour over OSN.

References

1. Benamara, F., Taboada, M., Mathieu, Y.: Evaluative language beyond bags of words: linguistic insights and computational applications. Comput. Linguist. **43**(1), 201–264 (2015)
2. Blei, D.M., Ng, A.Y., Jordan, M.I.: Latent Dirichlet allocation. J. Mach. Learn. Res. **3**, 993–1022 (2003)
3. Chen, Z., Liu, B., Hsu, M., Castellanos, M., Ghosh, R.: Identifying intention posts in discussion forums. In: Proceedings of the 2013 Conference of the North American Chapter of the Association for Computational Linguistics: Human Language Technologies, June 2013, pp. 1041–1050 (2013)
4. Ding, X., Liu, T., Duan, J., Nie, J.Y.: Mining user consumption intention from social media using domain adaptive convolutional neural network. In: Proceedings of the 29th AAAI Conference on Artificial Intelligence (AAAI 2015), pp. 2389–2395 (2015)
5. Indola, R.P., Ebecken, N.F.F.: On extending F-measure and G-mean metrics to multi-class problems. WIT Trans. Inf. Commun. Technol. **35**, 25–34 (2005)
6. Kim, Z.M., Jeong, Y.S., Hyeon, J., Oh, H., Choi, H.J.: Classifying travel-related intents in textual data. Int. J. Comput. Commun. Instrum. Eng. **3**(1), 96–101 (2016)
7. Liu, Y., Bi, J.W., Fan, Z.P.: Multi-class sentiment classification: the experimental comparisons of feature selection and machine learning algorithms. Expert Syst. Appl. **80**, 323–339 (2017)
8. Purohit, H., Dong, G., Shalin, V., Thirunarayan, K., Sheth, A.: Intent classification of short-text on social media. In: IEEE International Conference on Smart City/SocialCom/SustainCom (SmartCity), pp. 222–228 (2015)
9. Thelwall, M., Buckley, K., Paltoglou, G., Cai, D.: Sentiment in short strength detection informal text. J. Am. Soc. Inform. Sci. Technol. **61**(12), 2544–2558 (2010)
10. Tyshchuk, Y., Wallace, W.A.: Modeling human behavior on social media in response to significant events. IEEE Trans. Comput. Soc. Syst. **5**(2), 444–457 (2018)

Mixing Hetero- and Homogeneous Models in Weighted Ensembles

James Large and Anthony Bagnall[✉]

University of East Anglia, Norwich Research Park, Norwich, UK
{james.large,ajb}@uea.ac.uk

Abstract. The effectiveness of ensembling for improving classification performance is well documented. Broadly speaking, ensemble design can be expressed as a spectrum where at one end a set of heterogeneous classifiers model the same data, and at the other homogeneous models derived from the same classification algorithm are diversified through data manipulation. The cross-validation accuracy weighted probabilistic ensemble is a heterogeneous weighted ensemble scheme that needs reliable estimates of error from its base classifiers. It estimates error through a cross-validation process, and raises the estimates to a power to accentuate differences. We study the effects of maintaining all models trained during cross-validation on the final ensemble's predictive performance, and the base model's and resulting ensembles' variance and robustness across datasets and resamples. We find that augmenting the ensemble through the retention of all models trained provides a consistent and significant improvement, despite reductions in the reliability of the base models' performance estimates.

Keywords: Classification · Ensembles · Heterogeneous · Homogeneous

1 Introduction

Broadly speaking, there are three families of algorithms that could claim to be state of the art in classification: support vector machines; multilayer perceptrons/deep learning; and tree-based ensembles. Each has their own strengths on different problem types under different scenarios and contexts. However, our primary interest is when faced with a new problem with limited or no domain knowledge, what classifier should be used?

[10] introduced the Cross-validation Accuracy Weighted Probabilistic Ensemble (CAWPE), a weighted ensemble scheme [9] which demonstrates consistent significant improvements over its members, significant improvements over competing heterogeneous ensemble schemes, and, at worst, no significant difference to large homogeneous ensembles and heavily tuned state of the art classifiers. The key premise is that given a lack of domain knowledge to suggest one particular type of model over another, the best place to start on a new arbitrary problem is to heterogeneously ensemble over different kinds of classifiers instead

© Springer Nature Switzerland AG 2019
H. Yin et al. (Eds.): IDEAL 2019, LNCS 11871, pp. 129–136, 2019.
https://doi.org/10.1007/978-3-030-33607-3_15

of invest into a single type. This idea is nothing new of course, however a far larger share of the literature has typically been devoted to homogeneous ensembling or optimisation of individual models.

CAWPE cross-validates each member on the train data, generating error estimates, and then raises these estimates to the power $\alpha = 4$ [10] to generate accentuated weightings for the members' probability estimates when predicting. Models rebuilt on the full train data are used to form predictions for the ensemble. To this end, it needs to gather reliable error estimates. We do this through a ten-fold cross-validation (CV) [8]. During the CV process, however, models are made on each fold which are then discarded. A natural question is whether these can be retained and leveraged to improve predictive performance.

We investigate whether retaining these models, in addition to the models retrained to the full training set, can improve classification performance. We also assess whether accuracy can be maintained while skipping the retraining step on the full data, saving time in the training phase. While maintaining these models incurs no additional training time cost, prediction time and space requirements clearly increase in proportion to the number of CV folds. We further analyse the variance of the maintained classifiers and their effects on the resulting ensemble's variance.

Explicitly building homogeneous (sub-)ensembles from heterogeneous base classifiers is not a new idea. [7] builds forests of trees from different tree building algorithms and shows that larger purely-homogeneous forests can be matched or beaten by smaller mixed forests. Ensemble selection [4] (or pruning) can similarly be applied to purely hetero- or homogeneously generated model sets, or mixtures of the two [11]. Alongside these works, we specifically wish to study the effects of maintaining homogeneous models, with potentially lower-quality estimates of competency attached, on the CAWPE weighting scheme which relies heavily on the weightings applied.

We outline our experimental procedure in Sect. 2. Results are summarised in Sect. 3, and analysed further in Sect. 4. We conclude in Sect. 5.

2 Experimental Setup

The UCI dataset archive[1] is widely used in the machine learning literature. We have taken 39 real-valued, independent and non-toy datasets to use in our experiments, following feedback received on the superset of these datasets used in [10]. The datasets are summarised in Table 1.

Experiments are conducted by averaging over 30 stratified resamples of each dataset. Data, results and code can all be found at the accompanying website for this research[2]. For each resample, 50% of the data is taken for training (on which the cross validation process for each base classifier is performed), 50% for testing. We always compare classifiers on the same resamples, and these can be exactly reproduced with the published code.

---

[1] http://archive.ics.uci.edu/ml/index.php.
[2] http://www.timeseriesclassification.com/CAWPEFolds.php.

Table 1. A full list of the 39 UCI datasets used in our experiments. Full names saved for horizontal space: *[1] conn-bench-sonar-mines-rocks, *[2] conn-bench-vowel-deterding, *[3] vertebral-column-3clases.

Dataset	#Cases	#Atts	#Classes	Dataset	#Cases	#Atts	#Classes
bank	4521	16	2	page-blocks	5473	10	5
blood	748	4	2	parkinsons	195	22	2
breast-cancer-w-diag	569	30	2	pendigits	10992	16	10
breast-tissue	106	9	6	planning	182	12	2
cardio-10clases	2126	21	10	post-operative	90	8	3
sonar-mines-rocks*[1]	208	60	2	ringnorm	7400	20	2
vowel-deterding*[2]	990	11	11	seeds	210	7	3
ecoli	336	7	8	spambase	4601	57	2
glass	214	9	6	statlog-landsat	6435	36	6
hill-valley	1212	100	2	statlog-shuttle	58000	9	7
image-segmentation	2310	18	7	statlog-vehicle	846	18	4
ionosphere	351	33	2	steel-plates	1941	27	7
iris	150	4	3	synthetic-control	600	60	6
libras	360	90	15	twonorm	7400	20	2
magic	19020	10	2	vertebral-column*[3]	310	6	3
miniboone	130064	50	2	wall-following	5456	24	4
oocytes_m_nucleus_4d	1022	41	2	waveform-noise	5000	40	3
oocytes_t_states_5b	912	32	3	wine-quality-white	4898	11	7
optical	5620	62	10	yeast	1484	8	10
ozone	2536	72	2				

We evaluate three ensemble configurations that retain the models evaluated on CV folds of the train data against the original CAWPE, which ensembles only over the models retrained on the entire train set. These are to (a) (M)aintain all models trained on CV folds and add them to the ensemble alongside the fully trained models (CAWPE_M), (b) (M)aintain all models once more, but systematically (D)own-(W)eight them relative to the fully trained models due to their potentially less reliable error estimates (CAWPE_M_DW) (c) maintain *only* those models trained on the CV folds, and skip the retraining step on the full train data, (R)eplacing the original models (CAWPE_R).

All configurations of CAWPE tested use the same core base classifiers, those defined as the 'simple' set in [10] of logistic regression; C4.5 decision tree; linear support vector machine; nearest neighbour classifier; and a multilayer perceptron with a single hidden layer. These classifiers are each distinct in their method of modelling the data, and are approximately equivalent in performance on average. Because all dataset resamples and CV folds of the respective train splits are aligned, each ensemble configuration is therefore being built from identical (meta-) information and we are only testing the configuration's ability to combine the predictions.

For reference, we also compare to Random Forest (RandF) and eXtreme Gradient Boosting (XGBoost), each with 500 trees. This is to put the results into context, rather than to claim superiority or inferiority to them. XGBoost in particular would likely benefit from tuning, for example, which we do not perform for these experiments.

When comparing multiple classifiers on multiple datasets, we follow the recommendation of Demšar [5] and use the Friedmann test to determine if there are any statistically significant differences in the rankings of the classifiers. However, following recent recommendations in [1] and [6], we have abandoned the Nemenyi post-hoc test originally used by [5] to form cliques (groups of classifiers within which there is no significant difference in ranks). Instead, we compare all classifiers with pairwise Wilcoxon signed-rank tests, and form cliques using the Holm correction (which adjusts family-wise error less conservatively than a Bonferonni adjustment).

We assess classifier performance by four statistics of the predictions and the probability estimates. Predictive power is assessed by test set accuracy and balanced test set accuracy. The quality of the probability estimates is measured with the negative log likelihood (NLL). The ability to rank predictions is estimated by the area under the receiver operator characteristic curve (AUC). For multiclass problems, we calculate the AUC for each class and weight it by the class frequency in the train data, as recommended in [12].

3   Results

We summarise comparative results succinctly here in three forms: Fig. 1 displays CAWPE configurations and reference homogeneous ensembles ordered by average ranks in accuracy along with cliques of significance formed; Table 2 details the average scores of all four evaluation metrics; and Table 3 details pairwise wins, draws and losses between the original and proposed CAWPE configurations.

Maintaining the individual fold classifiers significantly improves over the original CAWPE. Within the three proposed configurations there is very little difference in performance. This is largely to be expected since they are working from the same meta-information, with the exception of CAWPE_R, which replaces the fully re-trained models only with those trained during CV. This does mean that training time can seemingly be saved by avoiding this final retraining step without a tangible reduction in predictive performance.

Note that while maintaining the fold classifiers improves performance with statistical significance, the average improvement in absolute terms is very small, roughly 0.3% in terms of accuracy, balanced accuracy, and area under the curve (Table 2). Meanwhile, XGBoost's average accuracy is a full 1.2% lower, but still significantly similar to the new CAWPE configurations. This is because the improvement found while being small, is very consistent. When looking at the paired wins, draws and losses between the configurations in Table 3, the contrast between the relatively balanced match-ups of the three new configurations, against the consistently beaten original configuration is clear to see.

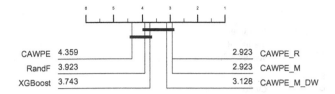

Fig. 1. Critical difference diagram displaying the average ranks of accuracy of the original CAWPE and three tested configurations and reference homogeneous ensembles. Classifiers connected by a solid bar are considered within the same clique and not significantly different from each other.

Table 2. Averages scores for four evaluation metrics of each of the CAWPE configurations and homogeneous ensembles tested.

Classifier	ACC ↑	BALACC ↑	AUC ↑	NLL ↓
CAWPE	0.861	0.787	0.915	0.53
CAWPE_M_DW	0.864	0.789	0.917	0.517
CAWPE_M	0.865	0.79	0.918	0.515
CAWPE_R	0.865	0.789	0.918	0.516
RandF	0.854	0.78	0.91	0.564
XGBoost	0.85	0.784	0.907	0.647

Table 3. Pairwise wins, draws and losses in terms of dataset accuracies between the ensemble configuration on the row against the configuration on the column.

	CAWPE_R	CAWPE_M	CAWPE_M_DW	CAWPE
CAWPE_R	-	17/4/18	23/0/16	32/0/7
CAWPE_M	18/4/17	-	23/0/16	31/0/8
CAWPE_M_DW	16/0/23	16/0/23	-	34/0/5
CAWPE	7/0/32	8/0/31	5/0/34	-

4   Analysis

CV is such a commonly used method of evaluating a model on a given dataset because of it's robustness and completeness relative to, for example, singular held-out validation sets [8]. A single fold of a CV procedure in isolation is of course simply the latter, and equivalent to a single subsample within a bagging context [2]; it is the repeated folding of the data that leads to each instance being predicted as a validation case once that makes the process complete.

All weighted ensembles rely to some extent on the reliability of the error estimates of their members, but CAWPE especially does given that it accentuates the differences in those estimates. We wish to analyse the extent to which the

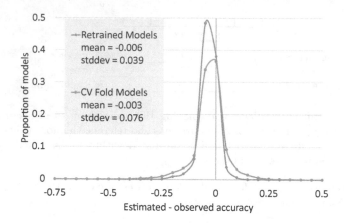

Fig. 2. Normalised counts of differences in estimated (on train data) and observed (on test data) accuracy for the retrained (blue) and individual CV fold (orange) models across all datasets and resamples. Positive x values indicate a larger estimated than observed accuracy, i.e. a classifier overestimating its performance. (Color figure online)

quality of error estimates suffers, and its effects on the ensemble's own performance and variance.

Figure 2 measures the counts of differences in estimated (on train data) and observed (on test data) accuracies and confirms expectations that completing the CV process and retraining models on the full dataset results in more accurate estimates of accuracy on average than the individual models on CV folds. Overall standard deviation almost doubles, but the number and degree of the outliers is perhaps the most important thing. The retrained models never have performance under-estimated by more than 0.3, and less than 2% of the models under estimate by more than 0.1.

Meanwhile, the individuals fold estimates have some extreme outliers in terms of underestimating in particular, with a small tail on Fig. 2 stretching all the way to −0.75. 7.6% of all fold models underestimate accuracy by more than 10%. Many of the extreme outliers were localised to two datasets, spread out across different learning algorithms. The breast-tissue dataset is a relatively balanced six class problem, while post-operative is a heavily imbalanced three class problem. These factors along with them being the datasets with the least instances likely lead to difficult folds to classify for certain models and seeds, which are of course averaged over when considering the remaining CV folds.

In context, however, the difference really is not too stark. The errors in estimates may double in variance, and these are being accentuated by CAWPE's combination scheme, but there are also fifty more models to average over. Figure 3 summarises the differences in variance across test performances between the configurations that maintain the fold models and the original CAWPE along two dimensions - variance in performance on arbitrary datasets, and variance in performance over formulations of the same dataset through resampling. Variance across resamples is reduced, while variance over datasets is less clear. It seems as though cases such as breast-tissue and post-operative affect this particular comparison as with the above, and this shows with variance in balanced accuracy still being clearly reduced.

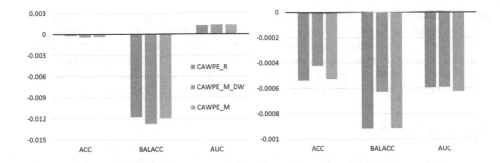

(a) Standard deviations of performances overs dataset

(b) Average standard deviations in performance over dataset resamples

Fig. 3. Standard deviations in performance metrics over (a) datasets and (b) dataset resamples of the three proposed CAWPE configurations, expressed as differences to the original. NLL is omitted due to the improper scaling factor brought about by it not being a measure in the range 0 to 1, however variance similarly drops for all three of the proposed configurations.

When there are only five members, erroneously discounting a classifier to the extent that it's outputs are effectively worthless is a large blow to the overall strength of the ensemble. In the case of ensembles with 50 or 55 members though, erroneously discounting one or two classifiers is not so harmful. Practitioners of homogeneous ensembles will of course be familiar with this, and it is the underpinning of the design of such an ensemble - averaging over high variance inputs to produce a low variance output [3].

5   Conclusions

We have experimentally evaluated the effectiveness of maintaining models used to estimate the accuracy of base classifiers in a weighted ensemble, in addition to or in place of the original models. The experiments show a minor but significant and very consistent improvement in performance across different evaluation metrics, even when skipping the retraining of the models on the full dataset. While variance in the estimates of performance that fuel the weightings within the ensemble increases, this is offset by the averaging effect of the greater number of models, as observed in typical homogeneous ensembles.

Further experimentation aims to discover a breaking point between the effectiveness of increased homogeneity versus heterogeneity. In these experiments, as in previous, a ten fold cross-validation process was used to evaluate the models and ultimately as the source of the expanded model set. Increasing or decreasing the number of folds relative to time and space requirements, or switching entirely to a randomised bagging approach with heterogeneous members are interesting routes to follow.

Acknowledgement. This work is supported by the UK Engineering and Physical Sciences Research Council (EPSRC) [grant number EP/M015807/1] and Biotechnology and Biological Sciences Research Council (BBSRC) Norwich Research Park Biosciences Doctoral Training Partnership [grant number BB/M011216/1]. The experiments were carried out on the High Performance Computing Cluster supported by the Research and Specialist Computing Support service at the University of East Anglia.

References

1. Benavoli, A., Corani, G., Mangili, F.: Should we really use post-hoc tests based on mean-ranks? J. Mach. Learn. Res. **17**, 1–10 (2016)
2. Breiman, L.: Bagging predictors. Mach. Learn. **24**(2), 123–140 (1996)
3. Breiman, L.: Random forests. Mach. Learn. **45**(1), 5–32 (2001)
4. Caruana, R., Niculescu-Mizil, A.: Ensemble selection from libraries of models. In: Proceedings of the 21st International Conference on Machine Learning (2004)
5. Demšar, J.: Statistical comparisons of classifiers over multiple data sets. J. Mach. Learn. Res. **7**, 1–30 (2006)
6. García, S., Herrera, F.: An extension on "statistical comparisons of classifiers over multiple data sets" for all pairwise comparisons. J. Mach. Learn. Res. **9**, 2677–2694 (2008)
7. Gashler, M., Giraud-Carrier, C., Martinez, T.: Decision tree ensemble: small heterogeneous is better than large homogeneous. In: 2008 Seventh International Conference on Machine Learning and Applications, pp. 900–905 (2008). https://doi.org/10.1109/ICMLA.2008.154. http://ieeexplore.ieee.org/lpdocs/epic03/wrapper.htm?arnumber=4796917
8. Kohavi, R.: A study of cross-validation and bootstrap for accuracy estimation and model selection. In: Proceedings of the 14th International Joint Conference on Artificial Intelligence, pp. 1137–1143. Morgan Kaufmann Publishers Inc. (1995)
9. Kuncheva, L., Rodríguez, J.: A weighted voting framework for classifiers ensembles. Knowl. Inf. Syst. **38**(2), 259–275 (2014)
10. Large, J., Lines, J., Bagnall, A.: A probabilistic classifier ensemble weighting scheme based on cross-validated accuracy estimates. Data Mining Knowl. Discov. (2019). https://doi.org/10.1007/s10618-019-00638-y
11. Partalas, I., Tsoumakas, G., Vlahavas, I.: A study on greedy algorithms for ensemble pruning. Aristotle University of Thessaloniki, Thessaloniki, Greece (2012)
12. Provost, F., Domingos, P.: Tree induction for probability-based ranking. Mach. Learn. **52**(3), 199–215 (2003)

A Hybrid Approach to Time Series Classification with Shapelets

David Guijo-Rubio[1,2(✉)], Pedro A. Gutiérrez[1], Romain Tavenard[3], and Anthony Bagnall[2]

[1] Department of Computer Sciences, Universidad de Córdoba, Córdoba, Spain
dguijo@uco.es
[2] University of East Anglia, Norwich NR47TJ, UK
[3] Univ. Rennes, CNRS, LETG/IRISA, Rennes, France

Abstract. Shapelets are phase independent subseries that can be used to discriminate between time series. Shapelets have proved to be very effective primitives for time series classification. The two most prominent shapelet based classification algorithms are the shapelet transform (ST) and learned shapelets (LS). One significant difference between these approaches is that ST is data driven, whereas LS searches the entire shapelet space through stochastic gradient descent. The weakness of the former is that full enumeration of possible shapelets is very time consuming. The problem with the latter is that it is very dependent on the initialisation of the shapelets. We propose hybridising the two approaches through a pipeline that includes a time constrained data driven shapelet search which is then passed to a neural network architecture of learned shapelets for tuning. The tuned shapelets are extracted and formed into a transform, which is then classified with a rotation forest. We show that this hybrid approach is significantly better than either approach in isolation, and that the resulting classifier is not significantly worse than a full shapelet search.

Keywords: Time series classification · Shapelets · Convolutional neural networks

1 Introduction

Shapelets [1] are discriminatory phase independent subsequences that form a basic primitive in many time series algorithms. For classification, shapelets are assessed using their distance to train set time series and the usefulness of these distances in discriminating between classes. Shapelet based features define a distinct form of discrimination which can be characterised as quantifying whether a particular shape exists in a series or not (at any location). Shapelets have proved an effective tool for classification [2] and have been a popular research topic. One key distinction between research threads is whether shapelets are extracted from the training data or whether the space of all possible shapelets is searched. The

© Springer Nature Switzerland AG 2019
H. Yin et al. (Eds.): IDEAL 2019, LNCS 11871, pp. 137–144, 2019.
https://doi.org/10.1007/978-3-030-33607-3_16

data driven approach was used in early work with shapelets [1,3,4] and has been employed to find effective classifiers [2,5]. The learned shapelet approach [6] was the first search based algorithm. Later work has bridged the gap between the learning shapelets and convolutional neural networks (CNN) through defining each shapelet as a variant of a convolutional filter [7]. Our aim is to investigate whether we can create a better classifier by hybridising data driven search and stochastic gradient descent learning. Our approach is to randomly sample shapelets in the data for a fixed time using the method described in [5], retain the best k found, then tune these shapelets with the learning shapelet algorithm [6] implemented using a neural network library. We test this on data from the UCR archive [8]. We demonstrate that, under controlled parameterisation, this approach is better than either algorithm in isolation. Furthermore, we show that a one hour search followed by tuning is not significantly worse than the reported results using the full shapelet transform [9].

The rest of the paper is structured as follows. In Sect. 2 we provide an overview of the shapelet transform and learning shapelets algorithm. In Sect. 3 we describe how we have hybridised the two approaches. Results are described in Sect. 4, and we conclude in Sect. 5.

2   Shapelet Background

We denote a vector in bold and a matrix in capital bold. A case/instance is a pair $\{\mathbf{x}, y\}$ with m observations $x_1, \ldots, x_m$ (the time series) and discrete class variable y with c possible values. A list of n cases with associated class labels is $\mathbf{T} = \langle \mathbf{X}, \mathbf{y} \rangle = \langle (\mathbf{x_1}, y_1), \ldots, (\mathbf{x_n}, y_n) \rangle$. A shapelet $\mathbf{s}$ is a time series $\langle s_1, \ldots, s_l \rangle$ where $l \leq m$. Shapelet based classification requires a method of finding and assessing shapelets, then an algorithm for using the selected shapelets for classification. All shapelet finding algorithms require the measuring of the distance between a candidate and a time series. This is done by sliding the candidate along the series and calculating the Euclidean distance at each position (after normalisation) to find the minimum. The distance between a shapelet $\mathbf{s}$ and a time series case is then given by Eq. 1,

$$sDist(\mathbf{s}, \mathbf{t}) = \min_{\mathbf{w} \in \mathbf{W}} (dist(\mathbf{s}, \mathbf{w})), \tag{1}$$

where $\mathbf{W}$ is the set of all subsequences which are the same length as $\mathbf{s}$ in $\mathbf{t}$, and $dist$ is the Euclidean distance between two equal length series. The original shapelet algorithm [1] constructed a decision tree by finding a shapelet at each node. Subsequent research [10] demonstrated it was better to use shapelets as a transformation. We can summarise the shapelet transform (ST) approach as: search for the best k shapelets in the data; transform the data so that each new attribute j represents distance between series i and shapelet j; construct a standard classifier on the resulting transformed data. It has been shown that there is little need to enumerate the possible space of shapelets in the data. Random sampling of a tiny proportion of the shapelet space does not lead to a significant decrease on accuracy [5]. The current shapelet transform algorithm can be configured so that it searches for shapelets for a maximum amount of time. We call this a contract classifier, since it is given a time contract it must fulfill.

The ST pipeline can be summarised as follows:

1. **Search**: randomly sample shapelets from the train data for a fixed amount of time, keeping the best.
2. **Transform**: create new train and test data where attributes are distances between instances and shapelets (see Eq. 1).
3. **Fit Model**: build the classifier on the train data.
4. **Predict**: estimate class values on the test data.

Another alternative to the enumeration of all possible shapelets is to use optimization techniques to learn good shapelets. Learning shapelets (LS [6]) algorithm involves learning shapelet values as model parameters rather than directly extracting them from the data. Hence, resulting shapelets are no longer subseries from the training set. The LS algorithm adopts an initialisation stage to find k shapelets through clustering shapelets observed in the data. It then jointly learns the weights for a regularised logistic regression and the shapelet set using a two stage iterative process for Shapelet Tranform representation to feed a final logistic regression classifier. The LS pipeline can be summarised as follows:

1. **Initialise**: find initial shapelets from the train data through clustering subsequences and initialise model weights.
2. **Fit Model**: For a given maximum number of iterations:
   (a) **Update Loss**: adjust loss function.
   (b) **Update Model Weights**: adjust weights to minimize loss.
   (c) **Update Shapelets**: adjust shapelets to mimize loss.
3. **Predict**: estimate class values on the test data.

An experimental comparison of these algorithms found ST to be significantly more accurate than LS [2]. LS suffers from three related problems that may have caused this difference. Firstly, LS is very sensitive to the initialisation of shapelets, and, secondly, the clustering algorithm adopted to partially overcome this problem is memory intensive. Finally, the implementation is very complex; despite communications with the LS author, we cannot rule out there being bugs in the Java implementation[1]. One way of possibly mitigating these problems is to use stable open source software.

3 Hybrid Shapelet Classifier

A shapelet distance $sDist$ is evaluated by sliding the shapelet to each series as would be done for a convolution filter (see Eq. 1). The LS model can be

---

[1] https://github.com/TonyBagnall/uea-tsc

be implemented as a variant of a CNN [7] within a standard neural network framework. A neural network model is defined that takes time series as inputs and outputs class probabilities.

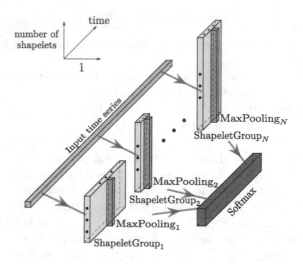

Fig. 1. Learning Time Series Shapelet (LS) model as a neural network architecture.

As shown in Fig. 1, this model is composed of a first layer (called the shapelet layer hereafter) that extracts a ST-like representation which then feeds into a logistic regression layer. In practice, the shapelet layer is made of several shapelet blocks (one block per shapelet length). Each shapelet block can be decomposed into:

1. a feature extraction step that computes pairwise distances $dist(\mathbf{s}, \mathbf{w})$ between the considered shapelets and all the subsequences with the same length as $\mathbf{s}$ in $\mathbf{t}$ and
2. a pooling step that retains the minimum of all distances.

Note that this is very similar in spirit to a convolutional layer that would compute dot products between filters and all the subsequences in $\mathbf{t}$ which would be typically followed by a (max-)pooling layer. Finally, the optimization procedure consists in tuning both the shapelet values and the parameters of the logistic regression through stochastic gradient descent.

There are two ways we could combine the approaches: we could use the transform to search for a better starting point for the LS classification algorithm or use the LS algorithm to tune the shapelets found by ST, but retain the classification method in ST. The first approach involves replacing the **initialise** stage in LS with a time constrained **search**, then proceeding with LS as normal. The second approach is to perform a tuning stage with LS' **fit model** between **search** and **transform** from ST. We evaluate both approaches and consider three classifiers:

1. **ST-RF**: Shapelet transform contracted for one hour or ten hours, then build and evaluate a rotation forest classifier on the transformed data
2. **Hybrid-LR**: Use the shapelets found for ST as an initialisation for the neural network (LS model), then use the final logistic regression classifier on the test data.
3. **Hybrid-RF**: As Hybrid-LR, but rather than use the logistic regression, use rotation forest as a classifier, as with ST.

We use the rotation forest classifier [11] with fixed parameters for ST and Hybrid-RF, since it has been shown to be very good at problems with continuous attributes [12].

4   Results

The rotation forest has 200 trees, uses a group size of 3 and class selection probability of 0.5. More details can be found in [12]. We set ST to find a maximum of 100 shapelets in either one hour or ten hours. The LS model is optimised using the Adagrad adaptive gradient optimisation algorithm [13]. The learning rate parameter is fixed to 0.1 and the training will be run for 2000 epochs for all datasets. Moreover the ℓ_2 regularisation parameter over classification weights is 0.01 and the batch size is fixed to 128. Note that for this first approximation, all the parameters are fixed to default values for simplicity and to reduce the computational cost.

We evaluate the shapelet approaches on a subset of 92 data of the 128 UCR data [8]. The data that are omitted are done so for practical and implementation reasons. 14 of the 128 data have missing values or are unequal length and our system is not able to handle this characteristic. The LS is unable to handle 22 of the remaining 114 data because of memory constraints; long shapelets require a large amount of memory. All reported results are on the standard train test splits. All learning and tuning are conducted exclusively on the train data, and accuracy is assessed on the unseen test split.

We are interested in testing whether tuning the shapelets after a data driven search significantly improves the overall classifier performance. To do this, we control other factors that cause variation. We have intentionally set up ST-RF and Hybrid-RF so that the only difference between the two experiments is the tuning stage. Any improvement can be attributed to this, since in all other ways ST-RF and Hybrid-RF are identical.

Figure 2(a) shows the scatter plot of test accuracy for ST-RF and Hybrid-RF for a one hour shapelet search. Tuning makes a significant difference: 51 datasets have improved accuracy, whereas just 32 have decreased performance (with 9 ties). The test of difference is significant with a binomial test, a paired T test and a Wilcoxon sign rank test. Figure 2(b) shows the same results for a 10 h shapelet search. The pattern of results is the same; tuning improves accuracy on 52 problems, makes things worse on 32 and makes no difference on 8. The difference is also significant.

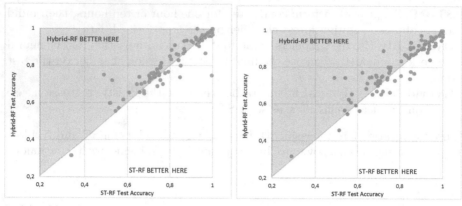

(a) 1 hour shapelet transform search   (b) 10 hours shapelet transform search

Fig. 2. Test accuracy for the shapelet transform before and after tuning with LS.

Whilst significant, it is worth noting that the improvement is not guaranteed. As a sanity check, it is worth checking whether it is better to use the logistic regression from LS as a classifier itself rather than extracting the shapelets for use with rotation forest. Figure 3 shows the critical difference diagrams [14] for the ST and Hybrid-RF transform in comparison to those found with Hybrid-LR. It demonstrates two things: firstly, tuning the shapelets then classifying with the LS model (Hybrid-LR) makes no significant difference when compared to ST alone; and secondly, that extracting shapelets from LS and using a rotation forest (Hybrid-RF) is significantly better than both a logistic regression classifier (Hybrid-LR) and using rotation forest with a transform generated by untuned shapelets (ST-RF).

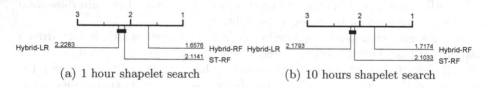

(a) 1 hour shapelet search   (b) 10 hours shapelet search

Fig. 3. Critical difference diagrams for three classifiers. The classifiers are: shapelet transform with rotation forest (ST); Learning Time Series Shapelets using the transform shapelets as a starting point (Hybrid-LR); and shapelet transform on shapelets extracted from LS with a rotation forest (Hybrid-RF).

For context, it is useful to compare the results to previously published results. ST is a key component of the meta-ensemble HIVE-COTE [9]. The ST results for HIVE-COTE were found through a computationally expensive enumeration of all possible shapelets. Figure 4 shows that we can achieve results that are not significantly worse than the enumerative search by simply tuning the one hour search shapelets.

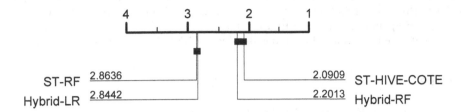

Fig. 4. Comparison of performance of three classifiers based on a one hour shapelet search with an exhaustive search based shapelet classifier presented in [9].

5   Conclusion

We have demonstrated that the core concept of tuning shapelets found in the data with a gradient descent algorithm has merit. Tuning significantly improved accuracy after both a one hour search and a ten hour search. Indeed, the improvement was enough to achieve results not significantly worse than the state-of-the-art shapelet approach. This is surprising and exceeded our expectation. However, these experiments have limitations. Firstly, the neural network implementation of learning shapelets is very memory intensive. This means we have not been able to compare against using more shapelets in the transform nor with all the datasets. Nor have we been able to evaluate on resamples rather than a single train test split. Secondly, whilst comparing on all problems is desirable to remove any suspicion of cherry picking, the presence of many smaller problems may mask the benefit of the full enumeration: many of the problems can be fully enumerated in one hour. Thirdly, we have not described the training time for the LS model in our results, because at the time of writing we do not have an integrated solution: we search for shapelets in the Java version [15] on CPU and train LS using a dedicated pytorch implementation on GPU. This makes timing comparisons problematic. A fully integrated version is in development using the sktime toolkit [16]. Results and code are available from the associated website [2]. Despite these limitations, these results are promising and support the central hypothesis than tuning can improve shapelets found in the data. The next stage is to evaluate the methods on larger problems and to assess the relative merits of greater search time, retaining more shapelets and selective tuning. Furthermore, an iterative search and tune algorithm may prove better than our current sequential model of search then tune.

---

[2] http://www.timeseriesclassification.com/hybrid.php.

Acknowledgement. This research has been partially supported by the Ministerio de Economía, Industria y Competitividad of Spain (Grant Refs. TIN2017-85887-C2-1-P and TIN2017-90567-REDT) as well as Agence Nationale de la Recherche through MATS project (ANR-18-CE23-0006). D. Guijo-Rubio's research has been supported by the FPU Predoctoral and Short Placements Programs from Ministerio de Educación y Ciencia of Spain (Grants Ref. FPU16/02128 and EST18/00280, respectively). Some experiments used a Titan X Pascal donated by the NVIDIA Corporation.

References

1. Ye, L., Keogh, E.: Time series shapelets: a novel technique that allows accurate, interpretable and fast classification. Data Min. Knowl. Disc. **22**(1–2), 149–182 (2011)
2. Bagnall, A., Lines, J., Bostrom, A., Large, J., Keogh, E.: The great time series classification bake off: a review and experimental evaluation of recent algorithmic advances. Data Min. Knowl. Disc. **31**(3), 606–660 (2017)
3. Mueen, A., Keogh, E., Young, N.: Logical-shapelets: an expressive primitive for time series classification. In: Proceedings of the 17th ACM SIGKDD International Conference on Knowledge Discovery and Data Mining (2011)
4. Lines, J., Davis, L., Hills, J., Bagnall, A.: A shapelet transform for time series classification. In: Proceedings of the 18th ACM SIGKDD International Conference on Knowledge Discovery and Data Mining (2012)
5. Bostrom, A., Bagnall, A.: Binary shapelet transform for multiclass time series classification. Trans. Large-Scale Data Knowl. Cent. Syst. **32**, 24–46 (2017)
6. Grabocka, J., Schilling, N., Wistuba, M., Schmidt-Thieme, L.: Learning time-series shapelets. In: Proceedings of the 20th ACM SIGKDD International Conference on Knowledge Discovery and Data Mining (2014)
7. Tavenard, R.: tslearn: a machine learning toolkit dedicated to time-series data (2017). https://github.com/rtavenar/tslearn
8. Dau, H., et al.: The UCR time series archive. arXiv e-prints arXiv:1810.07758 (2018)
9. Lines, J., Taylor, S., Bagnall, A.: Time series classification with HIVE-COTE: the hierarchical vote collective of transformation-based ensembles. ACM Trans. Knowl. Discov. Data **12**(5), 1–35 (2018)
10. Hills, J., Lines, J., Baranauskas, E., Mapp, J., Bagnall, A.: Classification of time series by shapelet transformation. Data Min. Knowl. Disc. **28**(4), 851–881 (2014)
11. Rodriguez, J., Kuncheva, L., Alonso, C.: Rotation forest: a new classifier ensemble method. IEEE Trans. Pattern Anal. Mach. Intell. **28**(10), 1619–1630 (2006)
12. Bagnall, A., Bostrom, A., Cawley, G., Flynn, M., Large, J., Lines, J.: Is rotation forest the best classifier for problems with continuous features? arxiv e-prints arXiv:1809.06705 (2018)
13. Duchi, J., Hazan, E., Singer, Y.: Adaptive subgradient methods for online learning and stochastic optimization. J. Mach. Learn. Res. **12**(July), 2121–2159 (2011)
14. Demšar, J.: Statistical comparisons of classifiers over multiple data sets. J. Mach. Learn. Res. **7**, 1–30 (2006)
15. UEA TSC: A weka compatible toolkit for time series classification and clustering (2019). https://github.com/TonyBagnall/uea-tsc
16. sktime: A toolbox for data science with time series (2019). https://github.com/alan-turing-institute/sktime

An Ensemble Algorithm Based on Deep Learning for Tuberculosis Classification

Alfonso Hernández[1], Ángel Panizo[1(✉)], and David Camacho[2]

[1] Computer Science Deparment, Universidad Autónoma de Madrid, Madrid, Spain
alfonso.hernandezs@estudiante.uam.es, angel.panizo@uam.es
[2] Information Systems Department, Technical University of Madrid, Madrid, Spain
david.camacho@etsisi.upm.es

Abstract. In the past decades the field of Artificial Intelligence, and specially the Machine Learning (ML) research area, has undergone a great expansion. This has been allowed for the greater availability of data, which has not been foreign in the field of medicine. This data can be used to train supervised Machine Learning algorithms. Taking into account that this data can be in form of images, several ML algorithms, such as Artificial Neural Networks, Support Vector Machines, or Deep Learning Algorithms, are particularly suitable candidates to help in medical diagnosis. This works aims to study the automatic classification of X-Ray images among patients who may have tuberculosis, using an ensemble approach based on ML. In order to achieve this, an ensemble classifier, based on three pre-trained Convolutional Neural Networks, has been designed. A set of 800 samples with chest X-Ray images will be used to carry out an experimental analysis of our proposed ensemble-based classification method.

Keywords: Deep Learning · Convolutional Neural Networks · Support vector machines · Image classification

1 Introduction

Machine Learning (ML) has undergone a great expansion in the last decade achieving superior performance in classification and regression tasks, specifically, supervised Machine Learning methods, like Artificial Neural Networks (ANNs). This leap in quality has been achieved thanks in part to the development of Deep Learning methods, which are a special type of ANN with a higher number of layers. Deep Learning popularity has increased in many fields including malware detection [1,2], health care & medicine [3–5] or speech recognition [6].

This work has been supported by the next research projects: DeepBio (TIN2017-85727-C4-3-P), funded by Spanish Ministry of Economy and Competitivity (MINECO), and CYNAMON (grant number S2013/ICE-3095) funded by Autonomous Goverment of Madrid (CAM), all of them under the European Regional Development Fund FEDER.

H. Yin et al. (Eds.): IDEAL 2019, LNCS 11871, pp. 145–154, 2019.
https://doi.org/10.1007/978-3-030-33607-3_17

The development of new techniques in the field of Health & Medicine has caused a demand for computing systems capable of managing high amounts of data and performing specific tasks automatically, like assisted surgery, clinical diagnosis, early illness detection or genomics among others. Some of these tasks require the analysis of images in the form of X-rays, CTs (computerized tomography) or live video. Deep Learning, especially Convolutional Neural Networks (CNNs), are the most suited ML techniques for image analysis. In recent years, CNNs have excelled in the image analysis field topping the lists of important related challenges like the ImageNet one [7].

The main goal of this article is to design and train an ensemble algorithm based on Deep Learning capable of diagnosing tuberculosis from chest X-Rays. To achieve this, first, we will compare the performance of several CNNs trained with a reduced dataset composed of 800 chest X-Ray images, similar to those used in medical projects. Second, the ensemble algorithm will be composed of the CNNs that have given the best results. When evaluating the performance of these methods we will favor those models that *minimize false negatives*. This is very important in the field of medical diagnosis since a false negative can lead to an erroneous treatment of the patient, or even worse no treatment at all, which could lead to serious health problems.

This paper is structured as follows: in Sect. 2, a brief background on the Machine Learning methods used in this work is briefly introduced, in Sect. 3 we describe the problem and the proposed approach for radiographical images classification based on ensembles, which are analyzed through an experimental evaluation in Sect. 4. Finally, in Sect. 5 the conclusions and the future lines of work are presented.

2   Background

CNNs are "biologically-inspired trainable architectures that can learn invariant features" [8,9]. Other image classification methods required hand-engineered filters in order to properly work. Contrary, CNNs have the ability to learn these filters by themselves with enough training. However, in order to achieve this, a large amount of labeled data is necessary. Usually, the amount of available labeled medical data is very small compared with the amount of data required to train a CNN. This makes applying CNNs to medical domains a big challenge and has generated a demand for algorithms capable of being trained with reduced or unbalanced datasets [10,11]. However, achieving this have not yet been solved and still remains an open problem.

Two main approaches can be found in the literature to tackle the lack of data problem. The first one consists of enlarging the dataset by applying methods such as mirroring, cropping or rotations to the images of the dataset, in a process called Data Augmentation [12]. The second approach is called Transfer Learning [13], and it consists on transferring the knowledge obtained in another task to solve a new one, letting us rebuild an existing model (pre-trained networks with other datasets) instead of creating a new one. However, even using these pre-trained networks, some hyperparameters, or the network architecture itself, must be modified in order to fit the model to each specific problem. For this purpose, some algorithms have been developed to help in this task [14].

Image analysis with CNNs is a hot research area and new architectures are continuously proposed. Several CNNs architectures have already been used for medical image analysis with good results [4]. Among these architectures it is worth mentioning the next ones: VGG16 and VGG19 [15], two networks composed of 16 or 19 weight layers of very small convolutional filters (3×3); ResNet50 [16], a network composed by an ensemble of residual nets, which are layers that have been reformulate as learning residual functions with reference to the layer inputs, instead of learning unreferenced functions. Each one of the residual nets has 152 layers, being much deeper than the VGG architectures, but with less complexity; and finally, InceptionV3 [17], a network composed by asymmetric blocks including convolutions, average pooling, max pooling, concats, dropouts, and fully connected layers.

Finally, another technique found in the literature to improve the results of an image classifier is building an ensemble with several models. This multi-model approach is based on the assumption that misclassified samples with the best algorithm can be correctly classified by other models, so combining several classifiers can obtain better results than the best stand-alone model. This approach has resulted in highly successful practical developments in several fields, including medicine and bio-inspired applications [18,19].

3   Algorithm Design

In this section, the ensemble algorithm proposed will be described. Our method is composed of three stages: Image Pre-processing, CNN selection, and Ensemble design. In the first stage, we use two different contrast tuning techniques and apply Data Augmentation to the X-Rays images. Then, in the second phase, several CNNs models will be build using different architectures and hyperparameters, each CNN will be combined with a Transfer Learning technique. Finally, in the last step, the previously made models will be joined together in groups of three along with one of the two proposed voting techniques in order to build ensembles. In Fig. 2 we can see the project architecture:

3.1   Image Pre-processing

Before building the models we will pre-process every image in the training set. First of all, we will duplicate every image in the training set by horizontally mirroring them. This improves the model's generalization by allowing them to find similar features on both sides of the thorax. Then, we will apply Histogram Equalization or Contrast Limited Adaptive Histogram Equalization (CLAHE) to every image. These methods allow to better highlight features in the lungs, Fig. 1 shows an example of a chest X-Ray and the results of applying both techniques. As we can see, while the histogram Equalization method makes the ribs less visible, the CLAHE method enhances the contrast between the ribs and the

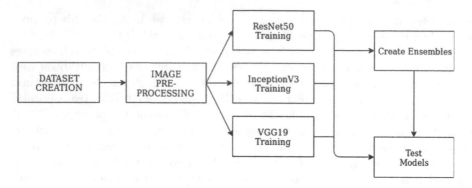

Fig. 1. In this flowchart we can see how the dataSet is first splitted in train, and test sets, and then preprocesd. Finally, it is used to rain ResNet50, VGG19 and InceptionV3 models and ensemblers of that models

background. The two methods are based on modifying the image by tweaking its histogram, so only one can be applied. Hence, the one that better fits our problem will be selected in the experimentation phase. Then, the images are resized to a shape of 224×224 using the bi-linear interpolation method. After this, we will add more images by adding random Gaussian noise to some of the images in the training set. This allows the model to better classify images with noise that, in medical images could be caused by an interference in the machine that takes the X-ray images [20]. Next, we will scale each pixel of each image by a factor of $1/255$ to normalize them, this avoids making the weights in the CNN grow too high. Finally, to allow the model to ignore the background, we will apply random cropping with a zoom range of 99 pixels. This is done with the Keras data structure *ImageDataGenerator*, which crops the images before training the model, cropping different images in each epoch, and leaving others without cropping.

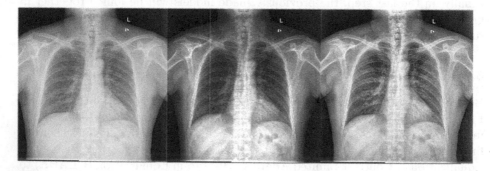

Fig. 2. Left: Raw image. Center: Image after histogram equalization. Right: Image after CLAHE with TileGridSize $= 20$

3.2   CNN Design

Due to the small size of the dataset, we will apply Transfer Learning from pre-trained models. The pre-trained models will be ResNet50, InceptionV3, VGG16, and VGG19 from the Keras library. These models have been trained with the 'imagenet' dataset, which has near 1.2 million images as the training set, another 50.000 images for validation and 100.000 for testing. To adapt the model to Tuberculosis classification, first, the last layer of the pre-trained models, which is a classifier that categorizes the input between 1000 classes, will be substituted by a binary classifier. Then, different models will be built by changing their hyperparameters: the activation function of the classification layer, the optimizer and it's learning rate. Finally, the transfer learning technique will be applied. We will re-train all layers of the network from 10 to 50 epochs. This will allow the model to look for the features in the images starting from the features extracted from the original dataset.

3.3   Ensemble Design

To build the ensembles, 3 CNN model will be selected taking into account that two models with the same architecture and different hyper-parameters will not be selected together. I.e, we will not choose 2 VGG19 models with different learning rates to be part of the same ensemble. Then, to decide the final prediction of the ensemble, the next two methods are proposed: **Voting by a majority (VM)**, where each image will be classified as the class that has been predicted by 2 (or more) stand-alone models; and **Sum of probabilities (SP)**, where each stand-alone model gives the probability of each sample to belong to each class. To classify an image, the probabilities of belonging to each class will be computed with the stand-alone model. The image will be classified as the class with a higher mean probability.

4   Experimental Results

4.1   Dataset

The dataset chosen for the project is the one published in [21], and it's formed by two different datasets. The first one is composed of 138 frontal chest X-rays from Montgomery County's Tuberculosis screening program, being 80 of them normal images and the other 58 from patients with TB (abnormal from now on). All of them have size either 4020×4892 or 4892×4020 pixels. The second dataset has 662 frontal chest X-rays, of which 326 are normal cases and 336 are abnormal cases, and their size varies, being close to 3000×3000 pixels. We used 90% of the dataset as the training set and the remaining 10% as the test set. We used 10% of the training set as validation split.

4.2   Performance Measurement Criterion

The goal of these ML methods is to remove or decrease the number of other tests needed to diagnose an illness. Generally, these other tests are longer, more expensive or invasive. Therefore, in our opinion, the overall accuracy of the model would not be a good criterion. If we want to decrease the number of tests performed, our model should be able to correctly classify at least one of the classes (has a disease/don't have a disease) with a very high probability. For example, is preferable having a model with lower overall accuracy and capable of telling if a patient has TB with a nearly zero error, than having a model with higher overall accuracy but a bigger TB positive error rate. Because if someone is classified as TB positive by this model, we can directly treat him with antibiotics without any further test. On the contrary, if the patient is classified as normal by this model further test can be done to check if the model was correct. This is not possible if the model has a non-despicable error for both classes. For that reason, our goal will be to make models that completely remove the false positives (healthy patient classified as unhealthy) or false negatives (patients with TB classified as healthy). The resulting models should have reduced to 0 one of the errors, hence, we will penalize the models which have both kinds of errors. In addition, it is more desirable to find a model that removes false negatives because, they can lead to an erroneous treatment of the patient, or even worse no treatment at all, which could lead to serious health problems. So the best model would be the one that has no false negatives and minimize the number of false positives. If we do not find a model with these characteristics, we will look for a model with no false positives and the less number of false negatives as possible.

4.3   Results

Table 1 shows the best hyperparameters configuration chosen for each of the candidate CNNs architectures. The hyperparameters are the *activation function*, the *model optimizer* and the *learning rate*, and the histogram equalization method will be also taken into account. The values of these hyperparameters have been tuned experimentally and will be used for the rest of experiments. The models shown in Table 1 have been trained from 10 to 50 epochs, testing and saving the weights after 10 epochs of training. Then, we have build ensembles using all possible combinations of CNN models and number of epochs, looking for the best results. Table 2 shows the four best results obtained using this approach. Looking at this table we can see that the ensemble with better accuracy is the one composed by the ResNet50 model trained for 50 epochs, the InceptionV3 model trained for 10 epochs and the VGG19 model trained for 50 epochs, achieving an overall accuracy of 0.8642. If we look at the confusion matrix, we can also see that this ensemble is the one that minimize the number of false-negatives, missclasifying only 9 abnormal cases. However, as explained in Sect. 4.2, this model has false positives and false negatives so it does not meet our performance criterion. As we can see in Table 2, there are no ensemble that reduces the false negatives to zero, but there are several ones that reduces the false positives to 0. From these models the one that has a higher accuracy, and better fits our performance criterion, is the one composed of: ResNet50 trained for 30 epochs, InceptionV3 trained for 10 epochs and VGG19 trained for 50 epochs. Both ensemble criterions, Sum of Probabilities (SP) and Voting for Majority (VM), give the same results: an overall accuracy of 0.8519, no false positives and 12 false negatives.

Table 1. Best CNN model's configuration

	ResNet50	InceptionV3	VGG19
Activation function	Linear	Sigmoid	Linear
Model optimizer	Adam ($\beta_1 = 0.9, \beta_2 = 0.999, nodecay$)		
Learning rate	0.0001	0.0001	0.001
Histogram type	Normal	CLAHE	Normal

Table 2. Best Ensemble models results, where VM stands for the *voting for majority* decision method and SP stands for the *sum of probabilities* one.

Decision method	CNN/Epochs		Confusion matrix			Total accuracy
VM	ResNet50	50		Normal	Abnormal	0.8642
	InceptionV3	10	Normal	39	2	
	VGG19	50	Abnormal	9	31	
VM & SP	ResNet50	40		Normal	Abnormal	0.7901
	InceptionV3	40	Normal	41	0	
	VGG19	50	Abnormal	17	23	
VM & SP	ResNet50	30		Normal	Abnormal	0.8519
	InceptionV3	10	Normal	41	0	
	VGG19	50	Abnormal	12	28	
SP	ResNet50	30		Normal	Abnormal	0.8272
	InceptionV3	40	Normal	41	0	
	VGG19	50	Abnormal	14	26	

Finally, we will compare the ensemble results with those obtained with the stand-alone models in order to verify that the ensembles improve the performance of any of its stand-alone parts. Table 3 shows the confusion matrix and accuracy achieved by each stand-alone model when trained during 10 to 50 epochs. As we can see, the stand-alone model which does not have false positives and has less false negatives is ResNet50 trained for 40 epochs. It has a total accuracy of 75.3%, significantly smaller than the same model trained for 30 epochs, which has an overall accuracy of **83.95%**. However, this model has almost the same False Positives than False Negatives, both bigger than zero. Finally, looking at the results of the ResNet50 model trained during 50 epochs we can see that it only has one false negative although it has less than 50% accuracy in normal cases. Comparing the stand-alone models used for the chosen ensemble (ResNet50 30 epochs, InceptionV3 10 epochs, and VGG19 50 epochs), we can see that the ensemble improves the overall accuracy of the 3 individual models, being higher than the ResNet50 model with 0.8495 against the 0.8519 of the ensemble. In addition, the only stand-alone model which reduces the false positives to zero is the InceptionV3 model but has 11 false negatives more than the selected ensemble.

Table 3. CNN best models confusion matrix.

Model	Epochs	10 Normal	Abnormal	20 Normal	Abnormal	30 Normal	Abnormal	40 Normal	Abnormal	50 Normal	Abnormal
ResNet50	Normal	13	28	40	1	34	7	41	0	18	23
	Abnormal	2	38	17	23	6	34	20	20	1	39
	Accuracy	0.6296		0.7778		0.8395		0.753		0.7037	
InceptionV3	Normal	41	0	41	0	41	0	41	0	40	1
	Abnormal	23	17	33	7	29	11	34	6	20	20
	Accuracy	0.716		0.5926		0.642		0.5802		0.7407	
VGG19	Normal	41	0	18	023	39	2	37	4	38	3
	Abnormal	28	2	20	20	15	25	14	26	14	26
	Accuracy	0.5308		0.4691		0.7901		0.7778		0.7901	

5   Conclusions and Future Work

This work presents an ensemble algorithm based on Deep Learning capable of diagnosing tuberculosis from chest X-Rays. We have proposed a problem in a medical diagnosis service, facing the most common problems in this kind of projects. Also, we have tried to find the ML model which gets the best results according to a performance criterion established according to the particularities of the medical image classification. The best model has been found to be an ensemble between 3 CNN models (ResNet50, VGG19, and InceptionV3), with an overall accuracy of **85%** and **no false positives**.

The future work will be focused on studying the ResNet50 models, because it was the model which was closest to remove false negatives, trying to improve its performance. Also, if the resulting model is going to be used in a real diagnosis service, it will necessary to find other datasets to validate the results, and to check the model robustness we could use k-fold cross validation. Finally, we must not forget that ML applications in medicine must meet certain ethical requirements [22], because it would be dangerous to trust them without any human validation. For this reason, some evidence of the prediction must be given to the doctor to let him validate the results, this can be achieved implemented an auto-explainable model, using techniques like the layer-wise relevance propagation (LRP) [23].

References

1. Martín, A., Lara-Cabrera, R., Camacho, D.: Android malware detection through hybrid features fusion and ensemble classifiers: the AndroPyTool framework and the OmniDroid dataset. Inf. Fusion **52**, 128–142 (2019)
2. Martín, A., Menéndez, H.D., Camacho, D.: MOCDroid: multi-objective evolutionary classifier for android malware detection. Soft Comput. **21**(24), 7405–7415 (2017)
3. Kermany, D.S.: Identifying medical diagnoses and treatable diseases by image-based deep learning. Cell **172**(5), 1122–1131.e9 (2018)
4. Litjens, G., et al.: A survey on deep learning in medical image analysis. Med. Image Anal. **42**, 60–88 (2017)
5. Bar, Y., Diamant, I., Wolf, L., Lieberman, S., Konen, E., Greenspan, H.: Chest pathology detection using deep learning with non-medical training. In: 2015 IEEE 12th International Symposium on Biomedical Imaging (ISBI). IEEE, April 2015
6. LeCun, Y., Bengio, Y., Hinton, G.: Deep Learn. Nature **521**(7553), 436 (2015)
7. Russakovsky, O., et al.: ImageNet large scale visual recognition challenge. Int. J. Comput. Vis. (IJCV) **115**(3), 211–252 (2015)
8. LeCun, Y., et al.: Handwritten digit recognition with a back-propagation network. In: Advances in Neural Information Processing Systems, pp. 396–404 (1990)
9. LeCun, Y., Kavukcuoglu, K., Farabet, C.: Convolutional networks and applications in vision. In: Proceedings of 2010 IEEE International Symposium on Circuits and Systems, pp. 253–256. IEEE (2010)
10. Mazurowski, M.A., Habas, P.A., Zurada, J.M., Lo, J.Y., Baker, J.A., Tourassi, G.D.: Training neural network classifiers for medical decision making: the effects of imbalanced datasets on classification performance. Neural Netw. **21**(2–3), 427–436 (2008)
11. Shin, H.-C., et al.: Deep convolutional neural networks for computer-aided detection: CNN architectures, dataset characteristics and transfer learning. IEEE Trans. Med. Imaging **35**(5), 1285–1298 (2016)
12. Wang, J., Perez, L.: The effectiveness of data augmentation in image classification using deep learning. Convolutional Neural Netw. Vis. Recognit. (2017)
13. Pan, S.J., Yang, Q.: A survey on transfer learning. IEEE Trans. Knowl. Data Eng. **22**(10), 1345–1359 (2010)
14. Martín, A., Lara-Cabrera, R., Fuentes-Hurtado, F., Naranjo, V., Camacho, D.: EvoDeep: a new evolutionary approach for automatic deep neural networks parametrisation. J. Parallel Distrib. Comput. **117**, 180–191 (2018)
15. Simonyan, K., Zisserman, A.: Very deep convolutional networks for large-scale image recognition. arXiv:1409.1556
16. He, K., Zhang, X., Ren, S., Sun, J.: Deep residual learning for image recognition. In: 2016 IEEE Conference on Computer Vision and Pattern Recognition (CVPR). IEEE, June 2016
17. Szegedy, C., Vanhoucke, V., Ioffe, S., Shlens, J., Wojna, Z.: Rethinking the inception architecture for computer vision. In: 2016 IEEE Conference on Computer Vision and Pattern Recognition (CVPR). IEEE, June 2016
18. Fierrez, J., Morales, A., Vera-Rodriguez, R., Camacho, D.: Multiple classifiers in biometrics Part 1: fundamentals and review. Inf. Fusion **44**, 57–64 (2018)
19. Del Ser, J., et al.: Bio-inspired computation: where we stand and what's next. Swarm Evol. Comput. **48**, 220–250 (2019)

20. Boyat, A.K., Joshi, B.K.: A review paper: noise models in digital image processing. Signal Image Process. Int. J. **6**(2), 63–75 (2015)
21. Antani, S., Wáng, Y.X., Lu, P.X., Thoma, G., Jaeger, S., Candemir, S.: Two public chest X-ray datasets for computer-aided screening of pulmonary diseases. Quant. Imaging Med. Surg. **4**, 252–477 (2014)
22. Vayena, E., Blasimme, A., Cohen, I.G.: Machine learning in medicine: addressing ethical challenges. PLOS Med. **15**(11), e1002689 (2018)
23. Montavon, G., Samek, W., Müller, K.-R.: Methods for interpreting and understanding deep neural networks. Digit. Signal Process. **73**, 1–15 (2018)

A Data-Driven Approach to Automatic Extraction of Professional Figure Profiles from Résumés

Alessandro Bondielli[1,2]([✉]) and Francesco Marcelloni[2]

[1] DINFO, University of Florence, Florence, Italy
`alessandro.bondielli@unifi.it`
[2] Department of Information Engineering, University of Pisa, Pisa, Italy
`francesco.marcelloni@unipi.it`

Abstract. The process of selecting and interviewing suitable candidates for a job position is time-consuming and labour-intensive. Despite the existence of software applications aimed at helping professional recruiters in the process, only recently with Industry 4.0 there has been a real interest in implementing autonomous and data-driven approaches that can provide insights and practical assistance to recruiters.

In this paper, we propose a framework that is aimed at improving the performances of an *Applicant Tracking System*. More specifically, we exploit advanced Natural Language Processing and Text Mining techniques to automatically profile resources (i.e. candidates for a job) and offers by extracting relevant keywords and building a semantic representation of résumés and job opportunities.

1 Introduction

For any company, it is of paramount importance to identify the right person for a job, especially in today's very dynamic market, where new technologies and specific professional figures considerably increase or decrease in popularity in a very short time span. Thus, it is crucial, especially in the context of Industry 4.0, to have data-driven tools which extract autonomously profiles of professional figures from both job opportunities and résumés, so as to capture the appearance of novel required competences and knowledge. Indeed, given the enormous amount of information and profiles on professional-oriented online platforms, such as *Linkedin*[1], these tools may help recruiters to identify the most suitable candidates for a given job offer. Further, by analyzing regularities and variations among data, they may allow predicting upcoming trends in the job market and identifying emerging and yet unseen professional figures.

Most of the current recruitment support approaches fall under the umbrella of so called *Applicant Tracking Systems* (ATSs). Available ATSs are comparable to Customer Relationship Management systems, with a focus on supporting

---

[1] https://www.linkedin.com.

© Springer Nature Switzerland AG 2019
H. Yin et al. (Eds.): IDEAL 2019, LNCS 11871, pp. 155–165, 2019.
https://doi.org/10.1007/978-3-030-33607-3_18

discovery and selection of candidates for job opportunities. Such systems are often based on manual collection and annotation of résumés and job opportunities by experienced recruiters, and on rule-based algorithms to identify suitable candidates.

In this paper, we present a preliminary version of a data-driven approach to the problem of automatically generating profiles of professional figures, by extracting relevant information from résumés and job opportunities. A classic and straightforward document similarity calculation approach cannot guarantee satisfactory performance in the ATS domain. This is due to the fact that documents in this domain are mostly similar to each other, and tend to be differentiated by minor aspects. For example, the two sentences "fluent in Python" and "fluent in Java" are extremely similar, but describe two different professional figures. Therefore, a simple document similarity calculation on whole documents could be unsuccessful since it may be prone to simply identify similarly written curricula. For this reason, we employ a composite approach which exploits Natural Language Processing (NLP) techniques, such as *word entropy* [4], for identifying relevant keywords, and recent distributional semantics techniques, such as word and document vectors [11], in order to represent similar profiles. Further, our approach exploits only the unstructured information contained in résumés. To the best of our knowledge, current ATSs do not support this kind of analysis and require a frequent interaction with the human experts. On the other hand, the development of ATSs has a distinctly commercial nature. Thus, the research community has not been particularly active on this specific topic. Today, with Industry 4.0, there is a growing interest in implementing autonomous and data-driven approaches that can provide insights on how the job market evolves over time, and practical assistance to recruiters. This is definitely a novelty for most commercial ATSs, especially for those targeted to small-to-medium sized companies that focus on a small domain of application.

This paper is the result of a collaboration with the recruitment agency IT Partner Italia[2]. The company mostly operates in the field of IT system integration. Due to the high volume of hiring processes managed on a daily basis, IT Partner Italia is aiming to build a next-generation ATS based, among other technologies, on NLP and Machine Learning techniques, in order to help recruiters in their most time consuming activities.

This paper is organized as follows. In Sect. 2 we discuss related work. Section 3 illustrates our proposed approach. Section 4 presents a preliminary evaluation of the approach on a subset of data. Finally, Sect. 5 draws some conclusion.

2   Related Work

Actually, very few researches on the impact of ATSs on the recruitment process have been proposed within the business economics field [5, 10]. In addition, a number of recent approaches have proposed the implementation of data mining

---

[2] http://www.itpartneritalia.com/.

and information retrieval techniques both in the recruitment process and in performing job recommendations [9, 16]. However, it is clear that extensive research in the direction of improving software such as ATS is still in its embryonic stage. Today's ATSs belong to two main categories: *direct recruitment* and *indirect recruitment* ATSs.

The target of the first category is the human resources department of a specific company, in order to directly handle the selection of candidates and the employment process. Thus, these ATSs are often developed to be easily integrated with other internal systems such as Enterprise Resource Planners.

Indirect recruitment ATSs, on the other hand, are developed for consulting, recruiting, and interim agencies. For this reason, such ATSs need to satisfy a wider range of criteria, concerning efficiency, autonomy, user-friendliness and speed of execution.

In the field of NLP, we have been observing an exponential growth of approaches aimed at representing words and documents in a semantically meaningful way. Most of such approaches are based on the observation originally proposed in [8] that words with similar meaning occur in similar contexts. Thus, the first steps along this direction were aimed at building implicit vector representation of words, by means of co-occurrences with other words.

More recently, deep learning has been extensively used for learning word representations [1, 12, 13]. The study of *word embeddings* has received in fact a lot of attention in the last few years thanks to the effectiveness of such representation in capturing a semantic quality of words. Recent literature has also focused on *character level embeddings* [2], that are able to represent *subword* information as well as word information, and *deep language models* [14], that focus on improving word embeddings models by taking into account contexts in order to model *polisemy* of words. On the contrary, a limited attention has been paid to build frameworks for representing sentences, paragraphs and entire documents embeddings, such as proposed in [11]. However, we believe that such kind of representations would have a great value in industrial environment, where both time and performances are important constraints.

3   The Proposed Approach

We aim to define a method to automatically extract the different profiles of professional figures from available résumés and job opportunities. Our approach consists of two main phases. In the first phase, we use NLP techniques to generate representations of résumés and job opportunities. In the second phase, we exploit these representations for generating profiles by exploiting clustering techniques. An overview of the entire approach is given in Fig. 1. This paper focuses mainly on the description and early evaluation of the first phase. In particular, we extract potential candidate keywords from each résumé and job opportunity by employing entropy as a weighting measure for their relevance. Then, we exploit the extracted keywords to generate a distributional semantic model of résumés and job opportunities by means of the *paragraph vector* algorithm [11].

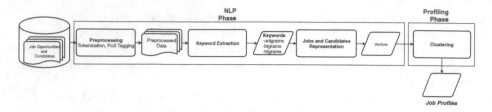

Fig. 1. Overview of the proposed approach.

3.1   Preprocessing

In order to improve the quality of our representations and to reduce the computational cost, we perform a preprocessing step. More specifically, we execute *sentence splitting, tokenization* and *Part-of-Speech (PoS) Tagging* by using SpaCy[3], a Python toolkit which provides deep learning based models to perform linguistic analysis for several languages, including Italian, out of the box. After tokenization and PoS Tagging, we also remove several stopwords (e.g. "Curriculum Vitae") that are not interesting in our analysis.

3.2   Keyword Extraction

Our keyword extraction algorithm has two main goals. First, we aim to identify keywords that can help us in profiling and describing résumés and job opportunities. Second, we want to be able to take into account the time, i.e. when a given résumé or job opportunity was produced and/or updated. The temporal aspect may in fact be crucial to enable the identification of regularities, i.e. recurrent terms and novelties in word usage. Such novelty in turn might be representative of previously unseen profiles.

Given the context of application, in addition to single words, we believe that *multiword expressions* should be taken into account as well. As in [6], we consider multiword expressions as "lexicalised word combinations as a theoretical, phraseological notion". For example, in our context of application, expressions representing technologies or software, such as "Dynamics AX", "Adobe Photoshop" or "SQL Server", should be considered as single semantic units of texts, i.e. multiword expressions. Such expressions may in fact provide a crucial insight in the process of identifying relevant keywords, as well as in the production of semantically relevant representation of résumés and job opportunities. For the sake of simplicity and readability, we will refer to both single words and multiword expressions as *n-grams*.

The keyword extraction algorithm is quite simple. Figure 2 shows its flowchart. We adopt a window-based strategy in order to identify relevant keywords for a given period of time. The time-window size is in the range $[1, n]$ days.

---

[3] https://spacy.io/.

For each time window, we compute the frequency of n-grams with respect to each document. We consider n-grams from unigrams to trigrams. In order to reduce complexity and noise, we consider only n-grams with specific PoS patterns. For instance, we want to identify n-grams such as "full stack developer", but have no interest in n-grams like "expert in", that would increase the computational complexity of the algorithm without extracting interesting terms. We then use the relative frequencies found in the time window to perform an on-line update of the overall frequency counts, and the weights for each n-gram. Once the weights are recomputed, we extract from current documents all n-grams whose weights are larger than a pre-fixed threshold WT.

This strategy allows us to identify both terms that occur frequently across all time windows, and previously not considered terms that appear as relevant only for a given window. In addition to the set of keywords extracted automatically from the documents, we also employ a set of keywords that have been manually selected as relevant by experienced recruiters. In our framework, both manually selected keywords and entropy-based keywords coexist. More specifically, we aim to introduce human experts in the loop, by enabling an *a-posteriori* selection of keywords considered as *relevant*, among those found by our algorithm, and marking as *noise* the least interesting ones. This is expected to improve performances when running the algorithm on novel data.

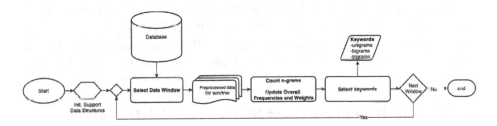

Fig. 2. Flowchart of the keyword extraction algorithm.

Concerning the selection criteria of n-grams, we consider specific content word PoS (e.g. *nouns, adjectives*) and specific PoS patterns that are often found in multiword expressions, such as *adjective-adposition-noun*, as in "fluent in Python". In order to decide whether an n-gram is relevant or not, we exploit classical association measures such as *Pointwise Mutual Information* (PMI) [3]. The PMI of two or more words quantifies the discrepancy between their joint probability and their individual marginal probability by assuming independence. More specifically, given two words x and y,

$$PMI(x; y) = \log \frac{p(x, y)}{p(x) * p(y)}$$

Probabilities are estimated by means of frequency. In particular, $p(x)$ (and $p(y)$) can be easily estimated as $f(x)/N$, where $f(x)$ is the frequency of x in the data, and N the total number of unigrams. Thus, $p(x, y) = f(x, y)/N_{bigrams}$.

In addition, since PMI is based on joint probabilities of events, it satisfies the *chain rule* (or general product rule) of probability. Thus, PMI for trigrams is computed as $PMI(x; yz) = PMI(x; y) + PMI(x; z|y)$.

We implement a slight variation of the PMI formula, namely *Positive PMI* (PPMI). PPMI is simply defined as $PPMI(x; y) = [PMI(x; y)]_+$.

We select as candidate keywords only multi-word terms that have a PPMI equal to or larger than 1.00. This ensures that we only select pairs or triplets of words that have a chance of co-occurring together higher than they were independent. Note that PPMI for unigrams is set as 1.00 by default.

Concerning the weighting of selected n-grams, several approaches for extracting relevant keywords from text documents have been proposed in the literature. We chose to adopt a weighting scheme based on term entropy. Term entropy [15] measures the average uncertainty of a given term in a collection of documents [4], thus providing an efficient method for estimating the relevance of such term for the collection of documents at hand. It is defined as:

$$e(t) = 1 + \sum_{j=1}^{D} \frac{p(t_j) \log(p(t_j))}{log(D)}$$

where D represents the overall number of documents in the collection, and $p(t_j)$ represents the probability of the term t for the document j. Again, such probability can be approximated by looking at term frequencies. More specifically, if $f(t_j)$ is the frequency of term t in document j, then $p(t_j) = \frac{f(t_j)}{\sum_{j=1}^{D} f(t_j)}$.

In summary, the proposed approach updates PPMI and entropy for all considered terms, and selects those with an entropy value equal to or larger than a pre-fixed threshold WT for each iteration (i.e., for each time window).

3.3   Semantic Representation of Résumés and Job Opportunities

In order to produce semantically relevant representation of résumés and job opportunities we explore the use of the *paragraph vector algorithm* proposed in [11]. This algorithm, also known as *doc2vec*, builds upon the notion of word embeddings by first learning word embedding representations, and then using them in order to generate a vector for the overall document.

As for *word2vec* [12,13], two architectures are available for computing the paragraph vector, namely *Distributed Memory* (PV-DM) and *Distributed Bag-of-Words* (PV-DBOW). The main difference is that, while PV-DM uses a concatenation of paragraph vectors and word vectors for predicting the next word in a window, PV-DBOW simply learns vectors by means of predicting words randomly sampled from the output paragraph.

Our goal is to identify semantic similarity among résumés and job opportunities. Given their semi-structured format, résumés and job opportunities are similar to each other, and use similar words. Therefore, an unsupervised model that considers word ordering such as PV-DM may be easily fooled in assigning a higher similarity score to different profiles due to very similar lexical and syntactic structures. Thus, we use the PV-DBOW for our goal. More specifically, we learn representations of résumés and job opportunities by exploiting only the keywords extracted by the Keyword Extraction algorithm. In particular, we adopt a join of all the keywords extracted by considering the different windows. The advantage is twofold. First, considering only relevant keywords allows learning a higher quality representation of texts. Second, as relevance of keywords is recomputed for each time window, this could ensure a greater ability to discriminate between different skills represented in résumés and job opportunities.

4 Preliminary Evaluation

We have performed several preliminary and exploratory experiments in order to assess the effectiveness of our proposed approach. In this section, we describe the available data, and illustrate results of such experiments.

We performed our experiments on a sample of the whole dataset. More specifically, we considered available résumés for 2016 and 2017, for a total of 12957 résumés. The data were anonymised and provided by IT Partner Italia following the General Data Protection Regulation (GDPR).

For each résumé, several additional information is provided, such as city of provenance and skills of the candidates (i.e. main skill, secondary skill, etc.), as well as manually inserted information concerning an actual evaluation of the candidate from a recruiter. On the one hand, we plan to exploit only information contained in the résumé to build our distributional model. This avoids the necessity of manual labour from recruiters. On the other hand, we exploit skill labels of résumés in order to assess the quality of our representation. Skills are extracted through proprietary algorithms by our industrial partner, and albeit not manually annotated, can provide a good approximation of the skills of a given candidate. We also observe that, due to different sources for résumés in our dataset (e.g. PDF, online forms), conversion into plain text may result in errors and inconsistencies.

We adopted the following parameters. For the PoS patterns, we included nouns, proper nouns and adjectives, and all combinations of three nouns, proper nouns, adjectives, adpositions and determiners, that form multiword patterns in Italian, e.g. "dispositivi Android" (Android devices), "reparto tecnico" (technical department), "sviluppatore Java" (Java developer). For the time window, we considered 180 days. We believe that using a too small time window would yield results that are too fine grained for our goal. Conversely, a window too large could hinder our ability to discover time-specific keywords. Concerning the entropy threshold, we performed several runs and finally selected a value of $WT = 0.7$. In fact, we found that a smaller threshold would yield noisy results, containing many uninteresting n-grams. Conversely, with a higher threshold the algorithm would extract a too small number of n-grams.

Second, we used the paragraph vector algorithm [11] to learn the representation of our data. More specifically, we used the Gensim implementation of the algorithm, known as Doc2Vec[4]. Doc2Vec allows fine tuning of the parameters.

---

[4] https://radimrehurek.com/gensim/models/doc2vec.html.

We performed several experiments in order to find the best parameters. In particular, we selected the PV-DBOW model. We trained it for 5 epochs, considering a minimum number of occurrences per word equal to 5. The size of the resulting vectors for representation was set to 200. All the other settings were left to default. In addition, we also considered a representation based on *pretrained word embeddings*. More specifically, we used FastText [7] 300-dimensional embeddings for Italian. Each résumé representation is computed as the average of embeddings for words in the text.

It is quite difficult to evaluate our approach in absence of a gold standard or labelled dataset. Thus, we manually selected two sets of résumés, based on their main skill, for an early evaluation. We trust that the skill can represent a reasonable approximation of a résumé. However, we also analysed the actual text in order to make sure we selected suitable candidates. Due to privacy concerns, no additional information over such data can be provided at this time.

In the first set, we selected similar résumés that have the same main skill "Java Junior". On the contrary, for the second set we selected résumés based on very different skill sets. In particular, we considered "Java Senior", "Help Desk intermediate", "Network engineer junior", "Web designer senior", and "Accounting employee". As it is common in many vector space models approaches, we exploit the *cosine distance* to determine similarity among learned representations (i.e. vectors). Results for the two sets and for the two representations (Doc2Vec and pretrained word embeddings) are reported in Table 1.

The analysis of the results shows that our approach is promising. We recall that the similarity is computed only by analysing the text of the résumés and not exploiting the skills. In fact, two different trends emerge for similar and dissimilar documents. More specifically, we observe that résumés for different "Java Junior" figures tend to have a high cosine similarity, while résumés for different professional figures obtain very low cosine similarity among them. On the contrary, when using the pretrained word embeddings representation, cosine similarity is high and very close to each other, thus highlighting that this representation is not very suitable for discriminating different profiles. This may be due to the fact that, whereas word embeddings are trained on general purpose corpora (i.e. Wikipedia), the Doc2Vec representation is learned directly from our domain-specific data. Moreover, in order to validate our approach, we manually evaluated our results, both for keyword extraction and similarity among résumés, with the help of experienced recruiters from the partner company. It appears that our approach can yield good results. However, it is clear that we presented a simple and tentative evaluation of our method, and further analysis is definitely needed. Nonetheless, the underlying idea appears to be promising.

Finally, with the aim of verifying whether the approach described so far is able to discriminate among curricula with different competences and skills, we applied a complete-linkage hierarchical clustering algorithm. We executed the clustering algorithm using both the Doc2Vec and the pretrained embeddings representations. Figures 3 and 4 show the two dendrograms obtained by the two representations, respectively. We can appreciate how actually different clusters

Table 1. Cosine similarities between *Java Junior* profiles and between different profiles using Doc2Vec (a–b) and pretrained word embeddings (c–d).

(b)

Main Skill	Java Sr.	Help Desk	Network Eng jr.	Web Design sr.	Accounting
Java Sr.	1	#	#	#	#
Help Desk	-0,01	1	#	#	#
Network Eng jr.	0,35	-0,01	1	#	#
Web Design sr.	0,43	-0,10	0,50	1	#
Account. Employee	0,60	-0,08	0,27	0,52	1

(a)

ID	1	2	3	4	5
1	1	#	#	#	#
2	0,65	1	#	#	#
3	0,66	0,96	1	#	#
4	0,70	0,95	0,94	1	#
5	0,51	0,53	0,58	0,61	1

(d)

Main Skill	Java Sr.	Help Desk	Network Eng jr.	Web Design sr.	Accounting
Java Sr.	1	#	#	#	#
Help Desk	0.96	1	#	#	#
Network Eng jr.	0.96	0,96	1	#	#
Web Design sr.	0,93	0,91	0,91	1	#
Account. Employee	0,93	0,97	0,94	0,91	1

(c)

ID	1	2	3	4	5
1	1	#	#	#	#
2	0,77	1	#	#	#
3	0,91	0,78	1	#	#
4	0,91	0,83	0,97	1	#
5	0,91	0,84	0,96	0,97	1

of curricula can be identified. Due to space limits, we cannot discuss in detail these clusters and how they can be characterized in terms of competences and skills. However, it is evident that the approach proposed is able to determine groups of curricula with similar characteristics. We can observe, however, that the partitions generated by using Doc2Vec are more homogeneous than the ones generated by employing pretrained embeddings representation. For instance, if we cut the dendrograms at distance 10, we can observe that for the pretrained embeddings representation several clusters contain less than 10 elements, and two macro clusters contain a number of résumés up to around 3500. Conversely, the clustering produced with Doc2Vec at the same distance is more homogeneous. Résumés and keywords are more evenly distributed across clusters.

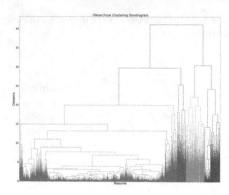 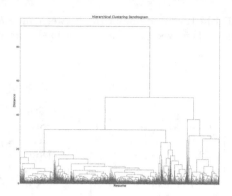

Fig. 3. Doc2Vec representation clustering.   **Fig. 4.** Pretrained embeddings clustering.

Further, we analysed a number of clusters for the two representations. We realized that the clusters generated by using pretrained word embeddings are generally less relevant to the recruiter. Just to give an example, using Doc2Vec we obtained for example a cluster with frequent keywords such as "web service", "TCP IP", "Active Directory", "MYSQL", "web application", that clearly describe the profile of a *web developer*. On the other hand, by analysing clusters obtained with pretrained word embeddings, they appear to be less descriptive. For example, one cluster contains the following frequent keywords: : "C", "Java", "Photoshop", "Google Analytics", "e-mail", "marketing", "assistente amministrativo" (administrative assistant). It is clear that such cluster cannot be considered as a proper representation for a given profile.

5   Conclusions

In this paper, we have presented a system for identifying and extracting time-relevant keywords from résumés and job opportunities, and for building semantically relevant representations of them. Our approach is data-driven and uses semi-supervised strategy for improvement. We have presented our approach and provided a preliminary evaluation of its quality. Although a lot of experimentation is still needed and further improvements are needed, we believe that our direction is promising to automatically extract professional profiles by adopting an unsupervised and data driven approach.

Our final goal is to develop tools that can be implemented in a next-generation ATS, that can take full advantage of information extraction and text mining techniques in order to help recruiters in their day-to-day job.

Acknowledgements. This work was partially supported by Tuscany Region in the context of the project TALENT, "FESR 2014-2020". We wish to thank Dr. Alessio Ciardini, CEO of IT Partner Italia, and Filippo Giunti, IT manager of IT Partner Italia, for providing us with the data and for the invaluable support and insight, both theoretical and practical, on the recruiting field from a company perspective.

References

1. Bengio, Y., Ducharme, R., Vincent, P., Janvin, C.: A neural probabilistic language model. J. Mach. Learn. Res. **3**, 1137–1155 (2003)
2. Bojanowski, P., Grave, E., Joulin, A., Mikolov, T.: Enriching Word Vectors with Subword Information. arXiv e-prints arXiv:1607.04606
3. Church, K.W., Hanks, P.: Word association norms, mutual information, and lexicography. Comput. Linguist. **16**(1), 22–29 (1990)
4. Dumais, S.T.: Enhancing Performance in Latent Semantic Indexing (LSI) Retrieval (1992)
5. Eckhardt, A., Laumer, S., Maier, C., Weitzel, T.: The transformation of people, processes, and IT in e-recruiting: Insights from an eight-year case study of a German media corporation. Empl. Relat. **36**(4), 415–431 (2015)
6. Evert, S.: Corpora and collocations. In: Lüdeling, S., Kytö, M. (eds.) Corpus Linguistics. An International Handbook, article 58, pp. 1212–1248 (2008)
7. Grave, E., Bojanowski, P., Gupta, P., Joulin, A., Mikolov, T.: Learning word vectors for 157 languages. In: Proceedings of the International Conference on Language Resources and Evaluation (LREC 2018) (2018)
8. Harris, Z.S.: Distributional structure. Word **10**, 146–62 (1954)
9. Heggo, I.A., Abdelbaki, N.: Hybrid information filtering engine for personalized job recommender system. In: Hassanien, A.E., Tolba, M.F., Elhoseny, M., Mostafa, M. (eds.) AMLTA 2018. AISC, vol. 723, pp. 553–563. Springer, Cham (2018). https://doi.org/10.1007/978-3-319-74690-6_54
10. Laumer, S., Maier, C., Eckhardt, A.: The impact of business process management and applicant tracking systems on recruiting process performance: an empirical study. J. Bus. Econ. **85**, 421–453 (2014)
11. Le, Q., Mikolov, T.: Distributed representations of sentences and documents. In: Proceedings of the 31st International Conference on Machine Learning (ICML 2014), vol. 32 (2014)
12. Mikolov, T., Chen, K., Corrado, G., Dean, J.: Efficient estimation of word representations in vector space. In: ICLR Workshop (2013)
13. Mikolov, T., Sutskever, I., Chen, K., Corrado, G., Dean, J.: Distributed representations of words and phrases and their compositionality. In: Proceedings of the 26th International Conference on Neural Information Processing Systems (NIPS 2013), vol. 2 (2013)
14. Peters, M.E., et al.: Deep contextualized word representations. arXiv e-prints arXiv:1802.05365
15. Shannon, C.E.: A mathematical theory of communication. Bell Syst. Tech. J. **27**, 379–423 (1948)
16. Shehu, V., Besimi, A.: Improving employee recruitment through data mining. In: Rocha, Á., Adeli, H., Reis, L.P., Costanzo, S. (eds.) WorldCIST 2018. AISC, vol. 745, pp. 194–202. Springer, Cham (2018). https://doi.org/10.1007/978-3-319-77703-0_19

Retrieving and Processing Information from Clinical Algorithm via Formal Concept Analysis

Aleksandra Vatian[1]([⊠]), Anna Tatarinova[2], Svyatoslav Osipov[1],
Nikolai Egorov[1], Vitalii Boitsov[1], Elena Ryngach[2],
Tatiana Treshkur[2], Anatoly Shalyto[1], and Natalia Gusarova[1]

[1] ITMO University, 49 Kronverksky pr., 197101 St. Petersburg, Russia
alexvatyan@gmail.com
[2] Almazov National Medical Research Centre,
2 Akkuratova str., 197341 St. Petersburg, Russia

Abstract. Information technologies play an invaluable role in improving the quality of medical care. New diagnostic technologies based on processing large medical data sets are being actively introduced into clinical practice, which, in turn, generates the new arrays of data. The natural solution here is the clinical information systems (CIS) which help the physicians in clinical decision making as well as in training and research. In this paper we propose FCA approach for retrieving and processing information suitable for using in CIS, first of all in the aspect of knowledge, from clinical algorithm. FCA in proposed form represents the clinical process as a sequence of decisions being made by the physician during the diagnosis (or treatment) in order to reach the desired diagnostic (or respectively treatment) state. In order to analyze the effectiveness of these decisions in relation to a specific clinical process we propose the necessary indices and metrics. The application of the proposed formalisms is illustrated by the example of a real clinical process.

Keywords: Clinical information systems · Formal concept analysis · Clinical process

1 Introduction

Information technologies play an invaluable role in improving the quality of medical care. New diagnostic technologies based on processing large medical data sets are being actively introduced into clinical practice, which, in turn, generates the new arrays of data. Because of the massive volume, variety, and continuous updating of medical data it becomes increasingly difficult to consider them fully when working with an individual patient both at the stage of diagnosis and at the stage of treatment.

The natural solution here is the clinical information systems (CIS) which help the physicians in clinical decision making as well as in training and research. Progress in the development of CIS is associated primarily with the involvement of high-quality information sources [1, 2]. As literature analysis shows, the division of these systems into knowledge-based and non-knowledge-based in recent years has shifted toward the

© Springer Nature Switzerland AG 2019
H. Yin et al. (Eds.): IDEAL 2019, LNCS 11871, pp. 166–174, 2019.
https://doi.org/10.1007/978-3-030-33607-3_19

second, i.e. machine-learning approach (see, for example, [3]). However, this approach does not cope well with such factors as accessibility, urgency, social factors, the specifics of the country and a particular medical institution. In this sense, taking into account the actual experience of practical doctors, which is accumulated in such documents as clinical guidelines, algorithms and protocols, can play an invaluable role. It can be said that these documents represent a definite path in the symptom space to the subset characterizing a particular disease, this path being optimal from the point of view of the medical community and confirmed by clinical practice.

The effectiveness of using clinical protocols, algorithms and guidelines in clinical practice is shown in numerous studies [4, 5]. It seems natural to use them as information sources for CIS. However, as practice shows, it is impossible to directly use them in decision support systems: they are written in natural language; they do not separate factual and conceptual information (data and knowledge); they are not checked for strict compliance with the rules of formal logic; they do not have a procedure for checking the performance of the relevant personnel.

Thus, there is the problem of retrieving and processing information suitable for using in CIS, first of all in the aspect of knowledge, from clinical algorithm. To solve the problem posed, we propose to use such formalism as a formal concept analysis (FCA). The problem is considered on a real example of the algorithm for the management of patients with stable coronary artery disease and high-grade ventricular arrhythmias [6].

2 Background and Related Works

The FCA formalism [7] represents the basic concepts of the domain using the theory of lattices, and its basic construct is the formal context, defined as a triple

$$K = (G, M, I), \tag{1}$$

where G is a set of objects, M is a set of attributes, and I is a binary relation such that $I \subseteq G \times M$. Given a context (3), one can define a formal concept (FC)

$$c = (A, B), \tag{2}$$

by a couple of the extent $A \subseteq G$, and the intent $B \subseteq M$, such that $A' = B$ and $B' = A$, where $A' = \{m \in M | \forall g \in A : (g, m) \in I\}$, $B' = \{g \in G | \forall m \in B : (g, m) \in I\}$. Thus, the FCA formalism is considered to represent knowledge as a lattice of concepts and conceptual relationships concerning the certain domain.

Although FCA was originally considered as a construct for displaying a purely taxonomic structure, at present its capabilities are expanded. One can describe such complex domain structures as the implication of features [8], the presence of characteristic patterns in graphs and sequences [9], as well as multivalued, fuzzy and probabilistic dependencies [10–12]. In order to select concept subsets within the whole concept lattice various indices are proposed based on thresholding of concept extent, on modifying the closure operator, on involving background knowledge etc. [13].

However, as emphasized in [13], their effectiveness depends critically on the content of the task, i.e. the developer is required a creative approach to their formation.

FCA has found quite wide application for describing dependencies in medical domain. In [11, 14, 15] FCA is used for mining association rules and analogical proportions among diseases based on processing the relevant datasets. [16] use FCA to identify and validate mappings across medical ontologies. In [12] for a better understanding of the structure of the subject area, it is proposed to build conceptual landscapes based on FCA. In [17], by means of FCA and big data approach, the accuracy of patient classification by disease is improved, which improves predictions about patient survival and mortality. Authors [9] seek the characteristic sequential pattern structures in hospitalization statistics for patients.

At the same time, FCA as a means of describing relationships between cognitive constructs, for example decisions, that reflect characteristic paths in the feature space and are of most interest from the point of view of knowledge representation, is not considered even at the level of problem statement. Thus, in the article we pose the following questions:

- how to form a conceptual lattice representing the clinical decision-making sequence from the description of the clinical process in the form provided by clinicians;
- which indices can help to analyze the effectiveness of these decisions in relation to a specific clinical process.

3   Proposed Approach

The motivation of this work, namely the algorithm for the management of patients with stable coronary artery disease and high-grade ventricular arrhythmias (VA) [6], is presented in Fig. 1. As our analysis [18] has shown, a system of standards [19 Its main entities are: health state (clinical state), HS; health condition (clinical symptoms), HC; healthcare activity (clinical activity), HA. In parentheses interpretations of the concepts as applied to a clinical process considered are given. Applying [19] to the algorithm we built a formalized description of the algorithm which was used as a source for constructing an FCA conceptual lattice. A fragment of the latter is presented in Table 1, and the full version is available by reference[1].

A content analysis of the clinical process shows that, of the above set of entities, HA best represent doctor's knowledge when making a decision and the HC form the information space in which these decisions are made. Since our task is to determine the sequence of decision-making by a doctor regarding treatment, we, with the expert help of doctors, have substantively identified those HA that relate to treatment; they are set apart into a subset of the final HA (FHA) located at the top of the Table 2. The rest of the activities refer to examinations and form a subset EHA.

---

[1] https://docs.google.com/spreadsheets/d/1G51nZpojzFp5dr8tay3YdLV71W1Vajdz-LVUOrYgF9M/edit?usp=sharing.

Table 1. Fragment of concept lattice

Health Activity (Final)	Prevalence time	HC1.1.0 no complaints	HC1.1.1 heartbeat complaints	HC1.1.2 heart dysfunctionality	HC1.1.3 pre-existing conditions/ Insensibility	HC1.1.4 anginal pains	HC1.2-1 Specialist's decision to continue	HC1.2-0 Specialist's decision not to continue	HC2.1-1 VA is detected	HC2.1-2 VA is not detected	HC2.2-1 CAD is detected	HC2.2-2 CAD is not detected	HC3-1 LVEF >= 40% is detected	HC3-2 LVEF >= 40% is not detected
FHA0 Occurence is not confirmed		1						1						
FHA1 Algorithm for noncoronary VA			1	1	1	1	1		1			1		1
FHA2 MR (Ischemic)			1	1	1	1	1		1		0.7		0.7	
FHA3 MR (Nonischemic)			1	1	1	1	1		1		0.7		0.7	
FHA4 CAD, AAT, CA, ICD			1	1	1	1	1		1		0.7		0.7	
FHA5 (Ischemic) BAB, AAT, CA, ICD			1	1	1	1	1		1		0.7		0.7	
FHA6 (Nonischemic) BAB, AAT, CA, ICD			1	1	1	1	1		1		0.7		0.7	
FHA7 (Ischemic) Psychotherapy			1	1	1	1	1		1		0.7		0.7	
FHA8 (Nonischemic) Psychotherapy			1	1	1	1	1		1		0.7		0.7	
Prevalence time		7000	2000	1000	2500	1500	2000	8000	1000	1000	500	500	200	800
Health Activity (Diagnostics)														
EHA1 Clinical history analysis	0.1	1	1	1	1	1	1	1						
EHA2.1 ECG, ECG monitoring	4						1	1	1	1	1			
EHA2.2 Laboratory tests, physical tests	24						1	1	1	1	1	1		
EHA3 Echocardiography	4												1	1
EHA4.1 ECG-test: TT, Bicycle testing	4												1	

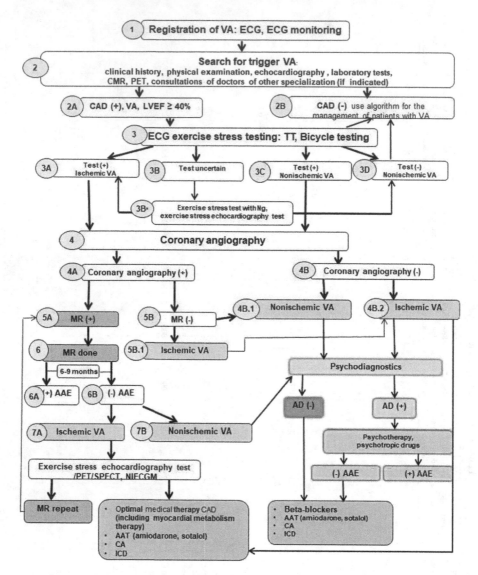

Fig. 1. The original form of algorithm for the management of patients with stable coronary artery disease and high-grade ventricular arrhythmias: AAE – Antiarrhythmic effect; AD – Anxiety disorder; CMR – Cardiac magnetic resonance; CA – Catheter ablation; CAD – Coronary artery disease; ECG – Electrocardiogram; ICD – Implantable cardioverter defibrillator; LVEF – Left ventricular ejection fraction; MR – Myocardial revascularization; Ng – Nitroglycerin; NIECGI – Noninvasive electrocardiographic mapping; PET – Positron emission tomography; SPECT – Single-photon emission computed tomography; TT - Treadmill test; VA – Ventricular arrhythmia

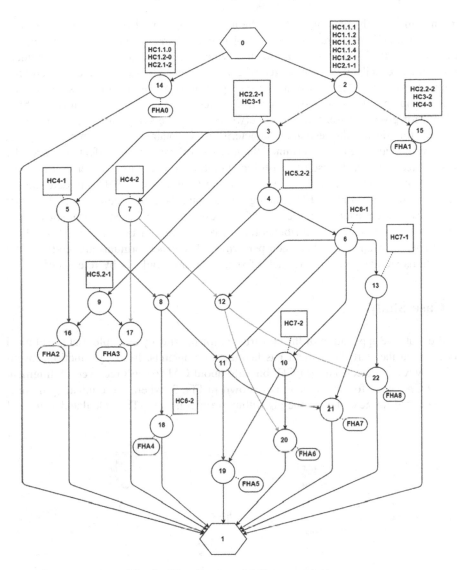

Fig. 2. Visualization of full concept lattice

So, the components of the formulas (1) and (2) acquire the following sense: G are elements of the FHA set, M are elements of the HC set, c is an inclusion-wise maximal subset of HC needed for diagnostics of given element of FHA set. Values of I are given in the range of (0...1), corresponding to the severity of c; the correspondence scale is established by clinician experts. The visualization of the constructed conceptual lattice is shown in Fig. 2.

Since the lattice is built for a specific disease, all the clinical symptoms included in it are required for this disease, but they can also characterize other diseases; in addition, a particular doctor may, when deciding to take into account or not to take into account certain symptoms thereby highlighting the active symptoms. Thus, in accordance with the approach [13], we have formed a set of indices characterizing the specifics of the work of a particular doctor regarding the typical sequence recorded in the clinical algorithm: Reconcilability - the degree of correspondence between active and required symptoms for given disease; Completeness - part of all required symptoms covered by active symptoms; Loss, showing in what degree active symptoms contradict those required for given disease; Surplus - overweight of active symptoms in relation to the required. Introducing the resource characteristics of obtaining each symptom (speed, price, availability), we also determined the metrics for assessing the effectiveness of the doctor's decisions on the use of a specific symptom: Probability - a metric that ranks examination activities EHA by the rate at which final activities FHA are determined; Benefit - a metric that ranks EHA taking into account both the rate at which FHA are determined and the examination costs; Wholeness - a metric based on statistics of the weighted joint occurrence of attributes for a given concept of FCA lattice, shows how fully a specific examination has been performed. The exact formulas and code for the calculations are not given due to lack of space and are available by reference[2].

4   Case Study

We illustrate the application of the constructed formalisms by example. Let the clinical process is at the start: the patient has heartbeat complaints, heart dysfunction, insensibility; VA was detected after ECG, but info about CAD wasn't received. A fragment of the FCA lattice for this situation is shown in Fig. 3, where the clinical symptoms (HC) selected for next step of the algorithm are highlighted. The calculated values of

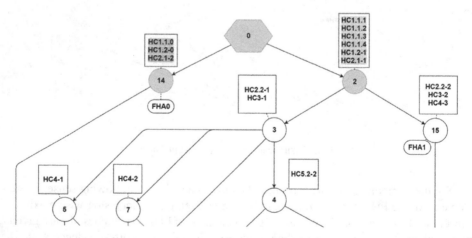

Fig. 3. Fragment of concept lattice with active symptoms highlighted

2 https://docs.google.com/document/d/1ID4Z8nAawXO1TPb1dAegjZ44WlNKBnmKGimtdctfW1s/edit?usp=sharing.

the indices and metrics for FHA and EHA are presented in Tables 2 and 3, respectively.

Analysis of Table 2 shows the following. For FHA0 Loss = 1 which means that the decision "VA is not confirmed" should be excluded from the further consideration. Large value of Surplus suggest that the existing set of clinical symptoms should be associated with other treatments - HAF1 or HAF2; this is also indicated by high Completeness values. At the same time, the high values of Reconcilability for the latest treatments suggest that the symptoms present are quite characteristic for them. Comparative evaluation of data Table 3 allows you to rank possible examination procedures so that they lead to real FHA more quickly and at lower cost.

Table 2. Index values for FHA

	Reconcilability	Completeness	Loss	Surplus
FHA1	1.0000	0.6586	0.0000	0.0000
FHA2	1.0000	0.6265	0.0000	0.0000
FHA0	0.0000	0.0000	1.0000	2.2985

Table 3. Metric values for DHA

	Probability	Wholeness	Benefit
EHA3	0.0899	0.2795	0.0481
EHA3.2	0.5397	0.2779	0.5771
EHA4.1	0.3175	0.2213	0.3395
EHA4.2	0.0529	0.2213	0.0354

5 Conclusion

In the article we have proposed FCA approach for retrieving and processing information suitable for using in CIS, first of all in the aspect of knowledge, from clinical algorithm. FCA in proposed form represents the clinical process as a sequence of decisions being made by the physician during the diagnosis (or treatment) in order to reach the desired diagnostic (or respectively treatment) state. In order to analyze the effectiveness of these decisions in relation to a specific clinical process we have proposed the necessary indices and metrics. The application of the proposed formalisms is illustrated by the example of a real clinical process.

The work was supported by the Government of Russian Federation, Grant 08-08 and Russian Science Foundation, Grant 19-19-00696.

References

1. Gardner, R.M.: Clinical information systems – from yesterday to tomorrow. Yearb. Med. Inform. **25**(Supp 1), S62–S75 (2016). https://doi.org/10.15265/iys-2016-s010

2. Wright, A., Bates, D.W.: Outpatient clinical information systems. In: Key Advances in Clinical Informatics: Transforming Health Care through Health Information, pp. 31–50. Academic Press (2018)
3. Chen, J., Li, K., Ronga, H., Bilal, K., Yang, N., Li, K.: A disease diagnosis and treatment recommendation system based on big data mining and cloud computing. arXiv:1810.07762v1 [cs.LG], 17 October 2018
4. Fischer, F., Lange, K., Klose, K., Greiner, W., Kraemer, A.: Barriers and strategies in guideline implementation—a scoping review. Healthcare 4, 36 (2016). https://doi.org/10.3390/healthcare4030036
5. Plummer, J.M., Newnham, M.S., Henry, T.: Improving the Quality of Care in Surgery: The Role of Guidelines, Protocols, Checklist and the Multidisciplinary Team, 27th March 2019. https://doi.org/10.5772/intechopen.84658. https://www.intechopen.com/onlinefirst/improving-the-quality-of-care-in-surgery-the-role-of-guidelines-protocols-checklistand-the-multidis
6. Ryngach, E.A., Treshkur, T.V., Tatarinova, A.A., Shlyakhto, E.V.: Algorithm for the management of patients with stable coronary artery disease and high-grade ventricular arrhythmias. Ther. Arch. 89(1), 94–102 (2017)
7. Ganter, B., Wille, R.: Formal Concept Analysis. Springer, Heidelberg (1999). https://doi.org/10.1007/978-3-642-59830-2
8. Neto, S.M., Song, M.A.J., Dias, S.M., Zárate, L.E.: Using implications from FCA to represent a two mode network data. In: SEKE (2015). https://doi.org/10.18293/seke201585
9. Buzmakov, A., Egho, E., Jay, N., Kuznetsov, S.O., Napoli, A., et al.: FCA and pattern structures for mining care trajectories. In: Workshop FCA4AI, "What FCA Can Do for Artificial Intelligence?", Beijing, China, August 2013 (2013)
10. Vityaev, E.E., Demin, A.V., Ponomaryov, D.K.: Probabilistic generalization of formal concepts. Program. Comput. Soft. 38, 219 (2012). https://doi.org/10.1134/S036176881205 0076
11. Kumar, C.A.: Fuzzy clustering-based formal concept analysis for association rules mining. Appl. Artif. Intell. Int. J. 26(3), 274301 (2012)
12. Haliţă, D., Săcărea, C.: Is FCA suitable to improve electronic health record systems? In: 2016 24th International Conference on Software, Telecommunications and Computer Networks (SoftCOM) (2016)
13. Kuznetsov, S.O., Makhalova, T.: On interestingness measures of formal concepts. arXiv: 1611.02646v2 [cs.AI], 19 April 2017
14. Keller, B.J., Eichinger, F., Kretzler, M.: Formal concept analysis of disease similarity. In: AMIA Joint Summits Translational Science Proceedings 2012, pp. 42–51, 19 March 2012
15. Săcărea, C., Şotropa, D., Troancă, D.: Using analogical complexes to improve human reasoning and decision making in electronic health record systems. In: Chapman, P., Endres, D., Pernelle, N. (eds.) ICCS 2018. LNCS (LNAI), vol. 10872, pp. 9–23. Springer, Cham (2018). https://doi.org/10.1007/978-3-319-91379-7_2
16. Zhao, M., Zhang, S., Li, W., Chen, G.: Matching biomedical ontologies based on formal concept analysis. J. Biomed. Semantics 9, 11 (2018). https://doi.org/10.1186/s13326-018-0178-9
17. Yuan, Y., Chen, W., Yan, L., Huang, B., Li, J.: A similarity-based disease diagnosis system for medical big data. J. Med. Imaging Health Inform. 7(2), 364–370 (2017)
18. Vatian, A., Dudorov, S., Chikshova, E., et al.: Design patterns for personalization of healthcare process. In: 2nd ICSSE-International Conference on Software and Services Engineering, 15–17 March 2019
19. ISO 13940-2016 Health informatics - System of concepts to support continuity of care (ISO 13940:2015)

Comparative Analysis of Approaches
to Building Medical Dialog Systems in Russian

Aleksandra Vatian[✉], Natalia Dobrenko, Nikolai Andreev,
Aleksandr Nemerovskii, Anastasia Nevochhikova,
and Natalia Gusarova

ITMO University, Kronverksky pr. 49, Saint Petersburg, Russia
alexvatyan@gmail.com

Abstract. Nowadays dialog systems have great promise in the field of medicine and healthcare. Only a few medicine dialog systems presented in the literature are accompanied by an experimental effectiveness evaluations. Moreover, they cover only English-speaken context. In this paper we set the task to conduct a comparative analysis of the effectiveness of Russian-language mixed-initiative medical dialog systems, depending on the chosen architecture and method of processing the users' intentions. We have developed and compared three types of chat-bots; Frame-based, ML-based and Ontology-based. As the metrics used the accuracy of the intent recognition as the percentage of intents correctly defined by the system relative to the total number of processed utterances. The results show that the accuracy of the intent recognition in all three approaches is quite the same, so finally we propose the architecture of s combined dialog system which covers all the needs for Russian-language medical domain.

Keywords: Dialog systems · Chatbot · Medical systems

1 Introduction

The rapid development of dialog systems is one of the phenomena of modern life. Smart conversational agents in different languages [1] such as Apple Siri, Google Now, Amazon Alexa, Yandex Alice were launched. Many large IT companies offer their services for creating chatbots based on natural language processing and machine learning, such as IBM Watson Conversation, Dialogflow etc.

Dialog systems have great promise in the field of medicine and healthcare [2]. They can perform not only routine tasks such as appointment to the doctor or tracking patient behavior, but also serve as valuable assistants, helping patients and elderly individuals in their living environments and in solving their healthcare problems [2–6]. An analysis of the literature [3, 7, 8] shows a large terminological variation in this area. In the following, the terms "dialog system" and "chatbot" are used equally.

Of particular interest in medicine and healthcare are the systems managing mixed-initiative [7, 8] dialog. The difficulties here are related to the fact that the dialog takes into account not only explicit, but also implicit interaction. A user interacting with a system can have many different kinds of intentions that can change as the dialog

© Springer Nature Switzerland AG 2019
H. Yin et al. (Eds.): IDEAL 2019, LNCS 11871, pp. 175–183, 2019.
https://doi.org/10.1007/978-3-030-33607-3_20

proceeds [4]. Besides, dialog organization is based on context, i.e. on its circumstances (external and internal), which may change during the course of discourse.

A qualitative leap in the construction of the mixed-initiative dialog systems occurred with the introduction of the machine learning paradigm into this area. However, if the mixed-initiative dialog system should work in a narrow domain and solve specific tasks, the advantages of such architecture change are not so obvious. According to the survey [3] of English-language medical dialog systems, the rule based and frame-based architectures retain their positions here. As far as the authors of this article are aware, similar studies for Russian-language medical dialog systems were not conducted at all. In this regard, comparative analysis of approaches to building medical dialog systems in Russian is of great interest.

2 Background and Related Works

The practice of modern chatbot developers [9–11] allows selecting the following basic artifacts that determine the processing of each utterance and the dialog as a whole: intent, entity, and dialog tree. Intent represents the intention, or the purpose of a user's utterance; entity represents a term or object that is relevant to utterance's intents and provides a specific context for an intent [12]. The dialog tree encodes conversation flow variants in the form of transitions between intents representing each utterance [9]. These definitions show how large a role in the creation of chatbot belongs to natural language processing (NLP). As for the field of medicine, a number of problems arise here, which are the subject of active study.

The first one is medical entities recognition and resolution. In recent years, various solutions based on the machine learning paradigm have been proposed in this area [13, 14]. However, depending on the specific task, well-known methods, such as dictionaries and conditional random fields, remain effective here [11, 15, 16]. Note that the results of solving the problem of medical entities recognition are language dependent. Our research [17] has shown that for medical texts in Russian, a sufficiently high quality of its solution is achieved without the use of external lexical resources or n-gram algorithms. At the same time, processing the text in English and its Russian-language translation leads to the selection of non-identical concepts.

One more problem characteristic for medical dialog systems is the following: many intents can be elicited from one utterance depending on the context interpretation. To solve it, a number of solutions have been proposed that, apply knowledge graphs and vocabularies [18] or use Deep Learning supervised classifiers [19]. Ready-made services[1] like LUIS or wit.ai for embedding in existing architectures are offered, but they remain black boxes for developers. Besides, LUIS does not support Russian.

An important role in building a dialog with the user in medical chatbots is played by tracking his emotional state, i.e. sentiment analysis. In order to do this different solutions have been proposed, starting with vocabulary techniques [20] and ending

---

[1] https://www.luis.ai, https://wit.ai/.

with NLP and machine learning capabilities of ready-made services like DialogFlow[2] [21, 22]. Nevertheless, as the study [23] shows, the proposed implementations are far from perfect: 72% of surveyed physicians believe that medical chatbots treat patients' emotions poorly.

A lot of chatbots for healthcare are announced in the market (see, for example, reviews [2, 5, 6, 24]). An overview of the solutions used shows that various types of architecture are represented in the zoo of medical dialog systems. But, from the point of view of NLP implementation, it makes sense to distinguish, on the one hand, the ontological approach [25, 26], and on the other hand, the approach of ready-made machine learning services, among which DialogFlow dominates. The only review found in the literature [3] analyzes cumulative effect of human users interacting with the system as a whole, which is useful for end users. In addition, it covers only English-speaking medical systems, and its results, as shown above, cannot be directly transferred to other languages, including Russian. At the same time, chatbot developers are much more interested in evaluating individual components of the medical dialog system, primarily language-dependent and concerning natural language understanding. Therefore the article sets the task to conduct a comparative analysis of the effectiveness of Russian-language mixed-initiative medical dialog systems, depending on the chosen architecture and method of processing the users' intentions.

The rest of the paper is organized as follows. Section 3 describes the architecture of the dialog systems designed for the investigation, as well as the methodology of the experiment. Section 4 presents and discusses the results of the experiment. Section 5 formulates the conclusion on the work.

3 Methods and Materials

To fulfill the tasks set in the article, we have created three medical dialog systems of mixed-initiative type designed to interact with users in the natural language, namely in Russian, implemented on the machine learning, frame-based and ontological approaches respectively. For the development of the ML-based dialog systems, the Dialogflow service owned by Google was chosen having the built-in NLP toolkit. Currently, support is provided for 20 languages, including Russian.

In addition to the purely framework approach, in the experiments we used the ontological approach, since they are the useful tool of NLP in a narrow domains.

The implementation of sentiment analysis of the utterance depends on the approach used. Namely, in Dialogflow, there are no tools for sentiment analysis of texts in Russian, and one has to create intents that characterize not only intention of the utterance, but it's tonality as well. With the frame-based approach, we used libraries[3] for sentimental analysis of texts in Russian.

One of the main problems in the development of chat bots is the way to form and store a dialog tree (that is, a script for the interaction between the bot and the user). We

---

[2] https://cloud.google.com/dialogflow/.

[3] https://github.com/thinkroth/Sentimental, https://github.com/bureaucratic-labs/dostoevsky.

use the approach of sparse trees [9], in which the dialog as a whole is formed from separate relatively independent mini-conversations that are combined into a common tree. In the frame-based and ontology-based approaches, the dialog trees are built by the developers themselves in the form of pre-written scripts. In the case of the ML approach, the selected Dialogflow platform allows, as mentioned earlier, to build context-based chains of intents that can be interpreted as dialog scenarios.

The selection of intents was made by experts. In order to do this, specialized forums on the Internet were selected, corresponding to the topics of each chatbot developed (see Table 1). The experts, based on the analysis of the content of the forums, formulated the basic intents that were directly used in the ML-based chatbot. For the frame-based and ontology-based chatbots, regular expressions were built on the basis of these words, which makes it possible to allocate several intents for each utterance and improves the effectiveness of the dialog system.

Table 1. Formation of content for experiments.

Chatbot	Topics	Number of forums explored	Number of intents highlighted
Frame-based	Advice on exercise for people with multiple sclerosis	3	7
ML-based	Support for people with bipolar affective disorder	6	14
Ontology-based	Advice on removing buccal fat pads	5	15

In order to highlight entities, in frame-based and ontology-based chatbots NLP tools were used: each statement of the utterance is divided into tokens, cleared from stop words and marked up using automatic morphological markup (POS tagging), and then each utterance is analyzed for the presence of entities of interest using a regular expression parser or processing unigram. For these purposes, we used the appropriate libraries from the NLTK package[4].

A diagram of the developed ML-based and frame-based dialog systems are presented in Figs. 1, 2, respectively. For the ontology-based dialog system, the ontology of the corresponding domain was constructed, which can be seen in Fig. 3. The ontology is presented in Russian with a translation into English for a more complete understanding. In this case, ontology is used as a repository of knowledge, but in general it can be associated with external ontologies for more complete and personalized answers.

In the literature, a number of metrics is proposed that can be used to assess the effectiveness of dialog systems in general [7] and medical dialog systems, in particular [3]. In accordance with the objectives of our work, we relied on metrics proposed in [27] measuring modifications of the dialog attributes that could cause the failure of the

---

[4] nltk.tokenize, nltk.tag, nltk.RegexpParser, nltk.UnigramTagger.

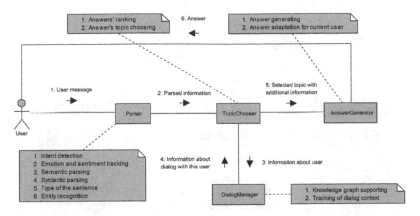

Fig. 1. The scheme of the ML-based dialog system.

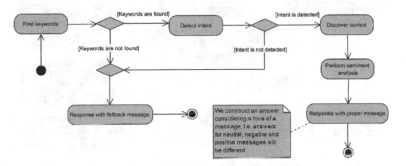

Fig. 2. The scheme of the frame-based dialog system.

dialog as a whole. In our case, the attribute of an individual utterance is considered as an attribute; accordingly, we measure the accuracy of the intent recognition as the percentage of intents correctly defined by the system relative to the total number of processed utterances.

The experiment was organized as follows. Similar to the stage of learning chat bots, specialized forums on the Internet were selected, corresponding to the topics of each chatbot. Posts of the forum containing at least 20 words were selected; each such post was considered as an utterance. For each utterance, intentions were marked up by two experts. In the case of a disagreement, a third expert was involved. In parallel, each utterance was processed by the corresponding chatbot. According to the results obtained, the above metric was calculated.

4 Results and Discussion

The results of the experiment are presented in the Table 2. As can be seen from the table, the accuracy of determining intentions in all cases is not so high, which is quite an expected result, given the experimental nature of the developed systems, the narrow domains processed and the complexity of NLP in Russian. At the same time, attention

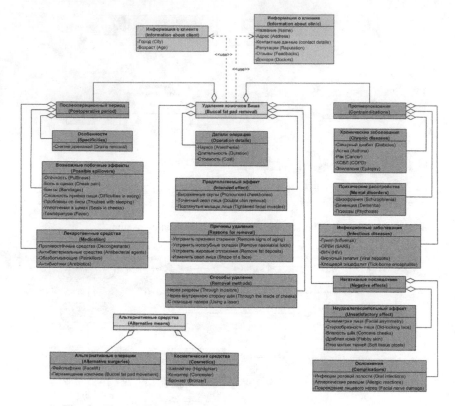

Fig. 3. The ontology scheme for the ontology-based dialog system.

Table 2. Accuracy of intent determination by different dialog systems.

Chatbot	Utterances in total	Number of correctly defined intents	Accuracy of intent recognition
Frame-based	23	16	69.6%
ML-based	50	32	64%
Ontology-based	64	43	67,2%

is drawn to the fact that none of the considered dialog systems showed a clear advantage, which is consistent with the conclusions made in [3].

A detailed analysis of the results made it possible to identify problems characteristic of each of the approaches. In particular, for DialogFlow approach in its pure form, the following shortcomings were identified:

- there is no possibility to set several intents in order to form a complex response for long utterances with compound intents which are characteristic for medical dialogs;
- there are no built-in tools for sentiment analysis in Russian, unlike English, which forces developers to create separate intents for each tonality;

- there is no possibility to make changes to the mechanism of the dialog system, except for adding new examples of utterances;
- due to a narrow domain there are fundamental data limitations and small datasets available which obstruct training the model up to the industrially needed level

On the other hand, the development of frame-based and ontology-based dialog systems also revealed a number of problems:

- there are linguistic-specific problems in Russian associated with linguistic analysis, in particular, with the resolution of the anaphora;
- a large variation of formulations is revealed, which requires the development of a large number of patterns to cover them

To solve the identified problems, it seems appropriate to combine the considered approaches in a single dialog system. The architecture of such a system developed by the authors is presented in Fig. 4.

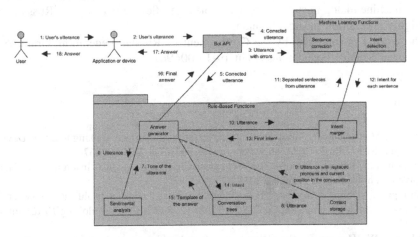

Fig. 4. The proposed architecture of combined dialog system

In this architecture there are two subsystems. The main subsystem, based on the rules, works directly with the utterance, namely: determines its tonality, compares the utterance with the current context, receives the intent of the entire utterance, restores the current place in the dialog tree and generates a suitable response. The auxiliary subsystem that uses machine learning corrects linguistic errors in the utterance that can lead to incorrect analysis, and determines the intent for individual sentences in the utterance for their further processing in the main subsystem.

The combined dialogue system created was experimentally tested on the entire content of Table 1 in accordance with the above described procedure. The average accuracy of intent recognition was $78 \pm 1, 8\%$ depending on topics, the best results were obtained for topics with more simple Russian vocabulary. The resulting increase in the efficiency of intents recognition for such a complex and flexive language as

Russian, and in a highly specialized medical domain seems quite convincing. Nevertheless, the authors plan to consider other combined architectures of dialog systems for this domain.

5 Conclusion

We have set the task to conduct a comparative analysis of the effective-ness of Russian-language mixed-initiative medical dialog systems, depending on the chosen architecture and method of processing the users' intentions. We have developed and compared three types of dialog systems; Frame-based, ML-based and Ontology based. As the metrics used the accuracy of the intent recognition as the percentage of intents correctly defined by the system relative to the total number of processed utterances. The results show that the accuracy of the intent recognition in all three approaches is quite the same – 64–70%. So based on these results we have developed combined dialog system which is based on two subsystems, that makes it possible to provide an effective human-machine dialog in such a rather complex and flexive language as Russian, and in a highly specialized medical domain.

The work was supported by the Government of Russian Federation, Grant 08-08 and Russian Science Foundation, Grant 19-19-00696.

References

1. McTear, M., Callejas, Z., Griol, D.: The Conversational Interface: Talking to Smart Devices. Springer, Switzerland (2016). https://doi.org/10.1007/978-3-319-32967-3
2. Farkash, Z.: Medical Chatbot—The 4 Greatest Challenges Medical Institutes Are Facing, Solved with Chatbots, 19 August 2018
3. Laranjo, L., Dunn, A.G, Tong, H.L.: Conversational agents in healthcare: a systematic review. J. Am. Med. Inform. Assoc. 25(9), 1248–1258 (2018). https://doi.org/10.1093/jamia/ocy072
4. White, R.W., Horvitz, E.: From health search to healthcare: explorations of intention and utilization via query logs and user surveys. J. Am. Med. Inform. Assoc. 21(1), 49–55 (2014)
5. Senaar, K.: Chatbots for Healthcare – Comparing 5 Current Applications. https://emerj.com/ai-application-comparisons/chatbots-for-healthcare-comparison/. Accessed 10 Feb 2019
6. Virtualspirits. Chatbot for Healthcare. https://www.virtualspirits.com/chatbot-for-healthcare.aspx. Accessed 07 June 2019
7. Jurafsky, D., Martin, J.H.: Speech and Language Processing: An Introduction to Natural Language Processing, Computational Linguistics, and Speech Recognition, 2nd edn. Prentice Hall, Englewood Cliffs (2008)
8. de Bayser, M.G., Cavalin, P., Souza, R., et al.: A Hybrid Architecture for MultiParty Conversational Systems (2017). https://arxiv.org/abs/1705.01214
9. Lambert, M.: Chatbot Decision Trees. Seriously, how hard can they be? https://chatbotsmagazine.com/chatbot-decision-trees-a42ed8b8cf32. Accessed 21 Apr 2018
10. Churilov, M. Conversational UI: Writing Chatbot Scripts Step by Step. https://chatbotslife.com/conversational-ui-writing-chatbot-scripts-step-by-stepa78b611a5eba. Accessed 8 Jan 2018

11. Paul, M.S.: How to build a Google Home App with DialogFlow. Conversation Design. https://medium.com/swlh/chapter-7-how-to-build-a-google-home-app-withdialogflow-conversation-design-63d6a1402ed0. Accessed 13 Aug 2018
12. Jakuben, B.: Intents, Entities, and Dialogs. https://teamtreehouse.com/library/intents-entities-and-dialogs. Accessed 3 June 2019
13. Liu, Z., et al.: Entity recognition from clinical texts via recurrent neural network. BMC Med. Inform. Decis. Mak. **17**(Suppl 2), 67 (2017)
14. Wei, Q., Chen, T., Xu, R., He, Y., Gui, L.: Disease named entity recognition by combining conditional random fields and bidirectional recurrent neural networks. Database (Oxford) (2016). https://doi.org/10.1093/database/baw140
15. Quimbaya, A.P., Munera, A.S., Rivera, R.A.G. et al.: Named entity recognition over electronic health records through a combined dictionary-based approach. Procedia Comput. Sci. **100**, 55–61 (2016)
16. Siu, A.: Knowledge-driven Entity Recognition and Disambiguation in Biomedical Text. Saarland University, Saarbrücken, May 2017. https://pure.mpg.de/rest/items/item_2475275_3/component/file_2475274/content. Accessed 03 June 2019
17. Vatian, A., et al.: Intellectualization of knowledge acquisition of academic texts as an answer to challenges of modern information society. Commun. Comput. Inf. Sci. **947**, 138–153 (2019)
18. Zhang, Ch., Du, N., Fan, W., Li, Y., Lu, C.-T., Yu, P.S.: Bringing Semantic Structures to User Intent Detection in Online Medical Queries. arXiv:1710.08015v1 [cs.CL], 22 October 2017
19. Amatriain, X.: NLP & Healthcare: Understanding the Language of Medicine. https://medium.com/curai-tech/nlp-hcalthcare-understanding-the-language-ofmedicine-e9917bbf49e7. Accessed 5 Nov 2018
20. Dharwadkar, R., Deshpande, N.A.: A medical ChatBot. Int. J. Comput. Trends Technol. (IJCTT) **60**(1), 39 43 (2018)
21. Fadhil, A.: Beyond patient monitoring: conversational agents role in telemedicine & healthcare support for home-living elderly individuals. https://arxiv.org/pdf/1803.06000
22. Oyebode, O.O., Orji, R.: Likita: a medical chatbot to improve healthcare delivery in Africa. www.hcixb.org/papers_2018/hcixb18-final-22.pdf
23. Palanica, A., Flaschner, P., Thommandram, A., Li, M., Fossat, Y.: Physicians' perceptions of chatbots in health care: cross-sectional web-based survey. J. Med. Internet Res. **21**(4), e12887 (2019). https://doi.org/10.2196/12887
24. Divya, S., Indumathi, V., Ishwarya, S., Priyasankari, M., Kalpana, D.S.: Survey on medical self-diagnosis chatbot for accurate analysis using artificial intelligence. Int. J. Trend Res. Dev. **5**(2)
25. Kazi, H., Chowdhry, B.S., Memon, Z.: MedChatBot: an UMLS based chatbot for medical students. Int. J. Comput. Appl. **55**(17), 1–5 (2012). (0975–8887)
26. Tielman, M., van Meggelen, M., Neerincx, M.A., Brinkman, W.-P.: An ontology-based question system for a virtual coach assisting in trauma recollection. In: Brinkman, W.-P., Broekens, J., Heylen, D. (eds.) IVA 2015. LNCS (LNAI), vol. 9238, pp. 17–27. Springer, Cham (2015). https://doi.org/10.1007/978-3-319-21996-7_2
27. Griol, D., Carbo, J., Molina, J.M.: An automatic dialog simulation technique to develop and evaluate interactive conversational agents. Appl. Artif. Intell. **27**(9), 759–780 (2013)

Tracking Position and Status of Electric Control Switches Based on YOLO Detector

Xingang Mou[1(✉)], Jian Cui[1], Hujun Yin[2], and Xiao Zhou[1]

[1] School of Mechanical and Electronic Engineering,
Wuhan University of Technology, Wuhan 430070, China
{sunnymou, cuijian123, zhouxiao}@whut.edu.cn
[2] School of Electrical and Electronic Engineering,
The University of Manchester, Manchester M13 9PL, UK
hujun.yin@manchester.ac.uk

Abstract. Tracking position and status of switches of electric control cabinets is the key to automatic polling and management systems for intelligent substation. It is a typical task of multi-target image detection and recognition. In this paper, we present an end-to-end switch position detection and state recognition system based on YOLO detector that can detect and recognize multiple targets in a single frame at one time. A four-category network based on YOLOv3-tiny is designed and optimized for real-time detection, then logistic regression is used to predict the probability of status of switches, and then an algorithm based on the prior information of the cabinet is developed to remove duplicate targets. Finally, the detected switches are sorted and numbered to compare with the information in the database. Experiments are reported to verify the proposed system.

Keywords: Convolutional neural network · YOLOv3 · State recognition · Power systems

1 Introduction

With the rapid growth in electricity consumption in recent years, the scale of power networks continues to expand. The number of control switches in substations has increased dramatically. Monitoring the states of these electric control cabinet switches is a major factor in the safe operation of the substations. Manual detection is time-consuming, laborious, and prone to errors and omissions. Therefore, it is important to realize automatic monitoring and tracking of switch status by means of machine vision.

Deep learning networks such as convolutional neural networks (CNNs) are often used for segmentation and recognition due to their remarkable performances. However, suitable CNN structures will need to be explored extensively for a specific application. In this paper, we propose a real-time multi-target switch position detection and recognition algorithm based on the YOLO detector, to detect position and recognize status of switches at any time. The algorithm is transplanted to an Android phone and has been deployed in actual inspection of substations.

© Springer Nature Switzerland AG 2019
H. Yin et al. (Eds.): IDEAL 2019, LNCS 11871, pp. 184–194, 2019.
https://doi.org/10.1007/978-3-030-33607-3_21

2 Related Work

Detection of electric switches requires information of switch positions and switch status. In [1] an algorithm based on image shape feature was proposed, which mainly uses projection method and template matching method.

Deep learning algorithms have achieved remarkable successes in large-scale image and video recognition in recent years [2,3]. Most deep learning algorithms use selective search as a regional nomination method [4]. R-CNN uses the selection search box to extract 2000 candidate region frames from the original image, and then performs feature extraction, classification and regression operations [5]. Fast R-CNN solves the problem of double counting in R-CNN and the speed is over 8 times faster than R-CNN [6]. Faster R-CNN uses CNN to region proposal and is even faster than Fast R-CNN [7]. These networks have two stages to realize detection and recognition.

YOLO [8] and SSD [9] are typical end-to-end target detection and recognition methods. YOLO combines the target detection and recognition tasks into one, greatly improving the running performance, reaching 45 frames per second. The SSD network has two parts. The first network is a standard network for image classification, and the latter network is a multi-scale feature mapping layer for detection, so as to detect targets of different sizes.

After SSD, the YOLO team designed YOLO9000, further improving detection and recognition performance, speed and robustness [10]. In 2018, the team designed YOLOv3, not only solving the problem of small target object detection in YOLO9000, but also streamlining the network, further improving the network speed [11].

3 Algorithm Design

We have designed an end-to-end target detection and recognition network. The network can operate on camera input picture, integrate target detection and target classification into one network, and realize real-time detection.

R-CNN, Fast R-CNN, and Faster R-CNN need to divide the target detection into two steps. First, the region proposal network (RPN) is used to extract the target information, and then the detection network is used to complete the location and category identification of the target [7]. These networks are slow and often cannot meet real-time requirements, and their training is also complex. SSD and YOLO do not need RPN, and can complete detection and identification in one step, so they are faster.

YOLOv3 is the latest version, which has better accuracy, faster speed, and better adaptability to targets of different sizes. Detecting the same picture with the same hardware, the speed of YOLOv3 can be 4 times faster than SSD, while the accuracy is also better. YOLOv3-tiny simplifies the number of layers of the convolutional neural network, increases the speed of the network by more than 4 times and can even up to 244 FPS. The algorithm will eventually be transplanted to the smartphone, the performance of the smartphone is not as good as the PC. So in order to ensure real-time computing, we choose the YOLOv3-tiny network [12].

3.1   Algorithm Overview

The network structure of yolov3 has 53 layers, and the real-time performance is still not good enough. In order to improve real-time performance, we must simplify the number of network layers. So, we refer to YOLOv3-tiny, reduce the number of layers in the network, optimize the network parameters, and further improve the real-time performance.

The specific process of multi-target switch position detection and state recognition algorithm is shown in Table 1.

Table 1. The process of the proposed algorithm

Algorithm	multi-target switch position detection and state recognition
Input: D	▷ Camera image of switch panel
Output: R	▷ Results of positions and status of switches
1: **Gamma correction**	▷ Enhance the quality of switch panel images of th e electrical control cabinet through Gamma correcti on.
2: **Use four-category netwo-rk to extract features**	▷ Refer to YOLOv3-tiny to design the four-category n etwork, and use the datasets we made to train this net work. Then use the trained network to extract feature s.
3: **Get bounding boxes and use logistic regression to recognize status of switches**	▷ Through the YOLO network, we obtain many bounding boxes, and then logistic regression is used to recognize status of switches.
4: **Non-maximum suppression**	▷ YOLO network use non-maximum suppression method to remove repetitive bounding-box
5: **Remove repetitive box using prior information**	▷ YOLO network can not remove all the repetitive bounding-box, so we design removing duplicates algorithm.
6: **Sort the coordinate of bounding-box and number switches**	
7: **Return R**	

3.2   Detection and Recognition Network Based on YOLOv3-Tiny

The YOLOv3-tiny network is still large. The computation is time consuming and cannot meet the real-time performance. Therefore, we have reduced the number of convolution kernels of per convolution layer to reduce the number of training parameters. Considering that the switch scale does not change much, only two scale switch detections were designed. The task of detection and identification of the switches state is considered as four-category task. Therefore, we designed four-category switch position detection and identification network models. The parameters of four-category network are shown in Table 2.

Table 2. Parameters of four-category switch position detection and state recognition network based on YOLOv3-tiny

Layer	Filters	Size	Input	Output
0 conv	16	$3 \times 3/1$	$416 \times 416 \times 3$	$416 \times 416 \times 16$
1 maxpool		$2 \times 2/2$	$416 \times 416 \times 16$	$208 \times 208 \times 16$
2 conv	32	$3 \times 3/1$	$208 \times 208 \times 16$	$208 \times 208 \times 32$
3 maxpool		$2 \times 2/2$	$208 \times 208 \times 32$	$104 \times 104 \times 32$
4 conv	64	$3 \times 3/1$	$104 \times 104 \times 32$	$104 \times 104 \times 64$
5 maxpool		$2 \times 2/2$	$104 \times 104 \times 64$	$52 \times 52 \times 64$
6 conv	128	$3 \times 3/1$	$52 \times 52 \times 64$	$52 \times 52 \times 128$
7 maxpool		$2 \times 2/2$	$52 \times 52 \times 128$	$26 \times 26 \times 128$
8 conv	256	$3 \times 3/1$	$26 \times 26 \times 128$	$26 \times 26 \times 256$
9 maxpool		$2 \times 2/2$	$26 \times 26 \times 256$	$13 \times 13 \times 256$
10 conv	256	$3 \times 3/1$	$13 \times 13 \times 256$	$13 \times 13 \times 256$
11 maxpool		$2 \times 2/1$	$13 \times 13 \times 256$	$13 \times 13 \times 256$
12 conv	512	$3 \times 3/1$	$13 \times 13 \times 256$	$13 \times 13 \times 512$
13 conv	128	$1 \times 1/1$	$13 \times 13 \times 512$	$13 \times 13 \times 128$
14 conv	256	$3 \times 3/1$	$13 \times 13 \times 128$	$13 \times 13 \times 256$
15 conv	27	$1 \times 1/1$	$13 \times 13 \times 256$	$13 \times 13 \times 27$
16 yolo				
17 route	13			
18 conv	128	$1 \times 1/1$	$13 \times 13 \times 128$	$13 \times 13 \times 128$
19 upsample			$2 \times 13 \times 13 \times 128$	$26 \times 26 \times 128$
20 route	19,8			
21 conv	256	$3 \times 3/1$	$26 \times 26 \times 384$	$26 \times 26 \times 256$
22 conv	27	$1 \times 1/1$	$26 \times 26 \times 256$	$26 \times 26 \times 27$
23 yolo				

YOLOv3-tiny network consists of 24 layers and is divided into two parts: the convolutional neural network and the YOLO network. Layers 0 through 12 are convolutional and pooling layers for extracting features of the target images from the input image. Each convolution layer is followed by a pooling layer, to abstract and reduce the size of the data and improve the speed of the entire network. In view of the large difference between the target switch pixel and the electrical cabinet panel background, the max pooling method in Yolov3-tiny can help our four-category network select the maximum value in the area to remove the influence of the panel background as much as possible. After the features are extracted by the convolutional layer, the original switch panel picture of 416×416 pixels generates a feature matrix of $13 \times 13 \times 512$.

The second part of YOLOv3-tiny is the feature interaction layer of the YOLOv3 network using only two scales. Each scale uses a convolution kernel to implement local feature interaction, while the fully connected layer implements global feature interaction. In this four-category network, the number of channels becomes 27 according to the number of target classifications. Position regression and classification of the target switch are performed on the feature maps of $13 \times 13 \times 27$.

On the YOLO network of the second scale, firstly, YOLOv3-tiny samples the feature map of the 13 layers of convolution of the entire network at the 19th layer to obtain a feature map of 26×26 size. Then, it merges the feature map obtained by sampling on the 19th layer with the feature map obtained from the 8th layer of the convolutional network. And a series of convolution operations are performed. The size of the feature map is unchanged, and the number of channels is changed according to the target classification number. In the proposed four-category network, the number of channels is changed to 27 according to the number of target classifications, and the positional regression of the target switch and the classification are performed on these feature maps of $13 \times 13 \times 27$ and $26 \times 26 \times 27$.

We use the sparsification method to compress the network [13]. The most computation in the network is convolution operation. The most data of the network are the weights of the filters. In order to compress the network, we can reduce the size of the filters or reduce the number of filters. When trying to change the 3×3 filter of any convolutional layer to 1×1, the recognition rate is seriously reduced. This method cannot work. So, we try to reduce the number of filters. When the number of filters of 0 conv, 2 conv, 4 conv, 6 conv or 8 conv is reduced, the recognition rate is seriously reduced. Because if the filters of the first few layers of conv are reduced, the network cannot extract enough information and the error will accumulate. The number of filters of 10conv, 12 conv, 13 conv and 14 conv are halved, causing the weight data is reduced from 8670384 to 5000368, compressed to 57.67%. At the same time, a high recognition rate is guaranteed.

3.3   Removing Duplicates and Sorting

The non-maximum suppression method used in the process of deep learning network detection cannot completely eliminate the duplicate target, and the output value of the deep learning network cannot be sequentially matched with the target switch on the actual switch panel. Therefore, we need to remove duplicates and sort the results. The process of removing duplicates is shown in Table 3.

After removing duplicates, we sort the results and make the results of the network identification correspondence with the actual switch positions. The Y coordinate values are first sorted and compared with the empirical threshold of line spacing that we've set it up before the experiment, and each line of the target switches is divided. Then, the X coordinate values of the target switches of each row are sorted, so that the correctly sorted target switches are obtained. The results are arranged from top to bottom in rows and uploaded to the substation back-end data management system.

Table 3. The process of removing duplicates

Algorithm	removing duplicates
Input: N	▷the number of bounding-box
Input: T	▷the experience threshold
Input: P[i]	▷the prediction probability of
	i bounding-box
Output C	▷ all the coordinate of bounding-box
1: **For** i=1 to N-1 **do:**	
2: E ← the euclidean distance betweeb i coordinate of i	
bounding-box and i+1 coordinate of i+1 bounding-box	
3: If E<T **then**	
4: If P[i]<=P[i+1] **then**	
5: remove i bounding-box	
6: **else**	
7: remove i+1 bounding-box	
8: If all the bounding-box is used **then**	
9: **return** C	
10: **return** C	

4   Experiments

In order to verify the validity of this network, we need to retrain the network using the actual data set, and use the weights obtained from this network to detect the target, and analyze the experimental data.

4.1   Datasets

Due to various different switch panel styles and backgrounds, a uniform standard needs to be established when labeling the dataset:

1. Accurately distinguish the target switch from the background.
2. The base needs to be framed into the target switch area, and the nameplate cannot be selected into the target switch area.
3. When the circular switch is off, the bounding box only selects the part of the base, and the blade part is not selected.

The data set should include the switch panel picture in various situations, such as switch panels with different degrees of tilt, different degrees of blur and different ages.

The circular and square switches can be divided into four categories for labeling. The open state of the circular switch is labeled as 0, the closed state of the circular

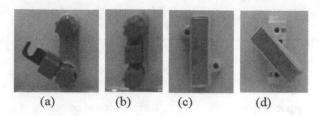

Fig. 1. (a) Open circular switch. (b) Closed circular switch. (c) Closed square switch. (d) Open square switch.

switch is labeled as 1, the closed state of the square switch is labeled as 2, and the open state of the square switch is labeled as 3. Examples of these switches are shown in Fig. 1.

The dataset contains 261 square switch panel images and 362 circular switch panel images, each of which contains approximately 40 switches. 131 square switch panel images and 203 circular switch panel images are front view, and the remaining 130 square switch panel images and 159 circular switch panel images have different shooting angles. Image quality, lighting and shadows are random. The ratio of train set to validation set is 8:1.

4.2   Training

The training of the proposed network was done on the Ubuntu1804 with Darknet. The Momentum method in Mini-Batch Gradient Descent was used to optimize the algorithm. This method can reduce the variance of the updated parameters, making the training convergence process more stable. The Momentum method can also improve the convergence speed of network training and save network training time.

First, we determine the hyperparameters based on the number of pictures in the dataset. If the memory is enough, you can increase the number of batch and decreases the number of subdivisions. Momentum is the momentum parameter of the optimization method in deep learning, which affects the speed of gradient descent. Decay is a weighted decay regular term that prevents overfitting during training. The angle, saturation, exposure, and hue parameters increase the training sample set by adjusting the angle, saturation, exposure, and hue of the image to enhance the training effect. Max_batches presets the maximum number of training iterations. Burn_in sets 1000 iterations to use the steps set by policy to adjust the learning rate. Scales sets the learning rate to 0.1 times the original learning rate. Then we load the network model, hyperparameters and datasets, and perform network feedforward convolution, calculate the loss function and update the network parameters. In the process of training, the model parameters are saved every 500 iterations, so that it is convenient to select the optimal model parameters according to the loss function. We do not load the pre-trained weights before the training and all the results are from our own datasets.

After the training iterations reach the preset maximum number, the entire network training process ends. Then, the training results can be tested in batches using the python script code. If the test results are not satisfactory, the hyperparameters are adjusted to continue the training.

The training platform configuration is Intel(R) Core (TM) i5-3570 CPU + NVIDIA 1080ti. Using the hyperparameters shown in Table 4, it takes about 4 h to complete 10,000 iterations.

Table 4. Hyperparameters of the network

Hyperparameter	Value
Batch	64
Subdivisions	2
Width	416
Heights	416
Channels	3
Momentum	0.9
Decay	0.0005
Angle	0
Saturation	1.5
Exposure	1.5
Hue	.1
Learning_rate	0.001
Burn_in	1000
Max_batches	10000
Policy	Steps
Steps	4000, 7000
Scales	.1,.1

The Intersection-over-Union (IoU) is the ratio of the area of intersection over the union area occupied by the ground truth bound and the candidate bound, and its formula is shown in Eq. (1).

$$IoU = \frac{area(C) \cap area(G)}{area(C) \cup area(G)} \tag{1}$$

where area (C) is candidate bound, area (G) is ground truth bound.

The loss function value and the average IoU during the entire network training process are shown in Fig. 2.

4.3   Results and Analysis

The experiment performance includes the inference time and recognition rate of the four-category network and comparative experiments on different platforms. The experiment uses the same set of test pictures consisting of 50 pictures of circular switches and 50 pictures of square switches. The image size of the single electrical

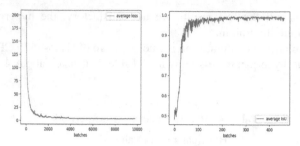

Fig. 2. Loss value and average IoU of the switch detection and identification network training.

cabinet switch panel is 1280 × 960 pixel. There are many target switches in each single electrical cabinet switch panel. The 100 test pictures contain appropriate tilt and shadow. Use the network obtained by the training for recognition. The number of correctly recognized switches is counted, and the average consumption time of single-frame image recognition is calculated.

Performance of Four-Category Network. The platform selected for this test uses Intel(R) Core (TM) i5-3570 CPU + NVIDIA 1080ti. The number of switches in the test pictures is indefinite, and including 2 × 9, 4 × 9, 6 × 9, 4 × 11, etc. Four-category network uses the optimized YOLOv3-tiny network. Table 5 below is the performance test result record of four-category network and original YOLOv3-tiny, and Fig. 3 is the test result interface.

Table 5. Performance of proposed four-category network

Model	Recognition rate	Average inference time per frame (ms)
YOLOv3-tiny	99.60%	4.1
Four-category network	99.47%	2.5

The results in Table 5 show that the recognition rate of the four-category network is very high, and the average inference time of the single-frame image on the same platform is short. This indicates that the target switch position detection and state recognition algorithm based on YOLO detector has strong robustness and good real-time capability.

Table 6 shows that recognition rates of this algorithm are not affected by tilt angles and shadows. The algorithm has strong anti-interference ability and good robustness and generalization and can cope with complex substation scenarios.

Different Test Platforms. When implementing automation switch status recognition for substations, it is convenient to use low-cost facilities such as mobile phones for detection. Table 7 shows the experimental results on the Intel(R) Core (TM) i5-3570 CPU + NVIDIA 1080ti and Huawei P9 Android smartphone.

Table 7 shows that the position detection and status recognition of the switch using the Intel(R) Core (TM) i5-3570 CPU + NVIDIA 1080ti PC are superior to the Huawei

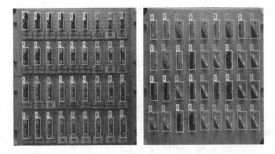

Fig. 3. Test results interface under different models

Table 6. Recognition rate under different picture quality.

Whether the mage is tilt	Whether there are shadows	Recognition rate
No	Yes	98.95%
Yes	No	98.95%
No	No	98.95%
Yes	No	98.95%

Table 7. Results under different test platforms.

Experiment platform	Recognition rate	Average time per frame (ms)
Intel Core i5-3570 NVIDIA 1080ti	99.47%	2.5
Huawei P9	97.66%	200

P9 Android smartphone in recognition rate and single-frame time consumption. The same algorithm has a slightly decreased recognition rate after transplanting to Android smartphones. This is because the CPU of the smartphone is ARM architecture and the CPU of PC is X86 architecture. The different instruction sets cause the results of calling the recognition network for calculation are slightly different, resulting in the decrease in the recognition rate.

5   Conclusion

In this paper, we have proposed an efficient, multi-target switch position detection and state recognition network based on YOLOv3. The images of the switch panel of the electrical cabinet in the substation were collected in the plant as the training dataset. We designed the parameters of the convolutional neural network and adjust the hyperparameters and use the dataset to train the streamlined network. In order to improve the robustness of the network and improve accuracy, we added deduplication and sorting algorithms. The algorithm can be transplanted to an Android smartphone. Experiments on the PC platform and Android smartphone platform show that the algorithm is fast, accurate and robust.

Acknowledgment. This work was funded by National Natural Science Foundation of China (61701357) and by China Scholarship Council.

References

1. Zhang, H., Wang, W., Xu, L., Qin, H., Liu, M.: Application of image recognition technology in electrical equipment on-line monitoring. Power Syst. Prot. Control **38**(6), 88–91 (2010)
2. Simonyan, K., Zisserman, A.: Very Deep Convolutional Networks for Large-Scale Image Recognition. arXiv preprint arXiv:1409.1556 (2015)
3. Sermanet, P., Eigen, D., Zhang, X.: OverFeat: Integrated Recognition, Localization and Detection using Convolutional Networks. arXiv preprint arXiv:1312.6229 (2014)
4. Uijlings, J.R.R., van de Sande, K.E.A.: Selective search for object recognition. Int. J. Comput. Vis. **104**(2), 154–171 (2013)
5. Girshick, R., Donahue, J., Darrell, T.: Rich feature hierarchies for accurate object detection and semantic segmentation. In: Proceedings of the IEEE International Conference on Computer Vision and Pattern Recognition (CVPR), pp. 580–599 (2014)
6. Girshick, R.: Fast R-CNN. arXiv preprint arXiv:1504.08083 (2015)
7. Ren, S., He, K., Girshick, R., Sun, J.: Faster R-CNN: towards real-time object detection with region proposal networks. In: Proceedings of the 28th International Conference on Neural Information Processing Systems (NIPS), pp. 91–99 (2015)
8. Redmon, J., Divvala, S., Girshick, R.: You only look once: unified, real-time object detection. In: Proceedings of the IEEE International Conference on Computer Vision and Pattern Recognition (CVPR), pp. 779–788 (2015)
9. Liu, W., Anguelov, D., Erhan, D.: SSD: Single Shot MultiBox Detector. arXiv preprint arXiv:1512.02325 (2015)
10. Redmon, J., Farhadi, A.: YOLO9000: better, faster, stronger. In: Proceedings of the IEEE Conference on Computer Vision and Pattern Recognition (CVPR), pp. 7263–7271 (2017)
11. Redmon, J., Farhadi, A.: YOLOv3: An Incremental Improvement. arXiv preprint arXiv: 1804.02767 (2018)
12. Bilel, B., Taha, K., Anis K., Adel, A., Kais, O.: Car Detection using Unmanned Aerial Vehicles: Comparison between Faster R-CNN and YOLOv3. arXiv preprint arXiv:1812. 10968 (2018)
13. Zhuang, L., Jianguo, L., Zhiqiang S., Gao, H., Shoumeng, Y., Changshui Z.: Learning Efficient Convolutional Networks through Network Slimming. arXiv preprint arXiv:1708. 06519 (2017)

A Self-generating Prototype Method Based on Information Entropy Used for Condensing Data in Classification Tasks

Alberto Manastarla[1(✉)] and Leandro A. Silva[2]

[1] Electrical Engineering and Computing, Mackenzie Presbyterian University,
Sao Paulo, SP, Brazil
manastarla@hotmail.com
[2] Computing and Informatics Faculty, Electrical Engineering and Computing,
Mackenzie Presbyterian University, Sao Paulo, SP, Brazil
leandroaugusto.silva@mackenzie.br

Abstract. This paper presents a new self-generating prototype method based on information entropy to reduce the size of training datasets. The method accelerates the classifier training time without significantly decreasing the quality in the data classification task. The effectiveness of the proposed method is compared to the K-nearest neighbour classifier (kNN) and the genetic algorithm prototype selection (GA). kNN is a benchmark method used for data classification tasks, while GA is a prototype selection method that provides competitive optimisation of accuracy and the data reduction ratio. Considering thirty different public datasets, the results of the comparisons demonstrate that the proposed method outperforms kNN when using the original training set as well as the reduced training set obtained via GA prototype selection.

Keywords: Prototype Selection (PS) · Data reduction · Data classification · Genetic Algorithm (GA)

1 Introduction

Data classification is one of the most popular supervised learning tasks used to predict classes based on the descriptive attributes of analysed instances. The K-nearest neighbour classifier (kNN) is a frequently used data classification algorithm, and despite being invented 50 years ago, [4] this classifier has been ranked one of the top 10 Data Mining methods [15]. When compared with other methods, kNN is considered more straightforward for data classification task benchmarks. Essentially, the method includes three operations: (i) An unlabelled instance (test sample) is compared to labelled instances stored in a training dataset through a similarity measure; (ii) The labelled instances are sorted

---

This study was financed in part by the Coordenacao de Aperfeicoamento de Pessoal de Nivel Superior - Brasil (CAPES) - Finance Code 001.

H. Yin et al. (Eds.): IDEAL 2019, LNCS 11871, pp. 195–207, 2019.
https://doi.org/10.1007/978-3-030-33607-3_22

based on their similarity to an unlabelled instance; (iii) Classification occurs when the unlabelled instance is given the majority class of the labelled instances' nearest neighbours. Due to these simple operations, as well as the reduced number of parameters (similarity measure and number of nearest neighbours, k), and accuracy of performance in different real datasets, this Instance-Based Learning algorithm is widely used in the data mining community as a benchmark algorithm [2, 16].

However, in terms of real application, kNN suffers several weaknesses. The optimal choice for the k parameter depends upon the dataset, mainly when the instance analysed is in a boundary region, meaning this parameter must be tuned to the specific application [5, 6, 11, 12]. Since the kNN algorithm does not have a model, it needs to compare the unlabelled instances, with all the labelled and stored instances in the training dataset, for each classification process. This exhaustive comparison process significantly increases classification time, especially when the dataset is large [5, 6, 11]. Also, the decision boundaries defined by the instances stored in the training set imply two additional weaknesses: First, even when the training set includes incorrect data, the algorithm exhibits low noise tolerance because the instances of the training set are considered relevant patterns, and second, it assumes that instances in the training set correctly delimit the decision boundaries among the classes [12].

The kNN algorithm has two distinct approaches for overcoming these weaknesses: Improve the calculation speed of the nearest neighbours, or reduce the training set by selecting only a small fraction of instances or features [9]. The object of focus in this paper is the latter approach, which encompasses variety data reduction techniques (DRTs), and among them we specifically explore prototype selection (PS) methods. DRTs aim to reduce the size of the training set by building a smaller representative set called a condensed set. These methods not only improve the speed of the classification process, but also build an enhanced training set by eliminating noisy instances (for example overlapped classes) and smoothing decision boundaries. Within the DRT group, PS methods have been compared regarding classification accuracy, and such studies have demonstrated that there is no clear best PS method nor evidence that supports the use of one method for all datasets [12]. Hence, choosing a primary optimisation goal, such as storage reduction, dataset structure, classification error rate, or noise tolerance, seems to be essential.

As PS methods have exhibited distinct results in previous literature, we are motivated to propose a new method that can be flexibly applied to a large number of datasets while maintaining competitive performance when compared to existing PS methods.

The proposed PS method in this paper allows for the customisation of parameters based on whatever primary optimisation goal is chosen. Titled the Self-Generating Prototype Entropy (SGPE), this method employs a new process for controlling the number of prototypes selected from the training set, which uses an information entropy function to decide when to create a new prototype based on the centre of gravity in a subset of the training set. The process enables precise control over the number of prototypes selected from the training set to form the condensed set that will be used in the training process. This new PS method offers an alternative to adapting the condensed set so that it appropriately represents the original dataset. Additionally, the introduced approach can minimise three main weaknesses related to kNN:

(i) Improve storage requirements by reducing the original training set through the adjustment of hyper-parameters in prototype generation;
(ii) Adjust the demand for powerful computational resources to calculate distances between a new instance and the condensed set;
(iii) Remove a certain level of noise instances through the prototype selection process.

The design of the proposed method includes the division of the training set into subsets; then the centre of gravity in these subsets must be found, and prototypes must be created. The method uses these prototypes to calculate the Euclidean distance between a prototype and instances from the training set to ultimately select instances from the training set and create the condensed set. The *modus operandi* of this proposed method is akin to Chen's algorithm [3], but with a fundamental improvement in the division subset operation. Our method uses an information entropy function to govern prototype generation and a threshold parameter that can control the number of prototypes used for selecting instances from the training set to form the condensed set.

The effectiveness of the proposed algorithm is examined for the K nearest neighbour classier (kNN) to compare accuracy and training processing time. Moreover, to verify the accuracy and data reduction achievements, a comparison between (SGPE) and a genetic algorithm (GA) for prototype selection was conducted using thirty public datasets. The limitations of our work include the use of only kNN classier; other variants of kNN such as the Informative kNN, Adaptive kNN, Bayesian kNN and SVM-kNN need to be verified. Also, there are still many other PS methods are available for comparison.

The remainder of this paper is organised as follows. Section 2 gives a brief explanation of prototype selection, while Sect. 3 describes the proposed methods. The methodology itself is elucidated in Sect. 4, while the experimental results and discussion are given in Sect. 5. In the last Section, conclusions are drawn.

2   Data Reduction Technique: Prototype Selection Methods

Prototype selection methods prove successful at simultaneously facing the computational complexity, storage requirements, and noise tolerance of the kNN classifier. When used, the PS method generates a pared down representative training set with similar or even higher classification accuracy compared to the original set [10].

More simply explained, PS methods are strategies for removing superfluous samples and can be divided into three broad groups [7]. First, *condensation* methods seek to maintain border samples and remove inner samples. The level of reduction is generally high for this group because there is commonly a higher proportion of inner samples, which has the potential to negatively impact accuracy. Second, *edition* methods remove border samples that are noisy or do not agree with their neighbours. The removal of such samples leaves a smoother decision boundary for the classifier. This group produces lower data reduction, and the improvements focus on increasing in accuracy. Finally, *hybrid* methods try to identify the smallest subset by either maintaining or increasing accuracy, which may involve the removal of both inner and border samples.

Formally, let TR be the original training set comprised of n instances. Every instance $\mathbf{I}_i$ is a pair $(\mathbf{x}_i, y_i)$ with i = 1,...,n, where $\mathbf{x}$ defines a vector of attributes and y defines the corresponding class label. Any input vector contains m input attributes comprised of quantitative or qualitative data that define the corresponding instance. The PS method produces a condensed set $\mathbf{S} \subseteq TR$.

The set $\mathbf{S}$ is used to classify new instances via the kNN classifier. Prototype selection can be considered as a multi-objective optimisation based on the following definition:

Definition 1. *PS Problem: The PS problem can be defined as a set of tuple (TR, A_{TR}, F, m), where TR is the original training set; A_{TR} is the set of all possible subsets $\mathbf{S}$ of TR; $F : A_{TR} \rightarrow \mathbb{R}^k$ is the vector-valued objective function defined as follows:*

$$F(\mathbf{S}) = [f_1(\mathbf{S}), f_2(\mathbf{S})]^T \quad \text{with} \quad \mathbf{S} \in A_{TR} \tag{1}$$

where $f_1 : A_{TR} \rightarrow \mathbb{R}$ is the first objective function defined as follows:

$$f_1(\mathbf{S}) = accuracy(TR, \mathbf{S}) \quad \text{with} \quad \mathbf{S} \in A_{TR} \tag{2}$$

In the above expression, *accuracy* is a function that computes the classification accuracy in terms of a percentage obtained by the kNN classifier by considering set TR minus $\mathbf{S}$ as the testing set and $\mathbf{S}$ as the training set. $f_2 : A_{TR} \rightarrow \mathbb{R}$ is the second objective function defined as follows:

$$f_2(\mathbf{S}) = reduce(TR, \mathbf{S}) = \frac{|TR| - |\mathbf{S}|}{|TR|}.100 \quad \text{with} \quad \mathbf{S} \in A_{TR} \tag{3}$$

In the above expression, *reduce* is a function that computes the reduction ratio of the set $\mathbf{S}$ concerning the original set TR. The m stands for maximisation or minimisation PS problem, which means the two defined objectives can be maximised or minimised.

The next section describes how we can overcome some of the disadvantages of the Condensation and Edition methods by applying a Hybrid algorithm called SGPE.

3   Self-generating Prototype Entropy

The term diameter of the set is used in this paper for describing a subset of the training dataset; it assumes the distance between its two farthest points.

To demonstrate how the proposed method operates, we demonstrate a one-dimensional example. The training set consists of eight points four from class one and class two, respectively. The two farthest points of the set have coordinates of 3 and 43. Before executing the first division, the thresholds must be set up to represent standard deviation tolerance, moving the average window and desired entropy level. In this simple example, the minimal values are used to set the entropy level, which shift the average window and standard deviation to 0.1, 3 and 1, respectively. These hyper-parameters enable control over the prototype generation for which entropy is the level of mixture class calculated for a certain subset. The moving average window and standard deviation operate on the outcome of the entropy function. When this outcome maintains the variation within a certain level, stability is achieved for the moving average window results based on the standard deviation values, and hence the method knows when to stop generating the prototype. The first division corresponds to the straight line that passes through point 23 (i.e., the middle between points 3 and 43) and generates two subsets. The next division is performed for the subset that contains a mixture of points from two classes. If more than one subset satisfies this condition, then the algorithm divides the subset with the largest diameter. In this case, left subset {3, 9, 14, 19} will be divided in the second step, while the subset {25, 28, 39, 43} will be divided in the third step. After the next step, the fifth division is performed. Here, no subset contains a mixture of points that differs from the entropy threshold. Hence, further division can only be realised for subsets that

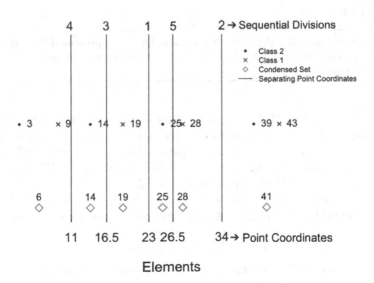

Fig. 1. A one-dimensional example of the SGPE operation.

meet the entropy threshold. Figure 1 displays the condensed set after five steps as well as the gravity centres created to obtain the subsets.

Algorithm 1. Pseudo-code: The Self-Generating Prototype Entropy (SGPE).

1: **function** SGPE_(X_{train} ,MAWindow, SDThreshold and EntropyThreshold)
2:     Assume $D \leftarrow X_{train}$, $i \leftarrow 1$, $N_d \leftarrow length(D)$, $SD \leftarrow$ SDThreshold, $MA \leftarrow$ MAWindow and $N_c \leftarrow 1$ (N_c is the current number of subsets of the training set);
3:     **while** $N_c < N_d$ **do**
4:         Find the two farthest points P_1 and P_2 in the set D;
5:         Divide the set D into two subsets D_1 and D_2, where
6:         $D_1 \leftarrow \{P \in D\colon \mathrm{d}(P, P_1) \leq \mathrm{d}(P, P_2)\}$;
7:         $D_2 \leftarrow \{P \in D\colon \mathrm{d}(P, P_2) \geq \mathrm{d}(P, P_1)\}$;
8:         Evaluates the subsets homogeneity using Entropy measure.
9:         **if** Entropy Outcome satisfies the EntropyThreshold in both subsets **then**
10:             $S(i) \leftarrow Subsets$;
11:         **end if**
12:         $N_c \leftarrow N_c + 1$, $C(i) \leftarrow D_1$ and $C(N_c) \leftarrow D_2$;
13:         Put $I_1 \leftarrow \{$i: $C(i)$ contains objects from two classes at least$\}$, $I_2 \leftarrow$ (i: i $\leq N_c$) $- I_1$;
14:         Put $I \leftarrow I_1$ if I_1 is non-empty else $I \leftarrow I_2$;
15:         Find the pair of two farthest points $Q_1(i)$ and $Q_2(i)$ in each $C(i)$ for $i \in I$;
16:         Find j such that $\mathrm{d}(Q_1(j), Q_2(j)) = \max(\mathrm{d}(Q_1(i),Q_2(i)))$ for $i \in I$, i.e. find a set $C(j)$ with largest diameter;
17:         Put $D \leftarrow C(j)$, $P_1 \leftarrow Q_1(j)$ and $P_2 \leftarrow Q_2(j)$;
18:         Evaluates the global homogeneity variability for each interaction.
19:         **if** Entropy Global Outcome for C satisfies MA and SD Thresholds **then**
20:             Break;
21:         **end if**
22:     **end while**
23:     find gravity centres $G(i)$ for each $S(i)$, $i = 1,2,...N_d$;
24:     Assign to each $G(i)$ the class that is most heavily represented in $S(i)$;
25:     The points $G(i)$, $i = 1,2,...N_d$, form the condensed training set;
26: **end function**

A predefined-user setting that allows for control over the number of prototypes is passed a hyper-parameter to the algorithm, as described in line 1 in the SGPE pseudo-code. Based on this hyper-parameter, the algorithm validates the prototype generation condition described in line 10 of the SGPE pseudo-code. The entropy Eq. (4) is then implemented to calculate the criteria and create or not create a new prototype according to the distribution inside each subset, where p is the probability of class A inside the subgroup and q of class B.

$$e = -plog_2(p) - qlog_2(q) \tag{4}$$

The moving average (5) and standard deviation (6) equations are used to calculate the stop condition of the prototype generation function based on the predefined-user settings, as described in line 1 and 21 of the SGPE pseudo-code.

$$\bar{e} = \frac{e_n + e_{n-1} + \cdots + e_n}{n} \tag{5}$$

$$s = \sqrt{\frac{\sum_{i=1}^{N}(e_i - \bar{e})^2}{N - 1}} \quad , \text{if } s < \text{predefined} - \text{userthreshold; then} \tag{6}$$

$$\text{stop prototype generation}$$

4   Methodology

The performance comparison among the algorithms is assessed by considering two widely used criteria: test-set accuracy (Atest) and training-set reduction (Rtrain). Since a successful algorithm should be able to significantly reduce the size of the training set without significantly reducing accuracy, it is opportune to assess algorithm performance through a quantitative metric that combines the criteria mentioned above.

This metric is denoted as the H-measure and was applied in a similar work described by [1], where a comparison was conducted between a GA and the Strength Pareto Evolutionary Algorithm. For the present paper, the performance of the algorithm is evaluated using the same metric.

Definition 2. *H-measure: Given a training dataset TR, a testing dataset TS, and a condensed set S produced by a PS method, the H-measure H is equal to the harmonic mean of the test-set.*

$$H = \frac{2.A_{test}.R_{train}}{A_{test} + R_{train}} \tag{7}$$

In the above equation, $A_{test}=accuracy(TS, \mathbf{S})$ and $R_{train}=reduce(TR, \mathbf{S})$.

Since the SGPE returns an uncontrollable set of solutions to compare its performance with GA, the best solution relative to the values of measure H is chosen as the final solution for the set of non-dominated solutions obtained by each execution. This process allows us to explicitly verify whether the SGPE can produce at least one solution of higher quality than GA. The performance achieved by the SGPE and GA algorithms is reported in Tables 3 and 4. The results are obtained by performing the typical ten-fold cross-validation procedure. In detail, each of the datasets considered is partitioned beforehand into ten training/test subsets. In general, during k-fold cross-validation, the dataset is divided into disjoint k subsets, which are used for training and testing.

At each iteration of the cross-validation procedure, a subset is used as a test set and all remaining subsets are used as the training set. This procedure runs k iterations until all the subsets have been used for testing. As the proposed algorithm and GA are non-deterministic approaches, they are executed three times at each iteration of the cross-validation procedure, and the overall comparison is performed based on the averages of the results from these races.

It is important to note that the kNN used is 1-NN. This choice is tied to the fact that setting k greater than one decreases the sensitivity of the PS method to noise, as stated by Wilson and Martinez in [14].

5   Experimental Results

This section explains the benefits provided by the application of SGPE to the prototype selection problem. Namely, we performed two experimental sessions to assess the capability SGPE has to (1) offer a better trade-off between classification and reduction compared to the traditional GA algorithm and (2) obtain better accuracy for the kNN applied without a prototype selection mechanism. Both experimental sessions involve thirty well-known datasets that belong to the UCI Machine Learning Database Repository. The characteristics of the datasets,

Table 1. Summary description of UCI Machine Learning Databases used in classification experiments.

#	Name	#Ex.	#Atts.	#Cl.	#	Name	#Ex.	#Atts.	#Cl.
1	penbase	10992	16	10	16	balance	625	4	3
2	ring	7400	20	2	17	wdbc	569	30	2
3	thyroid	7200	21	3	18	monk2	432	6	2
4	satimage	6435	36	7	19	bupa	345	6	2
5	optdigits	5620	34	10	20	haberman	306	3	2
6	texture	5500	40	11	21	cleveland	297	13	5
7	page_blocks	5472	10	5	22	heart	270	13	2
8	phoneme	5404	5	2	23	specftheart	267	44	2
9	winequality_white	4898	11	11	24	glass	214	9	7
10	spambase	4597	57	2	25	sonar	208	60	2
11	titanic	2201	3	2	26	wine	178	13	3
12	vehicle	846	18	4	27	hayes_roth	160	4	3
13	mammographic	830	5	2	28	iris	150	4	3
14	pima	768	8	2	29	appendicitis	106	7	2
15	wisconsin	683	9	2	30	hepatitis	80	19	2

Table 2. Parametrization of the algorithms in the experiments.

kNN	k:	1	
	Distance Measure:	Euclidian Distance	
SGPE	Entropy Value:	Small Dataset: 0.1	Large Dataset: 0.1 to 0.9
	Moving Average:	Small Dataset: 3	Large Dataset: 3 to 6
	Standard Deviation:	Small Dataset: 0	Large Dataset: 0.1 to 0.3
GA	Population size (N):	30	
	Crossover Operator:	0.5	
	Crossover rate (pc):	0.01	
	Mutation Operator:	Bit Flip	
	Selection Operator:	Binary Tournament	
	Termination condition:	Small Dataset: 5000	Large Dataset: 2000

including the number of instances, the number of the features, and the number of classes, are reported in Table 1. We intentionally selected datasets with different characteristics to prove the general applicability of the proposed approach. Hereafter, more detailed results are discussed in the next two experimental sections.

5.1   Experimental Section I

To demonstrate the improvement provided by the proposed method regarding data reduction and accuracy performance, a GA for prototype selection was selected, as it reported to have a high data reduction ratio [1]. The run setting of the algorithms is specified in Table 2 to perform a fair comparison. The fitness function parameter for GA is set to 0.5, which is the value typically used when no information is known about the dataset domain.

As shown in Table 3, the SGPE identifies at least one solution with higher quality in terms of H-measure than GA for all reference test cases, except for the Bupa, Hepatitis, Specftheart, Monk2, and Hayesboth datasets.

In terms of the average H-measure, the SGPE outperforms GA with an improvement of 12.33%. However, to statistically validate these promising results, a Wilcoxon signed-rank test was performed considering the sample data and H-measure values presented in Table 3. The Wilcoxon test [13] is a nonparametric procedure employed in hypothesis testing and involves a project with two samples. It is analogous to paired t-testing in non-parametric statistical procedures; therefore, it is a paired test that aims to detect significant differences between two sample means, in this case, the behaviour of two algorithms [8]. Based on the Wilcoxon signalled test performed on the values of the H measurement, our test output report exhibits that the SGPE statistically outperforms GA in a 99% confidence level.

In Table 3, it can be observed that the SGPE maintains a consistent data reduction ratio among the datasets around a mean of 85.45%, while that of the GA measures 72.77%.

5.2   Experimental Section II

After establishing the improvements of the data reduction ratio by applying this new approach, the second experiment exemplifies how the application of SGPE achieves a more consistent result than kNN without a prototype selection mechanism in terms of accuracy and training time. Table 4 presents the comparison in terms of classification accuracy between kNN and the kNN with the SGPE applied as a prototype selection method. In terms of average accuracy, the kNN with SGPE outperforms kNN with an improvement of the 2.6%. However, to statistically verify the validity of these results, the Wilcoxon's signed-rank test was performed considering the sample data and accuracy values presented in the Table 4.

The Wilcoxon's test results indicate that the application of SGPE as prototype selection method statistically verifies the accuracy results in aligning with the results obtained by the kNN with the original training set at a 95% confidence level. In terms of average training processing time, the kNN with SGPE outperforms kNN with an improvement of 0.060 s versus 0.422 s.

The tests also prove that the SGPE does not decrease the accuracy rate among the datasets when compared with the kNN with the original training set and offer a better trade-off between classification and reduction compared to the traditional GA algorithm (Fig. 2).

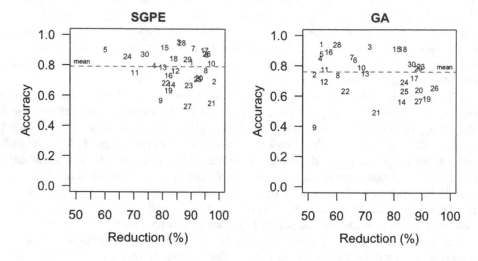

Fig. 2. Reduction effectiveness between SGPE and GA.

Table 3. Results for the compared kNN applied with prototype selection methods.

#	Datasets (Name and Dimension)	kNN(k = 1) (TR obtained via (SGPE))			kNN(k = 1) (TR obtained via (GA))		
		Acc	Red	H	Acc	Red	H
1	penbase-10992-16-10	83.41 ± 1.40	**90.03**	87	**95.71 ± 0.60**	54.43	69
2	ring-7400-20-2	69.88 ± 1.19	**98.24**	82	**75.57 ± 1.57**	52.16	62
3	thyroid-7200-21-3	**96.81 ± 0.90**	85.50	91	94.06 ± 1.22	71.39	81
4	satimage-6435-36-7	80.04 ± 3.06	77.09	79	**86.11 ± 2.23**	54.19	67
5	optdigits-5620-34-10	**90.96 ± 1.40**	59.96	67	88.99 ± 1.11	54.59	68
6	texture-5500-40-11	**88.07 ± 6.84**	95.45	92	85.09 ± 9.74	66.38	75
7	page_blocs-5472-10-5	**95.58 ± 3.27**	90.68	93	87.04 ± 5.62	65.19	75
8	phoneme-5404-5-2	**78.50 ± 1.74**	95.19	86	74.87 ± 2.12	60.31	67
9	winequality_white-4898-11-11	**57.44 ± 4.03**	79.26	67	40.70 ± 3.08	52.18	46
10	spambase-4597-57-2	**84.32 ± 8.04**	97.17	90	80.27 ± 10.54	88.54	74
11	titanic-2201-3-2	**79.27 ± 2.48**	70.29	75	78.91 ± 2.16	55.66	65
12	vehicle-846-18-4	**77.65 ± 3.48**	84.40	81	70.85 ± 4.31	55.44	62
13	mammographic-830-5-2	**79.28 ± 5.79**	80.00	80	77.47 ± 4.82	69.76	**80**
14	pima-768-8-2	**69.84 ± 4.44**	83.07	76	57.59 ± 4.86	82.68	68

(continued)

Table 3. (*continued*)

#	Datasets (Name and Dimension)	kNN(k = 1) (TR obtained via (SGPE))			kNN(k = 1) (TR obtained via (GA))		
		Acc	Red	H	Acc	Red	H
15	wisconsin-683-9-2	**93.24 ± 3.12**	80.67	**87**	92.50 ± 4.62	**80.82**	86
16	balance-625-4-3	73.71 ± 9.86	**82.08**	**78**	**90.48 ± 3.18**	57.12	70
17	wbdc-569-30-2	**94.84 ± 5.32**	**94.73**	**95**	73.29 ± 12.58	87.17	80
18	monk2-432-6-2	**86.74 ± 9.31**	**83.80**	85	92.33 ± 5.15	83.10	**87**
19	bupa-345-6-2	**62.94 ± 7.62**	82.03	71	59.71 ± 21.48	**91.59**	**72**
20	haberman-306-3-2	**73.48 ± 5.54**	**92.81**	**82**	65.23 ± 4.78	88.89	75
21	cleveland-297-13-5	**53.33 ± 9.86**	**97.31**	**69**	50.37 ± 6.96	73.74	60
22	heart-270-13-2	**70.37 ± 8.55**	**81.11**	**75**	64.44 ± 8.94	62.96	64
23	specftheart-267-44-2	72.02 ± 16.98	89.14	80	**81.07 ± 30.30**	**89.51**	**85**
24	glass-214-9-7	**86.19 ± 7.60**	67.76	**76**	70.48 ± 19.02	**83.64**	**76**
25	sonar-208-60-2	**67.72 ± 12.77**	**92.31**	**78**	64.55 ± 22.73	83.65	73
26	wine-178-13-3	**94.51 ± 5.86**	**95.51**	**95**	66.67 ± 38.19	94.38	78
27	hayes_roth-160-4-3	54.38 ± 13.19	**88.75**	67	**58.13 ± 18.41**	**88.75**	**70**
28	iris-150-4-3	**95.33 ± 8.34**	**96.33**	**91**	**95.33 ± 10.45**	60.00	74
29	appendicitis-106-7-2	**89.89 ± 17.47**	**88.68**	**89**	79.80 ± 40.09	**88.68**	84
30	hepatitis-80-19-2	**83.75 ± 13.24**	73.75	78	82.50 ± 10.54	**86.25**	**84**
Average		**79.48 ± 6.76**	**85.45**	**82**	76.00 ± 10.38	72.77	73

Table 4. Results for the compared kNN and kNN applied with Self-generating prototype method (SGPE). The results include the time taken to find the condensed set, training and test process.

#	Datasets (Name and Dimension)	kNN(k = 1) (Original TR)		kNN(k = 1) (TR obtained via (SGPE))		
		Acc	Time (Sec)	Acc	Time (Sec)	Red
1	penbase-10992-16-10	**94.39 ± 1.04**	1.714	83.41 ± 1.40	**0.230**	90.03
2	ring-7400-20-2	66.38 ± 1.64	0.971	**69.88 ± 1.19**	**0.051**	98.24
3	thyroid-7200-21-3	96.72 ± 0.99	0.987	**96.81 ± 0.90**	**0.184**	85.50
4	satimage-6435-36-7	79.81 ± 3.25	1.966	**80.04 ± 3.06**	**0.415**	77.09
5	optdigits-5620-34-10	81.48 ± 1.80	2.606	**90.96 ± 1.40**	**0.209**	59.96
6	texture-5500-40-11	76.98 ± 35.91	1.348	**88.07 ± 6.84**	**0.106**	95.45
7	page_blocs-5472-10-5	**94.13 ± 2.91**	0.315	95.58 ± 3.27	**0.057**	90.68
8	phoneme-5404-5-2	**90.96 ± 1.40**	0.209	78.50 ± 1.74	**0.027**	95.19
9	winequality_white-4898-11-11	45.46 ± 1.52	0.267	**57.44 ± 4.03**	**0.077**	79.26
10	spambase-4597-57-2	84.30 ± 8.18	1.251	**84.32 ± 8.04**	**0.106**	97.17
11	titanic-2201-3-2	78.82 ± 2.41	0.695	**79.27 ± 2.48**	**0.064**	70.29
12	vehicle-846-18-4	74.12 ± 5.76	0.035	**77.65 ± 3.48**	**0.021**	84.40
13	mammographic-830-5-2	74.47 ± 4.82	0.018	**79.28 ± 5.79**	**0.012**	80.00
14	pima-768-8-2	66.81 ± 4.69	0.021	**69.84 ± 4.44**	**0.014**	83.07
15	wisconsin-683-9-2	90.74 ± 5.05	0.020	**93.24 ± 3.12**	**0.014**	80.67
16	balance-625-4-3	**80.81 ± 4.13**	0.014	73.71 ± 9.86	**0.011**	82.08
17	wbdc-569-30-2	93.57 ± 4.21	0.039	**94.84 ± 5.32**	**0.024**	94.73
18	monk2-432-6-2	**96.05 ± 8.77**	0.014	86.74 ± 9.31	**0.010**	83.80

(*continued*)

Table 4. (*continued*)

#	Datasets (Name and Dimension)	kNN(k = 1) (Original TR)		kNN(k = 1) (TR obtained via (SGPE))		
		Acc	Time (Sec)	Acc	Time (Sec)	Red
19	bupa-345-6-2	55.29 ± 6.17	0.012	**62.94 ± 7.62**	0.011	82.03
20	haberman-306-3-2	65.59 ± 7.57	0.009	**73.48 ± 5.54**	0.009	92.81
21	cleveland-297-13-5	44.07 ± 8.62	0.017	**53.33 ± 9.86**	0.013	97.31
22	heart-270-13-2	61.11 ± 8.42	0.017	**70.37 ± 8.55**	0.014	81.11
23	specftheart-267-44-2	70.78 ± 26.09	0.010	**72.02 ± 16.98**	0.008	89.14
24	glass-214-9-7	81.43 ± 12.39	0.014	**86.19 ± 7.60**	0.015	67.76
25	sonar-208-60-2	58.20 ± 20.45	0.043	**67.72 ± 12.77**	0.035	92.31
26	wine-178-13-3	94.44 ± 11.11	0.016	**94.51 ± 5.86**	0.014	95.51
27	hayes_roth-160-4-3	**71.25 ± 10.70**	0.010	54.38 ± 13.19	0.010	88.75
28	iris-150-4-3	96.00 ± 6.44	0.010	**96.33 ± 8.34**	0.009	86.67
29	appendicitis-106-7-2	78.79 ± 16.39	0.010	**89.89 ± 17.47**	0.011	88.68
30	hepatitis-80-19-2	80.00 ± 14.67	0.016	**83.75 ± 13.24**	0.016	73.75
Average		77.43 ± 8.25	0.422	**79.48 ± 6.76**	0.060	85.45

6 Conclusions

This paper introduces a new PS method called SGPE (Self-Generating Prototype Entropy) to address the prototype selection problem. Two experimental sessions were conducted to verify the performance of SGPE for the kNN and kNN applied without a prototype selection method. After performing the Wilcoxon's signed-rank test, it is reasonable to assert that (1) SGPE statistically outperforms GA at a 99% confidence level in terms of H-measure; and (2) SGPE statistically maintains the same level of accuracy as kNN with the original training set at a 95% confidence level.

In the future, to further prove the benefits of this newly proposed method, we will perform additional experiments with new situations. For example, it is crucial to explore the behaviour of this new method when noisy data is presented into the training set and verify its implementation for parallel processing to further maximise the training processing time. Additionally, an extensive comparison with other algorithms will be made to garner a better understanding of the benefits and limitations provided by this new method.

References

1. Acampora, G., Tortora, G., Vitiello, A.: Applying SPEA2 to prototype selection for nearest neighbor classification. In: 2016 IEEE International Conference on Systems, Man, and Cybernetics (SMC), pp. 003924–003929. IEEE (2016)
2. Brighton, H., Mellish, C.: Advances in instance selection for instance-based learning algorithms. Data Min. Knowl. Disc. **6**(2), 153–172 (2002)
3. Chen, C., Jóźwik, A.: A sample set condensation algorithm for the class sensitive artificial neural network. Pattern Recogn. Lett. **17**(8), 819–823 (1996)

4. Cover, T., Hart, P.: Nearest neighbor pattern classification. IEEE Trans. Inf. Theory **13**(1), 21–27 (1967)
5. Deng, Z., Zhu, X., Cheng, D., Zong, M., Zhang, S.: Efficient KNN classification algorithm for big data. Neurocomputing **195**, 143–148 (2016)
6. Duda, R.O., Hart, P.E., Stork, D.G.: Pattern Classification. Wiley, New York (2001)
7. Garcia, S., Derrac, J., Cano, J., Herrera, F.: Prototype selection for nearest neighbor classification: taxonomy and empirical study. IEEE Trans. Pattern Anal. Mach. Intell. **34**(3), 417–435 (2012)
8. García, S., Molina, D., Lozano, M., Herrera, F.: A study on the use of non-parametric tests for analyzing the evolutionary algorithms' behaviour: a case study on the CEC'2005 special session on real parameter optimization. J. Heuristics **15**(6), 617 (2009)
9. Ougiaroglou, S., Evangelidis, G., Dervos, D.A.: FHC: an adaptive fast hybrid method for K-NN classification. Logic J. IGPL **23**(3), 431–450 (2015). https://doi.org/10.1093/jigpal/jzv015
10. Pękalska, E., Duin, R.P., Paclík, P.: Prototype selection for dissimilarity-based classifiers. Pattern Recogn. **39**(2), 189–208 (2006)
11. Torralba, A., Fergus, R., Freeman, W.T.: 80 million tiny images: a large data set for nonparametric object and scene recognition. IEEE Trans. Pattern Anal. Mach. Intell. **30**, 1958–1970 (2008). https://doi.org/10.1109/TPAMI.2008.128
12. Triguero, I., Derrac, J., Garcia, S., Herrera, F.: A taxonomy and experimental study on prototype generation for nearest neighbor classification. IEEE Trans. Syst. Man Cybern. Part C Appl. Rev. **42**, 86–100 (2012). https://doi.org/10.1109/TSMCC.2010.2103939
13. Wilcoxon, F.: Individual comparisons by ranking methods. Biometrics Bull. **1**(6), 80–83 (1945)
14. Wilson, D.R., Martinez, T.R.: Reduction techniques for instance-based learning algorithms. Mach. Learn. **38**(3), 257–286 (2000)
15. Wu, X., Kumar, V., Quinlan, J.R., et al.: Top 10 algorithms in data mining. Knowl. Inf. Syst. **14**(1) (2008). https://doi.org/10.1007/s10115-007-0114-2
16. Wu, X., Kumar, V.: The Top Ten Algorithms in Data Mining. CRC Press, Boca Raton (2009)

Transfer Knowledge Between Sub-regions for Traffic Prediction Using Deep Learning Method

Yi Ren[✉] and Kunqing Xie

Key Laboratory of Machine Perception (Ministry of Education),
School of Electronics Engineering and Computer Science,
Peking University, Beijing 100871, China
renyi_89@pku.edu.cn

Abstract. In modern society, traffic is a very important aspect of our social lives, but a lot of problems arise such as traffic accident, congestion, air pollution, etc. The emergence of ITS (Intelligent Transportation System) brings a range of scientific ways to tackle these problems, among which there is a crucial need to study prediction methods in modern traffic systems. Fortunately, big data in ITS and advanced machine learning technologies motivate us to implement extensive researches in this area. In this paper, we introduce an innovative method for traffic prediction. We split the whole traffic system into separate sub-regions. Then, each of them would be modeled by deep graph neural networks, which could extract unique characters of the region as well as preserve the topology property of it. Moreover, we transfer the useful knowledge between these regions and share the basic information in order to improve the performance of our model. Finally, we conduct experiment on real world data set and achieve the best performance, which proves the effectiveness of our method.

Keywords: Data mining · Deep learning · Big data · Traffic prediction · Intelligent Transportation System · Smart city

1 Introduction

In recent years, as rapid construction of traffic infrastructures and increasing travel demands of our daily lives, there is an urgent requirement to conduct scientific researches on traffic systems in order to make them smarter and more efficient. Under this background, ITS (Intelligent Transportation System) is invented as a combination of cutting-edge technologies and traffic management measures. As an active and important application in ITS, the study of traffic prediction is to explore quantitative patterns of passengers, vehicles or other traffic participants in various locations and time periods. It gives us more foresights and instructions to alleviate traffic pressures, optimize our transportation choices and avoid traffic accidents, and has already become an important part to build our smart cities.

Nowadays, the development of data mining technologies and machine learning models gives us unprecedented opportunities to implement our research. In this paper,

H. Yin et al. (Eds.): IDEAL 2019, LNCS 11871, pp. 208–219, 2019.
https://doi.org/10.1007/978-3-030-33607-3_23

we create the TKGNN (Transfer Knowledge Graph Neural Network) approach which is applicable for most traffic systems even if they are large, complex and spatially imbalanced. We divide the whole traffic system into different sub-regions. Each region is modeled individually to extract its own features. Deep neural networks with graph convolutional functions are used to deal with nonlinearity in the system as well as capture structural relationship in the traffic network. Besides, we adopt techniques of transfer learning concept, in order to transfer useful messages between different models and improve generalization ability of our method. In the end, we experiment our method on real world data set and compare it with multiple baselines to verify its utility. The result demonstrates that our approach achieves the best performance in different metrics.

2 Related Work

Traffic prediction has been researched for many years due to its importance to our daily activities. Some researchers use the average value of history data to predict future ones in the same location and time period [1]. It is a plain and straightforward way but losses too much information of traffic conditions, which has great influence to prediction results. Another kind of methods takes the traffic forecasting as a time series problem. For example, the model of ARIMA (Autoregressive Integrated Moving Average) calculates autocorrelation and partial autocorrelation factors of the series, taking drift series and error terms into regression function to make better results [2]. Some general machine learning methods such as SVR (Support Vector Regression) [3] and BN (Bayesian Network) [4] are also used for traffic prediction. Recently, more and more studies start paying attention to logical relationships between different locations in traffic system, and build it as a network or graph model to improve performances. Also, deep learning methods are put into use for many problems. For example, [5] adopts SAE (Sparse Auto-encoders) trained by greedy layer-wise algorithm to extract generic features for traffic flow. In [6], a structure of SBU-LSTM (Stacked Bidirectional and Unidirectional LSTM) is proposed, learning features of traffic series from both directions in order to boost the prediction accuracy. Researchers in [7] use CNN (Convolutional Neural Network) and ResNet architecture to capture spatial dependencies of gridding areas in city map. But unfortunately, most methods just adopt the commonly used machine learning models but not seriously consider the complexity and heterogeneity in traffic system, which could undermine predictive ability and aggravate the model instability.

3 Proposed Method

3.1 System Partitioning

In traditional studies, some researchers treat traffic prediction as a kind of time series problems. But in advanced system, locations (such as roads, stations, intersections, etc.) are usually well connected into a typical traffic network and have nonnegligible relationships with each other. So, it is beneficial to research traffic problems in network-level perspective.

But there are still lots of challenges to overcome. Since our societies are becoming increasingly diversified, the traffic functionalities of locations various from each other, which causes different traffic patterns in different areas. It is hard to build one versatile model to capture all kinds of attributes in different conditions. Furthermore, as the development of transportation infrastructures, the sizes and complexities of traffic networks are expanding more than ever, which means the model should grow dramatically to deal with this issue. This usually causes the curse of dimensionality in machine learning and the data will not be sufficient most of the time.

Hence, in order to make our model more applicable for all kinds of traffic problems, we adopt divide-and-conquer concept which partition the whole traffic system into smaller but more compact sub-regions, conducting forecast separately by building the specific model for each of them. To properly divide the main task and make our method effectual, we need to keep the integrity of each sub-regions.

There are three factors that will influence the partition criterion. Here we use notation N as the number of locations over the entire traffic system, integers $i, j \in [1, N]$ as the indices of the locations, and L_i as the ith location. Then we define three $N \times N$ matrices $\mathbf{W}^G$, $\mathbf{W}^F$ and $\mathbf{W}^D$, and they will be detailed in the following paragraphs.

$\mathbf{W}^G$ contains geographic information of the traffic network. In most scenarios, traffic attributes are greatly affected by the distinct functionalities of locations (such as commercial or residential areas, entertainment venues, etc.). Generally, locations with strong geographic connections will have similar attributes, which is a heuristic instruction to put them together. The element w_{ij}^G in $\mathbf{W}^G$ means the geographic connectivity between L_i and L_j, which is defined as follows:

$$w_{ij}^G = \begin{cases} dist(L_i, L_j) & \text{location } i \text{ and } j \text{ are linked} \\ 0 & \text{otherwise} \end{cases} \quad (1)$$

$\mathbf{W}^F$ represents the similarities of flow patterns, which reflects temporal correlations between different locations. Its element w_{ij}^F is defined as follows:

$$w_{ij}^F = exp\left(-\lambda \cdot cor\left(fl_i, fl_j\right)\right) \quad (2)$$

where fl_i denotes the flow patterns of L_i during a time period (such as a day or a week). It is calculated by the average of history flows in this location. The function $cor(\cdot, \cdot)$ is used to measure the correlation between two series. Here we employ DTW (Dynamic Time Warping) algorithm to do this job. The parameter λ is to adjust the sensitivity of this function.

$\mathbf{W}^D$ is the map for mutual dependencies between different locations. Every travel behavior has its origin and destination, and the element w_{ij}^D counts the average number of travels from L_i to L_j in a time window. These origin-destination pairs reflect causal relationships in the traffic system, which means the inflows of some locations are dependent on outflows of other ones.

Now we have three matrices that contain different aspects of information of our traffic network. We use them as adjacency matrices of three graphs G^G, G^F and G^D, the nodes of which are corresponding to our locations. Then, we employ network embedding method [8, 9] to get dense vectors of the nodes for each graph, and the node2vec algorithm is adopted in the following equations.

$$NetEm^X(\cdot) = node2vec(G^X)$$

$$v_i^X = NetEm^X(i) \tag{3}$$

$$X \in \{G, F, D\}$$

Next, three vectors for each node (or location) are concatenated into one after normalization operation which is conducted in order to adjust them to the same scale.

$$v_i = Concat(Norm(v_i^G), Norm(v_i^F), Norm(v_i^D)) \tag{4}$$

Finally, we choose the partitioning number k, and cluster the locations into k sub-regions based on their representing vectors.

3.2   Deep Learning Model for Graph

Most traffic networks can be abstracted as graph structure with nodes for locations and edges for linkages of them. Although deep learning models have the ability to cope with various kinds of features, it is not easy for them to represent general graph structures. Because unlike image or text series data, the components and their relationships in graphs are irregular and non-Euclidean, which means they cannot be sufficiently learned by standard deep learning approaches such as SEA, CNN, RNN, etc. To overcome this problem, graph neural networks (GNNs) are developed to make deep learning capable of dealing with these kinds of data [10–12].

In our case, we keep the multi-layer architecture of neural networks to extract features in different abstraction levels, and in each layer a special convolution kernel is deployed to aggregate information from the graph structure, as is shown in Fig. 1.

We let $\mathbf{x} \in \mathbb{R}^N$ be the input of our model and N is the number of nodes, each of which represents a location. The convolution operation is defined as follows:

$$GraphCov(\mathbf{x}; \boldsymbol{\theta}) = \mathbf{U}g(\boldsymbol{\Lambda}; \boldsymbol{\theta})\mathbf{U}^T\mathbf{x} \tag{5}$$

$$\mathbf{L} = \mathbf{I}_N - \mathbf{D}^{-\frac{1}{2}}\mathbf{A}\mathbf{D}^{-\frac{1}{2}} = \mathbf{U}\boldsymbol{\Lambda}\mathbf{U}^T \tag{6}$$

where $\mathbf{L} \in \mathbb{R}^{N \times N}$ is the normalized graph Laplacian matrix with adjacent matrix $\mathbf{A}$ and degree matrix $\mathbf{D}$, and $\mathbf{I}_N$ is an $N \times N$ identity matrix. $\mathbf{U}$ is the matrix of eigenvectors of $\mathbf{L}$. $\boldsymbol{\Lambda}$ is a diagonal matrix containing eigenvalues corresponding to eigenvectors of $\mathbf{U}$. $g(\boldsymbol{\Lambda}; \boldsymbol{\theta})$ is a function parameterized by $\boldsymbol{\theta}$, which can be approximated as [13]:

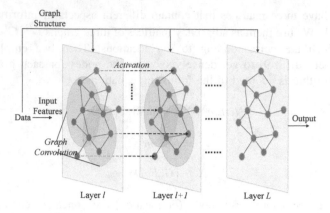

Fig. 1. Deep convolutional neural network for graph structure.

$$g(\Lambda; \boldsymbol{\theta}) \approx \sum_{m=0}^{M} \theta_m \times \mathbf{T}_m\left(\frac{2}{\lambda_{max}}\Lambda - \mathbf{I}_N\right) \tag{7}$$

where λ_{max} is the largest value of Λ. The terms of $\mathbf{T}_m(\cdot)$ are the Chebyshev polynomials which are defined as $\mathbf{T}_m(\mathbf{x}) = 2\mathbf{x}\mathbf{T}_{m-1}(\mathbf{x}) - \mathbf{T}_{m-2}(\mathbf{x})$, with $\mathbf{T}_0(\mathbf{x}) = 1$ and $\mathbf{T}_1(\mathbf{x}) = \mathbf{x}$. Therefore, the convolution function can be approximated as:

$$GraphCov(\mathbf{x}; \boldsymbol{\theta}) \approx \sum_{m=0}^{M} \theta_m \times \mathbf{T}_m\left(\frac{2}{\lambda_{max}}\mathbf{L} - \mathbf{I}_N\right)\mathbf{x} \tag{8}$$

Until now the function is still complicated, so measurements for simplification are need to reduce computational overhead and avoid the problem of overfitting. To be specific, constrains are made by letting $M = 1$, $\lambda_{max} = 2$ and $\theta = \theta_0 = -\theta_1$. Then we obtain the simplified equation:

$$GraphCov(\mathbf{x}; \boldsymbol{\theta}) \approx \theta\left(\mathbf{I}_N + \mathbf{D}^{-\frac{1}{2}}\mathbf{A}\mathbf{D}^{-\frac{1}{2}}\right)\mathbf{x} \tag{9}$$

Next, we put the function into practical deep learning model.

$$\mathbf{H}^{(l+1)} = \sigma\left(\left(\mathbf{I}_N + \mathbf{D}^{-\frac{1}{2}}\mathbf{A}\mathbf{D}^{-\frac{1}{2}}\right)\mathbf{H}^{(l)}\boldsymbol{\Theta}^{(l)} + \mathbf{b}^{(l)}\right) \tag{10}$$

$$\mathbf{H}^{(0)} = \mathbf{X} \tag{11}$$

Here we extend $\mathbf{x}$ to matrix $\mathbf{X} \in \mathbb{R}^{N \times D}$, which contains D-dimensional feature vectors (such as history flows, statistic information, etc.) for each traffic location. $\mathbf{H}^{(l)}$ denotes the hidden state of layer l, aggregating features from lower layer and abstracting them as its output. $\sigma(\cdot)$ is the activation function of the neural unit. But the term $\mathbf{I}_N + \mathbf{D}^{-\frac{1}{2}}\mathbf{A}\mathbf{D}^{-\frac{1}{2}}$ is unstable during the backpropagation process, which may cause gradients exploding or vanishing problems. We can replace it with $\tilde{\mathbf{D}}^{-\frac{1}{2}}\tilde{\mathbf{A}}\tilde{\mathbf{D}}^{-\frac{1}{2}}$ to alleviate this deficiency, where $\tilde{\mathbf{A}} = \mathbf{A} + \mathbf{I}_N$ and $\tilde{\mathbf{D}}_{ii} = \sum_j \tilde{\mathbf{A}}_{ij}$.

$$\mathbf{H}^{(l+1)} = \sigma\left(\tilde{\mathbf{D}}^{-\frac{1}{2}}\tilde{\mathbf{A}}\tilde{\mathbf{D}}^{-\frac{1}{2}}\mathbf{H}^{(l)}\mathbf{\Theta}^{(l)} + \mathbf{b}^{(l)}\right) \tag{12}$$

3.3 Knowledge Transfer

Now we have already described how to divide a traffic system into different parts and the deep learning method to extract high level features of them. In this section, we will introduce learning techniques that can coordinate multiple related models and transfer messages between them so as to acquire helpful knowledge across these sub-regions.

As mentioned above, the traffic system is partitioned into different sub-regions which are modeled by individual graph convolutional neural networks. Although be trained by different data, these models still contain common characters and inter-region connections in traffic system. If we can transfer these knowledges between the them, not only the generalization capability will be improved but also more supportive information can be leveraged to conduct prediction. So, in our method, we involve mechanisms to accomplish this goal and the overall framework of our method is depicted in Fig. 2.

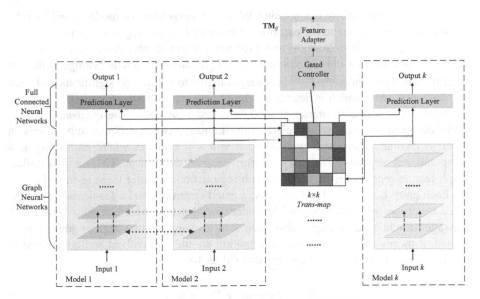

Fig. 2. Overview of the architecture of TKGNN.

The dotted lines between corresponding layers of separate models indicate that their weight matrices are relevant and some intrinsic characters can be shared or influenced by each other. It is proved that in deep neural networks, the abstraction degrees of features are not the same across multiple layers [14]. More specifically, the shallow layers are apt to extract generic features which are not constrained by particular domains, and this kind of features could be conducively transferred to related tasks. Whereas the features in deep layers are more task-specific and have lower transferability than the shallow ones [15, 16].

In our case, the lower level parameters of neural networks are connected between different models, which means the knowledge for basic features are compatible and can be referred by overall counterparts. This improves the generalization and robustness of our method. While as the layer goes deeper, this restriction would be gradually loosened. Each model relies more on its sub-region data to update parameters, making it more suitable for particular prediction task. The optimization functions are presented as follows:

$$\mathbf{W}_i = \arg \min_{\mathbf{W}} \mathcal{L}(\mathbf{Y}_i, f(\mathbf{X}_i; \mathbf{W})) + \sum_{j \neq i}^{k} \sum_{l=1}^{L} \alpha^{(l)} \|\mathbf{W}^{(l)} - \mathbf{W}_j^{(l)}\|_F \quad (13)$$

$$\alpha^{(l)} = \begin{cases} \beta & l \leq L1 \\ \beta \cdot \frac{L2-l}{L2-L1} & l \in (L1, L2) \\ 0 & l \geq L2 \end{cases} \quad (14)$$

In above equations, we use notation $\mathbf{W}_i$ as the parameters of model i, and its total number is corresponding to the partitioning number k of sub-regions. Operators $\mathcal{L}(\cdot, \cdot)$ and $\|\cdot\|_F$ represent loss function and Frobenius norm respectively, which are optimization objectives for our tasks. Superscript (l) means the variable belongs to the l th layer of neural networks. And $\alpha^{(l)}$ is a weight factor to control the significance of each regularization term, which is determined by hyperparameters β, $L1$ and $L2$.

Next, we describe an important module called *trans-map*. The *trans-map* is specially designed for transferring related information between different sub-regions in order to facilitate the overall prediction. It is a $k \times k$ matrix where each entry is a computational structure. The entry in *trans-map* is denoted as $\mathbf{TM}_{ij}$, which collect graph feature represented by model of sub-region i and transfer them to model j.

Each entry is composed of two parts: a gated controller and a feature adapter. The gated controller of $\mathbf{TM}_{ij}$ contains dependent information which decides how much message should pass through sub-region i to j. Suppose two regions are strongly related, it means much inter-region information should be involved and considered in this prediction task. The functions are defined as follows:

$$\mathbf{Z}_{ij} = sigmoid\left(\mathbf{F}_i \mathbf{W}_{ij}^Z + \mathbf{b}_{ij}^Z\right) \quad (15)$$

$$\mathbf{O}_{ij}^C = \mathbf{Z}_{ij} \otimes \left(\mathbf{F}_i \mathbf{W}_{ij}^C + \mathbf{b}_{ij}^C\right) \quad (16)$$

where $\mathbf{F}_i$ is the graph feature extracted by model i, also as the input of $\mathbf{TM}_{ij}$. $\mathbf{W}_{ij}^Z$, $\mathbf{b}_{ij}^Z$, $\mathbf{W}_{ij}^C$, $\mathbf{b}_{ij}^C$ are trainable parameters and $\mathbf{O}_{ij}^C$ is the output of the controller. Operator $\otimes$ is Hadamard product, which means element-wise product of matrices. The feature adapter of $\mathbf{TM}_{ij}$ is used to make the feature of model i adaptive to model j.

$$\mathbf{O}_{ij}^A = \sigma\left(\mathbf{O}_{ij}^C \mathbf{W}_{ij}^A + \mathbf{b}_{ij}^A\right) \tag{17}$$

In this equation, $\sigma(\cdot)$ is the activation function for feature transformation. $\mathbf{W}_{ij}^A$, $\mathbf{b}_{ij}^A$ are trainable parameters and $\mathbf{O}_{ij}^A$ is the output.

By employing *trans-map* module, we can get our prediction result more reliable. Another important thing is that the interpretability of our method is greatly improved by knowing dependent relationships between sub-regions, which is valuable but often missing in deep learning models.

In the end, prediction layers are built for each model to forecast the final result. This part of model is composed by stacked full connected neural networks that collects features from the graph neural network and entries in *trans-map* for its corresponding task.

4 Experiment

4.1 Data Set and Evaluation Metrics

Our data set is collected by Beijing subway system, which is one of the largest and most crowded subway systems in China. The range of dates is from 1st April 2015 to 29th June 2015. The records are automatically generated by the swiping card actions when people get in or pull out of the subway stations. Each record contains items of card id, card type, station number, line number, expense, deal time, etc. Here we use 80% of the data to train our model and the other 20% are used as testing set.

In the experiment, we use three metrics to test the performance of our method, which are MAS (Mean Absolute Error), RMSE (Root Mean Square Error) and MAPE (Mean Absolute Percentage Error). The equations of them are defined as follows:

$$MAE(\mathbf{T}, \mathbf{Y}) = \frac{1}{N_s} \sum_{i=1}^{N_s} |t_i - y_i| \tag{18}$$

$$RMSE(\mathbf{T}, \mathbf{Y}) = \sqrt{\frac{1}{N_s} \sum_{i=1}^{N_s} (t_i - y_i)^2} \tag{19}$$

$$MAPE(\mathbf{T}, \mathbf{Y}) = \frac{1}{N_s} \sum_{i=1}^{N_s} \left| \frac{t_i - y_i}{t_i} \right| \tag{20}$$

where $\mathbf{T} = \{t_i | i = 1, 2, \ldots N_s\}$ denotes the target values we need to predict, and $\mathbf{Y} = \{y_i | i = 1, 2, \ldots N_s\}$ is the set of results generated by the model. The number N_s is the total amount of samples.

4.2 Result Analysis

We test our model with 5 baselines: HA, ARIMA, SVR, MLP (Multi-Layer Perception) and Single-TKGNN. The last one has the same structure as our single model for each sub-region, but lack of traffic system partition and knowledge transfer.

Table 1 illustrates the overall comparison of results from different models. It shows that our approach achieves the best performance with regard to all three metrics, which proves the effectiveness of our research. Let's take a view of the Single-TKGNN, this model performs well in most cases comparing with other baselines, but still could not compete with the final method. This demonstrate that every piece of our procedures is meaningful, suggesting that we should implement them together and take them as an entire work.

Table 1. Results comparison of different methods.

Method	MAE	RMSE	MAPE
HA	27.05	50.75	0.2047
ARIMA	25.32	47.22	0.1834
SVR	23.28	41.36	0.1747
MLP	21.02	37.72	0.1617
Single-TKGNN	20.83	37.15	0.1632
TKGNN	**19.60**	**34.71**	**0.1476**

The comparison between real traffic flow and predicted results are shown in Fig. 3 (time has been discretized into interval numbers). For brevity, we only display two results. One is from our proposed approach, the other is from the best of the baseline methods. The figure shows that although both results have the similar trend with the ground truth flow, the curve of our approach fits the real flow more proximately than the contrast result. What is more, in the unstable and fluctuant situations, the accuracy of TKGNN is obviously better than the baseline method. These phenomena illustrate that our model can sharply capture the variation signals of different traffic conditions, and adjust the prediction result rapidly according to these messages. This is because TKGNN has the strong ability to extract intrinsic identities from different regions as well as exploit the relationships of the flows distributed in the traffic network.

Next, we would like to observe the change of prediction errors with regard to different partitioning number k. A small k implies that the sizes of sub-regions are relatively large. We can see from Fig. 4 that the errors are going down as we increase the number of k. This result is as expected, because as the number increases, multiple models are getting enough power to capture comprehensive characters of the entire traffic system. Then, as the number k increases to a threshold value, the method starts to lose its precision gradually. The reason could be that the intra-region features are

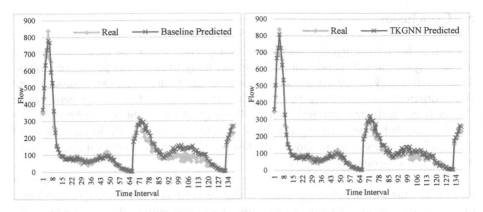

Fig. 3. Ground truth flow fitted by predicted result.

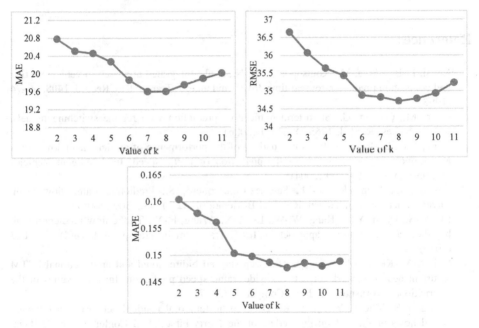

Fig. 4. Change of performance w.r.t. partitioning number.

getting meaningless and lack of information when the traffic structure in this region becomes too simple. The outcome suggests that we should choose a proper number of our sub-regions, which is neither too big nor too small, in order to achieve the best result. The appropriate choice depends on the size, complexity and heterogeneity of the traffic system in practice.

5 Conclusion

The researches on intelligent transportation is important due to the rapid development of traffic system and its strong correlation to our daily lives. In this paper, we propose an innovative way to predict traffic conditions in the future time. Our approach considers different kinds of characters in traffic system, using deep neural networks to extract latent attributes and deal with nonlinearity of the problem. Furthermore, we transfer knowledge between different parts of traffic system to take full advantage of information. The experiment shows our method has the best prediction result among multiple comparisons.

What we proposed is a general prediction approach which can work in different cases with little modification. For future work, we are going to discover various kinds of spatial and temporal patterns, representing them and putting them into new traffic solutions with deep learning models. In addition, further research will be done for the usage of high dimensional tensors in spatiotemporal problems.

References

1. Kaysi, I., Benakiva, M., Koutsopoulos, H.-N., et al.: Integrated approach to vehicle routing and congestion prediction for real-time driver guidance. J. Transp. Res. Record. **1408**, 66–74 (1993)
2. Cetin, M., Comert, G.: Short-term traffic flow prediction with regime switching models. J. Transp. Res. Record. **1965**(1), 23–31 (2006)
3. Vanajakshi, L., Rilett, L.-R.: A comparison of the performance of artificial neural networks and support vector machines for the prediction of traffic speed. In: Intelligent Vehicles Symposium, pp. 194–199 (2004)
4. Castillo, E., Menéndez, J.-M., Sánchez-Cambronero, S.: Predicting traffic flow using Bayesian networks. J. Transp. Res. Part B-Methodol. **42**(5), 482–509 (2008)
5. Lv, Y.-S., Duan, Y.-J., Kang, W.-W., Li, Z.-X., Wang, F.-Y.: Traffic flow prediction with big data: a deep learning approach. J. IEEE Trans. Intell. Transp. Syst. **16**(2), 865–873 (2015)
6. Cui, Z.-Y., Ke, R.-M., Wang, Y.-H.: Deep stacked bidirectional and unidirectional LSTM recurrent neural network for network-wide traffic speed prediction. In: Proceedings of the International Workshop on Urban Computing (2017)
7. Zhang, J.-B., Zheng, Y., Qi. D.-K.: Deep spatio-temporal residual networks for citywide crowd flows prediction. In: Proceeding of the Thirty-First AAAI Conference on Artificial Intelligence (2017)
8. Perozzi, B., Al-Rfou, R., Skiena, S.: Deep walk: online learning of social representations. In: Proceedings of the 20th ACM SIGKDD International Conference on Knowledge Discovery and Data Mining, pp. 701–710. ACM, New York (2014)
9. Grover, A., Leskovec, J.: node2vec: scalable feature learning for networks. In: Proceedings of the 22nd ACM SIGKDD International Conference on Knowledge Discovery and Data Mining (2016)
10. Kipf, T.-N., Welling, M.: Semi-supervised classification with graph convolutional networks. In: Proceedings of International Conference on Learning Representations (ICLR 2017) (2017)

11. Niepert, M., Ahmed, M., Kutzkov, K.: Learning convolutional neural networks for graphs. In: Proceedings of the 33rd International Conference on Machine Learning, PMLR 48, pp. 2014–2023 (2016)
12. Velickovic, P., Cucurull, G., Casanova, A., Romero, A., et al.: Graph attention networks. In: Proceedings of ICLR (2018)
13. Hammond, D.-K., Vandergheynst, P., Gribonval, R.: Wavelets on graphs via spectral graph theory. J. Appl. Comput. Harmonic Anal. **30**(2), 129–150 (2011)
14. Zeiler, M.D., Fergus, R.: Visualizing and understanding convolutional networks. In: Fleet, D., Pajdla, T., Schiele, B., Tuytelaars, T. (eds.) ECCV 2014. LNCS, vol. 8689, pp. 818–833. Springer, Cham (2014). https://doi.org/10.1007/978-3-319-10590-1_53
15. Yosinski, J., Clune, J., Bengio, Y., Lipson, H.: How transferable are features in deep neural networks? In: Advances in Neural Information Processing Systems, vol. 27, pp. 3320–3328 (2014)
16. Long, M., Cao, Y., Wang, J., et al.: Learning transferable features with deep adaptation networks. In: 32nd International Conference on Machine Learning (ICML 2015) (2015)

Global Q-Learning Approach for Power Allocation in Femtocell Networks

Abdulmajeed M. Alenezi[1,2]([⊠]) and Khairi Hamdi[1]

[1] School of Electrical and Electronic Engineering, The University of Manchester, Manchester, UK
abdulmajeed.alenezi@postgrad.manchester.ac.uk, k.hamdi@manchester.ac.uk
[2] Islamic University of Medinah, Madinah, Kingdom of Saudi Arabia

Abstract. In dense femtocell network, the complexity of the resource allocation increases significantly as the network becomes denser, which limits the performance of the network. The usage of reinforcement learning to solve the resource allocation problem showed promising results compared to conventional methods. In this work, we use global Q-learning approach on the macro base station to solve the resource allocation problem in a dense and complex network. We propose a new reward function that can be implemented on a centralized Q-learning and achieve good results in terms of maintaining the quality of service for the macro user and maximizing the sum capacity of the femtocell users. In comparison to other reward functions, the proposed reward function maintained both the QoS for the macro user and fairness among all femtocell users.

Keywords: Femtocell · Q-learning · Resource allocation

1 Introduction

Due to the high demand for wireless data transmission and the dramatic growth in the number of wireless users, researchers are trying to enhance the wireless network to maintain the required quality of service (QoS) and maximizing the capacity for each user. Several studies stated that the current techniques are not adequate to satisfy the high demand in the future as the mobile traffic will increase over thousands of times in the next decade [3].

One possible approach researchers try to investigate to satisfy the demand for high capacity is to use femtocells [1]. Femtocell is a small base station with low transmitted power that can be deployed by the end user. To extend the coverage in the building, femtocell is a promising solution for any indoor scenarios that are out of coverage [5]. One of the greatest advantages of femtocells is that it does not need a new spectrum as it allows the user to reuse the same spectrum assigned to the nearest macro user. However, implementing femto base stations in the same coverage of the macro user while using orthogonal frequency division multiple access (OFDMA) creates co-channel and cross-channel interference [8].

© Springer Nature Switzerland AG 2019
H. Yin et al. (Eds.): IDEAL 2019, LNCS 11871, pp. 220–228, 2019.
https://doi.org/10.1007/978-3-030-33607-3_24

This interference increases proportionally to the number of deployed femto base stations in the same area and significantly impacts the QoS for each user [11].

Several techniques were suggested and investigated to solve the resource allocation problem in the femtocell network. Most of the work done was either using frequency selective or power allocation techniques between the femto base stations [7,10]. However, as the number of femto base stations (FBS) increase, the current techniques can not solve the optimization problem while maintaining both high capacity and QoS. To solve the resource allocation problem in dense heterogeneous network (HetNet), reinforcement learning has been implemented recently in wireless communications [4].

In this paper, we propose a centralized Q-learning approach on the macro base station that maintains the QoS for the macro user and maximizes the sum capacity of the femto users equipment. Our contribution can be categorized into two main points.

- A new global Q-learning approach is presented to solve the resource allocation problem in a femtocell network. The proposed approach was able to achieve similar results to the cooperative Q-learning approach.
- A new reward function that can be implemented on the global Q-learning to maintain the QoS of the macro user and maximize the sum capacity of the femto users' equipment for dense femtocell network.

The rest of the paper is organized as follows. In Sect. 2, the system model is presented. We formulate the problem and propose a solution in Sect. 3. Section 4 presents simulation results, while Sect. 5 concludes the paper.

2 System Model

A HetNet scenario is considered, which consists of a macro base station (MBS) serving only one macro user, and L femto base stations. Each base station serves only one user at the same time. All base stations operate at the same spectrum, which creates interference in the downlink as the network's density increases. We focus on the power allocation problem in the downlink.

As the received signal at the macro user equipment (MUE) and the femto user equipment (FUE) contains co-channel and cross-channel interference from the other base stations, the signal to interference plus noise ratio can be calculated using the following equations

$$SINR_{MUE} = \frac{P_{MBS}h_{MBS,MUE}}{\sum_{i=1}^{L} P_i h_{FBS_i,MUE} + \sigma^2} \tag{1}$$

$$SINR_{FUE_i} = \frac{P_i h_{FBS_i,FUE_i}}{P_{MBS}h_{MBS,FUE_i} + \sum_{j=0,j\neq i}^{L} P_j h_{FBS_j,FUE_i} + \sigma^2}, \tag{2}$$

The parameters in Eq. (1) can be illustrated as follow: P_{MBS} is the transmitted power of the macro base station, $h_{MBS,MUE}$ is the channel gain from the MBS to the MUE, P_i is the transmitted power of FBS_i, $h_{FBS_i,MUE}$ is the channel gain from FBS_i to the MUE, σ^2 is the variance of the additive White Gaussian Noise (AWGN).

The parameters in Eq. (2) can be illustrated as follow: P_i is the transmitted power from FBS_i, h_{FBS_i,FUE_i} is the channel gain from FBS_i to the FUE_i, P_j is the power transmitted by FBS_j, h_{MBS,FUE_i} is the channel gain from MBS to the FUE_i ,and h_{FBS_j,FUE_i} is the channel gain from FBS_j to the FUE_i. Similar to the prior work in [2], all channel parameters are assumed to be known by the FBS. The normalized capacity for any user is calculated using the following equations:

$$C_{MUE} = log_2(1 + SINR_{MUE}) \tag{3}$$

$$C_{FUE_i} = log_2(1 + SINR_{FUE_i}), i = 1, .., L \tag{4}$$

where $SINR$ is the signal-to-interference-plus-noise ratio, which can be calculated using Eqs. 1 and 2.

3   Problem Formulation and Proposed Solution

The main goal of the optimization problem is to maximize the sum capacity for all FUE while maintaining the MUE capacity above a certain threshold. Each FBS has the same set of transmit powers $\overline{p} = (p_1, p_2, ..., p_{max})$. The optimization problem can be defined as:

$$\underset{\underset{\sim}{p}}{maximize} \sum_{k=1}^{M} C_{FUE_k} \tag{5}$$

To ensure high QoS, Eq. 5 needs to satisfy the following constraints:

$$P_i \leq P_{max}, i = 1, 2, ..., M$$
$$C_{MUE} \geq \Gamma_{MUE} \tag{6}$$
$$C_{FUE_i} \geq \Gamma_{FUE}, i = 1, 2, .., M$$

where Γ_{MUE} refers to the threshold capacity of the MUE, Γ_{FUE} refers to the threshold capacity of the FUE. By ensuring a limited power to each FBS, the goal is to maximize the sum capacity of all FUE without affecting the QoS for the MUE specified by the threshold Γ_{MUE}. To solve the optimization problem, we focus our attention on the Q-learning technique.

3.1   Q-Learning Approach

In most of the previous works, most of the researchers consider the FBS as an agent and apply the Q-learning algorithm at the FBS [2]. After comparing cooperative and non-cooperative agents, the best result is achieved by sharing the information at each iteration to achieve a faster learning process. In recent works, some papers suggested sharing only partial of the Q-table at each iteration. In all cases, agents need to communicate at each iteration through the backhaul network. According to [11], low-frequency sharing does not benefit the learning process and might reach similar results to the independent Q-learning. Thus agents need to share their information at each iteration to help each other to learn faster. In this way, more overhead communication is added to the network. The learning process can be improved at the cost of high communication.

Another aspect that needs investigation is the reward function [4,9]. In [2], the reward function achieved the best results in maximizing the FUE capacity while maintaining the MUE capacity close the threshold. However, after adding eight FUE near the macro user, the reward function failed to maintain the MUE capacity above the threshold.

3.2 Proposed Solution

To avoid the coordination and communication between agents in Q-learning, we can apply a centralized Q-learning at the macro base station. Assuming that the macro base station knows the location of the FBS, Q-Learning can be implemented using a controller at the macro base station. In femtocell network scenario, the Q-learning parameters can be defined as follows:

Agents: The macro base station acts as an agent. It uses the ϵ-greedy policy for exploration. The agent chooses an action with a probability of $1 - \epsilon$, and act randomly with a probability of ϵ.

Actions: The macro base station can choose a transmit power between P_{min} and P_{max} for each FBS from a set $A = (a_1, a_2, ..., a_{N_{power}})$. All actions have the same probability of occurrence, which can be applied using equal step sizes between P_{min} and P_{max}.

States: States are chosen based on the location of the FBS to both MUE and MBS. $S_t^i \in (D_{MUE}, D_{MBS})$. D_{MUE} defines how far is the FBS from the MUE, and D_{MBS} defines how far is the FBS from the MBS. By defining the states in this way, each FBS has a specific state as long as the location is fixed. In dense environment where multiple FBS cause interference to the MUE, they will share the same state and this can help in improving the learning speed as all nearby FBS will only use one state.

Reward Function: To ensure a sufficient reward function that achieves good results in maximizing the sum capacity of the FUEs, we need to include all the constraints in the reward function. To do that, we propose a new reward function that can be implemented in the centralized Q-learning approach. The reward function at time step t can be defined as

$$
R_t = \begin{cases} \prod_{i=1}^{L}(B_i C_{FUE_i})(C_{MUE})^2 - \frac{1}{k}(C_{MUE} - \Gamma_{MUE}) \\ - \sum_{i=1}^{L}(C_{FUE_i} - \Gamma_{FUE}), & \text{if } C_{MUE} \geq \Gamma_{MUE}, \\ \\ \prod_{i=1}^{L}(B_i C_{FUE_i})(C_{MUE})^2 - \frac{K_p}{k}, & \text{if } C_{MUE} < \Gamma_{MUE}. \end{cases}
$$

$$(7)$$

Unlike the reward function in [2], this reward function guarantees the QoS for the MUE above certain threshold and can maximize the sum capacity for all FUEs. To do that, we added a penalty K_p to the reward function whenever the MUE capacity is below the threshold. The rest of the parameters are illustrated as follow: B_i is the normalized distant from FUE_i to the MUE, k is the average normalized distance between all connected FUE's and the MUE. In Eq. 7, the first term implies high reward value when the capacity of either MUE or FUE is high. The MUE capacity is squared to imply a higher reward for the MUE. Note that Bi is normalized which reduces the reward value if the distance between the MUE and FUE is less than the reference distance. We use the $\prod$ of all connected FUEs to provide fairness among all the FUEs. When one of the FUE's capacity is below the threshold, this would affect the reward value as the total value is multiplied by a number less than one. The second and third terms are used to reduce the overall reward value. We apply the reward function in the Q update equation shown below

$$Q(s,a) \leftarrow Q(s,a) + \alpha[r + \gamma\max_a * Q(s^{'},a) - Q(s,a)] \tag{8}$$

where α is the learning rate, γ is the discount rate, and r is the reward function. For the optimal Q-value, the agent selects action a and receives a reward r affected by discount factor γ of performing the policy. Algorithm 1 shows the learning procedural.

Algorithm 1.Q-Learning algorithm

initialize $Q(x_t, a_t)$ arbitrarily
for all episodes **do**
Initialize x_t

for all steps of episode **do**
    Choose a_t for all FBS from set of actions
    Take action a_t, observe R_t, x_{t+1}
    $Q(x_t, a_t) \leftarrow (1-\alpha)Q(x_t, a_t) + \alpha\max_a(R_t + \gamma Q(x_{t+1}, a))$
    $x_t \leftarrow x_{t+1}$
end for
end for

---

4   Simulation Results

In this section, both simulation setup and results are presented to show the performance of the system.

4.1   Simulation Setup

The environment is simulated using a single macro base station (MBS) severing one MUE, and 10 FBSs. Each FBS is serving only one FUE. To simulate a dense environment where multiple FBS interfere with a MUE, we placed all the FBS in the coverage of the MUE. All FUEs are located within 10 m from its serving FBS. Figure 1 shows the location of the MBS, MUE, FBSs, and FUEs. The rings around the MBS and MUE defines the states. The total number of rings for both MUE and MBS is equal to 3. Defining the states is essential in the case of dense environment. For example, the FBSs that are located in the third ring of the MBS and the second ring of the MUE will use the same state.

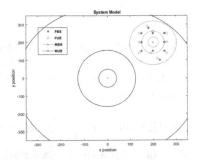

Fig. 1. Locations of MBS, MUE, FBS, and FUEs.

The MBS and all the FBSs are assumed to be operating over the same channel bandwidth at 2.4 GHz. The channel gain is assumed to be depending only on the path loss. The path loss model for the link between the MBS and its associated user is the same as the path loss model for the FBS's and its associated FUEs. The path loss is calculated using the following equation [6]:

$$L_p = (\frac{4\pi d_1}{\lambda})^2 (d/d_1)^n \tag{9}$$

where d_1 is the reference distance, d is the distance between the transmitter and receiver, n is the path loss exponent, and λ is the wavelength. The path loss exponent is set to 3 in this scenario.

The required QoS for both MUE and FUEs are defined in this simulation as $\Gamma_{MUE} = 1$ (b/s/Hz) and $\Gamma_{FUE_i} = 1$(b/s/Hz). It is specified as the minimum capacity that the user needs to support his application. In the Q-learning algorithm, the agent uses ϵ-greedy policy with $\epsilon = 0.1$. The maximum number of iterations is set to 100,000. For each number of connected FBS, the algorithm runs through all iterations using the algorithm in (Sect. 3.2). The learning rate in Eq. 8 is set to be $\alpha = 0.5$ where the discount factor is set to be $\gamma = 0.9$. For each FBS, the total number of actions is set to be 3. The FBS can select a transmit power from $P_t \in (-20 \text{ dBm}, 0 \text{ dBm}, 10 \text{ dBm})$. The rest of the parameters are illustrated in Table 1.

To investigate the effectiveness of the proposed algorithm, we run the algorithm using two methods. We want to know which method is the best in term of maintaining the QoS for the MUE while maximizing the FUEs capacity.

In the first method we applied, we want to investigate the implementation of one of the existing reward functions in the centralized Q-learning. To do that, we

Table 1. Simulation parameters

Parameter	Value	Parameter	Value
$d1_{MBS}$	50 m	$d1_{MUE}$	15 m
$d2_{MBS}$	150 m	$d2_{MUE}$	50 m
$d3_{MBS}$	400 m	$d3_{MUE}$	125 m
d_{th}	25 m	K_p	10^{-12}

use the reward function in [2]. Instead of updating the Q-value for each FBS, we changed the algorithm to include the sum of all reward values for each connected FUE as shown in Eq. 10. In this way, The Q-value at each iteration depends on all the connected FUE reward values.

$$Q(x_t, a_t) \leftarrow (1 - \alpha)Q(x_t, a_t) + \alpha\max_a(\sum R_t^i + \gamma Q(x_{t+1}, a)) \tag{10}$$

where $\sum R_t^i$ is the summation of the reward value for all connected FUEs.

In the second method which we proposed, we modified the reward function to serve all the connected FUEs. We applied the proposed reward function (RF) in Eq. 7 to the same algorithm. In the rest of the simulation, we call the proposed method (Proposed RF), and the RF in method 1 as (RF1). To have a fair comparison, we also simulate the RF in [2] and called it RF2.

4.2   Results

In this section, we compare the propose RF to the other two RFs discussed above. For each RF, we plotted the measurement of the MUE capacity, the sum capacity of the FUEs, and the capacity of each FUEs.

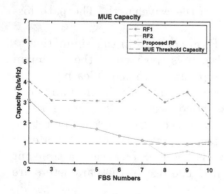

Fig. 2. MUE capacity

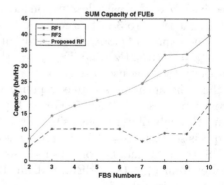

Fig. 3. Sum capacity of FUEs

Figure 2 shows the MUE capacity for the three RFs. We can notice that RF1 has the highest MUE capacity for both low and dense users environment. When

2–7 users are connected, all RFs satisfied the MUE capacity constraint which is above 1 b/s/Hz. RF1 outperformed the other two RFs. When we compare RF2 with the proposed RF, we can notice that both RFs succeed in maintaining the MUE capacity above the threshold up to 7 users connected. However, once the number of connected users exceeds 7, RF2 fails to satisfy the QoS for the MUE, which is similar to the finding in [2]. The proposed RF, however, maintains the MUE capacity above the threshold all the time. Figure 3 shows the sum capacity of FUEs for the three RFs. RF1 has the lowest sum capacity of FUEs while both RF2 and the proposed RF have similar results in maximizing the sum capacity of the FUEs. RF2 shows better results when the number of users exceeded 7.

To have a fair comparison between the three RFs, we need to consider both MUE capacity and the sum of FUEs. As shown in Figs. 2 and 3, RF1 purely depends on the MUE capacity as the sum capacity of FUEs was not high enough compared to the other RFs. The main reason for that is when using $\sum R_t^i$, the algorithm does not care about the users with the low reward value as long as there are other users with high reward values. Thus, this algorithm fails to maintain fairness among the FUEs. When we compare RF2 and the proposed RF, we can notice that both have similar results when the number of connected users does not exceed 7. As the number of users increases, we can notice that RF2 maximizes the sum capacity of FUEs better than the proposed RF but at the expense of not maintaining the MUE capacity above the threshold. Thus, the proposed RF showed better overall results as it maintains the QoS for MUE while maximizing the sum capacity of the FUEs.

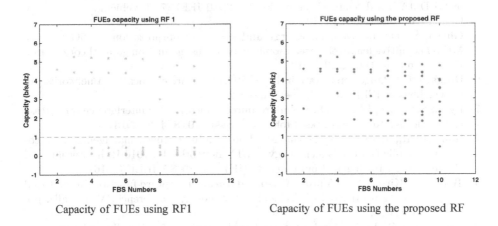

Capacity of FUEs using RF1   Capacity of FUEs using the proposed RF

Fig. 4. Performance of centralized Q-Learning

Figure 4 shows the capacity of the FUEs for RF1 and the proposed RF. RF1 failed to maintain the fairness among all FUEs as there is a significant gap between the users' capacity while the proposed RF maintained the fairness among all users as all of the users' capacity above the threshold except one user when ten users are connected.

5   Conclusion

In this paper, we applied a centralized Q-learning to solve the resource allocation problem in dense femtocell network. By reducing the number of used action, we can implement a centralized Q-learning and achieve similar results compared to the distributed approach, which significantly reduce the communication cost between the base stations. The results showed that with a proper reward function, we can achieve better results in term of maintaining the QoS for the macro user and fairness among all users, wheres the RF in [2] serves up to only 8 users. Future work will explore both modifying the reward function and the use of deep Q-Learning for the centralized approach.

References

1. Abdelnasser, A., Hossain, E., Kim, D.I.: Clustering and resource allocation for dense femtocells in a two-tier cellular ofdma network. IEEE Trans. Wirel. Commun. **13**(3), 1628–1641 (2014)
2. Amiri, R., Mehrpouyan, H., Fridman, L., Mallik, R.K., Nallanathan, A., Matolak, D.: A machine learning approach for power allocation in hetnets considering QOS. In: 2018 IEEE International Conference on Communications (ICC), pp. 1–7. IEEE (2018)
3. Bleicher, A.: A surge in small cells [2013 tech to watch]. IEEE Spectr. **50**(1), 38–39 (2012)
4. Galindo-Serrano, A., Giupponi, L.: Distributed q-learning for interference control in OFDMA-based femtocell networks. In: 2010 IEEE 71st Vehicular Technology Conference, pp. 1–5. IEEE (2010)
5. Ghosh, S., De, D., Deb, P.: Energy and spectrum optimization for 5G massive MIMO cognitive femtocell based mobile network using auction game theory. Wirel. Pers. Commun. **106**, 1–22 (2019)
6. Heegard, C.: Range versus rate in IEEE 802.11 G wireless local area networks. In: Proceedings of IEEE, vol. 802, September 2001
7. Pyun, S.Y., Lee, W., Jo, O.: Uplink resource allocation for interference mitigation in two-tier femtocell networks. Mobile Inf. Syst. **2018**, 1–6 (2018)
8. Raheem, R., Lasebae, A., Aiash, M., Loo, J.: Interference management for co-channel mobile femtocells technology in LTE networks. In: 2016 12th International Conference on Intelligent Environments (IE), pp. 80–87. IEEE (2016)
9. Tefft, J.R., Kirsch, N.J.: A proximity-based q-learning reward function for femtocell networks. In: 2013 IEEE 78th Vehicular Technology Conference (VTC Fall), pp. 1–5. IEEE (2013)
10. Yu, J., Han, S., Li, X.: A robust game-based algorithm for downlink joint resource allocation in hierarchical ofdma femtocell network system. IEEE Trans. Syst. Man Cybern. Syst. **99**, 1–11 (2018)
11. Zhang, H., Li, H., Lee, J.H., Dai, H.: Qos-based interference alignment with similarity clustering for efficient subchannel allocation in dense small cell networks. IEEE Trans. Commun. **65**(11), 5054–5066 (2017)

Deep Learning and Sensor Fusion Methods for Studying Gait Changes Under Cognitive Load in Males and Females

Abdullah S. Alharthi[✉] and Krikor B. Ozanyan

School of Electrical and Electronic Engineering, The University of Manchester,
Sackville Street, Manchester M1 3BU, UK
{abdullah.alharthi, k.ozanyan}@manchester.ac.uk

Abstract. Human gait is the manner of walking in people. It is influenced by weight, age, health condition or the interaction with the surrounding environment. In this work, we study gait changes under cognitive load in healthy males and females, using machine learning methods. A deep learning model with multi-processing pipelining and back propagation techniques, is proposed for cognitive load gait analysis. The IMAGiMAT floor system enabling sensor fusion from plastic optical fiber (POF) elements, is utilized to record gait raw data on spatiotemporal ground reaction force (GRF). A deep parallel Convolutional Neural Network (CNN) is engineered for POF sensors fusion, and gait GRF classification. The Layer-Wise Relevance Propagation (LRP), is applied to reveal which gait events are relevant towards informing the parallel CNN prediction. The CNN differentiates between males and females with 95% weighted average precision, and cognitive load gait classification with 93% weighted average precision. These findings present a new hypothesis, whereas larger dataset holds promise for human activity analysis.

Keywords: Convolutional Neural Networks (CNN) · Cognitive load gait · Ground Reaction Force (GRF) · Layer-Wise Relevance Propagation (LRP)

1 Introduction

In recent years, studies demonstrated that gait is no longer to be considered as an automated motor activity that receives minimal cognitive input [1, 2]. Purposeful gait is crucial for individual's aims and environment's burden, to concurrently execute cognitive demanding tasks while walking. Gait cannot be taken as a strict sequential recurrence, with each step and exact copy of the preceding one, due to the unavoidable surrounding environment variability. The progressive adjustment of gait, is considered as a way for humans to minimize the energy cost during the gait cycle, while navigating within often complex surroundings to successfully reach a desired destination [3, 4]. Humans continuously optimize energy cost in real-time, even if the energy savings while walking are small [5]. Thus, the energy conservation drives changes in gait dynamics, based on cognitive processing of information from the surrounding environment.

© Springer Nature Switzerland AG 2019
H. Yin et al. (Eds.): IDEAL 2019, LNCS 11871, pp. 229–237, 2019.
https://doi.org/10.1007/978-3-030-33607-3_25

Furthermore, the cognitive function of males and females is a long-discussed issue [6]. Typically, females show preeminence in verbal fluency, perceptual speed, accuracy and fine motor skills; while males outperform females in visuospatial perception processing, mathematical abilities and working memory [7, 8]. A recent study on gait kinematics for males and females at self-selected walking speeds, has concluded that the subject's gender can assist in pathology diagnosis, and treatment recommendations in several clinical applications [9].

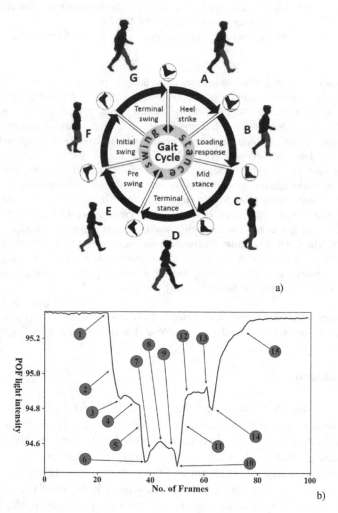

Fig. 1. (a) Gait cycle [12]; (b) Spatial average of gait GRF signals). Gait events recorded by the sensors based on figure (a): 1 - Heel strike, 2 - Foot-flattening, 3 - Single support, 4 - Opposite Heel strike, 5 - Opposite Foot-flattening, 6 - Double support, 7 - Toe-off, 8 - Foot swing, 9 - Heel strike, 10 - Double support, 11 - Toe-off, 12 - Foot swing, 13 - Opposite Heel strike, 14 - Single support, 15 - Toe-off.

Consistent with this, the gender difference in the response to cognitive load is a focus in this study, based on observing the impact on males and females of a cognitively demanding task while walking. In the longer term, this approach could contribute to improving the quantification of gait decline in studies of cognitive changes (under e.g. comparable stage of Alzheimer's disease) manifested in men and women [10, 11].

The complex gait events and intervals in Fig. 1(a), are typically captured using a well-established sensing technology e.g. cameras to record video sequence of the body motion, wearable sensors to acquire the legs' trajectories and body posture, and sensors under the foot to capture the vertical GRF. Bearing in mind that human walking involves the recurrent contact of the feet with the surface, to deliver a change in the body center-of-mass position. It can be a valuable source of information to study potential differences in the cognitive load impact on males and females. Therefore, floor sensors based on plastic optical fiber (POF) technology [13], is utilized to obtain gait data under the influence of cognitive load. A parallel CNNs network, is engineered to fuse and process the POF sensors signals for cognitive load gait classification. The LRP techniques, is proposed to interpret the CNNs prediction and back-propagate the impact of cognitively demanding tasks. It identifies the event that responsible for influencing the gait cycle.

2 Methodology

2.1 iMAGiMAT System

The gait cycle depends on the humans' recurrent contact with the ground. Thus several platforms are proposed to capture the footsteps while walking, using sensors fusion to capture walking patterns [14, 15]. A photonic guided-path tomography sensor head (the iMAGiMAT footstep imaging system [13]), is utilized in this experimental work. An illustration of iMAGiMAT system gait GRF spatial average signal is shown in Fig. 1(b). The spatial average of the POF signals is calculated (See Eq. (1)) to show the temporal activity on the surface of the carpet. As shown further, the processing of the calculated temporal GRF signal, allows interpretation as to which gait event occurred on the carpet surface, using gait cycle descriptors [12].

$$SA[t] = \frac{1}{116} \sum_{i=1}^{116} (R_i[t]) \tag{1}$$

R_i is the number of sensors in the iMAGiMAT system (116 sensors arranged in three plies, a lengthwise ply with 22 POF at 0° and two 47 POF arranged diagonally at 60° and 120°). t is the number of frames (20 frames per sec).

2.2 Data Acquisition

Ethical approval was received through the Manchester University Research Ethics Committee (MUREC) and data acquisition was performed in accordance with the board guidelines. In advance of each data recording trial, each subject answered short question to check over mental state and health conditions. Healthy males and females were invited to walk (while performing cognitive demanding tasks) along the 2 m length of the iMAGiMAT system. Gait spatiotemporal GRF signals were recorded in real time. Five healthy subjects (aged 20 to 31 years, 3 male and 2 females) participated in this experiment. Each subject was asked to walk several steps before and after stepping on the iMAGiMAT system for each experiment, so the captured gait data is unaffected by walking start and stop. 100 frames, each with duration of 5 s, were set for each experiment consistent with a capture frame rate of 20 frames per second. Two dual tasks experiments, as well as normal gait, were acquired for each subject; 20 gait trials for each manner of walking in a single assessment session for male and female, and 10 gait trials for two males and one female. Each manner of walking is identified as a class, as follows:

- Class 0, Normal Gait: walking at normal self-selected gait speed.
- Class 1, gait with serial 7 subtractions: normal gait speed, while at the same time performing serial 7 subtractions (count backwards from 100 by sevens e.g. (93, 86, 79...72, 65,) or by count from given random number, stop after completing the task).
- Class 2, gait while texting: normal gait attempted self-selected gait speed while typing in text on a smartphone.

2.3 Convolutional Neural Network (CNN)

A parallel CNNs model (see architecture in Fig. 2) is engineered using Keras libraries. The model hyper-parameters are selected based on extensive trial way (hyper parameter optimization). The stacked layers classify subjects' gait under cognitive load alongside with gait identification for males and females. The network fuses the POF sensors signals in the network deep layers, by mapping the signals to a gait class in supervisory training. The model is trained and validated using a batch size of 100 samples for each iteration; 250 echoes are optimal to train the model. The training and validation sizes are set to be 70% and 10% respectively, where 20% is held for testing the model accuracy. An ADAM (A Method for Stochastic Optimization) is used to train the model, where the parallel layer weight and bias shared and updated simultaneously in each iteration. The loss is computed using categorical cross-entropy in every iteration to minimize the network error. To improve the model performance a regularization method is utilized: a dropout with the size of 0.5 after the last MaxPooling layer was flattened, and an additional dropout with the size of 0.2 was added before the output layer. A *softmax* classifier is placed at the final layer to classify gait GRF signals.

2.4 Layer-Wise Relevance Propagation (LRP)

For most data classification cases a nonlinear deep neural network model acts as a black box, since the reasons as to why the models reached such decisions cannot be traced to physical phenomena. LRP [16] is an emerging machine learning method. It is a tool for understanding and interpreting nonlinear deep neural networks decisions. The aim of using LRP in this study is to understand the contribution of individual gait components in the input x (x: an iMAGiMAT sensor signal at a specific time frame), to the prediction $f(x)$ ($f(x)$: gait dual task class) made by the parallel CNN as a classifier f. The gait class prediction is redistributed to each intermediate node via backpropagation as far back as

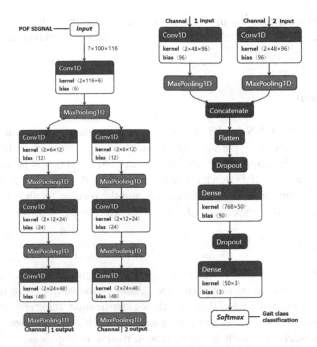

Fig. 2. Parallel CNNs architecture with 2 input and 8 parallel layers, followed by 5 layers.

the input layer. The output of the LRP backpropagation processes, is a heat map with identical size of the original input. The LRP signal highlights spatiotemporal gait pattern relevant for the model prediction. To understand this methodology, we first note that a CNN network consists of multiple computational units of the following form [17]:

$$(x)_i^s = l(0, \sum_j x_j^{(s-1)} \omega_{ji}^{(s-1,s)} + b_i^{(s+1)}) \tag{2}$$

where l is a non-linear (e.g. ReLU) activation function, i indexes neurons in layer s, j is computed over all neurons joined to neuron i and $\omega_{ij}^{(s-1,s)}$, $b_i^{(s+1)}$ are weight and bias parameters of the CNN, learned during the network training. Many such computational units join to form the entire network. The output $f(x)$ is evaluated in a forward-pass and the parameters are updated by back-propagating using the model error. As shown in [16], using the same network-graph architecture, we can redistribute the total relevance $f(x)$ at the output to input-layer relevance using local redistribution rules (αβ-rule) [16]:

$$R_j^{(s-1)} = \sum_i (\alpha \frac{x_j^{(s-1)} \omega_{ji}^{(s-1,s)+}}{\sum_j x_j^{(s-1)} \omega_{ji}^{(s-1,s)+}} - \beta \frac{x_j^{(s-1)} \omega_{ji}^{(s-1,s)-}}{\sum_j x_j^{(s-1)} \omega_{ji}^{(s-1,s)-}}) R_i \tag{3}$$

where ()+ and ()− denote the positive and negative parts, and the parameters α and β are chosen so that $\alpha - \beta = 1$ and $\beta \geq 0$ to avoid divisions by zero; here i indexes a feature map for layer s, j computed over all neurons joined to neuron i. This rule is applied in a backward pass through the network starting at the output layer to produce scores which are called the relevance map. It satisfies the relevance conservation property as per [16]: $\sum_j R_j^s = f(x)$.

3 Experiments

The proposed parallel CNN model in Fig. 2, is trained, validated, and tested on K $\times$ M $\times$ N signals; where K is the number of samples, M is number of frames, and N number of sensors. The training and validation size is 168 $\times$ 100 $\times$ 116, and the testing size is 42 $\times$ 100 $\times$ 116 GRF signals. The model is trained, validated, and tested three times, to avoid the random weight initialization, and the median result is reported using confusion matrix. Further experimental work is reported in the following sections.

3.1 Experiment 1

The aim of this experiment is to categorize gait GRF signals for males and females. The CNN model is trained on identifying the gender of the subject gait, by setting the data labels as a male or female. This two class classification achieved 95% weighted average precision, and 94% true positive for males and 96% for females (on 20% data unseen by the model) as shown in the confusion matrix in Fig. 3(a).

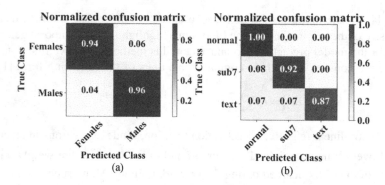

Fig. 3. Gait classification; (a) Males and females classification confusion matrix; (b)

3.2 Experiment 2

This experiment focuses on the difference between males and females, in classifying gait changes under cognitive load. The CNN model classified the 5 subjects' gait as normal gait, walking while performing serial 7 subtraction, or walking while texting. The confusion matrix in Fig. 3(b), is illustrating the classification accuracy for cognitive demanding tasks. The model achieved 93% weighted average precision with 100% true positives for normal gait, 92% true positives for gait with subtraction 7, and 87% true positives for walking while typing a text on a smartphone.

3.3   Experiment 3

This section addresses cognitive load gait deterioration difference between males and females by deploying the LRP methods of perturbations. The CNN is frozen and transferred using transfer learning methods to perform LRP with $\alpha = 2$, $\beta = 1$ using iNNvestigate library [18]. This method shows which gait event contributed positively and negatively in the CNN prediction, (Fig. 4 shows the GRF signals with the corresponding LRP scores for each signal). The model prediction $f(x)$ is back-propagated to gait GRF signals space to outline which gait event (mentioned in Fig. 1(b), using gait events in Fig. 1(a)). The patterns that contributed the most to predict gait class is shown in Fig. 5. The gait spatial average is calculated from the LRP relevance scores, to visually observe the temporal activity on the surface of the iMAGiMAT sensors at a

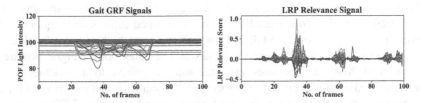

Fig. 4. Raw IMAGiMAT POF signals left, corresponding LRP scores signals right.

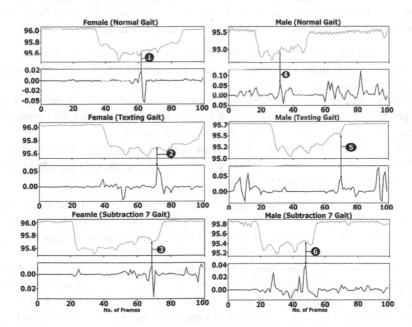

Fig. 5. LRP perturbation applied on males and females data for identifying gait events relevant for the CNN prediction to classify the cognitive load impact on gait. Spatial average of gait GRF signals: red, spatial average for LRP relevance signals over GRF temporal period: blue. Consistent with gait events analysis of GRF spatiotemporal signal in Fig. 1(a) and (b), the cognitive load difference in male and female is as follow: 1 - Toe-off, 2 - Opposite Heel strike, 3 - Foot-flattening, 4 - Foot swing, 5 - Single support, 6 - Opposite Heel strike. (Color figure online)

specific time. Taken into account that the POF sensors disturbance start when a subject step on the carpet, and the sensed signal will always be initiated by the heal strike followed by the gait events in Fig. 1(a). Further explanation of gait events is in [12].

As observed in Fig. 5, there are positive and negative LRP scores (the positive are numbered from 1 to 6). It shows the relevant gait event used towards informing the model's final classification outcome, or the gait event that contributed negatively to reduce the model accuracy. It reveals the difference between gait under cognitive load for males and females, by highlighting the gait events with the highest LRP scores.

4   Discussion and Conclusion

We presented a deep learning method to fuse and classify spatiotemporal GRF sensors signals. The gait experimental results for 3 males and 2 females show that gender is distinguishable by 95%. Similar results have been obtained using video sequence [19]. However, in this study, using the GRF of the five subjects, the high classification rate could be resulted from the difference in males and females body weight, posture or muscle strength. Males and females share the same cognitive load, and the classification result is 93%. However, according to the LRP analysis, gait events used for the CNN prediction are different for males and females as shown in Fig. 5. There are high relevance scores when there is no activity on the surface of the carpet. This information is used to inform the model that in some gait classes that the subject speed, was higher than the other classes, or the number of steps on the carpet is less than the other. An explanation of the CNN classification for the 3 classes is as follows.

- Normal gait: the event in the female gait cycle most influencing the classification is the toe-off, while in males it is the transition between the toe-off and foot swing (see Fig. 5, numbers 1 and 4).
- Walking while texting: the most significant for classification in female gait is the opposite heel strike, while in male gait it is the transition between the opposite heel strike and single support (see Fig. 5, numbers 2 and 5).
- Walking while subtracting in 7 s: in female gait most affected is the foot-flattening while in males it is the transition between foot-flattening and the opposite heel strike (see Fig. 5, numbers 3 and 6).

The results indicate that males and females do not differ in the accuracy of gait classification under cognitive load; however, LRP analysis suggests that in males the portion of the gait cycle, responsible for the classification of normal gait as well as gait under cognitive load, is slightly lagging behind that of females.

The main limitation of this paper is the small size of the data. Furthermore, the main hypothesis is not well researched in literature, yet it has substantial potential applications in biometrics, healthcare and various other aspects of human activity. The future direction will focus on securing larger data samples of gait GRF for males and females for further in-depth analysis.

References

1. Yogev, G., Hausdorff, J.M., Giladi, N.: The role of executive function and attention in gait. Official J. Mov. Disord. Soc. **23**(3), 329–342 (2008)
2. Woollacott, M., Shumway-Cook, A.: Attention and the control of posture and gait: a review of an emerging area of research. Gait Posture **16**(1), 1–14 (2002)
3. Bertram, J.E.A., Ruina, A.: Multiple walking speed–frequency relations are predicted by constrained optimization. J. Theor. Biol. **209**(4), 445–453 (2001)
4. Holt, K.G., Jeng, S.F., Ratcliffe, R., Hamill, J.: Energetic cost and stability during human walking at the preferred stride frequency. J. Motor Behav. **27**(2), 164–178 (1995)
5. Selinger, J.C., O'Connor, S.M., Wong, J.D., Donelan, J.M.: Humans can continuously optimize energetic cost during walking. Curr. Biol. **25**(18), 2452–2456 (2015)
6. Adenzato, M., Brambilla, M., Manenti, R., et al.: Gender differences in cognitive theory of mind revealed by transcranial direct current stimulation on medial prefrontal cortex. Sci. Rep. **7**(41219) (2017). https://doi.org/10.1038/srep41219
7. Sherwin, B.B.: Estrogen and cognitive functioning in women. Endocr. Rev. **24**(2), 133–151 (2003)
8. Zaidi, Z.F.: Gender differences in human brain: a review. Open Anat. J. **2**, 37–55 (2010)
9. Bruening, D.A., Frimenko, R.E., Goodyear, C.D., Bowden, D.R., Fullenkamp, A.M.: Sex differences in whole body gait kinematics at preferred speeds. Gait Posture **41**(2), 540–545 (2015)
10. Laws, K.R., Irvine, K., Gale, T.M.: Sex differences in cognitive impairment in alzheimer's disease. World J. Psychiatry **22**(1), 54–65 (2016)
11. McPherson, S., Back, C., Buckwalter, J.G., Cummings, J.L.: Gender-related cognitive deficits in alzheimer's disease. Int. Psychogeriatr. **11**(2), 117–122 (1999)
12. Alharthi, A.S., Yunas, S.U., Ozanyan, K.B.: Deep learning for monitoring of human gait: a review. IEEE Sens. J. (2019). https://doi.org/10.1109/JSEN.2019.2928777
13. Cantoral-Ceballos, J., et al.: Intelligent carpet system, based on photonic guided-path tomography, for gait and balance monitoring in home environments. IEEE Sens. J. **15**(1), 279–289 (2015)
14. Suutala, J., Fujinami, K., Röning, J.: Gaussian process person identifier based on simple floor sensors. In: Roggen, D., Lombriser, C., Tröster, G., Kortuem, G., Havinga, P. (eds.) EuroSSC 2008. LNCS, vol. 5279, pp. 55–68. Springer, Heidelberg (2008). https://doi.org/10.1007/978-3-540-88793-5_5
15. Yun, J., Woo, W., Ryu, J.: User identification using user's walking pattern over the ubiFloorII. In: Hao, Y., et al. (eds.) CIS 2005. LNCS (LNAI), vol. 3801, pp. 949–956. Springer, Heidelberg (2005). https://doi.org/10.1007/11596448_141
16. Montavon, G., Lapuschkin, S., Binder, A., Samek, W., Müller, K.R.: Explaining nonlinear classification decisions with deep taylor decomposition. Pattern Recogn. **65**, 211–222 (2017)
17. Bharadhwaj, H.: Layer-wise relevance propagation for explainable deep learning based speech recognition. In: 2018 IEEE International Symposium on Signal Processing and Information Technology (ISSPIT), pp. 168–174 (2018)
18. Maximilian, A., et al.: iNNvestigate neural networks (2018). https://github.com/albermax/innvestigate
19. Isaac, E.R.H.P., Elias, S., Rajagopalan, S., Easwarakumar, K.S.: Multiview gait-based gender classification through pose-based voting. Pattern Recogn. Lett. **126**, 41–50 (2018)

Towards a Robotic Personal Trainer
for the Elderly

J. A. Rincon[1][✉], A. Costa[2][✉], P. Novais[2][✉], V. Julian[1][✉],
and C. Carrascosa[1][✉]

[1] Institut Valencià d'Investigació en Intelligència Artificial (VRAIN),
Universitat Politècnica de València, Valencia, Spain
{jrincon,carrasco}@dsic.upv.es, vjulian@upv.es
[2] ALGORITMI Centre, Universidade do Minho, Braga, Portugal
{acosta,pjon}@di.uminho.pt

Abstract. The use of robots in the environment of the elderly has grown
significantly in recent years. The idea is to try to increase the comfort
and well-being of older people through the employment of some kind of
automated processes that simplify daily work. In this paper we present
a prototype of a personal robotic trainer which, together with a non-
invasive sensor, allows caregivers to monitor certain physical activities
in order to improve their performance. In addition, the proposed system
also takes into account how the person feels during the performance of
the physical exercises and thus, determine more precisely if the exercise
is appropriate or not for a specific person.

Keywords: Assistant robot · Emotion detection · Elderly

1 Introduction

The substantial growth in the average age of the world's population leads to a
larger number of older people than the rest of the age groups [1]. One of the
major problems afflicting this population is its high level of sedentary life [2].
According to this, the development of cognitive assistants at home, and especially
oriented to health and well-being of older people, is a very interesting field of
study because it not only allows a daily monitoring of the health in a comfortable
and fast way but also greatly lightens the workload that usually have hospitals
and professionals [3]. These assistants usually provide telemedicine service, video
calls, allowing the patient to be in contact with his doctor [4]. The advantage of
this type of tasks performed on a daily basis at home is that they would allow a
constant and exhaustive monitoring. So that by using models based on artificial
intelligence techniques, with the large amount of available data, it could be
possible to detect and anticipate possible problems or to suggest improvements
in lifestyle habits.

Other characteristic that cognitive assistants currently may possess is the
classification the emotional state of the human with whom they interact. As

H. Yin et al. (Eds.): IDEAL 2019, LNCS 11871, pp. 238–246, 2019.
https://doi.org/10.1007/978-3-030-33607-3_26

well as the ability to express an emotional state of their own, depending on the information they obtain from the surrounding environment. This, along with the rest, allows the users to feel a certain affinity with cognitive assistants. The information obtained by the assistant from this kind of interactions in combination with the large amount of data available, allow the assistants to classify and anticipate possible problems, and also to make personalized suggestions, in order to improve the well-being of the users.

This paper presents a cognitive assistant, specifically an assistant robot, that actuates as a personal trainer capable of recommending, detecting and classifying the physical activities [5, 6] performed by an elderly person. The robot is capable of interacting with humans, using voice and emotion detection as communication elements, recommending personalised physical activities. Moreover, while the elderly person is performing the exercises using a non-invasive sensor, the robot monitors whether the exercises are being done in an appropriate manner.

The rest of the paper is structured as follows: Sect. 2 presents a related work section; Sect. 3 explains the proposed approach; finally, Sect. 4 gives some conclusions and possible future works.

2   Related Work

EmIR is inserted in mainly three domains: assistive robotics, and human activity recognition. There are several novel advances in these domains, either separately or in combination, contributing to the advancement of the state of the art. In this section we present some of these projects, serving as a comparative to EmIR features.

Assistive Robots. Martinez-Martin et al. [7] provides a clear idea and several examples of assistive robotics and their features. Apart from them, we highlight three other projects. Caic et al. [8] presents the Vizzi robot. This robot has a very friendly aspect and is designed to interact with elderly people, proposing Exergames. The robot is able to navigate in flat surfaces and has an "emotion" displaying head. It also has speakers and microphones to capture and transmit voice commands. Currently it is not able to autonomously interact with the users or to interact with them, having an visual interface for a human caregiver to interact. In the same line is the work of [9], having the same physical features, it is more advanced on the interaction level and the unassisted movements. The robot is able to receive voice commands and perform autonomously certain tasks (that were pre-programmed) and navigate around a house, recording the paths and optimising them for future use. Lastly there is the SocialRobot [10], yet an advancement over the [9], being the most notable feature the ability to recognise human emotions and faces, and produce an empathetic interaction. It is also able to navigate in an home environment and visually identify users and objects.

Human Activity Recognition. Martinez-Martin et al. [11–13] presents an rehabilitation system using a humanoid robot to be used at home. They use the robot's cameras to visually identify the user's physical movements, using deep learning methods. The robot is able to navigate around the house and identify the user. The information is made available caregivers so they can adjust the exercises accordingly. Using wearables, the work of Kwapisz [14] shows how to identify activities using just a mobile phone. This environment is perfect for a real application due to the low effort that users have in "wearing" these devices. While this specific work did not present a very high accuracy, newer works like [15] or [16] that using state of the art approaches, like Deep Learning methods, achieve over 90% accuracy in detecting physical activities, using just small wearable devices (like wristbands) with accelerometers and gyroscopes. The issue with these projects is that the methods of training the system are rigid and tightly coupled to the person(s) that performed the activities, being unable to tolerate minor faults.

EmIR tries to overcome the presented issues of current projects and push the state of the art forward. In the next section we present an insight of the robot and cognitive assistant structure and operation.

3   Recommending and Following Activities

This section describes the main work developed for this paper. First of all, we detail the different activities that can be recommended to the users. Next, the new version of our low-cost robot, EmIR 3.0, is presented as the interlocutor for the user that is going to recommend the activities. After that, we show how the system makes the following of the activities checking if they have been properly done. First, the ad-hoc hardware to make it so is described and then the artificial neural network to identify the activities is detailed.

3.1   Activities' Description

These activities have been selected by physiotherapists, who have helped determine which exercises involve the chest as the main movement. The physiotherapist has recommended five exercises:

1. **Wall Push Up.** This is a strength-training exercise. Stand at arm's length in front of a wall. Lean forward a little and place the palms of your hands on the wall at shoulder width. Keep your feet still as you slowly move your body toward the wall. Gently push back so your arms are straight (Fig. 1).

Fig. 1. Wall push up exercise.

2. **Sit to stand.** This exercise is good for leg strength. Sit on the edge of the chair, feet apart at the hip. Lean forward slightly. Stand up slowly, using your legs, not your arms. Keep looking forward, not down. Stand up before sitting slowly, from the bottom down (Fig. 2).

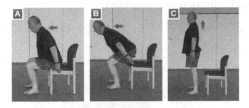

Fig. 2. Sit to stand exercise.

3. **Mini squats** Rest your hands on the back of the chair for stability and stand with your feet spread across your hips. Slowly bend your knees as far as is comfortable, keeping them facing forward. Aim to put them on your big toe. Keep your back straight at all times. Gently stand up, squeezing your buttocks as you do so (Fig. 3).

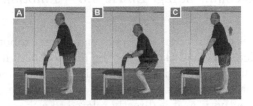

Fig. 3. Mini squats exercise.

4. **Back Leg Raises.** Slowly lift your right leg backwards, without bending the knees. The chest moves slightly toward the front. Hold this position for one second and then gently lower the leg (Fig. 4).

Fig. 4. Back leg raises.   Copyright © 004-2019 Koninklijke Philips N.V. All rights reserved.

5. **Clock Reach.** Just imagine that you are standing in the center of a clock. The number 12 is directly in front of you and the number 6 is at your back. Hold a chair with your left hand and raise your right leg and extend your right arm so that it points to the number 12. The return process is done by placing the arm back on the number three and then on the number 12. This process is done with the two arms (Fig. 5).

Fig. 5. Clock reach exercise. Copyright © 004-2019 Koninklijke Philips N.V. All rights reserved.

3.2   Activities Recommending: *EmIR 3.0 – Emotional Intelligent Robot*

This section describes our proposal for a personal trainer robot (Fig. 6), capable of recommending, detecting and classifying the physical activities carried out by patients. With the increase of elderly people living alone, and, in some cases, incurring in sedentarism, we see the need to develop systems that recommend any kind of physical activity, as it is directly related to welfare, healthy lifestyles, as well as the improvement of health and quality of life.

EmIR is a low cost robot divided into two parts. The first is a low level stage using an Arduino Mega 2560[1], which allows motor control and access to ultrasonic sensors for obstacle detection, allowing the robot to have a reactive behaviour. The second stage is performed using a raspberry pi 3 b+. This allows us to have a higher behaviour as a 7-inch LCD screen, on which it is possible to visualise not only the face of the robot, but also the different physical activities that the robot can recommend. The robot has been built in a modular way, facilitating the incorporation of new elements such as lidars, environmental sensors, etc. . . The robot also has a camera that allows you to identify people and their emotions [17].

---

[1] https://store.arduino.cc/mega-2560-r3.

Fig. 6. Prototype of the personal trainer robot.

3.3 Following Activities

Hardware Developed. In recent years, different types of internet of things (IoT) devices have been appearing. One of these devices is the one presented by AVNET[2] that we have used to detect if the activities are properly done.

These devices incorporate a wide range of sensors which can be used to acquire various types of data such as accelerometers, or gyroscopes, among others. Using the measurements of acceleration and rotation, it is possible to calculate the linear and angular velocities. These measurements, together with the raw acceleration and gyro measurements, allow the identification of the activities.

As some of the activities to be performed involve chest movement, the sensor was placed in a harness which is used by the patient as can be seen in Fig. 7.

Activities Classification. Once defined the exercises to classify, the following is to create our dataset. This was necessary because there is no public dataset, which measures these exercises. Ten patients were used to which a total of six repetitions per exercise were captured during 10 sessions, creating a database of 25.000 data. These data were then partitioned into three sub-datasets, one for training with 80% of the data, 10% for teas and, finally, 10% for validation.

Once the dataset was created, it was normalised between $[-1, 1]$ so that the dataset is on a common scale. This allows us to avoid distortions, since the characteristics of each measure have different ranges. Once the data was

---

[2] https://www.avnet.com/wps/portal/us/solutions/iot/building-blocks/smartedge-agile/.

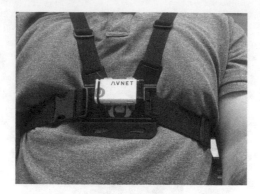

Fig. 7. Sensor and harness.

normalised, the next thing to do was to train our network. To do this, the data from our dataset were restructured. Converting them into matrices of 10 columns by 50 rows, where the 10 columns represent the data acquired by the sensor {Acceleration (X, Y, Z) Rotation (X, Y, Z) and Linear Speed (W, X, Y, Z)}, while the 50 rows represent the captured samples. This last value can be modified, changing the number of samples to train. In the various experiments that were performed, using 50 rows was the configuration that delivered the best result. This result of our classification of activities can be seen in the confusion matrix (Fig. 8). The matrix columns represent the number of predictions for each class, while each row represents the instances in the real class.

It can be observed that we have a success rate per exercise between 50 to 60%. this is due to small movements generated by patients, movements such as

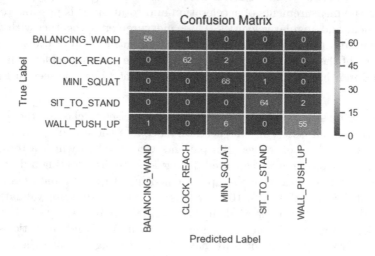

Fig. 8. Confusion matrix obtained from classification.

moving to the sides during the performance of exercises. As well as small spasms or involuntary movements. To try to solve this problem, we are getting more samples and we will try to filter or eliminate those involuntary or voluntary movements of patients.

4 Conclusions and Future Work

In the last decade, the use of robots in home environments has reached a very important role in our society, which no longer perceives the robot as a tool used only in the industrial environment. In this way, robots are acquiring an important role in different scenarios and one of the most important is the care of the elderly. In this paper we have presented a prototype of a robot applied to the care of the elderly. Specifically, the robot is responsible for monitoring the performance of physical exercises, also using a small sensor connected to the robot. In addition, the robot has been added new capabilities for recognition of emotions and recommendation of activities taking into account the result of monitoring the exercises.

As future works, the assistant will be tested by patients and workers of a daycare centre. Specifically, the *Centro Social Irmandade de São Torcato* in the north of Portugal. The validation will be performed through simple exercises with the patients under the supervision of caregivers. The goal is to improve the physical activities and tasks performed at the centre.

Acknowledgements. This work was partly supported by the Spanish Government (RTI2018-095390-B-C31) and FCT—Fundação para a Ciência e Tecnología through the Post-Doc scholarship SFRH/BPD/102696/2014 (A. Costa) and UID/CEC/00319/2019.

References

1. United Nations, Department of Economic and Social Affairs, Population Division. World population ageing 2015. (ST/ESA/SER.A/390) (2015)
2. Lachman, M.E., Lipsitz, L., Lubben, J., Castaneda-Sceppa, C., Jette, A.M.: When adults don't exercise: behavioral strategies to increase physical activity in sedentary middle-aged and older adults. Innov. Aging 2(1), igy007 (2018)
3. Costa, A., Julian, V., Novais, P. (eds.): Personal Assistants: Emerging Computational Technologies. ISRL, vol. 132. Springer, Cham (2018). https://doi.org/10.1007/978-3-319-62530-0
4. Marino, D., Miceli, A., Carlizzi, D.N., Quattrone, G., Sancin, C., Turchi, M.: Telemedicine and impact of changing paradigm in healthcare. In: Calabrò, F., Della Spina, L., Bevilacqua, C. (eds.) ISHT 2018. SIST, vol. 100, pp. 39–43. Springer, Cham (2019). https://doi.org/10.1007/978-3-319-92099-3_5
5. Aggarwal, J.K., Ryoo, M.S.: Human activity analysis: a review. ACM Comput. Surv. 43(3), 16:1–16:43 (2011)
6. Lara, O.D., Labrador, M.A.: A survey on human activity recognition using wearable sensors. IEEE Commun. Surv. Tutor. 15(3), 1192–1209 (2012)

7. Martinez-Martin, E., del Pobil, A.P.: Personal robot assistants for elderly care: an overview. In: Costa, A., Julian, V., Novais, P. (eds.) Personal Assistants: Emerging Computational Technologies. ISRL, vol. 132, pp. 77–91. Springer, Cham (2018). https://doi.org/10.1007/978-3-319-62530-0_5

8. Čaić, M., Avelino, J., Mahr, D., Odekerken-Schröder, G., Bernardino, A.: Robotic versus human coaches for active aging: an automated social presence perspective. Int. J. Soc. Rob. (2019). https://doi.org/10.1007/s12369-018-0507-2

9. Mateus, A., et al.: A domestic assistive robot developed through robot competitions. In: IJCAI 2016 Workshop on Autonomous Mobile Service Robots (2016)

10. Portugal, D., Santos, L., Alvito, P., Dias, J., Samaras, G., Christodoulou, E.: SocialRobot: an interactive mobile robot for elderly home care. In: 2015 IEEE/SICE International Symposium on System Integration (SII). IEEE, December 2015

11. Martinez-Martin, E., Cazorla, M.: A socially assistive robot for elderly exercise promotion. IEEE Access **7**, 75515–75529 (2019)

12. Martinez-Martin, E., Cazorla, M.: Rehabilitation technology: assistance from hospital to home. Comput. Intell. Neurosci. **1–8**, 2019 (2019)

13. Cruz, E., et al.: Geoffrey: an automated schedule system on a social robot for the intellectually challenged. Comput. Intell. Neurosci. **2018**, 1–17 (2018)

14. Kwapisz, J.R., Weiss, G.M., Moore, S.A.: Activity recognition using cell phone accelerometers. ACM SIGKDD Explor. Newsl. **12**(2), 74 (2011)

15. Vepakomma, P., De, D., Das, S.K., Bhansali, S.: A-wristocracy: deep learning on wrist-worn sensing for recognition of user complex activities. In: 2015 IEEE 12th International Conference on Wearable and Implantable Body Sensor Networks (BSN). IEEE, June 2015

16. Cao, L., Wang, Y., Zhang, B., Jin, Q., Vasilakos, A.V.: GCHAR: an efficient group-based context—aware human activity recognition on smartphone. J. Parallel Distrib. Comput. **118**, 67–80 (2018)

17. Rincon, J.A., Martin, A., Costa, Â., Novais, P., Julián, V., Carrascosa, C.: Emir: an emotional intelligent robot assistant. In: AfCAI (2018)

Image Quality Constrained GAN
for Super-Resolution

Jingwen Su[(⊠)], Yao Peng, and Hujun Yin

Department of Electrical and Electronic Engineering, The University of Manchester,
Manchester M13 9PL, UK
jingwen.su@postgrad.manchester.ac.uk,
{yao.peng,hujun.yin}@manchester.ac.uk

Abstract. As one of the most important research topics in image processing, super-resolution aims to estimate high resolution images from single or multiple low resolution images taken from the same scene. With the advent of deep learning techniques, generative adversarial networks are widely adopted for solving various image processing problems including super resolution. We investigate the effect of introducing image quality constraints into the training objective function of a generative adversarial network for super resolution. Experiment results demonstrate that network performance has great potential to be improved with such constraints.

Keywords: Image quality constraints · Super-resolution · Generative adversarial networks

1 Introduction

Image super-resolution (SR) techniques have been extensively studied and developed since 1990 [8], because high resolution (HR) images that have better perceptual quality with finer details are more useful than their low resolution (LR) counterparts in practice. Considering resolution limitations due to embedded sensor chips [20], SR techniques are more flexible and cost-efficient to enhance image resolution for LR imaging systems. Thus they have been broadly adopted in various fields, including surveillance [22,36], medical imaging [27], range imaging [31], astronomy [29] and so on. In general, if multiple LR images of the same scene are available, such methods are categorized as multiple image super-resolution (MISR) and the methods reconstruct HR images by utilizing relative shifts between related LR images [15]. While for single image super-resolution (SISR), since acquisition of several observed LR images is not available, during the upscaling process for enhancement, details that originally are absent and incomplete in the LR images will be predicted and refined to estimate HR images. SISR is more challenging than MISR, as much less prior knowledge regarding the real HR images and the mapping between LR and HR images is available,

© Springer Nature Switzerland AG 2019
H. Yin et al. (Eds.): IDEAL 2019, LNCS 11871, pp. 247–256, 2019.
https://doi.org/10.1007/978-3-030-33607-3_27

although both attempt to solve the same inverse problem of enhancing resolution. This also makes SISR a research topic being investigated intensively by experts and researchers because in practical situations auxiliary information such as several perspectives of the scene may not be easy to obtain.

The development of SISR techniques starts from the frequency domain methods [1,5] which are directly based on theoretical observation models, but the progress is hindered due to sensitivity and lack of practicability. Then attentions are shifted to the spatial domain such as non-uniform interpolation [19], bicubic interpolation [10] and regularized reconstruction methods [23,34]. Interpolation-based methods are straightforward but have a shortcoming in terms of accuracy. While reconstruction-based methods require prior knowledge to enforce constraints in the solution space and produce details accordingly, low computation efficiency and unstable performance when scaling factor changes are the obstacles to overcome. In the last decade, deep learning neural networks, as the most representative learning-based methods, become more and more popular and make a great difference in both methodology for solving ill-posed problems and improving image quality and accuracy [21,35]. Combining deep learning neural networks in image restoration presents a promising direction. Since generative adversarial networks (GANs) were proposed in [6], they have attracted a great deal of attention because of their potentials in simulating high dimensional and complex mapping functions between sample data distributions and target distributions.

In this paper, we attempt to integrate two common image-quality measures, the kurtosis and the structural similarity (SSIM), into the objective function of a generative adversarial network for SISR. The aim is to investigate the possibility of improving image quality by involving statistical image quality measures as part of optimization objectives. These two image quality measures are modified such that by minimizing the overall objective function of a super-resolution GAN, they are maximized in order to encourage distributions of the generated samples transform towards non-Gaussian and increase the similarity between generated samples and ground truth HR images.

The rest of this paper is structured as follows. Section 2 provides a succinct review of related work. The proposed method is described in Sect. 3, followed by experiments and results in Sect. 4. Section 5 draws conclusions.

2   Related Work

2.1   Deep Learning Based Image Super-Resolution

For SISR, many deep learning based methods have been proposed and they differ in network architecture, loss function, learning strategy, etc. [32]. Network architecture design determines capacity and generalization of the network, also the capability of training for very deep and wide networks. For example residual learning proposed in [11], batch normalization removal used in [16], and deep back projection [7]. Loss function decides the optimization direction of training, whether it is driven by accuracy or perceptual quality, and corresponding measures of generated images need to be selected for fair evaluations. However, according to [2] distortion measures such as peak signal-to noise ratio (PSNR) and structural similarity measure are not entirely consistent with human subjective evaluation over images, though rise in distortion measure means perceptual index drops in most cases. Thus it concludes that the loss functions varies and depends on specific applications. As for learning strategies, supervised learning is commonly adopted, upon which various loss functions have been proposed to obtain better results, for example content loss [9] and cycle-consistency loss [37]. Unsupervised learning [24] and weakly-supervised learning [3] have also been developed for solving SR problems. Except for deep neural networks, there are other existing approaches that adopt learning-based methods such as Markov network [4], manifold [25], and compressive sensing in sparse signal processing [30]. These topics remain active for SISR as well.

Generative Adversarial Networks. As one of the most popular deep learning models, GANs have made impressive progress in many applications of image generation [17,37]. As introduced in [6], GANs are comprised of two deep neural networks, a generator and a discriminator. The generator G takes random noise as input and is optimized to generate fake samples such that it is difficult for the discriminator D to distinguish fake samples from authentic ones. While the discriminator is trained to maximally distinguish between real images and fake images. Overall training objective function is a two-player minimax game expressed as following,

$$\min_{G} \max_{D} V(D,G) = \mathbb{E}_{x \sim p_{data}}[\log(D(x))] + \mathbb{E}_{z \sim p_z}[\log(1 - D(G(z)))] \quad (1)$$

where x represents a sample from the true data distribution p_{data}, and z is a noise vector sampled from prior distribution p_z. The generator is trained to minimize its objective function (2nd term), while the discriminator aims to maximize the entire function. Competition between these two networks motivates the improvement of their respective functionalities.

As proposed in [14], the super-resolution GAN (SRGAN) achieves the goal of generating realistic natural images by 4 times upscaling factors, through a perceptual loss function that contains an adversarial loss and a content loss. Also the residual convolutional blocks are used in the proposed network. For performance evaluation, mean opinion score (MOS) testing is adopted as qualitative measure, based on subjective scores given by human raters, and quantitative measures PSNR and SSIM are also included. In [14], because the goal of SRGAN is to improve perceptual quality of generated images, realism in human visual system is valued more importantly than accuracy and computation efficiency. However, high frequency artefacts are observed when the network architecture goes deeper in the SRGAN [14]. Further improvements can be made by incorporating image quality measures in the loss function.

2.2   Image Quality Measures

Kurtosis Measure. Kurtosis is a statistical measure of sharpness or outliers of a signal distribution and normalised kurtosis is defined as

$$kurtosis = \frac{\sum_{i=1}^{N}(x_i - \bar{x})^4}{\sigma} - 3 \tag{2}$$

where $\bar{x}$ is the mean, σ is the standard deviation, N is the number of samples. Equation (2) also serves as a similarity measure to Gaussian distribution. Distributions that are described as sub-Gaussian have negative kurtosis values, while super-Gaussian distributions have positive kurtosis values. Therefore the absolute or squared value of normalised kurtosis can be considered as a non-Gaussianity measure.

In digital image restoration, there are some existing studies using kurtosis measure for image deconvolution [28,33]. While super-resolution can be regarded as a special case of deconvolution, little has been explored on using kurtosis measures. As mentioned in [13], kurtosis measure can be interpreted as an indicator of noise measurement and image resolution. Maximizing absolute value of kurtosis enhances image quality in such a way that generated images are less smooth and sharp edges are preserved. For example, average absolute kurtosis measure of HR images of DIV2K dataset is 2.66, LR images have average value of 2.55.

Structural Similarity Measure. SSIM is a full reference metric that measures the similarity between a generated image and its ground truth image, thus in order to employ SSIM the prerequisite of available reference images must be met. SSIM is formulated as,

$$SSIM(x_1, x_2) = \frac{(2\mu_{x_1}\mu_{x_2} + c_1)(2\sigma_{x_1 x_2} + c_2)}{(\mu_{x_1}^2 + \mu_{x_2}^2 + c_1)(\sigma_{x_1}^2 + \sigma_{x_2}^2 + c_2)} \tag{3}$$

where x_1 and x_2 are two windows, μ_{x_1} is the average of x_1, μ_{x_2} is the average of x_2, $\sigma_{x_1}^2$ is the variance of x_1, $\sigma_{x_2}^2$ is the variance of x_2, $\sigma_{x_1 x_2}$ is the covariance of x_1 and x_2, c_1 and c_2 are set conventionally to 0.0001 and 0.0009 to stabilize the division. Average SSIM between upsampled LR images and ground truth HR images from DIV2K dataset is 0.8647.

3   Proposed Method

The aim is to investigate the effect of involving image quality constraints as regularisation terms into the objective function of the generator given in SRGAN [14]. We propose a regularised loss function to the generator with absolute kurtosis in exponential form and a modified SSIM measure, specified in the following sections.

3.1 Kurtosis Measure Based Constraint

The loss function of kurtosis constraint is defined as,

$$l_{kurt} = w_{kurt} \times e^{-|kurt(G(x))|} \tag{4}$$

We employ exponential function on negative absolute kurtosis such that the loss l_{kurt} can be constrained between 0 and 1, in order to lower the potential risk of divergence. During training, l_{kurt} is minimized, absolute kurtosis of generated images $|kurt(G(x))|$ is maximized, hence the goal of enhancing image quality can be achieved as expected. Besides, the weight w_{kurt} is defined to balance with other losses in the objective function.

3.2 Structural Similarity Measure Based Constraint

Equation 3 is modified to enforce a suitable constrained loss function as following,

$$l_{ssim} = w_{ssim} \times [1 - \frac{(2\mu_{I^{HR}}\mu_{G(I^{LR})} + c_1)(2\sigma_{I^{HR}G(I^{LR})} + c_2)}{(\mu_{I^{HR}}^2 + \mu_{G(I^{LR})}^2 + c_1)(\sigma_{I^{HR}}^2 + \sigma_{G(I^{LR})}^2 + c_2)}] \tag{5}$$

As we know, the higher value of SSIM, the more similar a generated HR image is to its ground truth HR image, and the range of SSIM index is $[0, 1]$. Therefore, minimizing the loss l_{ssim} in the form defined as above encourages the similarity to be increased as training proceeds and the effect of introducing l_{ssim} is controllable by the weight w_{ssim}.

3.3 Full Objective Function

The full objective function is formulated as follows,

$$Loss = l_{MSE} + l_{VGG} + l_{ad} + l_{ssim} + l_{kurt}$$

$$= w_{pixel} \times \frac{1}{r^2WH} \sum_{x=1}^{rW} \sum_{y=1}^{rH} (I_{x,y}^{HR} - G(I_{x,y}^{LR}))^2$$

$$+ w_{fea} \times \frac{1}{W_{i,j}H_{i,j}} \sum_{x=1}^{W_{i,j}} \sum_{j=1}^{H_{i,j}} (\phi_{i,j}(I^{HR})_{x,y} - \phi_{i,j}(G(I^{LR}))_{x,y})^2$$

$$+ w_{ad} \times \sum_{n=1}^{N} -logD(G(I^{LR}))$$

$$+ l_{ssim} + l_{kurt}$$

where l_{MSE}, l_{VGG} and l_{ad} are the same as stated in [14], l_{ssim} and l_{kurt} are defined in the above sections. r is the downsampling factor, W and H are the width and height of low resolution image I^{LR}. $\phi_{i,j}$ indicates the feature maps derived by $j - th$ convolution before the $i - th$ maxpooling layer of VGG19 network of the discriminator, $W_{i,j}$ and $H_{i,j}$ are the dimensions of the feature maps. μ represents the average, σ^2 the variance, $\sigma_{I^{LR}G(I^{LR})}$ the covariance of the input low resolution images I^{LR} and the generated images by the generator $G(I^{LR})$, c_1 and c_2 are the variables to stabilize division. $D(G(I^{LR}))$ denotes the probability predicted by the discriminator that the generated image $G(I^{LR})$ is natural and realistic, N is the number of training samples.

4   Experiments

4.1   Training Details

Training was performed on the DIV2K dataset [26], a benchmark dataset for image restoration that contains 800 high quality images. To obtain LR images, a 128×128 patch was randomly cropped out from a HR image and then downsampled by bicubic function with a scaling factor of 4. Training set was augmented by random flips and rotations. The batch size was set to 16 initially. The learning rate was set to 0.001 for both generator and discriminator, and decreased to half using multi-step learning rate scheme at 50,000, 100,000, 200,000, and 300,000 iterations. Adam solver [12] with $\beta_1 = 0.9, \beta_2 = 0.999$ was adopted as the optimizer. The number of total iterations was 500,000. Generator network has the architecture of SRResNet with 16 residual blocks and employs the least square loss function, while discriminator uses identical architecture of the VGG128 network. For the weights in the objective function, we assigned fixed values of 0.01 for w_{pixel}, 1 for w_{fea} and 0.05 for w_{ad}, while w_{ssim} and w_{kurt} are variables to be adjusted according to different settings of experiments. For performance evaluation, the trained networks were tested on DIV2K dataset, BSD100 dataset [18] and Set14 [35], quantitatively measured by the peak signal-to-noise ratio (PSNR). All experiments are implemented using PyTorch framework on NVIDIA Titan V GPUs.

Table 1. Performance of involving image quality constraints in SRGAN on DIV2K dataset, compared with baseline SRGAN.

Average PSNR (dB)	SRGAN[a]	SRGAN_$kurt$_$SSIM$
DIV2K	24.67	24.77

[a]The baseline SRGAN was implemented by using the basic super-resolution toolbox: https://github.com/xinntao/BasicSR

Table 2. Performance of involving image quality constraints in SRGAN on BSD100 and Set14 dataset, compared with baseline SRGAN.

Average PSNR (dB)	SRGAN[a]	SRGAN_$kurt$_$SSIM$
BSD100	24.19	25.37
Set14	25.80	25.77

[a]The baseline SRGAN was implemented by using the basic super-resolution toolbox: https://github.com/xinntao/BasicSR

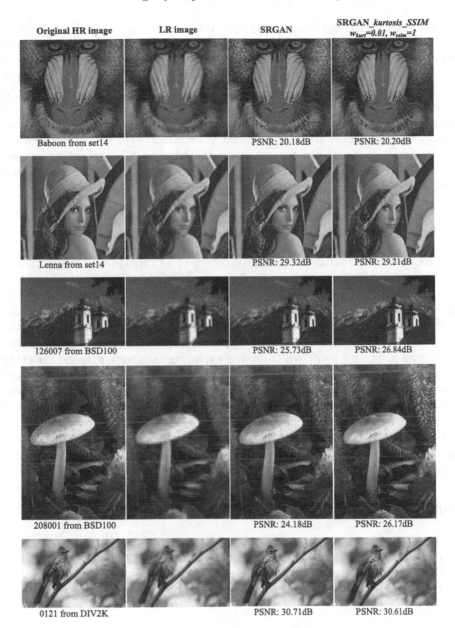

Fig. 1. Example test results generated by baseline SRGAN and the SRGAN trained with image quality constraints.

4.2 Results

We investigated the effect of image quality constraints in training GANs for image super-resolution by comparing the baseline SRGAN and the proposed network with constrained loss functions. Variants of SSIM and kurtosis were incorporated in the objective function for evaluating and matching the perceived quality of the generated and ground truth HR images. Quantitative results of typical runs are given in Tables 1 and 2, respectively showing the usefulness of the proposed approach on DIV2K dataset, as well as when generalizing to the BSD100 and Set 14 datasets. For visual comparison, Fig. 1 presents example images generated by SRGANs trained with and without image quality constraints. Weights for kurtosis and SSIM constraints were set experimentally as 0.01 and 1, respectively. Compared to the baseline SRGAN, the proposed image quality constrained GAN generates promising results both visually and quantitatively.

5   Conclusions

This paper presents an investigation of the effect of introducing image quality constraints into the training objective function of SRGAN. The results show that, with the constraints of kurtosis and SSIM, although the network performance has not been drastically improved, it exhibits the potential for further improvements after fine-tuning hyperparameters such as weights of various loss functions in the objective function. Therefore the approach is promising and worth further investigation. Future work will continue exploring constraints based on other image quality measures including full-reference and non-reference measures, and their applicability in boosting performance of the generative adversarial networks without disturbing training stability and convergence. Feasible combinations of image quality constraint terms will also be considered. In this case, balancing each loss term is crucial and requires extensive trials to assign suitable values for weighting individual terms. Multi-objective optimization schemes could also be considered.

References

1. Bergen, J.R., Anandan, P., Hanna, K.J., Hingorani, R.: Hierarchical model-based motion estimation. In: Sandini, G. (ed.) ECCV 1992. LNCS, vol. 588, pp. 237–252. Springer, Heidelberg (1992). https://doi.org/10.1007/3-540-55426-2_27
2. Blau, Y., Michaeli, T.: The perception-distortion tradeoff. In: Proceedings of the IEEE Conference on Computer Vision and Pattern Recognition, pp. 6228–6237 (2018)
3. Bulat, A., Yang, J., Tzimiropoulos, G.: To learn image super-resolution, use a GAN to learn how to do image degradation first. In: Proceedings of the European Conference on Computer Vision, pp. 185–200 (2018)
4. Freeman, W.T., Pasztor, E.C.: Learning to estimate scenes from images. In: Proceedings of the International Conference on Advances in Neural Information Processing Systems, pp. 775–781 (1999)
5. Gerchberg, R.: Super-resolution through error energy reduction. Opt. Acta: Int. J. Opt. **21**(9), 709–720 (1974)

6. Goodfellow, I., et al.: Generative adversarial nets. In: Proceedings of the International Conference on Advances in Neural Information Processing Systems, pp. 2672–2680 (2014)
7. Haris, M., Shakhnarovich, G., Ukita, N.: Deep back-projection networks for super-resolution. In: Proceedings of the IEEE Conference on Computer Vision and Pattern Recognition, pp. 1664–1673 (2018)
8. Irani, M., Peleg, S.: Super resolution from image sequences. In: Proceedings of the International Conference on Pattern Recognition, vol. 2, pp. 115–120 (1990)
9. Johnson, J., Alahi, A., Fei-Fei, L.: Perceptual losses for real-time style transfer and super-resolution. In: Leibe, B., Matas, J., Sebe, N., Welling, M. (eds.) ECCV 2016. LNCS, vol. 9906, pp. 694–711. Springer, Cham (2016). https://doi.org/10.1007/978-3-319-46475-6_43
10. Keys, R.: Cubic convolution interpolation for digital image processing. IEEE Trans. Acoust. Speech Signal Process. 29(6), 1153–1160 (1981)
11. Kim, J., Kwon Lee, J., Mu Lee, K.: Accurate image super-resolution using very deep convolutional networks. In: Proceedings of the IEEE Conference on Computer Vision and Pattern Recognition, pp. 1646–1654 (2016)
12. Kingma, D.P., Ba, J.: Adam: a method for stochastic optimization. arXiv preprint arXiv:1412.6980 (2014)
13. Kumar, V., Gupta, P.: Importance of statistical measures in digital image processing. Int. J. Emerg. Technol. Adv. Eng. 2(8), 56–62 (2012)
14. Ledig, C., et al.: Photo-realistic single image super-resolution using a generative adversarial network. In: Proceedings of the IEEE Conference on Computer Vision and Pattern Recognition, pp. 4681–4690 (2017)
15. Li, X., Hu, Y., Gao, X., Tao, D., Ning, B.: A multi-frame image super-resolution method. Sig. Process. 90(2), 405–414 (2010)
16. Lim, B., Son, S., Kim, H., Nah, S., Mu Lee, K.: Enhanced deep residual networks for single image super-resolution. In: Proceedings of the IEEE Conference on Computer Vision and Pattern Recognition Workshops, pp. 136–144 (2017)
17. Ma, L., Jia, X., Sun, Q., Schiele, B., Tuytelaars, T., Van Gool, L.: Pose guided person image generation. In: Proceedings of the International Conference on Advances in Neural Information Processing Systems, pp. 406–416 (2017)
18. Martin, D., Fowlkes, C., Tal, D., Malik, J., et al.: A database of human segmented natural images and its application to evaluating segmentation algorithms and measuring ecological statistics. In: Proceedings of the IEEE Conference on Computer Vision (2001)
19. Nguyen, N., Milanfar, P., Golub, G.: A computationally efficient superresolution image reconstruction algorithm. IEEE Trans. Image Process. 10(4), 573–583 (2001)
20. Park, S.C., Park, M.K., Kang, M.G.: Super-resolution image reconstruction: a technical overview. IEEE Signal Process. Mag. 20(3), 21–36 (2003)
21. Schulter, S., Leistner, C., Bischof, H.: Fast and accurate image upscaling with super-resolution forests. In: Proceedings of the IEEE Conference on Computer Vision and Pattern Recognition, pp. 3791–3799 (2015)
22. Segall, C.A., Molina, R., Katsaggelos, A.K.: High-resolution images from low-resolution compressed video. IEEE Signal Process. Mag. 20(3), 37–48 (2003)
23. Shen, H., Peng, L., Yue, L., Yuan, Q., Zhang, L.: Adaptive norm selection for regularized image restoration and super-resolution. IEEE Trans. Cybern. 46(6), 1388–1399 (2015)
24. Shocher, A., Cohen, N., Irani, M.: "zero-shot" super-resolution using deep internal learning. In: Proceedings of the IEEE Conference on Computer Vision and Pattern Recognition, pp. 3118–3126 (2018)

25. Su, K., Tian, Q., Xue, Q., Sebe, N., Ma, J.: Neighborhood issue in single-frame image super-resolution. In: Proceedings of the IEEE International Conference on Multimedia and Expo, pp. 4–pp (2005)
26. Timofte, R., Agustsson, E., Van Gool, L., Yang, M.H., Zhang, L.: Ntire 2017 challenge on single image super-resolution: methods and results. In: Proceedings of the IEEE Conference on Computer Vision and Pattern Recognition Workshops, pp. 114–125 (2017)
27. Trinh, D.H., Luong, M., Dibos, F., Rocchisani, J.M., Pham, C.D., Nguyen, T.Q.: Novel example-based method for super-resolution and denoising of medical images. IEEE Trans. Image Process. **23**(4), 1882–1895 (2014)
28. Vural, C., Sethares, W.A.: Blind deconvolution of noisy blurred images via dispersion minimization. In: Proceedings of the International Conference on Digital Signal Processing Proceedings, vol. 2, pp. 787–790 (2002)
29. Willett, R., Jermyn, I., Nowak, R., Zerubia, J.: Wavelet-based superresolution in astronomy. Astronomical Society of the Pacific (2003)
30. Yang, M.C., Chu, C.T., Wang, Y.C.F.: Learning sparse image representation with support vector regression for single-image super-resolution. In: Proceedings of the IEEE International Conference on Image Processing, pp. 1973–1976 (2010)
31. Yang, Q., Yang, R., Davis, J., Nistér, D.: Spatial-depth super resolution for range images. In: Proceedings of IEEE Conference on Computer Vision and Pattern Recognition, pp. 1–8 (2007)
32. Yang, W., Zhang, X., Tian, Y., Wang, W., Xue, J.H., Liao, Q.: Deep learning for single image super-resolution: a brief review. IEEE Trans. Multimedia (2019). https://doi.org/10.1109/TMM.2019.2919431
33. Yin, H., Hussain, I.: Independent component analysis and nongaussianity for blind image deconvolution and deblurring. Integr. Comput.-Aided Eng. **15**(3), 219–228 (2008)
34. Yuan, Q., Zhang, L., Shen, H.: Multiframe super-resolution employing a spatially weighted total variation model. IEEE Trans. Circuits Syst. Video Technol. **22**(3), 379–392 (2011)
35. Zeyde, R., Elad, M., Protter, M.: On single image scale-up using sparse-representations. In: Boissonnat, J.-D., et al. (eds.) Curves and Surfaces 2010. LNCS, vol. 6920, pp. 711–730. Springer, Heidelberg (2012). https://doi.org/10.1007/978-3-642-27413-8_47
36. Zhang, X., Tang, M., Tong, R.: Robust super resolution of compressed video. Vis. Comput. **28**(12), 1167–1180 (2012)
37. Zhu, J.Y., Park, T., Isola, P., Efros, A.A.: Unpaired image-to-image translation using cycle-consistent adversarial networks. In: Proceedings of the IEEE International Conference on Computer Vision, pp. 2223–2232 (2017)

Use Case Prediction Using Product Reviews Text Classification

Tinashe Wamambo, Cristina Luca$^{(\boxtimes)}$ ⓘ, and Arooj Fatima ⓘ

Anglia Ruskin University, East Road, Cambridge CB1 1PT, UK
tinashe.wamambo@pgr.anglia.ac.uk,
{cristina.luca,arooj.fatima}@anglia.ac.uk

Abstract. Research into sentiment analysis and its capabilities at analysing product reviews has increased tremendously in recent years. In this paper, we propose an approach to classify product reviews and identify use cases. Several iterations showing the application of natural language processing techniques and machine learning classifications are depicted. A number of machine learning classifiers are trained/tested in various iterations, their performance and accuracy at predicting the existence of use cases in product reviews is evaluated.

Keywords: Machine learning · Natural language processing · Sentiment analysis · Ensemble methods · Text classification

1 Introduction

Research into sentiment analysis has increased tremendously in recent years because e-commerce has increased significantly in popularity. A vast number of users purchasing products online on retail websites refer to product reviews as part of their purchasing process to find out the quality of the product, the advantages and disadvantages of purchasing the product, and to find out other users' experiences with the product. However, they are not provided with details of the use cases other users used the product for, which means opinions expressed in product reviews do not detail the full experience.

The focus of this research is to perform text classification on product reviews retrieved from Amazon using sentiment analysis and machine learning. The goal is to identify use cases contained within the reviews.

The rest of the paper is organised as follows: Sect. 2 presents the related work which is the literature review carried out as part of this research. Section 3 introduces the text analysis iterations carried out to detect and identify use cases. Section 4 describes the machine learning experiments carried out for this research and evaluates the results. Section 5 concludes the paper.

2 Related Work

This section reviews the most relevant methods used to classify and analyse text. Details of the techniques and strategies applied by various researchers in their

© Springer Nature Switzerland AG 2019
H. Yin et al. (Eds.): IDEAL 2019, LNCS 11871, pp. 257–264, 2019.
https://doi.org/10.1007/978-3-030-33607-3_28

attempts to improve text classification will be explained, and the results of their approaches will be provided. Certain aspects of the reviewed literature will be included into this paper's approach to achieve the aim of this research.

Text classification is divided into 3 levels: document level, sentence level and entity/aspect level. According to Shivaprasad and Shetty [14] document level analysis expresses the entire document's sentiment by extracting it from the document's overall opinion. Sentence level analysis separates objective and subjective information. Entity/aspect level analysis classifies the opinion rather than paragraphs, sentences and document to identify sentiment expressed towards entities and their aspects.

Due to the unstructured nature of natural language, it is difficult for a machine/system to interpret the meaning of a sentence. Chinsha and Shibily [4] performed opinion mining at all three levels similarly to Shivaprasad and Shetty [14]. They discovered that current approaches detected overall polarity of text which lacked precision at aspect identification. They proposed a syntactic based approach that applied aspect identification, orientation detection and syntactic dependency to aggregate opinion words.

Part of Speech (POS) tagging has been used to extract product features. Devi et al. [5] focused on detecting words that were tagged as nouns because the aspect or feature is in general described by a noun or noun phrase. Alfrjani et al. [1] applied POS tagging to determine if each token in a review was a noun, verb, adjective, adverb, etc with the intention of adding the POS tag to each token as a feature [1]. A method based on frequent noun and noun phrase consisting of three components: frequent based mining and pruning, order based filtering, and similarity based filtering [7] was suggested to improve feature extraction using POS tags.

A key challenge with opinion mining is to predict the sentiment orientation of sentences accurately. Zhao et al. [16] highlighted that feature engineering that generates a "feature set suitable for one domain may not generate good performance for other domains" [16]. Deep neural network frameworks were applied to learn a high level representation of the data by handling star ratings of reviews as weak labels for training whilst classifying sentence sentiment.

Fang and Zhan [6] proposed an algorithm for negation phrase identification, a mathematical approach for sentiment score computation and a feature vector generation method [6] to tackle the fundamental problem of sentiment polarity categorisation. Results from their experiments showed that Random Forest performed best on manually-labelled sentences. Support Vector Machine (SVM) had the most significant enhancement on machine-labelled sentences as training data increased and Random Forest performed best for datasets on all scopes [6].

Bakiu and Guzman [2] focused on detecting user experience and satisfaction by analysing product features. A collocation algorithm was applied to extract features mentioned in user reviews and sentiment analysis techniques were used to map extracted sentiments and features. Support Vector Machine (SVM) classifiers were applied to automatically detect user experience sentiments associated to the extracted features.

Devi et al. [5] used Support Vector Machines (SVM) to determine overall polarity of product features enabling them to isolate users' perception of a product and its features. Results showed Support Vector Machines (SVMs) performed high level text classification on large and small datasets becaused they were robust and efficient.

Xia et al. proposed a three stage cascade model to solve the polarity shift problem that affects the classification performance of machine-learning based sentiment analysis [17]. First, they built a hybrid model that employed rules and statistical methods to detect explicit and implicit polarity shifts [17]. Second, they removed polarity shift in negations using an elimination method and thirdly, they trained SVM, Logistic Regression and Naïve Bayes classifiers using datasets with varying polarity shifts to classify sentiment at document level.

A statistical method that used Gini Index based feature selection method incorporated with a Support Vector Machine (SVM) classifiers [10] was proposed by Manek et al. [10] for sentiment classification in reviews. Results showed that the Support Vector Machine (SVM) linear algorithm had a 97.25% [10] accuracy at predictiong sentiment polarity in reviews compared to other feature selection methods such as syntactic dependency, negation and term frequency.

Khan et al. [8] proposed a semi-supervised approach that incorporated Support Vector Machine (SVM) classifiers with lexicon based methodology to improve sentiment analysis performance [8]. Information gain and cosine similarity mathematical models were employed to revise SentiWordNet sentiment scores. They emphasized the importance of nouns and employed them as semantic features [8] alongside other parts of speech. Results showed that nouns increased the coverage of sentiment lexicons as well as semantic value. Their proposed mathematical models improved SentiWordNet's sentiment scores in a domain independent context.

Chifu et al. [3] used an ant clustering algorithm to perform sentence level aspect-based analysis. They assumed sentences would refer to different aspects of the product. Similar sentences were grouped into clusters and aspects were extracted from each cluster. They identified opinions expressed towards features, enabling them to determine feature polarity. Results showed the labelling quality of the classifier did not "require annotated training data to be built for the given semantic domain" [3].

Saranya and Jayanthy [13] used machine learning to text classification using an ontology model with the aim of identifying user sentiment. They used Word-Net to extract emotional words and build a domain ontology. A classifier was trained using the ontology to classify sentiment based on polarity. Results showed that Support Vector Machine (SVM), Naive Bayes and K-Nearest Neighbours classifiers proved appropriate for sentimental analysis.

A dual sentiment analysis approach was applied by Raut and Barve [12] to train a Naive Bayes classifier to classify polarity and solve the polarity shift problem encountered in automated opinion classification methods [12]. They reversed the polarity of processed reviews to produce two training reviews from each review and expand their training dataset, eliminating the need to collect large amounts of reviews. Their algorithm was not dependent on dataset size for classification accuracy because system accuracy remained stable and did not shift as dataset size increased. However, reversed reviews did not increase the classifier's accuracy because the syntactic structure was damaged causing them to become partially incoherent.

For this research, we propose a rule based machine learning text classification approach to detect the existence of use cases within the content of product reviews. This approach will hold similarities to approaches proposed by Devi et al. [5], Alfrjani et al. [1], Zhao et al. [16], Chinsha and Shibily [4] and Shivaprasad and Shetty [14] in their pursuit to enhance text and sentiment classification with the main difference being the rules applied.

3   Text Analysis Iterations

This section describes the text analysis iterations carried out to detect and identify use cases using natural language processing techniques such as POS tagging, rule based word and phrase identification and feature engineering. The reviews were not pre-processed to apply tags identifying the use cases they contained because the iterations below were used to find out if it was possible to identify use cases using POS tagging, action words and extracted features seeing as use cases would most likely be activities undetaken by the reviewers using the product.

3.1   Iteration 1

Using TextBlob [9] and NLTK [11] as dependencies, a tool called Review-Analyzer was developed to perform text analysis on product reviews. Capabilities to preform text pre-processing and stop word removal using NLTKs corpus library were implemented into the Review-Analyzer tool. The ability to perform docu- ment and sentence level analysis on reviews in a similar manner to Chinsha and Shibily [4] was also implemented, but it did not include a syntactic based approach to identify aspects which Chinsha and Shibily [4] proposed. Parts of Speech (POS) tagging was implemented to identify nouns or noun phrases in reviews and extract them as features similar to Devi et al. [5]. Approximately 500 reviews retrieved from Amazon were used to test Review-Analyzer. A manual analysis of the results showed that the expected sentiment and polarity were successfully determined.

3.2   Iteration 2 (Use Case Identification)

Spelling correction was added to the pre-processing component of the Review-Analyzer tool to ensure spelling mistakes were corrected before performing text analysis. Further investigation of product reviews and the features extracted revealed that words useful for use case identification could be tagged as verbs. To capture this, Parts of Speech (POS) tagging for Review-Analyzer was expanded to include verb POS tags which meant the approach headed in a direction different to Devi et al. [5]. There was an increase in features extracted from reviews. A manual analysis of the results showed improved accuracy in use case identification, however, there was also an increase in false positives. This is because certain reviews were incorrectly determined to contain a use case due to ambiguity in the reviews. For example, the statement "I was running late for work" was incorrectly determined to contain the use case 'running'.

3.3   Iteration 3 (Feature Extraction)

Using Review-Analyzer's pre-processing, text analysis and POS tagging capabilities, more than 10,000 product reviews were analysed at both document and sentence level. Features were extracted using POS tagging with specific focus on nouns, noun phrases and verbs because as stated by Devi et al. [5], Alfrjani et al. [1] and Khan et al. [8] features are usually described as nouns and noun phrases. Words described as verbs were also extracted as features because it is anticipated that use cases will be described by action words e.g. running, climbing, etc. The features were used to apply boolean labels to the reviews, if the POS tags extracted for a review contained verbs it was labelled as containing a use case (true) otherwise it was labelled as not containing a use case (false). It is important to note that at this stage, the labels did not reveal the specific use cases identified within each product review, they only stated whether or not a use case was present. Inspection of the labelled reviews revealed that in certain instances reviews were incorrectly labelled as containing use cases.

4   Machine Learning Iterations

This section explains the machine learning iterations carried out for this research. This includes text classification, feature engineering and ensemble methods.

In the iterations to follow, experiments were carried out using Support Vector Machine (SVM), Logistic Regression and Naive Bayes to find out if classifying product reviews to detect use cases would yield good results as experienced by Xia et al. [17], Manek et al. [10], Devi et al. [5] and Khan et al. [8] in their research. K Nearest Neighbours was used for this research because Saranya and Jayanthy [13] showed that the classifier could perform sentiment analysis. Random Forest was used because Fang and Zhan [6] reported the classifier's good performance on labelled sentences. Decision Tree and Stochastic Gradient Descent (SGD) were used to find out if their performance compared to the above illustrated classifiers which have been used by various researchers and yielded positive results.

4.1   Iteration 1 (Document/Sentence Level Classification)

In this iteration, text classification was performed on the product reviews at document and sentence level to find out if there would be a significant difference in the performance of the classifiers at predicting use cases. SKLearn's LabelEncoder was used to convert the reviews and their labels into binary format coherent for machine learning classifiers to digest with 1 being a review with a use case and 0 being a review without a use case. The converted dataset was split into training and testing data with 75% being training data and 25% being testing data. As input to the classifiers in binary format, the reviews were placed in the X axis and the labels in the Y axis. The majority of the classifiers were trained and tested with default settings, except the Stochastic Gradient Descent classifier which used a maximum of 100 passes over the training instead of 1000 maximum passes and the Support Vector Machine classifier which used a linear kernel.

Table 1. Machine learning classifiers accuracy

Classifier	Document level	Sentence level
K nearest neighbours	92.187%	99.270%
Decision tree	93.604%	99.232%
Random forest	93.604%	99.232%
Logistic regression	93.566%	99.270%
Stochastic Gradient Descent (SGD)	93.604%	99.270%
Naive bayes	91.498%	91.498%
Linear Support Vector Machine (SVM)	93.604%	99.270%

Table 1 shows the accuracy of the classifiers at document and sentence level. All classifiers reported high accuracy with classifiers popular amongst other researchers i.e. SVM, Naives Bayes, Logistic Regression and K Nearest Neighbours yielding high accuracy to show why they have been championed for text classification. Sentence level classification yielded much higher accuracy compared to document level classification. This is because at document level text will most likely contain multiple opinions making it a complex to classify whereas at sentence level, opinions are granular and ambiguity is reduced leading to high confidence classification.

4.2 Iteration 2 (Document Level Ensemble Classification)

To improve classification at document level, a max voting ensemble method that combined the classifiers used in Sect. 4.1 was developed to make predictions for each data point. The final predictions were extracted from those obtained from the majority of the classifiers [15]. The dataset used in Sect. 4.1 was used in this iteration as well. The accuracy of the ensemble classifier was 93%, an unsignificant increase compared to the document level accuracies in Sect. 4.1. A potential reason for this is that the dataset used to test the ensemble classifier

Table 2. Classification report

Class	Precision	Recall
Use case predicted	94%	100%
Use case unknown	91%	27%

contained reviews that were very difficult to classify because they contained ambiguous opinions, as highlighted in Sect. 3.2.

Table 2 shows the classification report for the ensemble classifier. With an accuracy of 93%, the ensemble classifier had a high precision and high recall at predicting use cases which meant the classifier was very picky and correctly classified many reviews. However, it had high precision and low recall at predicting reviews not containing a use case meaning the classifier was very picky and correctly classified many reviews that did not contain a use case, but it incorrectly classified a lot of reviews as containing a use case meaning there is a high probability of false positives. Another possible cause for the low recall could be an imbalance in the classes, this requires further investigation.

5 Conclusion

This paper proposed a feature engineering approach predicated on POS tagging, rule based word and phrase identification to perform text classification on product reviews and predict use cases. Accuracy at sentence level was much higher than at document level because of granularity and reduced ambiguity in opinions within sentences. Document level ensemble classification did not increase accuracy at document level as expected. There is high possibility that imbalanced classes as well as the issue highlighted in Sect. 3.2 caused ambiguity in opinions expressed at document level impacting accuracy and recall at predicting use cases in reviews leading to a high probability of false positives.

The aim of the research is to accurately predict use cases contained in product reviews. Results from the iterations show that accuracy at document level needs improving to limit the prevalence of false positive to ensure output of accurate results. This will be the focus of further work to be carried out for this research.

References

1. Alfrjani, R., Osman, T., Cosma, G.: A new approach to ontology-based semantic modelling for opinion mining. In: UKSim-AMSS 18th International Conference on Computer Modelling and Simulation (2016). https://doi.org/10.1109/UKSim.2016.15
2. Bakiu, E., Guzman, E.: Which feature is unusable? Detecting usability and user experience issues from user reviews. In: IEEE 25th International Requirements Engineering Conference Workshops (2017). https://doi.org/10.1109/REW.2017.76

3. Chifu, E., Leţia, T., Chifu, V.R.: Unsupervised aspect level sentiment analysis using ant clustering and self-organizing maps. In: International Conference on Speech Technology and Human-Computer Dialogue (2015). https://doi.org/10.1109/SPED.2015.7343075

4. Chinsha, T.C., Shibily, J.: A syntactic approach for aspect based opinion mining. In: Proceedings of the 2015 IEEE 9th International Conference on Semantic Computing (2015). https://doi.org/10.1109/ICOSC.2015.7050774

5. Devi, D.V.N., Kumar, C.K., Prasad, S.: A feature based approach for sentiment analysis by using support vector machine. In: 2016 IEEE 6th International Conference on Advanced Computing (2016). https://doi.org/10.1109/IACC.2016.11

6. Fang, X., Zhan, J.: Sentiment analysis using product review data. J. Big Data **2**, 5 (2015). https://doi.org/10.1186/s40537-015-0015-2

7. Hemmatian, F., Sohrabi, M.K.: A survey on classification techniques for opinion mining and sentiment analysis. Artif. Intell. Rev. (2017). https://doi.org/10.1007/s10462-017-9599-6

8. Khan, F.H., Qamar, U., Bashir, S.: A semi-supervised approach to sentiment analysis using revised sentiment strength based on SentiWordNet. Knowl. Inf. Syst. **51**, 851–872 (2017). https://doi.org/10.1007/s10115-016-0993-1

9. Loria, S.: TextBlob: Simplified Text Processing. https://buildmedia.readthedocs.org/media/pdf/textblob/dev/textblob.pdf. Accessed 25 June 2019

10. Manek, A.S., Shenoy, P.D., Mohan, M.C., Venugopal, K.R.: Aspect term extraction for sentiment analysis in large movie reviews using Gini Index feature selection method and SVM classifier. J. World Wide Web Internet Web Inf. Syst. **20**, 135–154 (2017). https://doi.org/10.1007/s11280-015-0381-x

11. NLTK 3.4.3 documentation. https://www.nltk.org/. Accessed 25 June 2019

12. Raut, M.Y., Barve, S.S.: A semi-automated review classification system based on supervised machine learning. In: Proceedings of 1st International Conference on Intelligent Systems and Information Management (2017). https://doi.org/10.1109/ICISIM.2017.8122162

13. Saranya, K., Jayanthy, S.: Onto-based sentiment classification using machine learning techniques. In: Proceedings of International Conference on Innovations in Information, Embedded and Communication Systems (2017). https://doi.org/10.1109/ICIIECS.2017.8276047

14. Shivaprasad, T.K., Shetty, J.: Sentiment analysis of product reviews: a review. In: Proceedings of International Conference on Inventive Communication and Computational Technologies (2017). https://doi.org/10.1109/ICICCT.2017.7975207

15. Singh, A.: A Comprehensive Guide to Ensemble Learning (with Python codes). https://www.analyticsvidhya.com/blog/2018/06/comprehensive-guide-for-ensemble-models/. Accessed 6 Mar 2019

16. Zhao, W., et al.: Weakly-supervised deep embedding for product review sentiment analysis. IEEE Trans. Knowl. Data Eng. (2018). https://doi.org/10.1109/TKDE.2017.2756658

17. Xia, R., Xu, F., Yu, J., Qi, Y., Cambria, E.: Polarity shift detection, elimination and ensemble: a three-stage model for document-level sentiment analysis. J. Inf. Process. Manag. **52**, 36–45 (2016). https://doi.org/10.1016/j.ipm.2015.04.003

Convolutional Neural Network for Core Sections Identification in Scientific Research Publications

Bello Aliyu Muhammad$^{(\boxtimes)}$, Rahat Iqbal, Anne James,
and Dianabasi Nkantah

Coventry University, Coventry, UK
muhamm75@uni.coventry.ac.uk

Abstract. The overwhelming volume of data generated online continuous to grow at an exponential and unprecedented rate. Over 80% of such data is unstructured. Scientific research publications constitute a significant portion of such unstructured data. Systematic literature review (SLR) activity is a rigorous and challenging process. The key challenge in SLR is the automatic extraction of the relevant data from the sheer volume of research publications. Lack of a unified framework has been identified as the key problem. A canonical model, based on the structure of the papers was proposed as the framework for data extraction purposes in SLR. Implemented as a classification problem, traditional machine learning models were used to realise the canonical model. A good accuracy was reported in these traditional models. However, there is room for improvement. This paper presents the result of the work on the same problem using convolutional neural network (CNN), which is more sophisticated (deeper). The results show an improvement over the traditional machine learning models with an accuracy of 85%. Unlike the previous CNN NLP works, this work also demonstrates the application of CNN on a bigger NLP dataset such as the data from the scientific research publications. The result also shows that the CNN performs even better in NLP tasks with bigger datasets.

Keywords: Data mining · Natural language processing · Machine learning · Neural network · Systematic literature review (SLR)

1 Introduction

The sheer volume of data generated from several online sources is growing at an unprecedented rate every day [1]. For example, in the biomedical field alone, more than 1 million publications (papers) are dropped into the PubMed database every year; that is about two (2) papers per minute [2]. This leads to the information overload. Over 80% of this data is unstructured. Unstructured data is text heavy and does not have any predefined model [3]. Scientific research articles are made up of unstructured data [4]. The natural language processing (NLP) techniques seek to support the automatic text processing, including the review of scientific research publications (studies).

The systematic literature review (SLR) is a rigorous and intensive process that aims to integrate the empirical research in order to create one generalisation. It involves

© Springer Nature Switzerland AG 2019
H. Yin et al. (Eds.): IDEAL 2019, LNCS 11871, pp. 265–273, 2019.
https://doi.org/10.1007/978-3-030-33607-3_29

pulling out the relevant studies from many sources, evaluating the studies for inclusion in the review and synthesising the evidence. Automation/Support for the systematic review process has been an ongoing endeavour with some progress. However, the data extraction stage of the process of SLR is still not feasible [5]. Lack of a unified frame work has been identified as the key reason for this non-feasibility [6].

Reference [4] proposed a canonical model as a unified framework (approach) to enable the automatic data extraction from the research papers. The canonical model is a representation of the different sections in a typical the scientific research paper. In total, the canonical model consists six (6) candidate sections: *Introduction, Background, Methodology, Result, Conclusion and Discussion.*

The canonical model is required to automatically identify and extracts these sections for further processing. Previously, the problem was implemented as a text classification tasks using the traditional machine learning methods: SVM, Naive Bayes, Random Forest and Logistic Regression. Good results were obtained from these methods. However, the results from these traditional machine learning method leaves room for improvement. We expected better result from artificial neural networks because they are more sophisticated than the traditional machine learning models as they mimic the biological neuron model.

The convolutional neural network (CNN) has been a success in natural language processing (NLP) tasks. However, it has mostly been experiment with small datasets such as a *sentence or question or relation*. This work explored the scalability of the CNN on a bigger dataset such as the scientific research reports.

2 Model

The neural network based deep learning model maps the words in a text document to vectors. These vectors are then mapped into a fixed length representations. Different neural networks models exists such as the convolutional neural networks (CNN), recurrent neural network (RNN), Recursive neural networks (RecNN) etc. In this paper, CNN is chosen because they are computationally faster than the recurrent neural network (RNN). One reason why RNN are computationally slow is because each word in the string has to be processed sequentially. In contrast, CNN simultaneously processes elements. This speed up processing immensely [10].

The convolutional neural network (CNN) has initially been used for image processing and hence, been a major break-through for image classification [7, 8]. However, it has also been applied to natural language processing (NLP) tasks and is proven to be effective particularly in text classification tasks. References [11–13] reported an excellent result for CNN NLP tasks including sentence classification and relation classification. Reference [12] performed questions classification using the CNN. In general, questions are short text about a sentence or 2. Similarly, reference [12] performed relation classification in CNN. Also, Reference [13] developed a k-max pooling for sentence modelling. The Fig. 1 below shows the architecture of our CNN model used in this project.

The model consists of an input layer (involving a concatenated *glove* word embedding), a convolutional layers, a max pooling layer and an output. Our data,

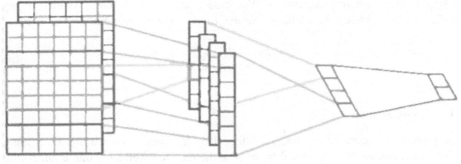

n x k representation of the   **Convolution filters of win-**   **Connected   output**
Words/text in the documents   **dow sizes and feature maps**   **layer with dropouts**

Fig. 1. Architecture of the CNN model.

discussed in Sect. 3 below, contains a bigger dataset (text) as much as the text from sections of a research papers. Each section consists of a number of paragraphs, and each paragraph in turn consists of a number of sentences. A section of length n is represented by

$$y = x1 \oplus x2 \oplus \ldots \oplus xn, \tag{1}$$

Where $\oplus$ is the concatenation function for the individual sentences $x1, x2 \ldots xn$ that make up the section/paragraph. Each convolution layer involves a filter operation $w \in R^{hk}$, applied to a window of k features to produce a feature map. For example, a feature Gi is generated by:

$$Gi = f\left(w \cdot \sum\nolimits_{i+k-1} x + b\right) \tag{2}$$

Here $b \in R$ is a bias term and f is a non-linear function such as the hyperbolic tangent. $\sum_{i+k-1} x$ is the summation of k words in every section (i.e. number of words in all the sentences in the section). This filter is applied to each possible window of words in the sentence $\{x_{1k}, x_{2:k+1}, \ldots, x_{n-k+1:n}\}$ to produce a feature map in each layer. Every layer in the network acts as a filter (to check) for the presence of specific features or patterns present in the original data using the formula in Eq. (2). The first layer in the network detects a large number of features. Increasingly, subsequent layers detect smaller features, which are more abstract and are usually present in the features detected by the other earlier layers. The final layer in the network is able to make the classification/prediction by combining all the specific features detected by the respective layers (filters) in the data.

The max-pooling operation is then applied over the feature map and the maximum value $\hat{G} = max\{g\}$ is taken as the feature corresponding to this particular filter [12]. This is done such that the highest value filter (the most important feature) is captured for each feature map. By default, this pooling scheme naturally deals with variable lengths of the respective documents. The model uses multiple filters (with varying window sizes) to obtain the multiple features used in the experiment.

$$g = [g1, g2, \ldots g_{n-k+1}]$$
$$\text{where } g \in R^{n-k+1}. \tag{3}$$

2.1 Regularisation

One problem in machine learning, including deep learning, is overfitting. Overfitting refers to a situation where a model trains with the data too well such that it affects the prediction on the new (unseen) data [14]. In deep learning, regularisation is a common and efficient method to deal with overfitting. Dropout is one such efficient regularisation technique. At every iteration it selects and removes some nodes in the network along with their incoming and outgoing connections to such nodes. The Fig. 2 below shows the dropuout in the network.

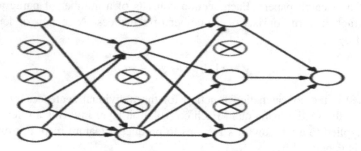

Fig. 2. The dropout from the network

So, each iteration has a different set of nodes. This is similar to the ensemble technique in machine learning. For our study, we employed the dropout technique of regularization which randomly drops out a nodes i.e. setting the nodes to zero during the forward propagation. For output y in forward propagation, dropout becomes:

$$y = w \cdot (z \circ r) + b, \tag{4}$$

Where $\circ$ is the element-wise multiplication operator and $r \in R$ is a 'masking' vector of Bernoulli random variables with probability p of being up to 1. Gradients are back propagated only through the unmasked units. At test time, the learned weight vectors are scaled by p such that $\hat{w} = pw$, and $\hat{w}$ is used (without dropout) to score (predict) unseen documents.

3   Dataset

The dataset used for the experiment has been collected as part of the research project at Coventry University. The data is collected from software engineering subjects. We considered a broad view of what constitutes software engineering subjects including software design, development and innovation. The subjects include: Machine learning, software cost estimation, natural language processing, software testing etc. Each data item is a full-text research article document. The data was obtained from the reputable online databases including the IEEEXplore, Science Direct, ACM digital library, etc. These sources have very rich software engineering subjects. The training data consists of six (6) classes each with a total of four hundred (400) documents, totaling two thousand four hundred (2,400) training documents. The data description is shown in Table 1 below.

Table 1. Data description.

S/N	Class	No. of documents	Doc size (words)
1	Introduction	400	242,680
2	Lit. review	400	309,338
3	Method	400	431,098
4	Result	400	358,256
5	Conclusion	400	91,609
6	Discussion	400	83,846

The labelled data is used to generate the bag of words model. So, given the set of classes and a set of terms associated with each class, the bag of words model is used to determine the appropriate class (category) to which any document belongs. The training and validation was done as described in the Sect. 3.1 below. The data amount to more than three and half million (3,500,000) parameters. Previously, most researches use a short-text data to work with CNN in NLP. Using this huge data, our research takes CNN NLP research to a whole new level.

3.1   Training

From the above data, the top 3,515,881 trainable parameters were used. 80:20 ratio was used for the training and validation. We also used a filter window (k) of 5 each with a filter size of 128. Maximum sequence length of 1,000 was also used. Dropout rate of 0.5 and batch size of 2 were also used. The accuracy and the loss function were used as the metric for measuring the e model.

3.2   Word Embedding

The word embedding is the conversion of the textual data into numbers that would be passed to the CNN. It is a kind of transfer learning for words. Rather than calculating the word embedding ourselves, we used *Globe* embedding, a global vectors for word representation by Stanford Group. It also ensures a semantic representation of the words. The *Globe* embedding consists of up to 400,000 word vectors with dimensionality of 100 and were trained using the continuous bag-of-words architecture [9]. Words not present in the *Globe* embedding are initialized by randomly sampling from uniform distribution in [−0.1, 0.1] [15]. The word embedding is fine-tuned during training to improve the performance of classification.

4   Result and Discussion

Table 2 below shows the result of the CNN model. It shows the epochs and the accuracy results of the training and validation. The difference between successive values is due to the improvement as the weight are updated in the feed forward network. Overall, a training accuracy of 93% and validation accuracy of 85% were respectively achieved by the model for each of the classes. 15 epochs were used in the experiment, the weights of the training and validation vectors were updated 15 times in a feed forward manner to ensure improvement of the model. After the 10th epoch, the validation accuracy of the model remains the same, hence, the final value of the validation accuracy is reported in this research. Figure 3 below visualised the results.

Table 2. Accuracy scores

Epoch	Training accuracy	Validation accuracy
1/15	0.3969	0.4425
2/15	0.5485	0.6235
3/15	0.6625	0.645
4/15	0.7250	0.6566
5/15	0.8250	0.6451
6/15	0.8240	0.6500
7/15	0.8975	0.7440
8/15	0.8625	0.7894
9/15	0.8750	0.7500
10/15	0.9250	0.8000
11/15	0.8875	0.8230
12/15	0.9250	0.8450
13/15	0.9750	0.867
14/15	0.9375	0.8720
15/15	0.9250	0.8451

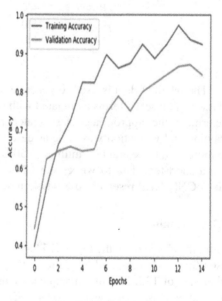

Fig. 3. Accuracies curve

4.1   Loss Function

The loss function determines the strength of our model. It depicts how our model evolves as the training progresses. The lower the loss values, the better our model becomes. Table 3 below shows the values for the training loss and validation loss.

Table 3. Loss values

Epoch	Training loss	Validation loss
1/15	1.4680	1.3072
2/15	1.1499	1.3001
3/15	1.1115	1.2919
4/15	0.9451	1.2873
5/15	0.7497	1.1414
6/15	0.7034	1.5273
7/15	0.6452	1.2248
8/15	0.4755	1.0078
9/15	0.4063	1.0055
10/15	0.4011	1.3590
11/15	0.3737	1.2219
12/15	0.2344	1.1734
13/15	0.1823	1.0924
14/15	0.1431	1.1138
15/15	0.1814	1.3792

As the number of epochs increases, the loss function (the inconsistency between the predicted and actual label) for the training decreases. The validation loss also decreases with the increase in the number of epochs. At the start of the computation, both the training loss and the validation loss were quite high. This is as a result of not enough learning for the model. However, as the weights are updated in the model, the losses begin to fall resulting in a more fitted model. As can be seen in the Table 3 above, the training loss is lower than the validation loss as the number of epochs climbs up. This is because the validation consist of the data which has not been used in training the model and the model tries to predict the class of the new data during the validation. The loss function is at acceptable level at the 10th epoch after which the validation accuracy stalls. Hence, the model is well fitted for the dataset.

4.2 Comparison

Table 4. Results of the various models.

S/N	Model	Accuracy
1	SVM	55%
2	Naïve Bayes	64%
3	Logistic regression	78.5%
4	CNN	85%

The same problem was implemented previously by [16] using the traditional machine learning algorithms such as the Naïve Bayes, Support Vector Machines (SVM) and the logistic regression using the same dataset. From the Results of machine learning experiment reported in reference [16] the CNN results outperforms the traditional models reported in the work. Having used a larger and more closely related labels, the application of CNN in NLP is an effective technique especially for the text classification. Table 4 above shows the results of the various algorithms with their various accuracies.

5   Conclusion and Future Work

In this work, we implemented a text classification problem involving a larger dataset on CNN NLP involving 6 classes built on top of *Globe* word embedding. The CNN model achieved an accuracy of 85%. This work is the first to try the CNN NLP task involving a large and complex data set such as the full text of research publication (journal/conference proceedings). The result also shows that the CNN performs even better in NLP tasks with bigger text sizes (dataset). The future work will consider the effect of different dataset types such as *Ngram, TF-IDF, word count* etc. on the CNN.

References

1. Melinat, P., Kreuzkam, T., Stamer, D.: Information overload: a systematic literature review. In: Johansson, B., Andersson, B., Holmberg, N. (eds.) BIR 2014. LNBIP, vol. 194, pp. 72–86. Springer, Cham (2014). https://doi.org/10.1007/978-3-319-11370-8_6
2. Landhuis, E.: Scientific literature: information overload. Nature **535**(7612), 457 (2016)
3. Blumberg, R., Atre, S.: The problem with unstructured data. DM Rev. **13**(42–49), 62 (2003)
4. Muhammad, A.B., Iqbal, R., James, A.: The canonical model of structure for data extraction in systematic reviews of scientific research articles. In: 2018 Fifth International Conference on Social Networks Analysis, Management and Security (SNAMS), pp. 264–271. IEEE (2018)
5. Jaspers, S., De Troyer, E., Aerts, M.: Machine learning techniques for the automation of literature reviews and systematic reviews in EFSA. EFSA Support. Publ. **15**(6), 1427E (2018)
6. Jonnalagadda, S., Goyal, P., Huffman, M.: Automating data extraction in systematic reviews: a systematic review. Syst. Rev. **4**(1), 78 (2015)
7. Krizhevsky, A., Sutskever, I., Hinton, G.E.: ImageNet classification with deep convolutional neural networks. In: Proceedings of NIPS 2012 (2012)
8. Graves, A., Mohamed, A.R., Hinton, G.: Speech recognition with deep recurrent neural networks. In: Proceedings of ICASSP 2013 (2013)
9. Mikolov, T., Sutskever, I., Chen, K., Corrado, G., Dean, J.: Distributed representations of words and phrases and their compositionality. In: Proceedings of NIPS 2013 (2013)
10. Dauphin, Y.N., Fan, A., Auli, M., Grangier, D.: Language modelling with gated convolutional networks. In: Proceedings of the 34th International Conference on Machine Learning, vol. 70, pp. 933–941. JMLR.org, August 2017
11. Kalchbrenner, N., Grefenstette, E., Blunsom, P.: A convolutional neural network for modelling sentences. arXiv preprint arXiv:1404.2188 (2014)
12. Kim, Y.: Convolutional neural networks for sentence classification. arXiv preprint arXiv: 1408.5882 (2014)

13. Zeng, D., Liu, K., Lai, S., Zhou, G., Zhao, J.: Relation classification via convolutional deep neural network. In: COLING, pp. 2335–2344 (2014)
14. Goodfellow, I., Bengio, Y., Courville, A.: Deep Learning. Adaptive Computation and Machine Learning Series. MIT Press, Cambridge (2016)
15. Zhou, P., Qi, Z., Zheng, S., Xu, J., Bao, H., Xu, B.: Text classification improved by integrating bidirectional LSTM with two-dimensional max pooling. arXiv preprint arXiv: 1611.06639 (2016)
16. Muhammad, A.B., Iqbal, R., James, A.: Machine learning based data analytics for automatic identification of core sections in research publications. Nat. Lang. Eng. J. (2019, under review)

Knowledge Inference Through Analysis of Human Activities

Leandro O. Freitas[✉][ID], Pedro R. Henriques[ID], and Paulo Novais[ID]

ALGORITMI Center, University of Minho, Braga, Portugal
leanfrts@gmail.com, {prh,pjon}@di.uminho.pt

Abstract. Monitoring human activities provides context data to be used by computational systems, aiming a better understanding of users and their surroundings. Uncertainty still is an obstacle to overcome when dealing with context-aware systems. The origin of it may be related to incomplete or outdated data. Attribute Grammars emerge as a consistent approach to deal with this problem due to their formal nature, allowing the definition of rules to validate context. In this paper, a model to validate human daily activities based on an Attribute Grammar is presented. Context data is analysed through the execution of rules that implement semantic statements. This processing, called semantic analysis, will highlight problems that can be raised up by uncertain situations. The main contribution of this paper is the proposal of a rigorous approach to deal with context-aware decisions (decisions that depend on the data collected from the sensors in the environment) in such a way that uncertainty can be detected and its harmful effects can be minimized.

Keywords: Activity analysis · Attribute grammar · Uncertainty handling

1 Introduction

Activity recognition systems have to monitor user's movements aiming to understand his actions to build models capable of being processed. Attribute Grammars (AGs) emerges as a good approach to define models to validate context data related to human actions. Despite having their origin coined to programming languages specification [8], they are already used in Intelligent Environments field to analyse the structure of human tasks [7]. AG provides a formal and strict structure, and specifies semantics through the definition of attributes for symbols and evaluation or validation rules associated with production rules.

Many times, uncertainty in context-aware systems is neglected. Explicit representation of it allows the quantification of its interference in the dataset [3]. Different sources of data, with a wider diversity of sensors, require optimized reasoning, to help decrease the negative impact on users [12].

In this paper, an approach to perform knowledge inference over human daily activities is presented. The validation of activities was done through the definition of an Attribute Grammar, ensuring the correctness of data structure. The proposal was validated through two public datasets, presented as case studies.

© Springer Nature Switzerland AG 2019
H. Yin et al. (Eds.): IDEAL 2019, LNCS 11871, pp. 274–281, 2019.
https://doi.org/10.1007/978-3-030-33607-3_30

The paper is organized as follows: Sect. 2 presents related works to this approach. The datasets and the attribute grammar developed to validate the context data are described in Sect. 3. Section 4 describes in details the approach proposed to deal with uncertainty in context-aware systems with its validation. Section 5 presents conclusions of the study and future research directions.

2   Related Work

Spacial-temporal aspects are considered in [5] to divide uncertainty into: *Point Of View*, including problems related to the conversion of sensor data into global position; *Temporal*, related to time lapses; *Transaction Attribute*, considering that attribute values may have different meanings depending on the source; *Location*, referring to problems with GPS logs with missing data, and; *Identity*, including problems with user's personal data. In [4], a framework for decision making in smart houses through voice recognition was presented. Hierarchic knowledge model for inference under situations with incomplete or outdated data was developed. Uncertainty is tackled through the learning process and the Markov model to the recognition of tasks execution.

These contributions differ from this paper once here the formal model is based on an attribute grammar. It validates the context data and provides a set of semantic rules that help the identification of dubious situations. Other papers like [1] and [10] can also be related to the proposal of this paper. However due to the sake of limited space, they will not be discussed in details.

3   Validation of Human Activities

This section firstly introduces the datasets that were used for the development of the work. In spite of using public datasets, the experiments were created based on an independent research problem, with independent goals. After that, the attribute grammar is described.

3.1   Dataset 1: Human Activities Recognition Dataset

This dataset of human activities in a smart house was created by [6], considered a reference in the field of context awareness, at the Washington State University, USA. The activities monitored were *phone calling, washing hands, cooking, eating* and *cleaning*. The dataset has six thousand and four hundred and ninety-seven registers, acquired through motion, item, water and burner sensors.

During the *phone call* activity the volunteers to receive information about the tasks they should execute after finishing the call. Besides that, all the participants were asked to cook the same meal (oatmeal), following the same recipe. This is important because one of the parameters taken into consideration for the experiments' analysis is the time spent in each activity. And, specifically for these two, the time may naturally vary. For the development of this work, it was assumed that the information from the dataset refers to the same user, i.e., supposing that only one person was monitored. This makes possible to simulate different behaviours of the same user, taking into consideration the context, preferences and needs.

No other information about the work developed by [6] was used in the research presented here.

3.2 Dataset 2: Activity Daily Living and Binary Sensors

The dataset was developed by [11] at the University of Madrid, Spain. It contains data captured from two users during a total of 35 days. For the development of this work it was decided to use data from only one user, since one of the goals is to identify patterns of behaviours and uncertainties related to it. Thus, the chosen dataset contains two thousand and three hundred and thirty five registers, collected during 22 days. The dataset was properly normalized and prepared.

The house used in the experiment have five rooms: Entrance, Living, Bedroom, Bathroom and Kitchen. The monitored activities were *Spare Time/TV*, *Grooming*, preparing *Breakfast*, *Lunch*, *Dinner* and *Snack*, *Sleeping*, *Toileting*, *Showering* and *Leaving*. To recognize the activities binary sensors were used (PIR - Passive Infra-Red, magnetic, flush, pressure and electric).

The registers contain information about the *Start* and *End* time of activities, *Location* of the sensor (object in the house that had a sensor installed), *Type* of the sensor, *Place* (room of the house) and the *Activity*'s name. No other information from the experiments developed by [11] was used.

3.3 Attribute Grammar for Context Validation

The main advantage of attribute grammars is that they allow lexical, syntactic and semantic data verification. According to [2], in the first step (lexical analysis), the data is transformed into terminal symbols and will represent leaves (terminal nodes) in an Abstract Syntax Tree (AST). Then, non-terminal symbols are organized into a hierarchical structure, creating the AST, (syntactic analysis). At last (semantic analysis phase), the actual data associated with the symbols in the AST nodes via inherited or synthesized attributes is processed, considering pre-established semantic rules, to extract (infer) new contextual information and ensure the correctness of the input as required to produce the desired output.

Below, the some of the grammar production rules (the first ones), that stablish the input data structure (define the language syntax), are presented.

p1: *set_of_activities → activity+*
p2: *activity → activity_name ,record, status, duration_status, measurements*
p3: *record → activity_time, sensor_ID+*
p4: *activity_time → date, hour*
p5: *sensor_ID → location, type∗*

The first production rule (**p1**) represents the axiom of the grammar. In this case, it refers to the definition of a set of activities. This symbol represents the root of the AST and comprises all the activities. This will be the reference for the output, after validating the tuples from the dataset. At last, it can be composed of one or more *activity* symbols. The production **p2** represents the structure of each activity in the dataset.

Rule **p3** (*record*) defines the details about sensors used and data collected. This includes the time the activity was performed and the sensors used for it. Each sensor has an identification, which may have different structure depending on the dataset. For instance, it can be an alphanumeric string in one dataset and in other it can be the composition of two or more values from different columns (**p5**). The symbol *status* keeps the information about the beginning and ending time of an activity. During the execution of an activity, the system can receive several data from the sensors, which are analysed by the symbol *measurements*. The symbol *activity_time*, according to **p4**, is composed of the activity's beginning date and hour. The symbols *activity_name*, *date*, *hour*, *location* and *type* store terminal values.

4   Reasoning over Context Data

Inference rules are applied to datasets aiming deduce new information and identify patterns of behaviour. From that, it is possible to tackle uncertainties which may be caused by sensor failures or unexpected changes of behaviour. This section presents two scenarios to illustrate the approach. Based on the grammar productions (syntatic rules) and taking into account the attributes associated with the symbols, it is possible to create any set of attribute evaluation rules. The following rules (to evaluate attributes *duration*, *durationStatus*, *record*, *status*) were created to be applied in the case studies.

1. activity.duration = getSystemTime(TEXT.value) - record.hour;
2. activity.durationStatus = durationAnalysis(activity.duration, expectedDuration[]);
3. activity.record = featureAnalysis(activity_time, activity.durationStatus);
4. activity.status = uncertaintyAnalysis(activity.record, expectedTotalAmount);

Rule 1 computes in the total time spent by the user to finish the activity. Rule 2 analyses if this duration is within an expected time interval. The total amount of activities performed in a specific day is calculated in rule 3. The result of it, is used in the *uncertaintyAnalysis* (rule 4), described in details in Subsect. 4.2.

4.1   Case Study: Abnormal Amount of Activities per Day

The experiment described in this section was conducted in both of the datasets. Details are provided in the paragraphs below.

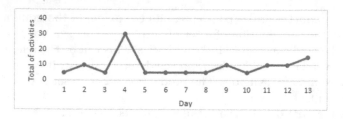

Fig. 1. Total of activities per day

Dataset 1. In this experiment it was analysed the total of activities performed per day. A sudden change of these values along the days may represent change related to behaviour patterns or to the user's health. Figure 1 presents the values computed when analysing dataset 1.

The graph shows good consistency regarding the number of activities executed per day. The majority of the days, 84.61%, the user performed between five and ten activities. Despite having an increased number of activities in the last day of monitoring, it is reasonable to say that it does not affect his standard behaviour, once this represents an increasing of five tasks (16.67%). Semantically, this graph says that the user has a regular routine, probably using a constant time to perform his tasks.

The uncertain situation is evidenced in *day 4* where the number of activities was around 66.66% more than what can be defined as the normal amount in a regular day. There are several reasons for that, but considering only the information from the dataset, it is impossible to choose one with a high level of confidence. For instance, it might refer to a weekend day where the user usually performs more tasks (e.g. receive visitors or clean the house).

Regardless, as the number of activities in this day has a high discrepancy when compared to other days it is imperative for the system to be capable of identifying it. This is the first step in order to minimize the impact of uncertainty in the domain. By the identification of its existence, it will be possible to circumvent the situation with security measures.

Dataset 2. The second dataset differs from the first in the sense that for the identification of the days it was used date format, instead of sequential numbers. However, this does not interfere in the analysis in any way. Figure 2 present the values computed when analysing dataset 2.

The graph shows that in most of the days the user performs between 88 and 120 daily activities (72.72%). Considering that this number represents the great majority of the days, it is reasonable to say that his daily routine follow a pattern. Besides that, in the days that he performed a lower amount of activities (75), this number is still close to the majority. The context-aware system could analyse this situation as abnormal or not, depending on its objectives. The first day has only one activity performed because the monitoring started close to midnight.

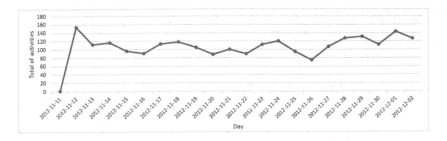

Fig. 2. Total of activities per day

Regarding the days where the user performed more activities than the average, those that present higher discrepancy were recorded in 2 days (2012-11-12, 154 activities and 2012-12-01, 143 activities). As aforementioned, to consider this difference as a problem, it is important to analyse the system's goals. Anyway, these two days correspond to the most critical situations.

Nevertheless, the graph shows that the user, for some reason, changed his routine. This can be related to a variety of reasons, including system failures. Cases like this generate uncertainty. The following section describes an approach to deal with uncertainty in Activity Daily Living environments.

4.2 Uncertainty Handling Approach

In spite of the two datasets used in this study have different structures, both of them represent the same type of domain: smart home. This means that they have similarities and these can be analysed following the same approach. This section tackles the problem of uncertainty, which was identified on the analysis of both datasets. Misinterpretation of sensor signs results in unreliable decision making. Hardware failures of sensors and communication problems may influence this. Thus, the goal is to analyse the total of activities performed in one day and answer questions like: why the user performed (or, why the system detected) a different number of activities than the usual? Algorithm 1 describes how this uncertainty is tackled.

The method *countActivities* calculates the total amount of activities performed in one day. This process is done for all registered days (lines 1 and 2). The results are stored in a list and represent the *expectedTotalAmount* of activities. They represent the normal amount of activities of a day. After that, the same process is executed for one specific day, which is the object of analysis. The result is stored in the variable *totalInCurrentDay* (line 4). The method *checkTotalActivities* (line 5) analyses the *totalInCurrentDay* against the list of expected values. A return *false* indicates a discrepancy between the amount of activities performed by the user and the expected total for him to perform. If a considerable difference between them is evidenced, some actions must be taken in order to prevent the system of using inconsistent or erroneous data. The first one is to run diagnosis tests aiming to find hardware failures (line 6). The method

---

Algorithm 1. uncertaintyAnalysis (activity.record)

---

1 **for** *allDays* **do**
2 | expectedTotalAmount[] = countActivities(activity.record)
3 **end**
4 *totalInCurrentDay* ← *countActivities(activity.record)*
5 **if** *!checkTotalActivities(totalInCurrentDay, expectedTotalAmount[])* **then**
6 | *SD* ← *runSensorsDiagnosis(totalActivities)*
7 | **if** *SD* **then**
8 | | messageAlert(SD)
9 | **end**
10 | *GD* ← *runGapsDiagnosis()*
11 | **if** *GD* **then**
12 | | userInteration(GD)
13 | **end**
14 **end**

---

runSensorDiagnosis starts communications with the sensors to ensure they are working as expected. If problems are found, the system send warnings to the user (method *messageAlert* of lines 7 and 8) about it for him to solve the problem (e.g. repair or replace the sensor). Then, the system runs diagnosis tests (line 10) seeking information gaps and their influence in decision making. The method *runGapsDiagnosis* searches for any type of data that is necessary to build contexts for this activity. These gaps may influence the composition of contexts, once the system will lack context data to build situations. One approach to deal with this is asking the user for help, to acquire more data or to eliminate ambiguities [9]. Thus, if the system detects gaps of information, it starts an interaction of such kind (method *userInteraction* of lines 11 and 12).

5 Conclusion

This paper presents a formalization of human daily activities through attribute grammars. Another contribution is related to how uncertain situations can be evidenced through the execution of inference rules. Uncertainty is naturally embedded in daily life. Thus, the challenge relies on finding ways of minimizing its negative impact on the system's behaviour and in the user's routine. Thus, the next steps of the research include the improvement of uncertainty analysis through the combination of machine learning algorithms and the validation provided by the grammar.

Acknowledgements. "This work has been supported by national funds through FCT – Fundação para a Ciência e Tecnologia within the Project Scope: UID/CEC/00319/2019."

References

1. Bobek, S., Nalepa, G.J.: Uncertainty handling in rule-based mobile context-aware systems. Pervasive Mob. Comput. **39**, 159–179 (2017). https://doi.org/10.1016/j.pmcj.2016.09.004
2. Burger, C., Karol, S., Wende, C.: Applying attribute grammars for metamodel semantics. In: ECOOP 2010 Workshop Proceedings - International Workshop on Formalization of Modeling Languages, FML 2010, January 2010. https://doi.org/10.1145/1943397.1943398
3. Camara, J., Peng, W., Garlan, D., Schmerl, B.: Reasoning about sensing uncertainty and its reduction in decision-making for self-adaptation. Sci. Comput. Program. **167**, 51–69 (2018). https://doi.org/10.1016/j.scico.2018.07.002
4. Chahuara, P., Portet, F., Vacher, M.: Context-aware decision making under uncertainty for voice-based control of smart home. Expert Syst. Appl. **75**, 63–79 (2017). https://doi.org/10.1016/j.eswa.2017.01.014
5. Chen, S., Wang, Z., Liang, J., Yuan, X.: Uncertainty-aware visual analytics for exploring human behaviors from heterogeneous spatial temporal data. J. Vis. Lang. Comput. **48**, 187–198 (2018). https://doi.org/10.1016/j.jvlc.2018.06.007
6. Cook, D.J., Schmitter-Edgecombe, M.: Assessing the quality of activities in a smart environment. Methods Inf. Med. **48**(5), 480–5 (2009)
7. Freitas, L.O., Henriques, P.R., Novais, P.: Attribute grammar applied to human activities recognition in intelligent environments. In: Novais, P., Lloret, J., Chamoso, P., Carneiro, D., Navarro, E., Omatu, S. (eds.) ISAmI 2019. AISC, vol. 1006, pp. 62–70. Springer, Cham (2020). https://doi.org/10.1007/978-3-030-24097-4_8
8. Knuth, D.E.: Semantics of context-free languages. Math. Syst. Theory **2**, 127–145 (1968)
9. Lim, B.Y., Dey, A.K.: Investigating intelligibility for uncertain context-aware applications. In: Proceedings of the 13th International Conference on Ubiquitous Computing, UbiComp 2011, pp. 415–424. ACM, New York (2011). https://doi.org/10.1145/2030112.2030168
10. Noor, M.H.M., Salcic, Z., Wang, K.I.K.: Enhancing ontological reasoning with uncertainty handling for activity recognition. Know.-Based Syst. **114**(C), 47–60 (2016). https://doi.org/10.1016/j.knosys.2016.09.028
11. Ordóñez, F.J., De Toledo, P., Sanchis, A.: Activity recognition using hybrid generative/discriminative models on home environments using binary sensors. Sensors **13**(5), 5460–5477 (2013). https://doi.org/10.3390/s130505460
12. Tian, W., et al.: A review of uncertainty analysis in building energy assessment. Renew. Sustain. Energy Rev. **93**, 285–301 (2018). https://doi.org/10.1016/j.rser.2018.05.029

Representation Learning of Knowledge Graphs with Multi-scale Capsule Network

Jingwei Cheng$^{(\boxtimes)}$, Zhi Yang, Jinming Dang, Chunguang Pan, and Fu Zhang

College of Computer Science and Engineering,
Northeastern University, Shenyang, China
chengjingwei@mail.neu.edu.cn

Abstract. Representation learning of knowledge graphs has gained wide attention in the field of natural language processing. Most existing knowledge representation models for knowledge graphs embed triples into a continuous low-dimensional vector space through a simple linear transformation. In spite of high computation efficiency, the fitting ability of these models is suboptimal. In this paper, we propose a multi-scale capsule network to model relations between embedding vectors from a deep perspective. We use convolution kernels with different sizes of windows in the convolutional layer inside a Capsule network to extract semantic features of entities and relations in triples. These semantic features are then represented as a continuous vector through a routing process algorithm in the capsule layer. The modulus of this vector is used as the score of confidence of correctness of a triple. Experiments show that the proposed model obtains better performance than state-of-the-art embedding models for the task of knowledge graph completion over two benchmarks, WN18RR and FB15k-237.

Keywords: Representation learning · Capsule network · Multi-scale · Dynamic routing · Knowledge graph completion

1 Introduction

Knowledge Graphs evolved from the Semantic Web, the essence of which is a directed graph composed of entities connected by relations. Each edge is a triple of the fact (*head entity, relation, tail entity*) (denoted as (h, r, t)). An example triple from Freebase looks like (Hamlet, story_by, William_Shakespeare). The academic community has proposed a set of translation-based knowledge representation learning models. However, the inherent shortcoming of translation-based models still exist, as shallow networks cannot adequately extract the relevant features of entities and relations. In view of this, Dettmers et al. proposed the ConvE model [3] which utilized deep convolution network to model entities and relations. Nguyen et al. [10] proposed to use capsule network [12] to model

The work is supported by the National Natural Science Foundation of China (61672139) and Project No. JCKY2018205C012.

H. Yin et al. (Eds.): IDEAL 2019, LNCS 11871, pp. 282–290, 2019.
https://doi.org/10.1007/978-3-030-33607-3_31

entities and relations. This model achieves the state-of-the-art in the task of knowledge graph completion. However, as the convolution layer in the capsule network use a single window size convolution kernel, the feature map obtained after a convolution operation contains only partial features represented by the head and tail entities and partial interaction features represented by relations.

Our main contributions in this paper are as follows:

- A knowledge graph representation model based on multi-scale capsule network (MCapsE) is proposed. MCapsE make use of multi-scale convolution kernels in the convolution layer, which helps to extract more semantic features of entities and relations, and thus improve the distinction between entities and relations.
- We evaluate MCapsE on WN18RR [3] and FB15k-237 [13]. MCapsE obtains the best mean rank and the highest Hits@10 on WN18RR and the highest mean reciprocal rank and the highest Hits@10 on FB15k-237.

2 Related Work

Existing knowledge representation models based on structured triples without introducing external knowledge can be divided into three categories: early shallow neural network models, translation-based models, and deep neural network models. We focus on the latter two categories that directly related to our work.

2.1 Translation-Based Model

The earliest translation-based model was TransE [1] model proposed by Bordes in 2013, which was inspired by the fact that the algebraic operation of word vectors in Word2vec model is still meaningful. However, TransE has some obvious shortcomings. It fails in modeling complex relations well such as 1-N, N-1 and N-N. Wang et al. thus proposed TransH [15] in 2014. The authors believe that the entity vector and the relation vector are not in the same hyperplane. They proposed to map the head entity vector and the tail entity vector to the hyperplane where the relation is located, and then perform translation operations. In 2015, Lin et al. proposed TransR [7], which considers that entities and relations are not in the same space. They first map the head entity vector and tail entity vector to the space where the relation vector is located through the transformation matrix, and then carry on the translation operation. This model significantly improves the ability of representation learning of knowledge graphs. Ji et al. proposed TransD [4] and argued that entities and relations are diverse, so the transformation matrixes should be related not only to relations, but also to entities. TransSparse [5] model took into account the problem of heterogeneity and imbalance of relations.

2.2   Deep Neural Network Model

Translation-based models belong to shallow neural network models, which are incapable of extracting correlation features between entities and relations. Dettmers et al. thus proposed ConvE [3] model in 2018. For the first time, this model uses deep convolution network to model entities and relations. Subsequently, Nguyen et al. [9] proposed to use capsule network to model entities and relations, which consists of convolution layers and capsule layers. Since the convolution layer in the capsule network only uses convolution kernels of the same size, feature maps obtained after convolution operations contains only partial features represented by head and tail entities and partial interactive features represented by relations. To obtain more interactive features of larger context, we propose to use convolution kernels of different sizes in the convolution layer so that we can extract features of different abstract levels.

3   MCapsE

The Multi-scale Capsule Network Embedding (MCapsE) contains two parts: structural representation based on TransE and entity and relation representations based on Multi-scale Capsule Network. Firstly, we use TransE as a pre-training model to encode the entities and relations in triples. The generated entity vectors and relation vectors are composed into an embedding matrix. Then, we designed three kinds of convolution kernels with different sizes of window, by which we encode semantic features of entities and relations to form the final representation in the capsule network. Figure 1 shows the overall framework of MCapsE. The function of each layer of the model is elaborated as follows.

3.1   Structure Representation Based on TransE Model

TransE aims to represent entities and relations as low-dimensional distributed vectors. Given a triple (h, r, t), the corresponding k-dimensional embeddings for h, r, t are respectively v_h, v_r and v_t. The vectors of a correct triple should satisfy the formula $v_h + v_r \approx v_t$, whereas a wrong triple should not. Therefore, TransE defines the following scoring function, as is shown in the Eq. (1).

$$f_r(h, t) = \|\mathbf{v_h} + \mathbf{v_r} - \mathbf{v_t}\|_{L_1 \| L_2}, \text{ where } \|\mathbf{v_h}\|_2^2 \leq 1, \|\mathbf{v_t}\|_2^2 \leq 1 \tag{1}$$

Equation (1) denotes the distance of L_1 or L_2 between vector $v_h + v_r$ and vector v_t, and serves as a measure of degree of correctness of a triple. The k-dimensional embedding vectors v_h, v_r and v_t of entities and relations are obtained by using the above-mentioned TransE pre-training model.

3.2   Representations Based on Multi-scale Capsule Network

Embedding Matrix: We treat each embedding triple $[v_h, v_r, v_t]$ as a matrix $A = [v_h, v_r, v_t] \in R^{k \times 3}$, where $A_{i,:} \in R^{1 \times 3}$ is the i-th row of A.

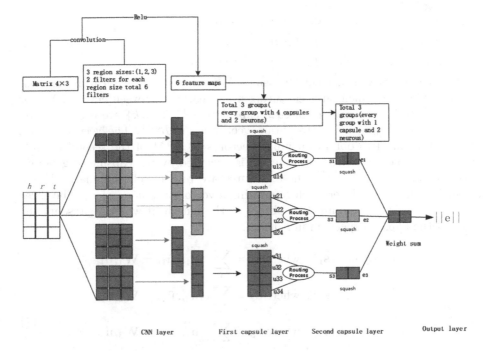

Fig. 1. The overall framework of Multi-scale Capsule Network Embedding (MCapsE) model

Convolutional Layer: The input of a convolutional layer is the embedding matrix A. We use three convolution kernels $\omega_j \in R^{j \times 3}$ with different window sizes, where $j \in \{1, 2, 3\}$. As is shown in Eq. (2), a convolutional operation is executed on each row of the matrix A by using each of the three convolution kernels to produce feature maps. The padding is set to SAME, that is, the output format is filled with zeros. The feature map lists q of the output of convolutional layer is thus obtained.

$$c_{j,i} = g\left(\omega_{\mathrm{j}} \cdot A_{i,:} + b\right)$$
$$q_j = [c_{j1}, c_{j2}, \dots, c_{jk}] \in R^k \tag{2}$$
$$q = [q_1, q_2, q_3]$$

where $\cdot$ is the dot product operation, $b \in R$ is the offset vector, and g is the non-linear activation function. In this paper, we use the linear rectification function ReLU as the activation function.

Capsule Layer: We use two capsule layers in MCapsE. In the first layer, we organize capsules in groups. Each group use as input a feature map list from the output of Convolutional layer (see Eq. (3)).

$$q_1 = [u_{11}, u_{12}, u_{13}..., u_{1k}]$$
$$q_2 = [u_{21}, u_{22}, u_{23}..., u_{2k}] \tag{3}$$
$$q_3 = [u_{31}, u_{32}, u_{33}..., u_{3k}]$$

In each group, we construct k capsules. We encapsulate each feature map list into capsule, which facilitates capturing more features between entities. Vector $u_{ji} \in R^{(N/3) \times 1}$ and vector u_i of each capsule $i \in \{1, 2, ..., k\}$ are multiplied by weight matrix $W_{ji} \in R^{d \times N}$ to obtain vector $\hat{u}_{ji} \in R^{d \times 1}$. The vectors $\hat{u}_i$ are weighted summed to obtain an input vector $s_i \in R^{d \times 1}$ of the second capsule layer. A nonlinear compression function is executed by the capsule to generate a vector output $e_i \in R^{d \times 1}$. These features are respectively summarized as a capsule in the second layer, each generating a vector output e_1, e_2, e_3. The three vectors are weighted summed to obtain e, the length of which represents the score of the triple as specified in Eq. (4).

$$\mathbf{e}_1 = squash\,(\mathbf{s}_1)\,, \text{ where } \mathbf{s}_1 = \sum_i c_i \hat{\mathbf{u}}_{1i}, \hat{\mathbf{u}}_{1i} = \mathbf{W}_{1i}\mathbf{u}_{1i}$$

$$\mathbf{e}_2 = squash\,(\mathbf{s}_2)\,, \text{ where } \mathbf{s}_2 = \sum_i c_i \hat{\mathbf{u}}_{2i}, \hat{\mathbf{u}}_{2i} = \mathbf{W}_{2i}\mathbf{u}_{2i}$$

$$\mathbf{e}_3 = squash\,(\mathbf{s}_3)\,, \text{ where } \mathbf{s}_3 = \sum_i c_i \hat{\mathbf{u}}_{3i}, \hat{\mathbf{u}}_{3i} = \mathbf{W}_{3i}\mathbf{u}_{3i} \tag{4}$$

$$\mathbf{e} = \sum_{j \in \{1,2,3\}} \mathbf{e_j}$$

In Eq. (5), we proposed a improved squash function for amplification when the modulus length is close to zero, rather than being compressed globally in the original squash function. c_i is the coupling coefficient determined by the routing process, as shown in Algorithm 1. Since there is one capsule in the second layer of each set of features, we use softmax for all capsules in the first layer to output vectors of proper length for each capsule in the next layer.

$$squash\,(\mathbf{s_i}) = \frac{\|\mathbf{s_i}\|^2}{0.5 + \|\mathbf{s_i}\|^2} \frac{\mathbf{s_i}}{\|\mathbf{s_i}\|} \tag{5}$$

we define the scoring function f in Eq. (6):

$$f(h, r, t) = \|MCapsnet\,(g\,([\mathbf{v}_h, \mathbf{v}_r, \mathbf{v}_t] * \mathbf{\Omega}))\| \tag{6}$$

where $\mathbf{\Omega}$ is the shared parameter in the convolutional layer and * represents the convolution operator.

We use the Adam Optimizer [6] to train MCapsE by minimizing the Loss Function [14] in Eq. (7):

$$\mathcal{L} = \sum_{(h,r,t) \in \{\mathcal{G} \cup \mathcal{G}'\}} \log\left(1 + \exp\left(-t_{(h,r,t)} \cdot f(h, r, t)\right)\right) \tag{7}$$

Algorithm 1. The routing process algorithm extended from Sabour et al. (2017) [12]

Input: parameters $\hat{u}_{ji}$
Output: e
1: some description
2: **for** j∈{1,2,3} **do**
3:     **for** all capsule i∈the first layer **do**
4:         $b_i \leftarrow 0$
5:     **for** iteration=$1, 2, \ldots, m$ **do**
6:         $\mathbf{c} \leftarrow softmax(\mathbf{b})$
7:         $\mathbf{s_j} \leftarrow \sum_i c_i \hat{u}_{ji}$
8:         $\mathbf{e_j} = squash(\mathbf{s_j})$
9:         **for** all capsule i ∈the first layer **do**
10:             $b_i \leftarrow b_i + \hat{u}_{ji} \cdot \mathbf{e_j}$
    $\mathbf{e} = \sum_{j \in \{1,2,3\}} \mathbf{e_j}$
11: **return e**

---

where $t_{(h,r,t)} = \begin{cases} 1, & \text{if } (h,r,t) \in \mathcal{G} \\ -1, & \text{if } (h,r,t) \in \mathcal{G}' \end{cases}$, $\mathcal{G}$ is the positive sample triple set, and $\mathcal{G}'$ is the negative sample triple set. Since the triples in knowledge graphs are positive samples, the negative samples need to be generated. We uses the method proposed by Wang et al. [15] to set different probabilities to replace the head or tail entities. The method divides the relation into 1-1, 1-N, N-1 and N-N according to the categories of relations. We will give more chance to replacing the head entity if the relation is 1-N and give more chance to replacing tail entity if the relation is N-1.

4   Experimental Results and Analysis

We evaluate the performance of MCapsE[1] on the task of knowledge graph completion.

We use two most commonly used datasets, WN18RR and FB15k-237. In MCapsE, the vector dimension is set to $d \in \{100, 200\}$, learning rate is set to $\lambda \in \{0.00001, 0.0001, 0.001, 0.1, 1\}$ with SGD, and the three convolution kernel window sizes in the convolutional layer is set to $\{1 \times 3, 2 \times 3, 3 \times 3\}$. The number of convolution kernels is $N \in \{50, 100, 200, 300, 400\}$, the number of training iterations epoches $epoch \in \{50, 100, 200\}$, $batch\_size \in \{128, 256\}$, the number of neurons in the capsule layer is set to $d \in \{10, 20\}$, and the times of dynamic path iterations $m \in \{1, 3, 5\}$. To speed up convergence, we use the results of the TransE model pre-training to initialize the vectors of entities and relations.

In MCapsE, for WN18RR dataset, the model optimal parameters are: $k = 100, N = 50, epoch = 100, m = 1, batch\_size = 128, \lambda = 0.00001, d = 10$. For FB15k-237, the optimal parameters of the model are $k = 100, N = 400, epoch = 100, m = 3, d = 10, m = 1, batch\_size = 128, \lambda = 0.00001$.

---

[1] Our code is available at: https://github.com/1780041410/McapsE.

Evaluation Protocol: knowledge graph completion is designed to predict missing head or tail entities in triples. We use three evaluation indicators: the average ranking of all correct entities (Mean Rank, MR), average reciprocal ranking of all correct entities (Mean Reciprocal Rank, MRR), and ratio of correct entities in the top 10 (Hits@10).

Table 1 compares the results of MCapsE with previous state-of-the-art results. MCapsE reaches the best in Hits@10, which shows MCapsE effectively capture the semantic information of entities and relations by using multi-scale window convolution kernels. It has certain advantages in extracting features at different levels, and further improves the discrimination of entity representation.

Table 2 shows Hits@10 and MRR results for predicting head and tail entities for each relation category on FB15k-237. McapsE works better than ConvKB and CapsE model in predicting entities, especially in the N-1 and N-N relations.

Table 1. Knowledge graph completion results

Method	WN18RR			FB15k-237		
	MR	MRR	Hits@10	MR	MRR	Hits@10
DISTMULT [16]	5110	0.425	49.1	254	0.241	41.9
ComplEx [14]	5261	0.444	50.7	239	0.247	42.8
ConvE [3]	4187	0.433	51.5	**244**	0.325	50.1
KBGAN [2]	_	0.213	48.1	-	0.278	45.8
M3GM [11]	1864	0.311	53.3	-	-	-
TransE [1]	743	0.245	56	347	0.294	46.5
ConvKB [8]	763	0.253	56.7	254	0.418	53.2
CapsE [9]	719	0.415	56	303	0.523	59.3
MCapsE	**710**	0.402	**57.2**	317	**0.532**	**60.5**

Table 2. Results on FB15k-237 by relation Category

Task	Head prediction (hits@10)				Tail prediction (hits@10)			
Relation types	1-1	1-N	N-1	N-N	1-1	1-N	N-1	N-N
ConvKB [8]	**51.56**	**57.15**	49.51	48.85	0.46	0.096	0.72	0.39
CapsE [9]	47.4	47.72	59.78	57.28	0.45	0.17	0.69	0.52
MCapsE	48.2	47.23	**59.95**	**59.63**	0.43	**0.18**	**0.73**	**0.56**

5   Conclusion

This paper proposed a knowledge graph representation learning method based on multi-scale capsule network, and models the relation between embedded vectors from a deep perspective to improve the representation of knowledge. Experiments show that on WN18RR and FB15k-237 benchmarks, the performance of proposed method on the task of knowledge graph completion is comparable to the state-of-the-art models, and the performance for N-N modeling of relation classification task is superior to other models.

In the future, we will consider the following methods to improve the representation performance of the model. By considering the neighboring entities of a given entity, the entity can be represented as the weighted mean of neighboring entities. By integrating external knowledge as a supplement, such as description information and/or categorical information of entities, to achieve a more fine-grained representation.

References

1. Bordes, A., Usunier, N., Garcia-Duran, A., Weston, J., Yakhnenko, O.: Translating embeddings for modeling multi-relational data. In: Advances in Neural Information Processing Systems, pp. 2787–2795 (2013)
2. Cai, L., Wang, W.Y.: KBGAN: adversarial learning for knowledge graph embeddings (2017)
3. Dettmers, T., Minervini, P., Stenetorp, P., Riedel, S.: Convolutional 2D knowledge graph embeddings In: Thirty-Second AAAI Conference on Artificial Intelligence (2018)
4. Ji, G., He, S., Xu, L., Liu, K., Zhao, J.: Knowledge graph embedding via dynamic mapping matrix. In: Proceedings of the 53rd Annual Meeting of the Association for Computational Linguistics and the 7th International Joint Conference on Natural Language Processing (vol. 1: Long Papers), pp. 687–696 (2015)
5. Ji, G., Liu, K., He, S., Zhao, J.: Knowledge graph completion with adaptive sparse transfer matrix. In: Thirtieth AAAI Conference on Artificial Intelligence (2016)
6. Kingma, D.P., Ba, J.: Adam: a method for stochastic optimization. arXiv preprint arXiv:1412.6980 (2014)
7. Lin, Y., Liu, Z., Sun, M., Liu, Y., Zhu, X.: Learning entity and relation embeddings for knowledge graph completion. In: Twenty-Ninth AAAI Conference on Artificial Intelligence (2015)
8. Nguyen, D.Q., Nguyen, T.D., Nguyen, D.Q., Phung, D.: A novel embedding model for knowledge base completion based on convolutional neural network (2018)
9. Nguyen, D.Q., Vu, T., Nguyen, T.D., Nguyen, D.Q., Phung, D.: A capsule network-based embedding model for knowledge graph completion and search personalization. arXiv preprint arXiv:1808.04122 (2018)
10. Nguyen, D.Q.: An overview of embedding models of entities and relationships for knowledge base completion (2018)
11. Pinter, Y., Eisenstein, J.: Predicting semantic relations using global graph properties (2018)
12. Sabour, S., Frosst, N., Hinton, G.E.: Dynamic routing between capsules. In: Advances in Neural Information Processing Systems, pp. 3856–3866 (2017)

13. Toutanova, K., Chen, D.: Observed versus latent features for knowledge base and text inference. In: Workshop on Continuous Vector Space Models & Their Compositionality (2015)
14. Trouillon, T., Welbl, J., Riedel, S., Gaussier, R., Bouchard, G.: Complex embeddings for simple link prediction (2016)
15. Wang, Z., Zhang, J., Feng, J., Chen, Z.: Knowledge graph embedding by translating on hyperplanes. In: Twenty-Eighth AAAI Conference on Artificial Intelligence (2014)
16. Yang, B., Yih, W.T., He, X., Gao, J., Deng, L.: Embedding entities and relations for learning and inference in knowledge bases (2014)

CNNPSP: Pseudouridine Sites Prediction Based on Deep Learning

Yongxian Fan[1($\boxtimes$)], Yongzhen Li[2], Huihua Yang[1,4],
and Xiaoyong Pan[3($\boxtimes$)]

[1] School of Computer and Information Security, Guilin University of Electronic
Technology, Guilin 541004, China
yongxian.fan@gmail.com
[2] School of Electronic Engineering and Automation,
Guilin University of Electronic Technology, Guilin 541004, China
1585691075@qq.com
[3] Institute of Image Processing and Pattern Recognition,
Shanghai Jiao Tong University, Shanghai 200240, China
xypan172436@gmail.com
[4] School of Automation, Beijing University of Posts and Telecommunications,
Beijing 100876, China
yhh@bupt.edu.cn

Abstract. Pseudouridine (ψ) is a kind of RNA modification, which is formed at specific site of RNA sequence due to the catalytic action of Pseudouridine synthase in the process of gene transcription. It is the most prevalent RNA modification found so far, and plays a vital role in normal biological functions. Several computational methods have been proposed to predict Pseudouridine sites, but these methods still do not achieve high accuracy. At present, deep learning has become a popular field of machine learning, and convolutional neural network (CNN) is one widely used algorithm. CNN can automatically dig into the hidden features of data and make accurate predictions, so a new algorithm based on CNN was proposed for extracting the features from RNA sequences with and without ψ sites. And a predictor called CNNPSP was developed to predict ψ sites in RNAs across three species (H. sapiens, S. cerevisiae and M. musculus). Both the rigorous jackknife test and independent test indicated that the new predictor outperformed the existing methods in this task.

Keywords: Convolutional Neural Network · Deep learning · Pseudouridine sites

This work was supported in part by the National Natural Science Foundation of China (No. 61462018, 61762026), Guangxi Natural Science Foundation (No. 2017GXNSFAA198278), Guangxi Key Laboratory of Trusted Software (No. kx201403), Guangxi Colleges and Universities Key Laboratory of Intelligent Processing of Computer Images and Graphics (No. GIIP201502).

H. Yin et al. (Eds.): IDEAL 2019, LNCS 11871, pp. 291–301, 2019.
https://doi.org/10.1007/978-3-030-33607-3_32

1   Introduction

Many RNAs are modified when the gene is being transcribed, and so far, more than one hundred RNA modifications have been found [1]. Of them, pseudouridine (ψ) is the first discovered modification, and is the most abundant RNA modification and widely spread in RNAs of many organisms. The nitrogen-carbon covalent bond between uracil and ribose in uridine is changed to carbon-carbon covalent bond, and then the uridine is converted into pseudouridine.

Pseudouridine can alter RNA secondary structure and tertiary structure, increase the number of base stacking and complementary base pairing, and rigidify the sugar–phosphate backbone. In addition, it is also directly or indirectly related to some neuropathic diseases, e.g. the Parkinson's and Alzheimer's disease. Due to its special structure and chemical properties, as well as biological and medical significance, pseudouridine sites have attracted more and more attentions. Aiming at identifying pseudouridine sites, high-throughput sequencing technology (ψ-SEQ) was proposed [2]. In this method, a comprehensive and high-resolution mapping for sequence was conducted to determine the location of the ψ site. But the technique is time-consuming and costly for genome-wide analysis [3]. Therefore, it is urgent to develop some convenient computational algorithms to extract the information of pseudouridine sites, and then identify the sites.

To date, computational methods formulate the pseudouridine sites identification as a binary classification problem. Li et al. [4] and Chen et al. [3] intercepted the gene sequence, then converted the sequence into digital features, and used support vector machine (SVM) algorithm to predict the pseudouridine modification sites of RNAs. Chen et al. integrated the chemical properties of nucleotides when encoding the sequences into features, which is further fed into SVM, their predictor covers one more species M. Musculus than Li et al. But both of their predictors have some disadvantages: the input features are manually curated according to domain knowledge. Thus, more discriminate features need to be automatically constructed from RNA sequences. In addition, the accuracy of SVM algorithm for classifying pseudouridine sites should be improved. To this end, new methods need to be proposed for extracting high-level features instead of manually handcrafting them, and classifying pseudouridine sites with high accuracy.

Deep learning automatically learns high-level features from raw inputs without manual curation for features or rules [5], and they have been shown to improve performance in computer vision analysis, speech recognition, natural language understanding and recently, even in computational biology [6, 7]. CNN is a typical deep learning framework, and it can extract hidden features in data effectively. Therefore, we propose a CNN based algorithm to extract sequence features, then predict the pseudouridine sites using these abstract features.

2 Materials and Methods

At first, we encode the sequences in benchmark dataset, and extract features from sequences by a CNN; Then we intercept the query sequences through a fixed length sliding window in the form of sample sequences, and encode the sequence fragment; Finally, we judge whether the intercepted sequences include pseudouridine site or not by the trained CNN model. With the window sliding to the right, we can yield the predicted results across the whole input sequence.

2.1 Data Set

In this paper, the CNN algorithm is used to extract abstract features and classify the sequences to contain pseudouridine sites or not. We evaluate our method on the benchmark datasets provided in [3]. In [4], we can only obtain the information about pseudouridine sites in the data set. Due to the database update, it is difficult to obtain sequences accurately. Thus, we do not test our model on datasets in [4]. Fortunately, both positive and negative sample sequences of benchmark datasets S(1), S(2), S(3) for H. sapiens, S. cerevisiae and M. musculus have been provided in Chen et al. [3], as well as independent test datasets S(4), S(5) for H. sapiens and S. cerevisiae. So in this paper, we used the same datasets S(1), S(2), S(3), S(4), S(5) and constructed a new independent dataset S(6) for M. musculus.

2.2 Evaluation Measurements

It is very important to choose the appropriate evaluation measurements to measure the effect of feature extraction and classification. We choose four metrics: 1. sensitivity (Sn); 2. specificity (Sp); 3. overall accuracy (Acc); 4. Mathews correlation coefficient (MCC). Among them, the overall accuracy (Acc) and the Mathews correlation coefficient (MCC) are the metrics that evaluate the overall performance of the classification. The above four metrics can be formulated as follows (as shown in formula (1)):

$$
\begin{cases}
Sn = \frac{TP}{TP+FN} \\
Sp = \frac{TN}{TN+FP} \\
Acc = \frac{TP+TN}{TP+FN+TN+FP} \\
MCC = \frac{(TP\times TN)-(FP\times FN)}{\sqrt{(TP+FP)(TP+FN)(TN+FP)(TN+FN)}}
\end{cases}
\tag{1}
$$

We treat the samples which containing pseudouridine sites as the positive samples, in contrast, samples without pseudouridine sites are considered as the negative samples. In formula (1), TP represents the number of positive samples that are correctly predicted, FP represents the number of positive samples that are incorrectly predicted to be negative samples, TN represents the number of negative samples that are correctly predicted, and FN represents the number of negative samples which are incorrectly predicted to be positive samples in true.

2.3   Validation Method

Choosing appropriate validation method is also very important. Jackknife test, k-fold cross validation and independent test are often used to evaluate the metrics values in an experiment. Of these three validation methods, Jackknife test is the most rigorous validation method, so we choose Jackknife test method in this study.

2.4   Sequence Encoding

We found sequences in dataset used in Chen et al. and extended the sequences from the source [8, 9], then intercepted sequences in the manner shown in formula (2). The intercepted sequence is represented by S (U) (as shown in formula (2)). U represents uridine or pseudouridine, and we treat it as a reference point for intercepted sequence. N stands for nucleotide; L, R represents the number of nucleotides at the left and right of U, respectively.

$$S(U) = N_{-L}N_{-(L-1)}N_{-(L-2)}\ldots N_{-2}N_{-1}UN_{+1}N_{+2}\ldots N_{+(R-2)}N_{+(R-1)}N_{+R} \qquad (2)$$

According to whether U is a pseudouridine or uridine, the intercepted sequence can be represented by S^+ (U) or S^- (U). As shown at formula (3).

$$S(U) = \begin{cases} S^+(U), \text{if } U \text{ is a pseudouridine site} \\ S^-(U), \text{otherwise} \end{cases} \qquad (3)$$

We use S^+ to represent all the positive samples which contain pseudouridine sites and S^- to represent all the negative samples without pseudouridine sites. Therefore, S^+, S^- are the set of $S^+(U)$ and $S^-(U)$, respectively (as shown at formula (4)).

$$\begin{cases} S^+ = \sum S^+(U) \\ S^- = \sum S^-(U) \\ S = S^+ \cup S^- \end{cases} \qquad (4)$$

$\sum$ is a summation notation, U stands for union of two sets. It can be seen from formula (2) that the length of sample sequence equals to L + R + 1.

The benchmark datasets of three species H. sapiens, S. cerevisiae and M. musculus, they are denoted as S (1), S (2) and S (3), respectively. For these three datasets, we take L = R = 10, thus the length of a sample equals to L + R + 1 = 21.

We adopt different ways to represent samples. (1) Encode samples based on the difference of nucleotides (see below); (2) Add its density; (3) Combine the difference of nucleotides, its density and its chemical properties.

First, we encode RNA sequences based on the difference of nucleotides. RNA sequence consists of four nucleotides A, U, G, and C. We randomly extract dinucleotides and sort them in order as a set, thus there are 16 sets. As shown in Fig. 1(a), each pair of combination is converted to a one-hot encoded vector with length 16.

Starting from the left of a sample, and shifted to the right with a step one nucleotide (Fig. 1(b)). When the sample length is L + R + 1 we yield a 16 × (L + R) matrix as shown in Fig. 1.

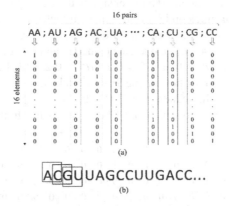

(a)

(b)

Fig. 1. Sequence is encoded in the difference of nucleotides.

The pseudouridine site in the RNA sequence is not only influenced by sequence and enzymes, but also associated with concentration and chemical properties of nucleotides. Thus we also integrate chemical properties into sequence encoding. The chemical properties of nucleotide [3] are shown in Table 1.

Table 1. The chemical properties of nucleotide.

Chemical property	Class	Nucleotides
Ring structure	Purine	A, G
	Pyrimidine	C, U
Functional group	Amino	A, C
	Keto	G, U
Hydrogen	Strong	C, G
	Weak	A, U

Second, the chemical properties and density of nucleotides should be taken into account on the basis of the above method. We had 16 dimension to encode the difference of nucleotides, we added another 3 dimension for chemical properties (Table 1) and 1 dimension for density.

(1) The 17th dimension represents the ring structure properties of the first dinucleotide, the purine is represented by '1', and the pyrimidine is represented by '0';

(2) The 18th dimension encodes functional group properties, the amino group is represented by '1' and the ketone group is represented by '0';

(3) The 19th dimension encodes hydrogen properties, '1' represents the strong covalent bond, and '0' represents the weak covalent bond.

(4) The 20th dimension represents the density of the nucleotide in the first dinucleotide. We can calculate the density of the four types of nucleotides in the preceding L + R nucleotides when the length of the sample is L + R + 1, respectively. So the value in 20th dimension is the density of nucleotide that is the same type as the nucleotide in the first dinucleotide. The RNA sequence "AGAUCU" remove the last nucleotide is "AGAUC" that contains two adenosine(A), one cytidine(C), one guanosine(G) and one uridine(U). The density of A, C, G, U is 2/5 = 0.4, 1/5 = 0.2, 1/5 = 0.2 and 1/5 = 0.2 in sequence "AGAUC". If the nucleotide in the first dinucleotide is A, the value in 20th dimension is 0.4. So the sequence "AGAUCU" encoding result can be denoted by R(AGAUCU) shown at formula (5). "T" denotes matrix transpose in formula (5).

$$
R(A\,G\,A\,U\,C\,U) = \begin{pmatrix} 0 & 0 & 1 & 0 & 0 & 0 & 0 & 0 & 0 & 0 & 0 & 0 & 0 & 0 & 0 & 0 & 1 & 1 & 0 & 0.4 \\ 0 & 0 & 0 & 0 & 0 & 0 & 0 & 0 & 1 & 0 & 0 & 0 & 0 & 0 & 0 & 0 & 1 & 0 & 1 & 0.2 \\ 0 & 0 & 1 & 0 & 0 & 0 & 0 & 0 & 0 & 0 & 0 & 0 & 0 & 0 & 0 & 0 & 1 & 1 & 0 & 0.4 \\ 0 & 0 & 0 & 0 & 0 & 0 & 0 & 1 & 0 & 0 & 0 & 0 & 0 & 0 & 0 & 0 & 0 & 0 & 0 & 0.2 \\ 0 & 0 & 0 & 0 & 0 & 0 & 0 & 0 & 0 & 0 & 0 & 0 & 1 & 0 & 0 & 0 & 1 & 1 & 0.2 \end{pmatrix}^{T} \quad (5)
$$

We take L = R = 10, we have the sample length 21, we can encode each sample into a matrix size of 20 × 20 instead of 16 × 20 after integrating chemical properties and density.

2.5   Convolutional Neural Networks

The rapid increase in biological data dimension is challenging conventional analysis strategies. CNN is one of the most famous algorithms in deep learning, and it can extract hidden features in data efficiently. In this study, we extract sample features and classify them by a CNN. The structure of CNN consists of five layers, including an input layer, a convolutional layer, a sub-sampling layer, a fully connection layer and an output layer. The architecture of CNN is constructed in our study is shown in Fig. 2.

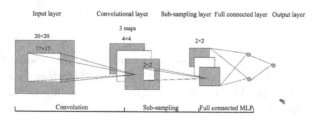

Fig. 2. Architecture of Convolutional Neural Network used in this study.

We take dataset S (1) of H. sapiens as an input example, as shown in Fig. 2. The length of the intercepted sequence varies in different species, resulting in different dimensions of the constructed input matrix. For H. sapiens, a sequence is constructed as a matrix size of 20 × 20, the matrix is used as an input to the CNN. After using three convolutional kernels size of 17×17 and sub-sampling, we obtained three feature matrices size of 2×2. Finally, the fully connection layer is used to reduce the dimension and eliminate the impurity, the results are treated as outputs.

Convolutional layer consists of three feature matrices, and the aim of the convolution operation is to enhance the original characteristics and remove interference information. After the samples are converted into a matrix as input, the elements of input matrix multiplied by corresponding elements of the convolutional kernel, then doing the sum of them and offset variable. Starting from the upper left of the input matrix, then the convolutional kernel moves a step to the right and repeating the above operation.

As shown in Fig. 2, convolutional layer consists of three feature maps, each feature map has 4×4 neurons, and each neuron in each feature map corresponds to a convolutional kernel size of 17×17 in the input layer. Therefore, there are $4 \times 4 \times 3 = 48$ neurons in this layer, and $(17 \times 17 + 1) \times 3 = 870$ weights (the "+1" is for the bias). However, the traditional neural network needs a unique weight for each connection, because each of the 48 neurons has $17 \times 17 + 1 = 290$ connections, thus $48 \times 290 = 13290$ weights are needed from this layer to the prior layer. However, only 870 weights are needed in the CNN because of the "shared weight" characteristic of the CNN. The aim of the subsampling is to reduce the amount of data processing while retaining the useful information.

3 Results

We first test the performance of different encoding methods. As shown in Fig. 3, the blue bar shows the accuracy is higher than the green one with only difference of nucleotides, and is lower than the yellow one with difference, density and chemical properties of nucleotides for three species. We mixed the difference of nucleotides, its density and its chemical properties to encode sequences.

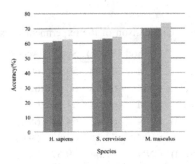

Fig. 3. Performance of different encoding methods. The green bar stands for the accuracy obtained by the model trained on the difference of nucleotides in identifying Ψ sites; The blue one is accuracy for adding the density of nucleotide; The yellow one is the accuracy for adding chemical properties and density of nucleotides. (Color figure online)

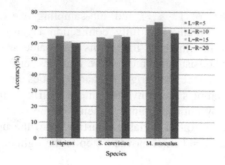

Fig. 4. The overall accuracy obtained by CNN algorithm in identifying Ψ site with different values of L and R.

In the article, we assigned different values to L and R, and then evaluated the overall accuracy in identifying pseudouridine sites for three species (Fig. 4).

The accuracy for H. sapiens or M. musculus reaches a peak when L = R = 10, after L and R value increase or decrease to 5, 15 and 20, respectively, the accuracy is lower than that obtained by CNN with L = R = 10. While the accuracy for S. cerevisiae is the highest when L = R = 15. Thus, we assign the same value 10 to L and R for H. sapiens and M. musculus, and the length of samples is L + R + 1 = 21 for them. While we set L = R = 15 for S. cerevisiae, and the length of samples is 31.

Chen et al. [3] combined the chemical properties of nucleotides when encoding the sequence, and achieved higher classification accuracy than Li et al. [4]. In addition, he added a new species M. musculus in his predictor. Therefore, we mainly compare the results with [3]. We created a new sequence encoding method and applied a new CNN algorithm instead of SVM algorithm to predict Pseudouridine sites. And we also obtained the experimental results by jackknife test on benchmark datasets S(1), S(2) and S(3) for H. sapiens, S. cerevisiae and M. musculus, respectively. The comparison with Chen et al., the experimental results are shown in Table 2.

Table 2. Comparison of the experimental results.

Species	Method	Sn(%)	Sp(%)	Acc(%)	MCC
H. sapiens	SVM	61.01	59.80	60.40	0.21
(Dataset S(1))	CNN	64.44	60.73	62.59	0.25
S. cerevisiae	SVM	64.65	64.33	64.49	0.29
(Dataset S(2))	CNN	65.92	64.34	65.13	0.30
M. musculus	SVM	73.31	64.83	69.07	0.38
(Dataset S(3))	CNN	77.66	69.57	73.62	0.47

The experimental results are shown in Table 2 by jackknife testing on datasets for three species. The Acc obtained by CNN improved by 2.19% and 4.55% than those obtained by SVM for H. sapiens and M. musculus, respectively, and the Acc also increased by 0.64% for S. cerevisiae. Sn and Sp have also improved for three species. And the overall measure parameter MCC values obtained by CNN are higher than that by SVM.

The predictor CNNPSP for predicting ψ sites in RNA sequences was developed in this study. PPUS [4] and iRNA-PseU [3] are so far the only two existing predictors available for predicting the Ψ sites in RNA sequences. It should be pointed out PPUS can be only used to identify the Ψ sites in the RNA sequences from H. sapiens and S. cerevisiae, but not from M. musculus. Three independent datasets S(4), S(5) and S(6) from H. sapiens, S. cerevisiae and M. musculus are used to compare the performance of three predictors. The comparison of prediction performance is shown in Table 3.

Table 3. Comparison of different predictors for predicting Ψ sites in RNA sequences.

Species	Predictor	Acc(%)	MCC	Sn(%)	Sp(%)
H. sapiens	PPUS	52.50	0.13	6.00	99.00
(Dataset S(4))	iRNA-PseU	65.00	0.30	60.0	70.00
	CNNPSP	70.00	0.40	72.00	68.00
S. cerevisae	PPUS	71.00	0.44	56.00	86.00
(Dataset S(5))	iRNA-PseU	73.00	0.46	81.00	65.00
	CNNPSP	75.50	0.51	81.00	70.00
M. musculus	iRNA-PseU	69.50	0.39	79.00	60.00
(Dataset S(6))	CNNPSP	74.00	0.48	81.00	67.00

As shown in Table 3, the Acc values obtained by CNNPSP improved by 2.50% and 4.50% than those obtained by iRNA-PseU predictor for S. cerevisiae and M. musculus, respectively, and increased by 5.00% for H. sapiens. The Acc values of CNNPSP are much higher than that of predictor PPUS. In addition, the overall measurement MCC values obtained by CNNPSP are higher than that of iRNA-PseU and PPUS for all three species.

Receiver operating characteristic (ROC) curve is another important performance measurement of a predictor, and the area under the ROC curve (AUC) represents a popular evaluation metric of a binary classifier. The larger the AUC is, the better performance a predictor yields. In Fig. 5, we plotted the ROC curves for CNNPSP and iRNA-PseU for three species. Obviously, the AUC value of CNNPSP is 0.0502 higher than that of iRNA-PseU for H. sapiens, and also increased about 0.0038 and 0.0312 for S. cerevisiae and M. musculus, respectively. Therefore, we can draw the conclusion that the performance of our predictor CNNPSP is better than that of predictor iRNA-PseU. From Table 3 and Fig. 5, we can conclude that CNNPSP is currently the best predictor for predicting ψ sites in RNA sequences from H. sapiens, S. cerevisiae and M. musculus.

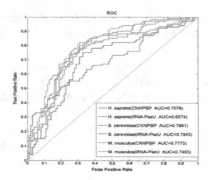

Fig. 5. The ROC curves to show the performance of two different predictors for three species. CNNPSP is a new predictor developed in this paper, iRNA-PseU is a state-of-the-art predictor developed by Chen et al.

4 Conclusion and Future Work

The CNN has shown good performance in many research fields due to its special algorithm structure. In this paper, we constructed new encoding strategy for sequences, and proposed a new CNN based method for extracting abstract features from RNA sequences, then constructed a new predictor CNNPSP to predicted ψ sites across three species H. sapiens, S. cerevisiae and M. musculus. Both the rigorous jackknife test and independent test indicated that the prediction performance of CNNPSP is better than that of existing predictors iRNA-PseU and PPUS. It also shows that the CNN algorithm could be well applied to biological sequences.

References

1. Czerwoniec, A., et al.: MODOMICS: a database of RNA modification pathways. 2008 update. Nucleic Acids Res. **37**, D118–D121 (2009)
2. Carlile, T.M., Rojas-Duran, M.F., Zinshteyn, B., Shin, H., Bartoli, K.M., Gilbert, W.V.: Pseudouridine profiling reveals regulated mRNA pseudouridylation in yeast and human cells. Nature **515**, 143 (2014)
3. Chen, W., Tang, H., Ye, J., Lin, H., Chou, K.-C.: iRNA-PseU: identifying RNA pseudouridine sites. Mol. Ther-Nucleic Acids **5**, e332 (2016)
4. Li, Y.-H., Zhang, G., Cui, Q.: PPUS: a web server to predict PUS-specific pseudouridine sites. Bioinformatics **31**, 3362–3364 (2015)
5. Pan, X., Rijnbeek, P., Yan, J., Shen, H.-B.: Prediction of RNA-protein sequence and structure binding preferences using deep convolutional and recurrent neural networks. BMC Genomics. **19**, 511 (2018)
6. Alipanahi, B., Delong, A., Weirauch, M.T., Frey, B.J.: Predicting the sequence specificities of DNA- and RNA-binding proteins by deep learning. Nat. Biotech. **33**, 831 (2015)
7. Pan, X., Shen, H.-B.: RNA-protein binding motifs mining with a new hybrid deep learning based cross-domain knowledge integration approach. BMC Bioinf. **18**, 136 (2017)

8. Sun, W.-J., et al.: RMBase: a resource for decoding the landscape of RNA modifications from high-throughput sequencing data. Nucleic Acids Res. **44**, D259–D265 (2016)
9. Speir, M.L., et al.: The UCSC genome browser database: 2016 update. Nucleic Acids Res. **44**, D717–D725 (2016)

A Multimodal Approach to Image Sentiment Analysis

António Gaspar[(✉)][iD] and Luís A. Alexandre[iD]

Instituto de Telecomunicações, Universidade da Beira Interior,
Rua Marquês d'Ávila e Bolama, 6201-001 Covilhã, Portugal
{antonio.pedro.gaspar,luis.alexandre}@ubi.pt

Abstract. Multimodal sentiment analysis is a process for the classification of the content of composite comments in social media at the sentiment level that takes into consideration not just the textual content but also the accompanying images. A composite comment is normally represented by the union of text and image. Multimodal sentiment analysis has a great dependency on text to obtain its classification, because image analysis can be very subjective according to the context where the image is inserted. In this paper we propose a method that reduces the text analysis dependency on this kind of classification giving more importance to the image content. Our method is divided into three main parts: a text analysis method that was adapted to the task, an image classifier tuned with the dataset that we use, and a method that analyses the class content of an image and checks the probability that it belongs to one of the possible classes. Finally a weighted sum takes the results of these methods into account to classify content according to its sentiment class. We improved the accuracy on the dataset used by more than 9%.

Keywords: Multimodal sentiment analysis · Image · Text · Deep learning

1 Introduction

Sentiment analysis is an important topic nowadays. With the advent of social media networks, the amount of data available is increasing exponentially which made sentiment analysis techniques and methods grow. These have many possible applications, among which are the anticipation of the behaviours and trends of the crowds. This kind of analysis is also used on opinion mining, which correlates the sentimental information with the influence of someone or something, that often has the purpose to convince or attract the individuals to do some action, for example, to buy a new product or to vote in a determinate candidate

This work was partially supported by Instituto de Telecomunicações under grant UID/EEA/50008/2019 and by project MOVES - Monitoring Virtual Crowds in Smart Cities (PTDC/EEI-AUT/28918/2017) financed by FCT - Fundação para a Ciência e a Tecnologia.

© Springer Nature Switzerland AG 2019
H. Yin et al. (Eds.): IDEAL 2019, LNCS 11871, pp. 302–309, 2019.
https://doi.org/10.1007/978-3-030-33607-3_33

for the elections. Although sentiment analysis is widely used for many tasks, its application has a high dependence on the textual content, which is present on the majority of comments that are published on social media networks.

Nonetheless and in spite of the textual content containing an objective message, which most of the times is clear to all of the participants when associated with an image, it can transform the image's natural meaning. A widely used expression is "A picture is worth a thousand words", meaning that an image can clearly transmit a message to the viewers that would otherwise require a large textual description to describe its contents. But unfortunately, the meaning of an image is not always clearly recognised because the viewers can have different cultures and life experiences, which means that they may have different ideas and perspectives about the interpretation of an image. This fact introduces subjectivity into the interpretation. Along this document we will present our proposal to resolve this challenging situation using current state-of-the-art methods on sentiment analysis, which are discussed in Sect. 2. Our contribution is a method that can help to reduce the subjectivity of image sentiment analysis. The next section, presents the related work. Our method is presented in Sect. 3. Section 4 contains the experiments and the last section the conclusions.

2   Related Work

2.1   Sentiments

In the psychology area, sentiments are different from emotions as is described by the authors of the papers [6,9]. Sentiments are the result of subjective experiences that were lived from an emotion. Emotions, in general, are the triggers for actions that can be positive or negative. An emotion can occur as a response to an internal or external signal of the environment context. For example, pain, which is considered an internal signal, can trigger in most of cases the sadness emotion that produces bad feelings. Playing a game, can trigger joy and pleasure which are positives feelings. Emotions are the base of sentiments. These can construct the history of the all feelings that are processed and memorised. This fact is important in sentiment analysis because through it is possible to reduce the subjectivity according to the culture where the analysed data belongs to. However, often the data cannot be organised by the culture. It is the case of the data collected from social media networks. For this reason, artificial intelligence may help to find the best features for classification. Next, we present some of these techniques related with the present theme.

2.2   Sentiments and Artificial Intelligence

Nowadays text, images, videos and all multimedia content can be processed and analysed. This process is supported essentially by models that run on computers. These models can obtain important information from raw data. To analyse the data, most models represent information using sets of features which in turn represent the classes of the target objects. This process can be done through many

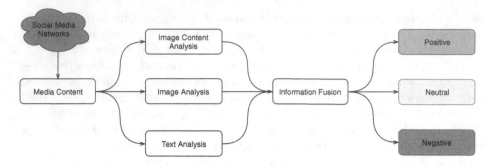

Fig. 1. Overview of the proposed method.

different approaches, but currently, deep neural networks such as Convolutional Neural Networks (CNNs) have been producing very good results when applied to image data.

There have been many proposals of methods for Image Sentiment Analysis. The authors of [10] studied the sentiment analysis process. They propose a method that is capable of classifying images at sentiment polarity level. The dataset they use is composed of 3 million tweets, which include text and images, and was constructed by them collecting the information on Twitter. For the classification, they propose a method that leverages the text classification and correlates it with the image. They conclude that text associated with image is often noisy and is weakly correlated with the image content, but it is possible to classify its sentiment using a model that is trained with the images classified with text labels. In another work described in [4], the authors explore four different architectures of convolutional neural networks to do sentiment analysis in visual media. This work was based on a labelled set that has the main categories of the description of the scene. With their results, the authors compose their own dataset and train a model that improves the results.

3   Proposed Method

Our method is based on a multimodal approach. It is composed of three parts that are fused in the end. Each part is designed for a specific task. The first part corresponds to the text analysis and the second part corresponds to the isolated image analysis. Both parts classify the sentiment analysis polarity, which is negative, neutral and positive. Next, in the third part, we studied the most common object class occurrence on the training and validation dataset, for each polarity class. Finally, we fuse all the parts using a weighted sum that is capable of predicting the polarity at the sentiment level. Figure 1 contains an overview of the proposed method.

3.1   Text Classification

Text sentiment analysis is a procedure derived from natural language processing (NLP). There are many proposed methods in this field, that are able to handle the job. These can be applied to problems like opinion mining and crowd influence through social networks. In this work we test two current state-of-the-art methods, the Vader [7] and the TextBlob [3] methods.

The Vader method (Valence Aware Dictionary and Sentiment Reasoner) [7] is a method composed by a list of lexical features, that are labelled according to their semantic orientation. Vader can produce four classifications, which are, negative, neutral, positive and compound score. Vader does not require to be trained because it is constructed through a standard sentiment lexicon.

The TextBlob [3] is a Python library that implements methods for processing textual data. It provides an API for processing NLP tasks such as part-of-speech tagging, and sentiment analysis. The sentiment analysis polarity, has a float range between −1 and 1, where values above 0.1 means positive, values below −0.1 means negative and values between −0.1 and 0.1 means neutral.

For the text analysis process, we tested both methods on the dataset described in Sect. 4.1. To compare both methods, we use the respective confusion matrices that take into consideration the results that the B-$T4SA$ dataset already provides on the validation set. With the analysis of these results, the TextBlob method is the chosen since it is the one that reveals a higher accuracy (64.271% vs 41.078%) in the B-$T4SA$ dataset.

3.2   Image Classification

The developed method for image analysis is based on a deep learning approach. This is implemented with Pytorch [8], which is a deep learning framework that supports several features and automatic differentiation. For this work, we explore three versions of the Resnet, which are, the ResNet18, the ResNet50, and the ResNet152. We use the ResNet topology because it is a state-of-the-art method that reduces significantly the vanishing gradient problem. To use these models we need to set them up and prepare the data. To do that we follow the next steps.

1. **Data Preprocessing.** One of the biggest challenges on deep learning approaches is the data quality. Any deep learning approach is hungry for data, because it is through it that the network extracts and learns the features used for classification. The dataset used has many images with different scales and sizes. This fact can slow the training process. The pre-processing method used scales and re-sizes each image to 224 × 224.
2. **Model Choice and its Adaptation.** Pytorch comes with several built-in models. In this work, we selected three of these models, the ResNet18, ResNet50, and ResNet152. All the models are set up with the same hyper-parameters. These are, the learning rate that starts with 0.001, the momentum with 0.9 and the gamma parameter with 0.1. We use an optimiser that

will hold the current state and will update the parameters based on the computed gradients. This is the SGD (Standard Gradient Descendent). We use a schedule that provides several methods to adjust the learning rate based on the number of epochs. This will adjust the learning rate in every seven epochs. We define 30 epochs to train.

3. **Model Training and Evaluation**. We train the models with a GeForce GTX 1080 TI, using the training set to train and the validation set to validate the training phase. ResNet18 uses 512 features from each image and achieves 50.3%, ResNet50 and ResNet152 use 2048 features and achieve better results. ResNet152 achieves the best result, 52.2%, and exceeds the result presented in the dataset paper [10] for only the image analysis, which is 51.3%. For this reason this is the model selected for the next phase.

3.3   Image Content Analysis

Image content analysis is a complex subject because an image might contain many objects. In this work we try to identify automatically the class of the object that an image can represent. To do this we use a pre-trained model with the ImageNet [5] to classify the data into its class through the ImageNet classifications (1000 possible object classes). All images on the ImageNet are quality-controlled and human-annotated. We use an InceptionResNetV2 trained model, which according to the author [1], has 80.17% of accuracy. This model comes from a python package that is called pretrained models [2]. In this work, our intention with the image content analysis is to build a probability distribution that makes it possible to classify an image according to its sentiment polarity, such as negative, neutral and positive. So, with the InceptionResNetV2, we built a model that was feed with the union of the training and validation sets to increase the number of the images. The InceptionResNetV2 classified the contents of each dataset image into one of the ImageNet classes. Each of these images contains a sentiment classification in the training and validation sets that we used to build a table with the probability distribution of the image sentiment for each ImageNet class. In the Table 1 we present an example of the full table (1000 rows). With this analysis, we can increase the information that we give to our proposal method. Next, we present the fusion of these three approaches to build our proposed method.

Table 1. An example of the results of our method with ImageNet.

Class ID	Class name	Negative	Neutral	Positive
445	bikini, two-piece	28.28%	33.02%	39.70%
700	paper towel	40.54%	27.03%	32.43%
966	red wine	27.59%	20.69%	51.72%

3.4   Information Fusion

We built a method where we join the three methods explained before. This is used to classify multimedia content at sentiment polarity level, without a high text dependence. To do that, we make a weighted sum where we attribute the normalised validation accuracy value of each method as a weight, which is used to balance the importance that we give to each method.

The accuracy in the validation set of the three methods was 64.27, 52.15 and 40.09, respectively. The sum of these values is 156.51 and it is used to normalise them and create the weights used for producing the final decision:

$$w_1 = 0.41, w_2 = 0.33, w_3 = 0.26.$$

The results of each method are now used in a voting system. Each of the three partial methods vote on its own decision class with a value equal to its normalised weight, that is, the text-based decision votes with 0.41, the image-based with 0.33 and the image content-based with 0.26. The image content-based decision is simply the sentiment that is more probable for the given image class. The final decision is the class that has the highest vote value.

4   Experiments

In this section, we present the dataset used to evaluate our method, as well as the results that we achieved in this work. The results that we present, were obtained on the test set.

4.1   The Dataset

The authors of [10] built a dataset with three million tweets. These tweets contain text and images. Nonetheless this huge amount of data, it has some problems, such as, duplicated entries and malformed images. Another problem is the occurrence number around the three possible classes, negative, neutral and positive. In this case, positive and neutral content occur more times than the negative content. These situations led the authors to build a subset that is composed by tweets that have images and text in their corpus, non duplicated and non malformed images, as well the same number of occurrences on the different classes.

The subset is called *B-T4SA*, and is divided into three partitions: the train part, the validation part and the test part. Each one of these subsets has three classes, negative, neutral and positive. Each class has the same number of images as the others. Figure 2 shows some example tweets of the three classes.

4.2   Results

The result which we obtained is 60.42% accuracy on the test set. Table 2 shows the confusion matrix of our method and Table 3 presents the results of the baseline and of our method. We can conclude that, with this results, it is possible to classify multimedia content using our method. Nonetheless there is space to improve it, for instance, by using all of the probabilistic information gathered for the image content-based decision instead of using only the most probable class.

(a) **Negative**: "Not only are we surrounded by clowns, fools and useful idiots,we now have 000s more rapists, parasites and predators."

(b) **Neutral**: "Moon Mission: Hakuto robot rover undergoes testing in Japan ahead of lunar travel."

(c) **Positive**: "Join us tomorrow for another incredible Sunday! We have three services catering for all."

Fig. 2. Examples of the three classes present in the dataset, negative, neutral and positive.

Table 2. Confusion matrix of the proposed method in the test set. The accuracy in the test set is 60.42%

		Predicted class		
		Negative	Neutral	Positive
True Class	Negative	8841	4960	3199
	Neutral	2980	9998	4022
	Positive	2338	2689	11973

Table 3. Results comparison between the method by the authors of the baseline paper [10] and the proposed method.

Method [10]	Proposed method
51.30%	60.42%

5 Conclusions

In this work, we explore the sentiment analysis of tweets that contain both text and image, focusing on images and their content. We achieve a result on the isolated method image that exceeds the baseline method for the same theme in the paper [10]. We built a probability distribution table, that is based on 1000 classes of the ImageNet, that summarise a probability of a given image being negative, neutral or positive according to its content. Finally, we built a method that can classify multimedia content with text and image and generate a sentiment classification based on the image content. This method improves the results presented in [10] by 9%. For future work we intend to further improve the method and make more tests with other datasets.

References

1. Pretrained Models GitHub pretrained models for pytorch github. https://github.com/cadene/pretrained-models.pytorch. Accessed 17 Jun 2019
2. Pretrained Models pretrained models for pytorch. https://pypi.org/project/pretrainedmodels/. Accessed 17 Jun 2019
3. TextBlob. https://textblob.readthedocs.io/en/dev/. Accessed 17 Jun 2019
4. Bonasoli, W., Dorini, L., Minetto, R., Silva, T.: Sentiment analysis in outdoor images using deep learning, pp. 181–188 (2018). https://doi.org/10.1145/3243082.3243093
5. Deng, J., Dong, W., Socher, R., Li, L.J., Li, K., Fei-Fei, L.: ImageNet: a large-scale hierarchical image database. In: CVPR 2009 (2009)
6. Hovy, E.H.: What are sentiment, affect, and emotion? applying the methodology of Michael Zock to sentiment analysis. In: Gala, N., Rapp, R., Bel-Enguix, G. (eds.) Language Production, Cognition, and the Lexicon. TSLT, vol. 48, pp. 13–24. Springer, Cham (2015). https://doi.org/10.1007/978-3-319-08043-7_2
7. Hutto, C., Gilbert, E.: Vader: a parsimonious rule-based model for sentiment analysis of social media text (2015)
8. Paszke, A., et al.: Automatic differentiation in PyTorch (2017)
9. Pawar, A.B., Jawale, M.A., Kyatanavar, D.N.: Fundamentals of sentiment analysis: concepts and methodology. In: Pedrycz, W., Chen, S.-M. (eds.) Sentiment Analysis and Ontology Engineering. SCI, vol. 639, pp. 25–48. Springer, Cham (2016). https://doi.org/10.1007/978-3-319-30319-2_2
10. Vadicamo, L., et al.: Cross-media learning for image sentiment analysis in the wild. In: 2017 IEEE International Conference on Computer Vision Workshops (ICCVW), pp. 308–317, October 2017. https://doi.org/10.1109/ICCVW.2017.45

Joining Items Clustering and Users Clustering for Evidential Collaborative Filtering

Raoua Abdelkhalek$^{(\boxtimes)}$, Imen Boukhris$^{(\boxtimes)}$, and Zied Elouedi$^{(\boxtimes)}$

LARODEC, Institut Supérieur de Gestion de Tunis,
Université de Tunis, Tunis, Tunisia
abdelkhalek_raoua@live.fr, imen.boukhris@hotmail.com, zied.elouedi@gmx.fr

Abstract. Recommender Systems (RSs) are supporting users to cope with the flood of information. Collaborative Filtering (CF) is one of the most well-known approaches that have achieved a widespread success in RSs. It consists in picking out the most similar users or the most similar items to provide recommendations. Clustering techniques can be adopted in CF for grouping these similar users or items into some clusters. Nevertheless, the uncertainty comprised throughout the clusters assignments as well as the final predictions should also be considered. Therefore, in this paper, we propose a CF recommendation approach that joins both users clustering strategy and items clustering strategy using the belief function theory. In our approach, we carry out an evidential clustering process to cluster both users and items based on past preferences and predictions are then performed accordingly. Joining users clustering and items clustering improves the scalability and the performance of the traditional neighborhood-based CF under an evidential framework.

Keywords: Recommender Systems · Collaborative Filtering · Uncertain reasoning · Belief function theory · Users clustering · Items clustering

1 Introduction

Collaborative Filtering (CF) is one of the most well-known approaches that have achieved a widespread success in Recommender Systems (RSs) [1]. It consists in finding out the most similar users (user-based) or the most similar items (item-based) to provide recommendations. Thus, the predictions of the users' preferences are performed based on the users or the items holding similar ratings. Although their simplicity and intuitiveness, these two CF methods disclose some limitations such as the scalability problem [2], since they require a lot of heavy computations to provide predictions. That is to say, in order to compute the user-user similarities and/or the item-item similarities, the entire ratings matrix needs to be searched which reveals a poor scalability performance. In this context, clustering techniques can be involved to first group users or items according to

© Springer Nature Switzerland AG 2019
H. Yin et al. (Eds.): IDEAL 2019, LNCS 11871, pp. 310–318, 2019.
https://doi.org/10.1007/978-3-030-33607-3_34

their available past ratings and predictions can be made independently within each cluster. Hence, clustering would be an ideal process that may increase the scalability of CF systems as well as maintaining a good prediction performance. While clustering users or items, they are likely to bear upon more than only one cluster which is known as soft clustering. Thus, the uncertainty spreading around the clusters assignment should be considered. In this paper, we embrace the belief function theory (BFT) [3–5] which is among the most used theories for representing and reasoning under uncertainty. We opt for the Evidential c-Means (ECM) [6], an efficient soft clustering method allowing to deal with the uncertainty at the clusters assignment level. When the uncertainty intervenes at clusters assignment, while using the belief function theory, the output is referred to as credal partition. The pertinence of reasoning under uncertainty in CF is also considered at the final prediction level. Indeed, we also tend to quantify and represent the uncertainty arising behind the provided predictions based on the Evidential k-Nearest Neighbors [7] formalism. Such evidential classifier improves the classification performance by allowing a credal classification of the objects. Therefore, the proposed approach allows us to provide more reliable and intelligible predictions by providing an evidential representation of the final results. This representation allows the users to get an overall information about their future preferences, which may increase their confidence and satisfaction towards the RS. Our aim then is not only to unify both user-based and item-based CF but also to represent and encapsulate the uncertainty appearing in both clusters assignment and prediction process under an evidential framework.

This paper is organized as follows: Sect. 2 recalls the concepts of the belief function theory. Section 3 introduces the related works on CF. Then, our proposed recommendation approach is emphasized in Sect. 4. Section 5 reports the experimental results. Finally, the paper is concluded in Sect. 6.

2   Belief Function Framework

The belief function theory [3–5], also referred to as evidence theory, represents a flexible framework for modeling and quantifying imperfect knowledge. Let Ω be the frame of discernment defined as a finite set of variables w. It refers to n elementary events such that: $\Omega = \{w_1, w_2, \cdots, w_n\}$. The basic belief assignment (bba) represents the belief committed to each element of Ω such that $m : 2^{\Omega} \to [0, 1]$ and $\sum_{E \subseteq \Omega} m(E) = 1$. The mass $m(E)$ quantifies the degree of belief exactly assigned to an element E of Ω. Evidence may not be equally trustworthy. Hence, a discounting operation can be used to get the discounted bba m^{δ} such that: $m^{\delta}(E) = (1 - \delta) \cdot m(E), \forall E \subset \Omega$ and $m^{\delta}(\Omega) = \delta + (1 - \delta) \cdot m(\Omega)$. where $\delta \in [0,1]$ is the discounting factor.

To make decisions, the *pignistic probability* denoted $BetP$ can be used, where $BetP$ is defined as follows: $BetP(E) = \sum_{F \subseteq \Omega} \frac{|E \cap F|}{|F|} \frac{m(F)}{(1 - m(\varnothing))}$ *for all* $E \subseteq \Omega$.

3   Related Works on Collaborative Filtering

Clustering-based CF approaches have been suggested in RSs to learn a cluster model and to provide predictions accordingly. For instance, the CF approach introduced in [2] consisted in a partition of the users in a CF system based on clustering and used the obtained partitions as neighborhoods. In the same way, authors in [8] have proposed a CF approach based on users' preferences clustering. In [9], a CF method has been presented where the training set has been used to generate users clusters and predictions are then generated. A different clustering based-CF has been proposed in [10], where, instead of users, authors proposed to group items into several clusters based on the k-METIS algorithm. These works mentioned above rely either on users clustering or items clustering to predict the users' preferences. In this work, we are rather interested in joining both users clustering and items clustering under the belief function theory [3–5]. In fact, a hybrid CF approach combining both user-based and item-based strategies under the belief function framework has been proposed in [11]. Such evidential approach showed that the fusion framework is effective in improving the prediction accuracy of single approaches (user-based and item-based) under certain and uncertain frameworks. However, it is not able to scale up to real data sets with large collections of items and users since a lot of heavy computations is required. This problem is referred to as the scalability problem which we tackle in our proposed approach.

4   E-HCBCF: Evidential Hybrid Clustering-Based CF

In this approach, we tend to cope with the scalability problem by performing a clustering process of both users and items where uncertainty is also handled.

4.1   Evidential Users Clustering

In this step, an evidential clustering process is performed for the users in the system using the Evidential c-Means (ECM) [6]. We define $\Omega_{clus} = \{w_1, w_2, \ldots, w_c\}$ where c is the numbers of clusters. Hence, the users clustering process bestows a credal partition that allows a given user to be associated to multiple clusters, or rather multiple partitions of clusters. We start by considering the ratings matrix to randomly initialize the cluster prototypes. Then, we compute the euclidean distance between the users and the non empty sets of Ω_{clus}. At the end, the convergence and the minimization of a given objective function [6] lead to the generation of the final credal partition. In order to make a final decision about the cluster of the current user, we transform each bba into a pignistic probability $BetP(w_k)$ based on the obtained credal partition of each user in the system. These values reflect the degrees of membership of the user u to the cluster w_k. A hard partition can be easily derived by assigning each user to the cluster holding the highest $BetP$ value.

4.2   Modeling Users' Neighborhood Ratings

According to the obtained users clusters, the set of the k-nearest neighbors of the active user U_a is extracted. We compute the distances between U_a and the other users in the same cluster as follows [7]: $dist(U_a, U_i) = \frac{1}{|I(a,i)|}\sqrt{\sum_{I \in I(a,i)}(r_{a,j} - r_{i,j})^2}$, where $|I(a,i)|$ is the number of items rated by the user U_a and the user U_i, $r_{a,j}$ and $r_{i,j}$ are respectively the ratings of the users U_a and U_i for the item I_j. The set of the k-most similar users, that we denote by Γ_k, is selected based on the computed distances. We define the frame of discernment $\Omega_{pref} = \{r_1, r_2, \cdots, r_L\}$ as a rank-order set of L preference labels (i.e ratings) where $r_p < r_l$ whenever $p < l$. Since we use the belief function theory in order to conveniently model uncertainty, we transform the rating of each user U_i belonging to Γ_k into a basic belief assignment m_{U_a,U_i} spanning over the frame of discernment based on the formalism in [7]. This bba reflects the evidence of U_i for the rating of U_a.

$$m_{U_a,U_i}(\{r_p\}) = \alpha_0 e^{-(\gamma_{r_p}^2 \times (dist(U_a,U_i))^2)} \tag{1}$$

$$m_{U_a,U_i}(\Omega_{pref}) = 1 - \alpha_0 e^{-(\gamma_{r_p}^2 \times (dist(U_a,U_i))^2)}$$

Where α_0 is fixed to the value of 0.95 and γ_{r_p} is the inverse of the average distance between each pair of users having the same ratings r_p. In order to evaluate the reliability of each neighbor, these bba's are then discounted such that:

$$m_{U_a,U_i}^{\delta_u}(\{r_p\}) = (1 - \delta_u) \cdot m_{U_a,U_i}(\{r_p\}) \tag{2}$$
$$m_{U_a,U_i}^{\delta_u}(\Omega_{pref}) = \delta_u + (1 - \delta_u) \cdot m_{U_a,U_i}(\Omega_{pref})$$

The discounting factor $\delta_u = \frac{dist(U_a,U_i)}{max(dist)}$, where $max(dist)$ is the maximum value of the computed distances between the users.

4.3   Generating Users' Neighborhood Predictions

Once the $bba's$ of the similar users are generated, they are fused as follows [7]:

$$m^{\delta_u}(\{r_p\}) = \frac{1}{N}(1- \prod_{U \in \Gamma_k}(1-\alpha_{r_p})) \cdot \prod_{r_p \neq r_q} \prod_{U \in \Gamma_k}(1-\alpha_{r_q}) \qquad \forall r_p \in \{r_1, \cdots, r_{Nb}\}$$

$$m^{\delta_u}(\Omega_{pref}) = \frac{1}{N} \prod_{p=1}^{Nb}(1 - \prod_{U \in \Gamma_k}(1 - \alpha_{r_p})) \tag{3}$$

Nb is the number of the ratings provided by the similar users, α_{r_p} is the belief committed to the rating r_p, α_{r_q} is the belief committed to the rating r_q $\neq r_p$ and N is a normalized factor defined by [7]:

$$N = \sum_{p=1}^{Nb}(1- \prod_{U \in \Gamma_k}(1-\alpha_{r_p}) \prod_{r_q \neq r_p} \prod_{U \in \Gamma_k}(1-\alpha_{r_q}) + \prod_{p=1}^{Nb}(\prod_{U \in \Gamma_k}(1-\alpha_{r_q})))$$

4.4   Evidential Items Clustering

Until now, only the user-based assumption has been adopted. In what follows, the item-side would also be considered. In this phase, we aim also to carry out an evidential items clustering process for the corresponding items in the system. The uncertainty about items assignments to clusters is considered where each item in the user-item matrix can belong to all clusters with a degree of belief. The ECM [6] is applied once again and the ratings matrix is explored in order to generate the final credal partition of the corresponding items. Once the credal partition corresponding to each item in the system has been produced, a pignistic probability $BetP(w_k)$ is then generated based on the obtained $bba's$. Thereupon, we apply at this level these pignistic probabilities with the aim of assigning each item to its appropriate cluster through the consideration of the highest values.

4.5   Modeling Items' Neighborhood Ratings

Given a target item I_t, we now consider the items having the same cluster membership as I_t. To this end, the distances between I_t and the whole items are computed. Formally, the distance between the target item I_t and each item I_j is computed as follows [7]: $dist(I_t, I_j) = \frac{1}{|U(t,j)|} \sqrt{\sum_{U \in U(t,j)} (r_{i,t} - r_{i,j})^2}$. $|U(t,j)|$ denotes the number of users that rated both the target item I_t and the item I_j where $r_{i,t}$ and $r_{i,j}$ correspond to the ratings of the user U_i for the target item I_t and for the item I_j. Finally, only the k items having the lowest distances are selected and the rating of each similar item is transformed into a bba defined as:

$$m_{I_t,I_j}(\{r_p\}) = \alpha_0 e^{-(\gamma_{r_p}^2 \times (dist(I_t,I_j))^2} \tag{4}$$

$$m_{I_t,I_j}(\Theta_{pref}) = 1 - \alpha_0 e^{-(\gamma_{r_p}^2 \times (dist(I_t,I_j))^2}$$

Following [7], α_0 is set to 0.95 and γ_{r_p} is defined as the inverse of the mean distance between each pair of items having the same ratings. Finally, these $bba's$ are discounted as follows:

$$m_{I_t,I_j}^{\delta_i}(\{r_p\}) = (1 - \delta_i) \cdot m_{I_t,I_j}(\{r_p\}) \tag{5}$$

$$m_{I_t,I_j}^{\delta_i}(\Theta_{pref}) = \delta_i + (1 - \delta_i) \cdot m_{I_t,I_j}(\Theta_{pref})$$

where δ_i is a discounting factor depending on the items' distances such as: $\delta_i = \frac{dist(I_t,I_j)}{max(dist)}$ where $max(dist)$ is the maximum value of the computed distances.

4.6   Generating Items' Neighborhood Predictions

The aggregation of the $bba's$ for each similar item is performed as follows:

$$m^{\delta_i}(\{r_p\}) = \frac{1}{Z}(1 - \prod_{I \in \Gamma_k}(1 - \alpha_{r_p})) \cdot \prod_{r_p \neq r_q} \prod_{I \in \Gamma_k}(1 - \alpha_{r_q}) \qquad \forall r_p \in \{r_1, \cdots, r_{Nb}\} \tag{6}$$

$$m^{\delta_i}(\Theta_{pref}) = \frac{1}{Z} \prod_{p=1}^{Nb}(1 - \prod_{I \in \Gamma_k}(1 - \alpha_{r_p}))$$

Note that Nb is the number of the ratings given by the similar items, α_{r_p} is the belief committed to the rating r_p, α_{r_q} is the belief committed to the rating $r_q \neq r_p$ and Z is a normalized factor defined by [7]:

$$Z = \sum_{p=1}^{Nb}(1 - \prod_{I \in \Gamma_k}(1 - \alpha_{r_p}) \prod_{r_q \neq r_p} \prod_{I \in \Gamma_k}(1 - \alpha_{r_q}) + \prod_{p=1}^{Nb}(\prod_{I \in \Gamma_k}(1 - \alpha_{r_q})))$$

4.7   Final Predictions and Recommendations

After performing an evidential co-clustering process, predictions incorporating both items' aspects and users' aspects have been generated under the belief function theory. Thus, we obtain, on the one hand, bba's from the k-similar items (m^{δ_i}) and, on the other hand, $bba's$ derived from the k-similar users (m^{δ_u}). The fusion of the $bba's$ corresponding to these two kinds of information sources can be ensured through an aggregation process using Dempster's rule of combination [3] such that: $m_{Final} = (m^{\delta_i} \oplus m^{\delta_u})$. The obtained predictions indicate the belief induced from the two pieces of evidence namely the similar users and the similar items.

5   Experiments and Discussions

MovieLens[1], one of the commonly used real world data set in CF, has been adopted in our experiments. We perform a comparative evaluation over our proposed method (E-HCBCF) and the evidential hybrid neighborhood-based CF proposed in [11], denoted by (E-HNBCF). Such hybrid approach achieved better results than state of the art CF approaches both item-based and user-based ones. Compared to E-HNBCF, our new approach incorporates an evidential co-clustering process of the users as well as the items in the system and provides predictions accordingly. We followed the strategy conducted in [12] where movies are first ranked based on the number of their corresponding ratings. Different subsets are then extracted by progressively increasing the number of the missing rates. Hence, each subset will contain a specific number of ratings leading to different degrees of sparsity. For all the extracted subsets, we randomly extract 10% of the users as a testing data and the remaining ones were considered as a training data. We opt for three evaluation metrics to evaluate our proposal: the *Precision*, the *Distance criteron* (Dist_crit) [13] and the *elapsed time* such that:

$Precision = \frac{IR}{IR+UR}$ and $Dist\_crit = \frac{\sum_{u,i}(\sum_{i=1}^{n}(BetP(\{p_{u,i}\}) - \theta_i)^2)}{\|\widehat{p_{u,i}}\|}$. IR indicates that an interesting item has been correctly recommended while UR indicates that an uninteresting item has been incorrectly recommended. $\|\widehat{p_{u,i}}\|$ is the total number of the predicted ratings, n is the number of the possible ratings that

---

[1] http://movielens.org.

Table 1. Comparison results in terms of Precision and Dist_crit

Evaluation metrics	Sparsity degrees	E-HNBCF	E-HCBCF
Precision	53%	0.750	0.833
Dist_crit		0.946	0.905
Precision	56.83%	0.726	0.767
Dist_crit		0.933	1.020
Precision	59.8%	0.766	0.761
Dist_crit		0.940	0.938
Precision	62.7%	0.814	0.805
Dist_crit		0.932	0.928
Precision	68.72%	0.864	0.846
Dist_crit		0.961	0.844
Precision	72.5%	0.730	0.805
Dist_crit		0.936	0.927
Precision	75%	0.739	0.683
Dist_crit		0.872	0.972
Precision	80.8%	0.620	0.675
Dist_crit		1.080	1.003
Precision	87.4%	0.633	0.555
Dist_crit		0.665	0.651
Overall Precision		0.737	**0.747**
Overall Dist_crit		0.918	**0.909**

can be provided in the system. θ_i is equal to 1 if $p_{u,i}$ is equal to $\widehat{p_{u,i}}$ and 0 otherwise, where $p_{u,i}$ is the real rating for the user u on the item i and $\widehat{p_{u,i}}$ is the predicted value. Note that the lower the Dist_crit values are, the more accurate the predictions are while the highest values of the precision indicate a better recommendation quality. Experiments are conducted over the selected subsets by switching each time the number of clusters c. We used $c = 2$, $c = 3$, $c = 4$ and $c = 5$. For each selected cluster, different neighborhood sizes were tested and the average results were computed. Finally, the results obtained for the different numbers of clusters used in the experiments are also averaged. More specifically, we compute the precision and the Dist_crit measures for each value of c and we report the overall results. For the evidential co-clustering process, required parameters were set as follows: $\alpha = 2$, $\beta = 2$ and $\delta^2 = 10$ as invoked in [6]. For the $bba's$ generation, we remind that α_0 is fixed to the value 0.95 and γ_{r_p} is computed as the inverse of the mean distance between each pair of items and users sharing the same rating values. Considering different sparsity degrees, the results are displayed in Table 1. Compared to the evidential hybrid neighborhood-based CF, it can be seen that incorporating an evidential co-clustering process that joins both items clustering and users clustering provides a slightly better performance

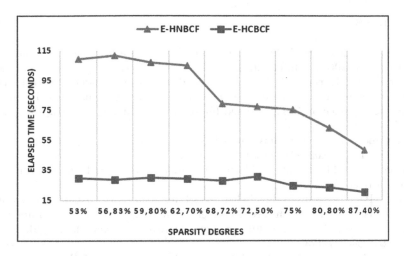

Fig. 1. Elapsed Time of E-HCBCF vs. E-HNBCF

than the other evidential approach. Indeed, the average precision of E-HCBCF is somewhat greater than the value obtained by the standard E-HNBCF (0.747 compared to 0.737). Besides, it acquires the lowest average value in terms of Dist_crit (0.909 compared to 0.918). These obtained results show that the evidential co-clustering process is effective to maintain a good prediction quality while improving also the scalability performance as depicted in Fig. 1. According to Fig. 1, the elapsed time corresponding to E-HCBCF is substantially lower than the basic E-HNBCF. This outcome is explained by the fact that the standard hybrid evidential CF method needs to search the closest neighbors of the active user as well as the similar neighbors of the target item by browsing the whole user-item ratings space, which results in a huge computing amount.

6   Conclusion

In this paper, we have proposed a new evidential co-clustering CF approach where both items' aspect and users' aspect come into play. In fact, a clustering model that joins both users clustering and items clustering has been built under the belief function theory based on the available past ratings. According to the obtained clusters, the k-nearest users and the k-nearest items are selected and predictions are then performed. Global evidence of the neighbors is finally aggregated to get an overall information about the provided predictions. Experiments on a real world CF data set proved the efficiency of the proposed approach where elapsed time has been significantly improved while maintaining a good recommendation performance.

References

1. Su, X., Khoshgoftaar, T.M.: A survey of collaborative filtering techniques. Adv. Artif. Intell. **2009**, 1–19 (2009)
2. Sarwar, B.M., Karypis, G., Konstan, J., Riedl, J.: Recommender systems for large-scale e-commerce: scalable neighborhood formation using clustering. In: International Conference on Computer and Information Technology, Dhaka, Bangladesh. IEEE (2002)
3. Dempster, A.P.: A generalization of Bayesian inference. J. Roy. Stat. Soc. Ser. B (Methodol.) **30**, 205–247 (1968)
4. Shafer, G.: A Mathematical Theory of Evidence, 1st edn. Princeton University Press, Princeton (1976)
5. Smets, P.: The transferable belief model for quantified belief representation. In: Smets, P. (ed.) Quantified Representation of Uncertainty and Imprecision. HDRUMS, vol. 1, pp. 267–301. Springer, Dordrecht (1998). https://doi.org/10.1007/978-94-017-1735-9_9
6. Masson, M.H., Denoeux, T.: ECM: an evidential version of the fuzzy c-means algorithm. Pattern Recogn. **41**(4), 1384–1397 (2008)
7. Denoeux, T.: A K-nearest neighbor classification rule based on Dempster-Shafer theory. IEEE Trans. Syst. Man Cybern. **25**, 804–813 (1995)
8. Zhang, J., Lin, Y., Lin, M., Liu, J.: An effective collaborative filtering algorithm based on user preference clustering. Appl. Intell. **45**(2), 230–240 (2016). https://doi.org/10.1007/s10489-015-0756-9
9. Xue, G.R., et al.: Scalable collaborative filtering using cluster-based smoothing. In: ACM SIGIR Conference on Research and Development in Information Retrieval, pp. 114–121. ACM (2005)
10. O'Connor, M., Herlocker, J.: Clustering items for collaborative filtering. In: ACM SIGIR Workshop on Recommender Systems, vol. 128. UC Berkeley (1999)
11. Abdelkhalek, R., Boukhris, I., Elouedi, Z.: Towards a hybrid user and item-based collaborative filtering under the belief function theory. In: Medina, J., et al. (eds.) IPMU 2018. CCIS, vol. 853, pp. 395–406. Springer, Cham (2018). https://doi.org/10.1007/978-3-319-91473-2_34
12. Su, X., Khoshgoftaar, T.M.: Collaborative filtering for multi-class data using Bayesian networks. Int. J. Artif. Intell. Tools **17**, 71–85 (2008)
13. Elouedi, Z., Mellouli, K., Smets, P.: Assessing sensor reliability for multisensor data fusion within the transferable belief model. IEEE Trans. Syst. Man Cybern. Part B Cybern. **34**, 782–787 (2004)

Conditioned Generative Model via Latent Semantic Controlling for Learning Deep Representation of Data

Jin-Young Kim and Sung-Bae Cho[✉]

Department of Computer Science, Yonsei University, Seoul, South Korea
{seago0828, sbcho}@yonsei.ac.kr

Abstract. Learning representations of data is an important issue in machine learning. Though generative adversarial network has led to significant improvements in the data representations, it still has several problems such as unstable training, hidden manifold of data, and huge computational overhead. Moreover, most of GAN's have a large size of manifold, resulting in poor scalability. In this paper, we propose a novel GAN to control the latent semantic representation, called LSC-GAN, which allows us to produce desired data and learns a representation of the data efficiently. Unlike the conventional GAN models with hidden distribution of latent space, we define the distributions explicitly in advance that are trained to generate the data based on the corresponding features by inputting the latent variables, which follow the distribution, into the generative model. As the larger scale of latent space caused by deploying various distributions makes training unstable, we need to separate the process of defining the distributions explicitly and operation of generation. We prove that a variational auto-encoder is proper for the former and modify a loss function of VAE to map the data into the corresponding pre-defined latent space. The decoder, which generates the data from the associated latent space, is used as the generator of the LSC-GAN. Several experiments on the CelebA dataset are conducted to verify the usefulness of the proposed method. Besides, our model achieves a high compression ratio that can hold about 24 pixels of information in each dimension of latent space.

Keywords: Generative model · Data representation · Latent space · Variational autoencoder · Generative adversarial nets

1 Introduction

Developing generative model is a crucial issue in artificial intelligence. Creativity is a human proprietary, but many recent studies have attempted to make machines to mimic it. There has been an extensive research on generating data and one of them, generative adversarial network (GAN), has led to significant achievements [1]. Many approaches to creating data as better quality as possible have been studied: for example, variational auto-encoder (VAE) [2] and GAN [1]. Both can generate data from manifold which is hidden to us so that we cannot control the kind of data that we generate.

© Springer Nature Switzerland AG 2019
H. Yin et al. (Eds.): IDEAL 2019, LNCS 11871, pp. 319–327, 2019.
https://doi.org/10.1007/978-3-030-33607-3_35

Generative models produce data from latent variable without any other information so that we cannot control what we want to generate. To cope with this problem, the previous research generated data first and found distributions of features on latent space by investigating the model with data, since the manifold of data is hidden in generative models [3]. This latent space is deceptive for finding an area which represents a specific feature of our interest; it would take a long time even if we can find that area.

To work out these problems, we propose a model which can generate the desired data and learn a data representation with a higher compression rate, as well. We predefine distributions corresponding to each feature and modify the loss function of VAE so as to generate the data from the latent variable which follows the specific distribution according to its features. However, this method makes the latent space to become a more complex multimodal distribution, resulting in an instability in training process. We prove that this problem can be solved and even made more efficiently by using an auto-encoder model with the theorem in Sect. 3.

2 Related Works

Some researchers have conducted to generate data such as text, grammar, and images [4–6]. Other researchers also proposed generative models of VAE and GAN [1, 2]. These are basis of the generative models, but they have different policies to construct density: explicitly and implicitly. There are lots of variations of these models. Radford et al. constructed deep convolutional GAN (DCGAN) with convolutional neural networks (CNN) for improving the performance [7]. Zhao et al. introduced energy-based GAN (EBGAN) using auto-encoder in discriminator [8]. Kim et al. proposed transferred encoder-decoder GAN for stabilizing process of training GAN and used it to classify the data [9, 10]. These studies focused on high productivity in generation so that they could not control the type of generated data.

Recently, some researchers began to set conditions on the data. Sohn et al. inputted data and conditions together into VAE and generated data whose type is what they want [11]. Van den Oord et al. set discrete embedding space for generating a specific data using vector quantized VAE (VQ-VAE) [12]. However, because of discrete space, they could not control latent space continuously. Larsen et al. used both VAE and GAN in one generative model [13]. As they just mixed two models and did not analyze a latent space, the manifold of data was hidden to us. Chen et al. used mutual information for inducing latent codes (InfoGAN) [14] and Nguyen et al. added a condition network that treated the information of data to be generated [15]. These two models needed an additional input to generate the desired data. These studies make us to generate data with condition, but we still do not figure out about latent space and it is hard to find the location of a specific feature in the latent space. Therefore, we propose a model that learns to generate concrete features that we want from the latent space determined when LSC-VAE is trained.

3   The Proposed Method

The proposed model is divided into two phases: initializing latent space and generating data. In the first phase, as the instability caused by the larger scale of latent space in this process, we use the LSC-VAE. As shown in Fig. 1(a), we train the LSC-VAE with $\mathcal{L}_{prior}$ for the data to be projected into the desired position on the latent space according to the characteristics of the data. The trained decoder of the LSC-VAE is used as a generator of LSC-GAN so that the LSC-GAN generates the data with the corresponding features by using latent variables sampled from a specific distribution. In the second phase, G and discriminator (D) are trained simultaneously so that G can produce data similar to real data as much as possible and that D can distinguish the real from the fake. The architecture of the generation process is shown in Fig. 1(b).

3.1   Initializing the Latent Space with LSC-VAE

VAE, one type of auto-encoders, is one of the most popular approaches to unsupervised learning of complicated distributions. Since any supervision is not in training process, the manifold constructed is hidden to us. Therefore, we allow LSC-VAE to learn a representation of data with supervision. It compresses data into a particular place on latent space according to its features. The proposed model consists of two modules that encode a data x_i to a latent representation z_i and decode it back to the data space, respectively.

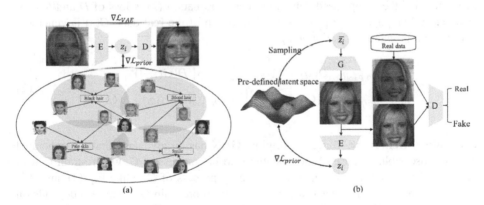

Fig. 1. (a) The process of pre-defining a latent space. The LSC-VAE is trained to project the data into the appropriate position on latent space. (b) Generating process of the proposed method. The latent space is pre-defined in the process of (a).

Let i be a feature which is included in data x and latent space z. The encoder is regularized by imposing a prior over the latent distribution $P(z)$. In general, $z \sim \mathcal{N}(0, I)$ is chosen, but we choose $z_i \sim \mathcal{N}(\mu_i, I)$ for controlling latent space. In addition, if we want to produce data which has multiple features i and j, we generate data from $z_{ij} \sim \mathcal{N}(\mu_i + \mu_j, I)$[1]. The loss function of LSC-VAE is as follows.

$$\mathcal{L}_{LSC-VAE} = -\mathbb{E}_{z_i \sim Q(z_i|x_i)}[\log P(x_i|z_i)] + \mathcal{D}_{KL}[Q(z_i|x_i)\|P(z_i)] = \mathcal{L}_{VAE} + \mathcal{L}_{prior} \quad (1)$$

where $\mathcal{D}_{KL}$ is the Kullback-Leibler divergence and Q is the encoder. The first term of Eq. (1) is related to reconstruction error and the second term is related to appropriate projection of data to the latent space. For example, when LSC-VAE projects the data with i- and j-features into the latent space, it is trained to map the data into the pre-defined latent space ($\mathcal{N}(\mu_i + \mu_j, I)$) with $\mathcal{L}_{prior}$ in Eq. (1).

3.2   Generating Data with LSC-GAN

The basic training process of GAN is to adversely interact and simultaneously train G and D. However, because the original GAN has a critical problem, unstable process of training [7], the Least Squares GAN (LS-GAN) is proposed to reduce the gap between the distributions of real data and fake data by Mao et al. [16]. The main difference of the proposed model with VAE-GAN and LS-GAN is that LSC-GAN is based on LSC-VAE for initializing a latent space to control it. To produce the desired data with i-feature, we just input latent variable $z_i \sim \mathcal{N}(\mu_i, I)$ to G. Besides, we add the encoder of LSC-VAE into LSC-GAN to make sure that the generated data actually have the desired features. The encoder projects the generated data back to latent space so as to be trained to minimize the difference between latent space where data is generated and the compressed data is projected. Equation (2) is loss of D and loss of encoder and G. p_{data} is the probability distribution of the real data. $G(z)$ is generated data from a probability distribution p_z.

$$\begin{aligned} \min_{D} &\mathbb{E}_{x_i \sim p_{data}(x_i)}[D(x_i) - 1)^2] + \mathbb{E}_{z_i \sim P_z(z_i)}[(D(G(z_i)))^2] \\ \min_{Q,G} &\mathbb{E}_{z_i \sim p_z(z_i)}\left[(D(G(z_i)) - 1)^2\right] + \mathcal{D}_{KL}[Q(z_i)|G(z_i)\| \mathcal{N}(\mu_i, I)] \end{aligned} \quad (2)$$

Pre-trained Generator. Since the original GAN has disadvantage that the generated data are insensible because of the unstable learning process of G, we pre-train G with decoder of LSC-VAE. The goal of the learning process of G is like Eq. (3), and it is equivalent to Eq. (4). However, it is not efficient to pre-train G, because it depends on the parameters of D. Therefore, we change this equation to Eq. (5) again. In this paper, to train G with Eq. (5), we use the decoder of LSC-VAE. The result of LSC-VAE is that $|p_{data}^{LSC-VAE} - p_G^{LSC-VAE}| \leq |p_{data} - p_G|$ so that it can reach a goal of GAN ($p_{data} \approx p_G$) stably, which is proved by Theorem 1.

$$\min_{G}(1 - D(G(z_i)))^2 \quad (3)$$

---

[1] More precisely, $\mathcal{N}(\mu_i + \mu_j, 2I)$ is correct but calculated as $\mathcal{N}(\mu_i + \mu_j, I)$ for convenience in computation and scalability.

$$\Leftrightarrow D(G(z_i)) \approx 1 \tag{4}$$

$$\Leftrightarrow G(z_i) \approx x \tag{5}$$

Validity and Efficiency of LSC-VAE. From the game theory point of view, the GAN converges to the optimal point when G and D reach the Nash equilibrium. Here, let p_G be the probability distribution of data generated from G. We show that if $G(z) \approx x$, i.e., $p_{data} \approx p_G$, the GAN reaches the Nash equilibrium. We define $J(D, G) = \int_x p_{data}(x)D(x)dx + \int_z p_z(z)(1 - D(G(z)))dz$ and $K(D, G) = \int_x p_{data}(x)(1 - D(x))dx + \int_z (D(G(z))dz^2$. We train G and D to minimize $J(D, G)$ and $K(D, G)$, respectively. Then, we can define the Nash equilibrium of the LSC-GAN as a state that satisfies Eqs. (6) and (7). Fully trained G and D are denoted by G^* and D^*, respectively.

$$J(D^*, G^*) \leq J(D^*, G) \forall G \tag{6}$$

$$K(D^*, G^*) \leq K(D, G^*) \forall D \tag{7}$$

Theorem 1. If $p_{data} \approx p_G$ almost everywhere, then the Nash equilibrium of the LSC-GAN is reached.

Before proving this theorem, we need to prove the following two lemmas.

Lemma 1. $J(D^*, G)$ reaches a minimum when $p_{data}(x) \leq p_G(x)$ for almost every x.

Lemma 2. $K(D, G^*)$ reaches a minimum when $p_{data}(x) \geq p_{G^*}(x)$ for almost every x.

The proof of Lemmas 1 and 2 was discussed by Kim et al. [10]. We assume that $p_{data} \approx p_G$. From Lemmas 1 and 2, if $p_{data} \approx p_G$, then $J(D, G)$ and $K(D, G)$ both reach minima. Therefore, the proposed GAN reaches the Nash equilibrium and converges to optimal points. By theorem 1, GAN converges when $p_d \approx p_g$, and it is done to some extent by the modified VAE, i.e. $\left| p_d - p_g^{LSC-VAE} \right| \leq |p_d - p_g|$ since a goal of modified VAE is $P(x|Q(z|x)) \approx x$. Therefore, the proposed method is useful to initialize the weight of the generative model.

4   Experiments

4.1   Dataset and Experimental Setting

To verify the performance of the proposed model, we use the CelebA dataset [17]. It is a large-scale face attributes dataset. We crop the initial 178×218 size to 138×138, and resize them as 64×64. We use 162,769 images in CelebA and 14 attributes: *'black hair', 'blond hair', 'gray hair', 'male', 'female', 'smile', 'mouth slightly open', 'young', 'narrow eyes', 'bags under eyes', 'mustache', 'eyeglasses', 'pale skin',* and *'chubby'.*

---

2 Since $\left(1_{p_{data}(x) > p_G(x)}\right)^2 = 1_{p_{data}(x) > p_G(x)}$ and $\left(1 - 1_{p_{data}(x) > p_G(x)}\right)^2 = (1 - 1_{p_{data}(x) > p_G(x)})$, we eliminate the square.

In the first phase, the LSC-VAE takes facial images as input and constructs a latent space. Next, the LSC-GAN exploits the LSC-VAE, resulting in the stabilized training process and the well-defined latent space. We assign 20 dimensions to each feature and set mean of the ith-20 dimensions as 1 for enough capacity to represent each characteristics of data. For example, if an image has i-feature, the elements of $i*20^{th}$ to $(i+1)*20^{th}$ of the image's latent variable are 1 in average and 0 in the remainder and we denote the latent variable as n_i.

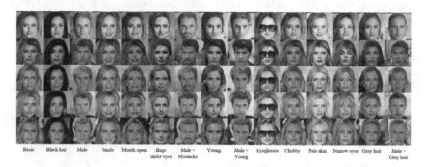

Fig. 2. The generated images. The images are shown in each column according to the features below the columns.

4.2 Generated Images

As shown in Fig. 2, we generate images from a specific latent space by using LSC-GAN. The images in the first column are generated to have 'female' and 'blond hair' features. We confirm that the condition works well. The images in the remaining columns are transformed using Eq. (8) for the features listed below. For example, if we generate an image x_i which has i-feature from the latent variable z_i, we add n_j to add j-feature into the image. We can confirm that the features of images are changed in the desired way across the row.

$$x_{ij} = G(z_i + n_j) \qquad (8)$$

where x_{ij} is an image which has i- and j-features, and z_i is the latent variable.

To show that the proposed model does not simply memorize data but understand features of data to generate them, we generate images from a series between two random images as Radford did [8]. As shown in Fig. 3, the change between images is natural so that we can say that LSC-GAN understands the image.

Unlike other GAN models, the LSC-GAN can handle features continuously, not discretely. We only train the model about the presence of the 'pale skin' feature, but the model has also learned about the reverse of 'pale skin' automatically as shown in the second column of Fig. 4. Besides, if we assign a value of 2 rather than 1 to the average of latent variable which is related to 'mustache', we can see that more mustaches are created in the last column in Fig. 4. Therefore, our model can automatically infer and generate the data with inverse-feature that do not exist in the training dataset.

Fig. 3. Interpolation between images in leftmost and rightmost columns.

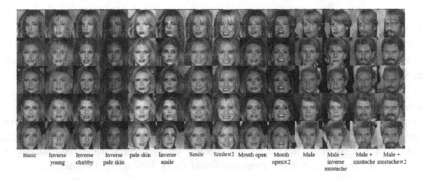

Fig. 4. The result of controlling features. Surprisingly, although the inverse features are not in training dataset, the proposed model infers them in unsupervised way.

This shows that the proposed model has the ability to deduce a negative feature by itself although only positive features are used in training

4.3 Compression Rate

Our model not only generates images according to input conditions, but also compresses efficiently. Besides, it is also related with the complexity of the model. We calculate the compression rate with $rate = size_{input\ data}/size_{bottleneck}/\#classes$ where the size of bottleneck is that of latent space. As shown in Table 1, the proposed model has the best compression rate compared to others. This proves experimentally that LSC-VAE, theoretically proven with theorem 1, is helpful in initializing the weights of the LSC-GAN, and it can achieve good performance even in small latent spaces.

Table 1. The compression rate of models with 14 classes.

Model	U-NET [18]	VQ-VAE [12]	Disco-GAN [19]	Cycle-GAN [20]	StarGAN [21]	Ours
Compression rate	0.036	3.429	2.926	0.027	0.188	24.546

5   Conclusions

In this paper, we address some of significant issues in generative models: unstable training, hidden manifold of data, and extensive hardware resource. To generate data whose type is what we want, we propose a novel model LSC-GAN which can control a latent space to generate the data that we want. To deal with a larger scale of latent space caused by deploying various distributions in one latent space, we use the LSC-VAE and theoretically prove that it is a proper method. Also, we confirm that the proposed model can generate data that we want by controlling the latent space. Unlike the existing generative model, the proposed model deals with features continuously and compresses the data efficiently.

Based on the present findings, we hope to extend LSC-GAN to more various datasets such as ImageNet or voice datasets. In the future, we plan to conduct more experiments with various parameters to confirm the stability of model. We will also experiment by reducing the dimension of the latent space to verify that the proposed model is efficient.

Acknowledgment. This work was supported by grant funded by 2019 IT promotion fund (Development of AI based Precision Medicine Emergency System) of the Korean government (Ministry of Science and ICT). J. Y. Kim has been supported by NRF (National Research Foundation of Korea) grant funded by the Korean government (NRF-2019-Fostering Core Leaders of the Future Basic Science Program/Global Ph.D. Fellowship Program).

References

1. Kingma, D.P., Welling, M.: Auto-encoding variational bayes. arXiv preprint arXiv:1312.6114 (2013)
2. Goodfellow, I., et al.: Generative adversarial nets. In: Neural Information Processing Systems, pp. 2672–2680 (2014)
3. Kim, J.Y., Cho, S.B.: Electric energy consumption prediction by deep learning with state explainable autoencoder. Energies 12(4), 739–754 (2019)
4. Yang, Z., Hu, Z., Salakgutdinov, R., Berg-Kirkpatrick, T.: Improved variational autoencoders for text modeling using dilated convolutions. In: International Conference on Machine Learning, pp. 3881–3890 (2017)
5. Kusner, M.J., Paige, B., Hernandez-Lobato, J.M.: Grammar variational autoencoder. In: International Conference on Machine Learning, pp. 1945–1954 (2017)
6. Denton, E.L., Chintala, S., Fergus, R.: Deep generative image models using a laplacian pyramid of adversarial networks. In: Neural Information Processing Systems, pp. 1486–1494 (2015)
7. Radford, A., Metz, L., Chintala, S.: Unsupervised representation learning with deep convolutional generative adversarial networks. arXiv preprint arXiv:1511.06434 (2015)
8. Zhao, J., Mathieu, M., LeCun, Y.: Energy-based generative adversarial network. arXiv preprint arXiv:1609.03126 (2016)
9. Kim, J.-Y., Cho, S.-B.: Detecting intrusive malware with a hybrid generative deep learning model. In: Yin, H., Camacho, D., Novais, P., Tallón-Ballesteros, Antonio J. (eds.) IDEAL 2018. LNCS, vol. 11314, pp. 499–507. Springer, Cham (2018). https://doi.org/10.1007/978-3-030-03493-1_52

10. Kim, J.Y., Bu, S.J., Cho, S.B.: Zero-day malware detection using transferred generative adversarial networks based on deep autoencoders. Inf. Sci. **460**, 83–102 (2018)
11. Sohn, K., Lee, H., Yan, X.: Learning structured output representation using deep conditional generative models. In: Neural Information Processing Systems, pp. 3483–3491 (2015)
12. Van den Oord, A., Vinyals, O.: Neural discrete representation learning. In: Neural Information Processing Systems, pp. 6309–6318 (2017)
13. Larsen, A.B.L., Sonderby, S.K., Larochelle, H., Winther, O.: Autoencoding beyond pixels using a learned similarity mertic. arXiv preprint arXiv:1512.09300 (2015)
14. Chen, X., Duan, Y., Houthooft, R., Schuman, J., Sutskever, I., Abbeel, P.: InfoGAN: interpretable representation learning by information maximizing generative adversarial nets. In: Neural Information Processing Systems, pp. 2172–2180 (2016)
15. Nguyen, A., Clune, J., Bengio, Y., Dosovitskiy, A., Yosinski, J.: Plug & play generative networks: conditional iterative generative of images in latent space. Comput. Vis. Pattern Recogn. **2**(5), 3510–3520 (2017)
16. Mao, X., Li, Q., Xie, H., Lau, R.Y., Wang, Z., Smolley, S.P.: Least squares generative adversarial networks. In: International Conference on Computer Vision, pp. 2813–2821 (2017)
17. Liu, Z., Luo, P., Wang, X., Tang, X.: Deep learning face attributes in the wild. In: International Conference on Computer Vision, pp. 2730–2738 (2015)
18. Ronneberger, O., Fischer, P., Brox, T.: U-Net: convolutional networks for biomedical image segmentation. In: Navab, N., Hornegger, J., Wells, W.M., Frangi, A.F. (eds.) MICCAI 2015. LNCS, vol. 9351, pp. 234–241. Springer, Cham (2015). https://doi.org/10.1007/978-3-319-24574-4_28
19. Kim, T., Cha, M., Kim, H., Lee, J., Kim, J.: Learning to discover cross-domain relations with generative adversarial networks. In: International Conference on Machine Learning, pp. 1857–1865 (2017)
20. Zhu, J.Y., Park, T., Isola, P., Efros, A.A.: Unpaired image-to-image translation using cycle-consistent adversarial networks. arXiv preprint arXiv:1703.10593 (2017)
21. Choi, Y., Choi, M., Kim, M., Ha, J., Kim, S., Choo, J.: StarGAN: unifired generative adversarial networks for multi-domain image-to-image translation. In: Computer Vision and Pattern Recognition, pp. 8789–8797 (2018)

Toward a Framework for Seasonal Time Series Forecasting Using Clustering

Colin Leverger[1,2]($\boxtimes$), Simon Malinowski[2], Thomas Guyet[3]ⒾD,
Vincent Lemaire[1], Alexis Bondu[1], and Alexandre Termier[2]

[1] Orange Labs, 35000 Rennes, France
`colin.leverger@orange.com`
[2] Univ Rennes, Inria, CNRS, IRISA, 35000 Rennes, France
[3] AGROCAMPUS-OUEST/IRISA - UMR 6074, 35000 Rennes, France

Abstract. Seasonal behaviours are widely encountered in various applications. For instance, requests on web servers are highly influenced by our daily activities. Seasonal forecasting consists in forecasting the whole next season for a given seasonal time series. It may help a service provider to provision correctly the potentially required resources, avoiding critical situations of over- or under provision. In this article, we propose a generic framework to make seasonal time series forecasting. The framework combines machine learning techniques (1) to identify the typical seasons and (2) to forecast the likelihood of having a season type in one season ahead. We study this framework by comparing the mean squared errors of forecasts for various settings and various datasets. The best setting is then compared to state-of-the-art time series forecasting methods. We show that it is competitive with them.

Keywords: Time series · Forecasting · Time series clustering · Naive Bayesian prediction

1 Introduction

Forecasting the evolution of a temporal process is a critical research topic, with many challenging applications. One important application is the forecast of future consumer behaviour in marketing, or cloud servers load for popular applications. In this work, we focus on the forecast of *time series*: a time series is a timestamped sequence of numerical values, and the goal of the forecast is, at a given point of time, to predict the next h values of the time series, with $h \in \mathbb{N}^*$. A practical example is the time series of outdoor temperature values for New York: sensors capture temperature values every hour, and the goal of forecasting can be to predict temperatures for the next 24 h. Many forecasting tasks can be reformulated as time series forecasting, making it an especially valuable research topic.

© Springer Nature Switzerland AG 2019
H. Yin et al. (Eds.): IDEAL 2019, LNCS 11871, pp. 328–340, 2019.
https://doi.org/10.1007/978-3-030-33607-3_36

Forecasting in time series is a difficult task, which has attracted a lot of attention. The most popular methods are Auto-Regressive, ARIMA or Holt-Winters, and come from statistics. These methods build a mathematical model of the time series and focus on predicting the next value ($h = 1$).

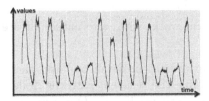

Fig. 1. Seasonal time series example borrowed from [9]. Two weeks of the Internet traffic data (in bits) from a private ISP with centres in 11 European cities. The whole data corresponds to a transatlantic link and was collected from 06:57 on 7 June to 11:17 on 31 July 2005. The time series is obviously seasonal, but the assumption of having one unique periodic pattern seems not suitable in this case.

Many time series, especially those related to human activities or natural phenomena, exhibit *seasonality*. This means that there is some periodic regularity in the values of the time series. In our New York temperature example, there is an obvious 24 h seasonality: temperatures gradually increase in the morning, and decrease for the night. Human activity behaviours often exhibit daily and weekly seasonality. Knowing or discovering that a time series is seasonal is a precious information for forecasts, as it can restrict the search space for the mathematical model of the time series. The classical approach to deal with seasonality is called STL [5], it builds a model while taking into account three components: seasonality, trend (long-term evolution of the time series: increase or decrease) and residual (deviation from trend and seasonality). It assumes that the time series exhibits a single periodic behaviour (ex: daily periodicity of temperatures). Holt-Winters and ARIMA models have been extended to deal with seasonality. But they also assume a single and clear periodic behaviour.

However, this assumption is often violated in practical cases. Consider the time series in Fig. 1, which shows an Internet traffic measurement for two weeks. While there is indeed a daily periodicity, there are two types of daily patterns: weekday patterns and weekend patterns. This cannot be well captured by the STL framework and the associated statistical methods.

In this article, we address the task of forecasting the next season of a seasonal time series, with regularities in the seasonal behaviours. We propose a framework for that task, in the light of the work in [12]. This framework is based on two steps: (1) the identification of typical seasonal behaviours and (2) the forecasting of the season. It only requires as input the time frame of the seasonality (ex: daily, weekly, etc.), and it determines the main seasonality patterns from the analysis of past data (without any assumption on the number of these patterns).

In our previous example, it would automatically identify the weekday pattern and the weekend pattern.

Our contributions are the following:

1. we propose a general framework for seasonal time series forecasting,
2. we provide extensive experiments of our framework with various settings and various datasets,
3. we compare our framework with alternative strategies for time series forecasting.

In the remainder, we present in detail our approach, which is based on a clustering step to determine the seasonal patterns, followed by a classification step to build a "next pattern predictor". Our thorough experiments on real data determines which are the best combinations of clustering and classification method, and show that our approach compares favourably to state-of-the-art forecasting methods.

2   Related Work

Time series analysis [3] has become a recent challenge since more and more sensors collect data with high rates. In statistical approaches, statistical models are fitted on the data to predict future values. The nature of the statistical model is chosen depending on the data characteristics (noise, trends, stationarity, etc.). For a wide literature on statistical methods for time series forecasting, we refer the reader to Brockwell's book [3]. Autoregressive methods (AR, ARIMA, etc.) demonstrate a great success in lots of applications. The advantages of those models are the hypothesis simplicity, the computational efficiency during the training phase and the handling of the noise. For instance, in [10], authors use several ARIMA models to predict day-ahead electricity prices. Kavasseri et al. [11] use fractional-ARIMA to generate a day-ahead and two days ahead wind forecasts. They show that the method is much better than the persistence model. But, in these works, methods do not directly handle the seasonal dimension of the data. Indeed, ARIMA expects data that is either not seasonal or has the seasonal component removed, e.g. seasonally adjusted via methods such as seasonal differencing [2]. The SARIMA model [8] extends the autoregressive model to deal with seasonality. Such model exhibits autocorrelation at past lags of multiple of season period for both autoregression and moving average components.

In the machine learning communities, one of the objectives is to create data analytic tools that would require fewer modelling efforts for data scientists. Concerning time series forecasting task, machine learning has been used in mainly two ways: use of neural network techniques and of ensemble methods. Neural networks models can be efficient for forecasting tasks, especially when the data is more complex and when the process is non-linear. One drawback of those models is their tendency to over-fit [19], which may cause lower performances. LSTM [7] is a classical neural network architecture used for time series forecasting, but not dedicated to seasonal time series. In [18], the authors use ANN and fuzzy clustering for creating daily clusters that are afterward being used for forecasts. Ensemble forecasting methods and hybrid models are created from several state of the art, independent models, that are mixed to create more complex chains. This strategy is the one proposed in the Arbitrated Dynamic Ensemble [4]. This metalearning method combines different models, regarding to their specifies against target datasets. In [15], authors use both ARIMA and SVMs models to forecast stock prices problem, to tackle the non-linearity of some datasets.

3   Seasonal Time Series Forecasting

A time series Y is a temporal sequence of values $Y = \{y_1, \ldots, y_n\}$, where $y_i \in \mathbb{R}^d$ ($d \in \mathbb{N}^*$) is the value of Y at time i and n denotes the length of the time series. If $d = 1$, Y is said univariate. Otherwise, Y is said multivariate. We assume here that time series are univariate, regularly sampled and that there is no missing data. We also assume that there is no trend component in the time series[1].

The problem of time series forecasting is a classical problem: given the knowledge of Y up to sample n, we want to predict the next samples of Y, *i.e.* $y_{n+1}, \ldots, y_{n+h}$, where h is called the prediction horizon. Predictions are denoted $\hat{y}_{n+1}, \ldots, \hat{y}_{n+h}$. Quality of the predictions is related to the difference between real and predicted values. Different measures can be used to evaluate the quality of predictions. We will introduce some measures in Sect. 5.

Let s be the seasonal periodicity of the considered time series. A typical season is a time series $\tilde{y} = \{\tilde{y}_1, \tilde{y}_2, \ldots, \tilde{y}_s\}$ of length s. Let $\mathcal{Y} = \{\tilde{y}^{1 \ldots k}\}$ be a collection of k typical seasons, then a seasonal time series $Y = \{y_1, y_2, \ldots, y_n\}$ is a univariate time series of length $n = k \times s$ such that:

$$y_i = \tilde{y}^{k_{\sigma(i)}}_{i - s \times \sigma(i)} + \epsilon, \ \forall i \in [n]$$

where $\sigma(i) = \lfloor \frac{i}{s} \rfloor$ is the season index of the i-th timestamp of the series, k_i is type of i-th typical seasonal time series and $\epsilon \sim \mathcal{N}(0, 1)$ is a gaussian noise.

The seasonal time series forecasting is a forecasting of a seasonal time series at an horizon s, *i.e.*, the prediction of the time series values for the whole next season ahead.

4   Framework for Seasonal Time Series Forecasting

The general sketch of our approach is composed of two different stages: one for learning the predictive models (see Fig. 2) and the other to use this model to predict the next season (see Fig. 3). The learning stage is composed of three steps: (i) data splitting, (ii) clustering of the seasons, (iii) training a classification algorithm. The predictive stage is composed of two steps: (i) predict a cluster

---

[1] In practice, it is not a problem, as most time series could be easily decomposed and detrended. Trend components can then be re-applied on the forecasted values.

index, (ii) forecast the next season using the predicted cluster index. We describe in this section all these steps in detail.

Let $Y = \{y_1, \ldots, y_n\}$ be an observed time series with m seasons of length s ($n = m \times s$). We assume that the time series in our possession are equally sampled and that there is always the same number of points s per season. We assume that s is known beforehand: it could be daily, weekly, monthly, or even yearly; and that it does not change over time. Finding the seasonality s is not in the scope of our study.

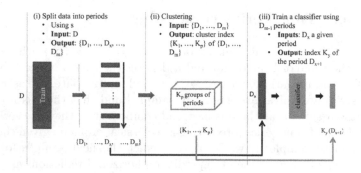

Fig. 2. Illustration of the learning steps.

4.1   Learning Stage

The learning stage is composed of three steps (see Fig. 2):

(i) The data splitting step consists in constructing a set of m seasons, $\mathcal{D} = \{D_1, \ldots, D_j, \ldots, D_m\}$, where D_i represents the subseries corresponding to the ith season of measurements. We assume that observations of the $\mathcal{D}$ are independent.

(ii) The m elements, seasons, of $\mathcal{D}$ are given to a clustering (or co-clustering) algorithm that gathers similar seasons into p groups of typical seasonal time series.

(iii) A probabilistic classifier is then trained to estimate, using the knowledge of a current season D_X, the expected group of next season (season $X + 1$ denoted the season just after the season X).

The data splitting step is straightforward (since input datasets are considered to be regularly sampled), for the other step we give just below details on them.

Clustering Step: Details and Algorithms - In this step, a clustering algorithm is used to group the m seasons of $\mathcal{D}$ into p groups. The choice of p will be discussed later at the end of this section. A representative series is computed inside each group. These series represent typical seasons that occur in the dataset. Hence, at the end of this step, every season of $\mathcal{D}$ can be assigned a label (that represents in which group it has been clustered) and p typical seasons are computed. We consider in this paper four different clustering algorithms:

K-Means for time series [13] K-means algorithm aims at creating a partitioning of the data in k clusters by minimising the intra-cluster variance. The use of K-means implies the use of a distance measure between two time series. Two of the major distance measures available for time series are Euclidean, and Dynamic Time Warping (DTW) [17].

K-Shape for time series[16] K-Shape algorithm creates homogeneous and well separated clusters and uses a distance measure based on a normalised version of the cross correlation (invariant to time series scale or shifting). It can be seen as a K-means algorithm but that uses a shape-based similarity measure.

Global Alignment Kernel K-means (GAK) [6] Kernel k-means is a version of k-means that computes the similarity between time series by using kernels. It identifies clusters that are not linearly separable in the input space. The Global Alignment Kernel is a modified version of DTW.

Co-clustering with MODL [1] MODL is a nonparametric method that uses a piecewise constant density estimation to summarise similar curves. Curves are partitioned into clusters and the curve values are discretised into intervals. The cross-product of these discretisation is an estimation of the joint density of the curves and points.

The choice of the number of clusters is of particular importance in our prediction framework. We use a tuning approach: for each candidate number of clusters, the training set is used to build the overall model. This model is then used to predict the seasons of the validation set. The number of cluster that leads to the lowest error on the validation set is selected. For K-means, K-shape and GAK algorithms, candidates number of cluster are systematically chosen in a pre-defined range $[2, 300]$. Partitions with empty clusters are discarded. On another hand, MODL co-clustering estimates in a nonparametric way the best number of clusters for each input time series. This estimated number usually leads to the best description of input time series, regarding to the coclustering task. However, this model can be simplified to reduce the number of clusters. From now on, the procedure is similar: different number of clusters are evaluated and the overall model that leads to the lowest error on the validation set is kept.

Predicting the Cluster Index of the Next Season - A probabilistic classifier is then trained to estimate, using the knowledge of the current season D_X, the expected group of next season. To train this classifier, we first apply a clustering model as described above to create groups of similar seasons. A learning set is then created to feed the classifier: each line of the learning set corresponds to time series values of a season D_X (explanatory variables) and a target variable which corresponds to the group of the next season (the group of season D_{X+1}). We think that including more data in the classifier would probably be beneficial: data about few past days, and not only the last one, data about national holidays, all sorts of exogene data we can find related to the problem treated; this last point will be studied in future work.

Any classifier could be chosen at this step, the only constraint is that the output of the classifier has to be probabilistic (not only a decision but a vector of conditional probabilities of all possible groups given the input time series). In the Fig. 3, $\widehat{K_Y}$ denotes a vector of probabilities (*i.e* $\{\widehat{K_1}, ..., \widehat{K_p}\}$). In this paper, three different types of probabilistic classifiers are investigated: a Naive-Bayes classifier, a Decision Tree and a Logistic Regression.

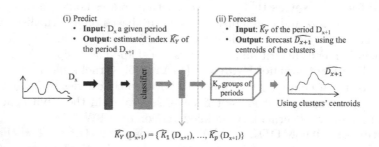

Fig. 3. Illustration of the forecasting steps.

4.2 Forecasting the Next Season

To forecast the next season, we use the classifier learned above to predict the group index of the next season (using the knowledge of the current season), and representative series of the different groups to generate the prediction of the next season. The forecasting of the next season is done in two steps (see Fig. 3). First, the current season (D_X) is given to the classifier. This classifier computes the probabilities of the next season to belong to each group ($\{\widehat{K_1}, ..., \widehat{K_p}\}$). Then, next season is predicted as a weighted combination of the groups centroids, the weights corresponding to the probabilities output by the classifier.

5   Experiments

We have performed different experiments to assess the performance of the proposed method.[2] We were curious about the impacts of the choice of both the clustering and the classification algorithms on the performance of the overall method. We also aimed to find out how our approach compares with classical time series prediction approaches.

We have used nine datasets to assess the performance of the proposed approach. Each dataset is a seasonal time series monitored over many seasons. Eight datasets are open source datasets: five are from the time series data library [9], one is from the City of Melbourne [14] (Pedestrian Counting System) and one

---

[2] For reproducible research, code and data are available online https://github.com/ ColinLeverger/IDEAL2019-coclustering-forecasts.

Table 1. Performances of all the possible combination of clustering and classifiers algorithms for our chain, in terms of MSE and MAE. Bold Figures are best MSE/MAE.

	MSE			MAE		
	Naive Bayes	Tree	Logistic Reg.	Naive Bayes	Tree	Logistic Reg
MODL	0.480	0.581	**0.479**	**0.462**	0.494	0.487
K-means	0.799	0.583	0.492	0.564	0.526	0.492
GAK	0.571	0.638	0.608	0.538	0.576	0.590
K-shape	0.745	0.824	0.852	0.575	0.677	0.693

is provided by Kaggle[3]. The two other datasets have been provided by Orange company via a project named Orange Money. The first time series provided by Orange represents the number of people browsing a website at a time of the day. The second one is a technical metric collected on one server (CPU usage). These nine time series have different seasonalities (daily, weekly, monthly, quarterly). All the time series have been z-normalised beforehand.

Each dataset is separated into a training set, a validation set and a test set. The training set is composed of 70% of the seasons, while both validation set and test set are composed of 15% of the seasons. However, no shuffle is done during the separation, as it is important to learn the relations between contiguous seasons. The shuffle would remove those relations, which is not acceptable for a "time split". The validation set is used to select the appropriate number of clusters of the different clustering algorithms. The test set is used to assess the performance of the different approaches.

Two metrics are used in this paper to evaluate the quality of the predictions: the Mean Square Error (MSE) and the Mean Absolute Error (MAE):

$$MSE(y, \hat{y}) = \frac{1}{s} \times \sum_{j=1}^{s} (y_j - \hat{y}_j)^2, \quad MAE(y, \hat{y}) = \frac{1}{s} \times \sum_{j=1}^{s} |y_j - \hat{y}_j|$$

where $y = y_1, \ldots, y_s$ is the ground truth and $\hat{y} = \hat{y}_1, \ldots, \hat{y}_s$ is the prediction. As these two measures represent prediction errors, the lower they are, the more accurate the predictions.

5.1 Overall Performance of the Approach

As explained in Sect. 4, our proposed approach is based on a clustering algorithm and a classifier. For comparison purposes, we made use of 4 different clustering algorithms and 3 different classifiers. We want to analyse the performance of these different algorithms. Table 1 gives the average MSE and MAE over the 9 datasets for every possible combination of clustering and classifier.

According to this table, MODL outperforms the other clustering algorithms. Combined with a Naive Bayes classifier, it reaches the best performance for 3 of

---

[3] See: https://www.kaggle.com/robikscube/hourly-energy-consumption.

the 4 settings. On the other hand, K-shape is the worse clustering algorithm for this task. A detailed comparison of the clustering algorithms and the classifiers will be made later in this section.

5.2 Performance of the Different Clustering Algorithms

Figure 4-(a), (b) and (c) compare the performance of MODL against the three other considered clustering algorithms. Each point in these figures represents a combination of experimental settings: clustering algorithms, classifiers, datasets, and metrics. For each of these combinations, the performance is computed according to the considered metrics and inserted in the figure. It shows that MODL outperforms the other clustering algorithms. For each comparison (MODL versus another clustering algorithm), a Wilcoxon test has been carried out. The p−value of these tests is depicted in each figure. For each comparison, we can see that this probability is less than 0.05, meaning that the performance of MODL is significantly better than the one of the other clustering algorithm. We have also carried out a Nemenyi test to compare the average ranks of the different approaches. The critical diagrams associated with these tests are given in Fig. 5. The left-hand side represents the comparison of the average rank according to the MSE measure, while the right-hand side is associated with the MAE measure. It shows that MODL outperforms the three other algorithms, significantly for the MSE measure.

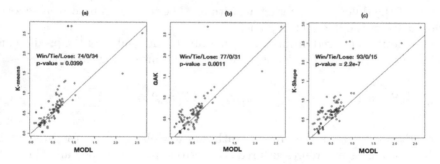

Fig. 4. Comparison between MODL and the other clustering algorithms (K-means, GAK and K-shape). The axis represents an error measure (either MSE or MAE) associated with the clustering algorithm of the axis labels. A point above the diagonal is in favour of the MODL approach.

5.3 Performance of the Different Classifiers

As explained above, an important part of our method is to predict the group (or cluster) of the next season. This prediction is made using a classifier. Once the group is predicted, a weighted combination of different centroids is used to make the forecast. A centroid of a cluster is the Euclidean mean of all its curves. In this section, we first make a performance comparison of the three considered classifiers. Let us compare the performance of the three considered classifiers: a Naive-Bayes classifier, a decision tree, and a logistic regression. The critical diagrams that compare the performance of the three classifiers are given in Fig. 5. The left diagram is for the MSE measure and the right one for the MAE measure. We can see that for the MSE measure, the performance of the classifiers are very close (no significant difference between the ranks). For the MAE, the Naive Bayes classifier is better than the two others.

5.4   Comparison with Other Prediction Methods

In this section, we aim at comparing the performance of the proposed approach (PA) with other competitive prediction methods. Following the analysis above, we use for our approach the following setting: MODL algorithm is used as the clustering algorithm and the classifier is chosen according to the validation set. We compare the performance of our approach with the following four other prediction methods: mean season, Markov model (MM), autoregressive models (AR) and an algorithm called Arbitrated Ensemble for Time Series Forecasting (AETSF). The mean season is a naive approach that consists in predicting the next season as the average season of the training set. The MM approach has been proposed in [12]. It follows the same framework as the one proposed here. However, the classifier is replaced by a Markov model, whose transition probability matrix is used to predict the cluster of the next season. Auto-regressive models have widely been used for time series forecasting. The order of the regressive process is set to the length of the seasonality of the time series. Finally, the AETSF

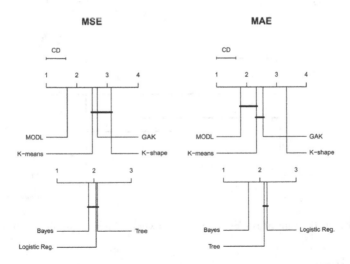

Fig. 5. Critical diagram, comparison of the average ranks of the difference clustering algorithms (on the top) different classification algorithms (on the bottom) in terms of MSE (left) and MAE (right).

method has been proposed in [4]. It is an ensemble method that combines prediction of several algorithms based on meta-learners. This method has been shown to reach state-of-the-art performance for time series prediction. We have used four base learners: generalised linear models, support vector regression, random forest and feed forward neural networks.

Table 2 compares the performance of all these approaches for the nine considered datasets and the MSE measure. In terms of average MSE and mean rank, the AETSF is the best method, just in front of the proposed approach. The critical diagram associated with the Nemenyi test is given in Fig. 6. The difference between AETSF and our approach is not significant on these nine datasets.

Table 2. Comparison of the proposed approach with other prediction methods.

Dataset	Mean season	MM	AR	AETSF	PA
1	1.2349	0.1800	0.7981	0.1771	**0.1750**
2	3.2345	0.7698	1.7460	**0.3988**	0.5172
3	1.7163	1.1057	0.9060	**0.1189**	0.1924
4	0.9428	0.5701	1.1262	**0.5392**	0.6359
5	1.4766	1.1515	0.9817	**0.1062**	0.2599
6	0.7424	**0.4260**	0.4407	0.6106	0.5633
7	1.2194	1.3411	2.1578	1.3028	**1.0181**
8	1.070	1.5949	0.7294	**0.6226**	0.7187
9	2.0271	0.9542	1.3424	0.2953	**0.2811**
Average MSE	1.5183	1.1365	0.8993	**0.4635**	0.4846
Average rank	4.44	3.67	3.22	**1.78**	1.89

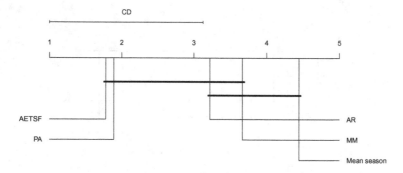

Fig. 6. Critical diagram of the comparison between different prediction approaches.

6   Conclusion

We have proposed in this paper a study of a framework for seasonal time series prediction. This framework involves three steps: the clustering of the seasons into groups, the prediction of the group of the next season using a classifier, and finally the prediction of the next season. One advantage of such a framework is that it is able to produce predictions at the horizon of one season in one shot (*i.e.*, there is no need to build different models for different horizons). We have provided a comparison of different clustering algorithms that can be used in the first step, and a comparison of different classifiers for the second step. Experiment results show that our proposed approach is competitive with other time series prediction methods. For future work, an important point will be to focus on the performance of the classifier. We think that including more exogenous data (example: holidays, social events, weather, etc.) might improve the performance of the classifier, thus the global performance of our framework.

References

1. Boullé, M.: Functional data clustering via piecewise constant nonparametric density estimation. Pattern Recogn. **45**(12), 4389–4401 (2012)
2. Box, G.E., Jenkins, G.M., Reinsel, G.C., Ljung, G.M.: Time Series Analysis: Forecasting and Control. Wiley, Hoboken (2015)
3. Brockwell, P.J., Davis, R.A., Calder, M.V.: Introduction to Time Series and Forecasting. STS, vol. 2. Springer, New York (2002). https://doi.org/10.1007/b97391
4. Cerqueira, V., Torgo, L., Pinto, F., Soares, C.: Arbitrated ensemble for time series forecasting. In: Ceci, M., Hollmén, J., Todorovski, L., Vens, C., Džeroski, S. (eds.) ECML PKDD 2017. LNCS (LNAI), vol. 10535, pp. 478–494. Springer, Cham (2017). https://doi.org/10.1007/978-3-319-71246-8_29
5. Cleveland, R.B., Cleveland, W.S., McRae, J.E., Terpenning, I.: STL: a seasonal-trend decomposition. J. Off. Stat. **6**(1), 3–73 (1990)
6. Cuturi, M.: Fast global alignment kernels. In: Proceedings of the International Conference on Machine Learning (ICML), pp. 929–936 (2011)
7. Gers, F.A., Schmidhuber, J., Cummins, F.: Learning to forget: continual prediction with LSTM (1999)
8. Ghysels, E., Osborn, D.R., Rodrigues, P.M.: Chapter 13 forecasting seasonal time series. In: Handbook of Economic Forecasting, vol. 1, pp. 659–711 (2006)
9. Hyndman, R.: Time series data library (TSDL) (2011). http://robjhyndman.com/TSDL
10. Jakaša, T., Androček, I., Sprčić, P.: Electricity price forecasting - ARIMA model approach. In: Proceedings of the International Conference on the European Energy Market (EEM), pp. 222–225 (2011)
11. Kavasseri, R.G., Seetharaman, K.: Day-ahead wind speed forecasting using f-ARIMA models. Renewable Energy **34**(5), 1388–1393 (2009)
12. Leverger, C., Lemaire, V., Malinowski, S., Guyet, T., Rozé, L.: Day-ahead time series forecasting: application to capacity planning. In: Proceedings of the 3rd Workshop on Advanced Analytics and Learning of Temporal Data (AALTD) (2018)
13. Lloyd, S.: Least squares quantization in PCM. Trans. Inf. Theory **28**(2), 129–137 (1982)
14. Melbourne, C.O.: Pedestrian counting system (2016). http://www.pedestrian.melbourne.vic.gov.au/
15. Pai, P.F., Lin, C.S.: A hybrid ARIMA and support vector machines model in stock price forecasting. Omega **33**(6), 497–505 (2005)

16. Paparrizos, J., Gravano, L.: k-shape: efficient and accurate clustering of time series. In: Proceedings of the International Conference on Management of Data (SIGMOD), pp. 1855–1870 (2015)
17. Petitjean, F., Ketterlin, A., Gançarski, P.: A global averaging method for dynamic time warping, with applications to clustering. Pattern Recogn. **44**(3), 678–693 (2011)
18. Vahidinasab, V., Jadid, S., Kazemi, A.: Day-ahead price forecasting in restructured power systems using artificial neural networks. Electr. Power Syst. Res. **78**(8), 1332–1342 (2008)
19. Zhang, G.P.: Time series forecasting using a hybrid ARIMA and neural network model. Neurocomputing **50**, 159–175 (2003)

An Evidential Imprecise Answer Aggregation Approach Based on Worker Clustering

Lina Abassi[✉] and Imen Boukhris

LARODEC, Institut Supérieur de Gestion de Tunis, Université de Tunis,
Tunis, Tunisia
lina.abassi@gmail.com, imen.boukhris@hotmail.com

Abstract. Crowdsourcing has become a popular and practical tool to gather low-cost labels from human workers in order to provide training data for machine learning applications. However, the quality of the crowdsourced data has always been an issue mainly caused by the quality of the contributors. Since they can be unreliable due to many factors, it became common to assign a task to more than one person and then combine the gathered contributions in order to obtain high quality results. In this work, we propose a new approach of answer combination within an evidential framework to cope with uncertainty. In fact, we assume that answers could be partial which means imprecise or even incomplete. Moreover, the approach includes an important step that clusters workers using the k-means algorithm to determine their types in order to effectively integrate them in the aggregation of answers step. Experimentation on simulated dataset show the efficiency of our approach to improve outcome quality.

Keywords: Answer aggregation · Crowd · Belief function theory · Imprecision · Worker clustering

1 Introduction

The expanding popularity of crowdsourcing services has contributed in providing on-demand easy access to human intelligence to perform a wide range of tasks [1,2] that are usually trivial for humans but not for computers. Although hiring experts is reliable, it can often be expensive especially when a large volume of data is involved. Crowd labelling can then be a budget efficient way to obtain information. Information most of the time consists in training data that are used to build machine learning models. Crowdsourcing presents an affordable and fast way to solve tasks, yet it can be subject to poor quality performance. This is caused by unintentional or intentional mistakes introduced by unreliable participants. To overcome this, crowdsourcing platforms made using redundancy [3] possible for requesters which basically allows them to hire multiple workers to accomplish one task. Thus, for each task, multiple answers are collected which

© Springer Nature Switzerland AG 2019
H. Yin et al. (Eds.): IDEAL 2019, LNCS 11871, pp. 341–349, 2019.
https://doi.org/10.1007/978-3-030-33607-3_37

led researchers to propose methods to combine these contributions in order to
get satisfactory results that turn up to be as accurate as those provided by
professionals [2]. The easiest way is to consider the answer with the most votes
[24]. More advanced probabilistic models are based on the general Dawid-Skene
model [10] which uses the Expectation-Maximization (EM) algorithm [15] to
infer both worker reliability and the tasks true labels. The method proposed in
[16] includes the estimation of task complexity as an additional parameter in the
model. As for the work in [12], it assumes that each worker admits two metrics
namely specificity and sensitivity to express his reliability. Using some experts
labelled tasks, the work in [11] proposes a frequentist approach that estimates the
true labels besides of worker expertise and question difficulty. Other works [2,9]
used experts labels to identify low quality workers in order to filter them. On the
other hand, many works adapted the belief function theory to ensure a better
quality. The theory of belief functions is a flexible mathematical framework for
uncertainty modelling and reasoning and has a variety of operations and rules
to manage and combine evidence. Among works embracing this theory, the work
in [17] suggesting the estimation of the abilities of workers based on the vote of
majority, whereas in [18,20,22] abilities are induced based on few experts data.
The work proposed in [19] is inspired by the EM technique and it generates
the true answers as well as labeller expertise and task difficulty jointly. As for
[14], the introduced approach focuses on recognizing experts in a crowdsourcing
scenario.

In this work also based on the evidence theory, our main assumptions include
relying on only observed labels (i.e no experts data is given) and allowing answering partially on single response tasks to express uncertainty towards the accuracy
of their labels (i.e accepting imprecise answers). The choice of the theory of belief
functions is explained by the fact that, unlike the probabilistic theory, it enables
to support and manage uncertainties as well as imprecision and ignorance. The
work proposed in [21] has the same assumptions. However, the key distinction
of our proposed approach is that it aims to identify workers abilities by clustering them in regard to their exactitude and precision levels and then revise
their responses accordingly to finally aggregate them and get the potentially
true ones.

An outline of the reminder of this paper is as follows: We present the necessary background of the belief function theory in Sect. 2. Section 3 describes our
proposed approach. We finally present our experimental findings in Sect. 4 and
conclude reporting some potential future work in Sect. 5.

2   The Theory of Belief Functions

The theory of belief functions, also called theory of evidence or Dempster-Shafer
theory [4,5] is one of the most adopted theories for representing and reasoning
under uncertainty. It was interpreted by Smets under the name Transferable
Belief Model (TBM) [13].

Representation of Evidence. Let Ω be the frame of discernment representing a set of n elementary and mutually exclusive events such that $\Omega = \{\omega_1, \omega_2, \ldots, \omega_n\}$. The set of all possible values that each subset of Ω can take is the power set of Ω denoted by 2^{Ω} defined as $2^{\Omega} = \{ E : E \subseteq \Omega \}$. A basic belief assignment (bba), also called a mass function, is a function $m^{\Omega} : 2^{\Omega} \to [0, 1]$ that satisfies: $\sum_{E \subseteq \Omega} m^{\Omega}(E) = 1$. It presents the part of belief committed strictly to event E.

If an event E of Ω have a strictly positive mass $m^{\Omega}(E) > 0$, it is called a focal element of the bba. A bba is called vacuous if $m(\Omega) = 1$ and $m(E) = 0$ for $E \neq \Omega$. It represents a state of total ignorance. A bba is called categorical if it has only one focal element E such as $m(E) = 1$ for $E \subseteq \Omega$. A bba is called certain if it is a categorical but its focal element E is a singleton. A bba is called simple support function if it has at most two focal elements: Ω and a subset E.

Discounting Evidence. When the reliability of a source of information can be quantified, it can be taken into consideration by the discounting procedure as follows:

$$\begin{cases} m^{\alpha}(E) = (1 - \alpha)m(E), \quad \forall E \subset \Omega, \\ m^{\alpha}(\Omega) = (1 - \alpha)m(\Omega) + \alpha. \end{cases} \quad (1)$$

$\alpha \in [0, 1]$ is the discounting factor and $(1 - \alpha)$ is the reliability of the source.

Combination of Evidence. There is a lot of combination rules in the belief functions framework. They are used to obtain one mass function in order to easily make a decision. Let m_1 and m_2 be two bbas defined on Ω provided by two distinct and cognitively independent reliable sources.

- Conjunctive rule of combination
  It is introduced by Smets [8] and characterized by allowing a mass on the empty set. It is noted by $\bigcirc\!\!\!\cap$ and defined as:

$$m_1 \bigcirc\!\!\!\cap m_2(E) = \sum_{F \cap G = E} m_1(F) m_2(G) \quad (2)$$

- Dempster's rule of combination
  It is introduced by Dempster [5] and it is the normalized version of the conjunctive rule since it does not support a mass on the empty set $(m(\varnothing) = 0)$. It is denoted by $\oplus$ and defined as:

$$m_1 \oplus m_2(G) = \begin{cases} \dfrac{m_1 \bigcirc\!\!\!\cap m_2(G)}{1 - m_1 \bigcirc\!\!\!\cap m_2(\varnothing)} & \text{if } E \neq \varnothing, \forall G \subseteq \Omega \\ 0 & \text{otherwise.} \end{cases} \quad (3)$$

- The Combination With Adapted Conflict
  The Combination With Adapted Conflict (CWAC) rule [7] is an adaptive weighting between the conjunctive and Dempster's rules. It has the first rule behaviour if mass functions are contradictory and the second one's if they are not. It is denoted by $\ominus$ and defined as follows:

$$m_{\ominus}(E) = (\ominus m_i)(E) = D m_{\bigcirc\!\!\!\cap}(E) + (1 - D) m_{\oplus}(E) \quad (4)$$

with $D = \max [d(m_i,m_j)]$ is the maximum Jousselme distance [6] between m_i and m_j. This distance measures the dissimilarity between *bbas*. It is defined as follows:

$$d(m_1, m_2) = \sqrt{\frac{1}{2}(m_1 - m_2)^t D(m_1 - m_2)}, \qquad (5)$$

where D is the Jaccard index defined by:

$$D(E, F) = \begin{cases} 0 & \text{if } E = F = \varnothing, \\ \frac{|E \cap F|}{|E \cup F|} & \forall E, F \in 2^\Omega. \end{cases} \qquad (6)$$

Decision Process. To make decisions, Smets [13] proposes to transform *bbas* into a pignistic probability, *BetP*, computed as follows:

$$BetP(\omega_i) = \sum_{E \subseteq \Omega} \frac{|E \cap \omega_i|}{|E|} \cdot \frac{m(E)}{(1 - m(\varnothing))} \quad \forall \omega_i \in \Omega \qquad (7)$$

The decided event is the one that has the highest value of *BetP*.

3   CIMP-BLA: The Clustering Approach of the Imprecise Belief Label Aggregation

Our work introduces a new approach of imprecise answer aggregation. It is a solution to improve the quality of consensus labels in a scenario where workers are given the possibility to submit more than one answer in case they have doubts about the exact solution. To cope with such imprecise labels and more, we resort to the belief function theory that has a greater expressive power than other theories. Also, as the quality of workers has a crucial impact on the accuracy of results, we first include a clustering step in order to separate three main types of labellers; pro, good and bad. Their answers are then altered accordingly and aggregated in a second step. Figure 1 presents the different steps of our proposed approach. We detail each step in what follows.

3.1   Answer Modelling

In the first step, each answer of the answers matrix is transformed under the belief function theory into a *bba*. Since we deal with partial responses, the evidence theory enables us to affect belief on composite hypotheses. So, if we suppose that our frame of discernment is $\Omega = \{1, 2, 3\}$ which means that each true answer is either 1, 2 or 3, the answers submitted by workers belong to $\{\{1\}, \{2\}, \{3\}, \{1,2\}, \{1,3\}, \{2,3\}, \{1,2,3\}\}$.

For instance, if worker 1 answers 1 to question 2, his response will be modelled as a certain *bba* $m_{21}(\{1\}) = 1$. If he hesitates and decided to give two answers 1 and 3, these responses will be changed to a categorical *bba* $m_{21}(\{1,3\}) = 1$. And if gives all possible answers, it means that he ignores completely the exact label, thus his response will be a vacuous *bba* $m_{21}(\{1,2,3\}) = 1$.

3.2 Worker Clustering Process

In the second step, workers are partitioned into three clusters according to two main features:

Exactitude Level. The exactitude level denoted as e_j is calculated using the average distance between a worker answer m_j and all the rest as follows:

$$e_j = 1 - \frac{1}{A_j} \sum_{i=1}^{N} d_j \tag{8}$$

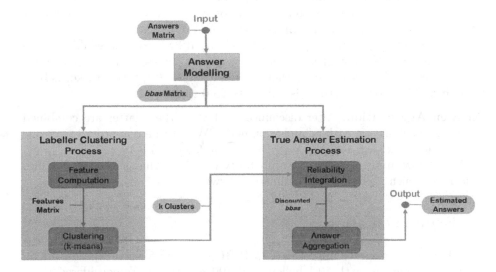

Fig. 1. CIMP-BLA

where N is the total number of tasks and d_j is the distance of Jousselme [6] between the answer of worker j and the averaged remaining answers computed as follows:

$$m_{\Sigma_{T-1}}(X) = \frac{1}{T-1} \sum_{j=1}^{T-1} (X) \tag{9}$$

where T is the total number of workers.

Precision Level. The precision level denoted as p_j is calculated using the average of the specificity levels [23] relative to each label as follows:

$$p_j = \frac{1}{A_j} \sum_{i=1}^{N} \delta_j \tag{10}$$

where δ_j is the specificity level computed as follows:

$$\delta_j = 1 - \sum_{X \in 2^{\Omega}} m_j(X) \frac{log_2(|X|)}{log_2(|\Omega|)} \tag{11}$$

After calculating both attributes for each worker, we proceed to cluster all workers using the k-means algorithm. We want to differentiate between three categories of workers notably the pro, good and bad labellers. Therefore k is fixed to 3. Once we obtain our groups, we calculate the average of both features to discover the nature of each cluster.

3.3 True Label estimation process

Reliability Integration. This next step consists in including the reliability of workers using the discounting operation (Eq. 1) in order to improve the estimation of true labels in the final step. The discounting factor α_j of each worker is adjusted regarding his type determined previously:

If the worker is pro, we keep his *bbas* unchanged by fixing his reliability level to 1 hence α_j=0. If the worker is good, we weaken his answers using the average of exactitude and precision levels thus $\alpha_j = 1 - (e_j + p_j/2)$. If the worker is bad, we remove his answers setting his reliability to 0 thus α_j=1.

Answer Aggregation. After discounting all *bbas*, these latter are combined using the combination with adapted conflict (CWAC) rule 4 in order to get one *bba* for each task such as $\bigominus_{j=1}^{T} m_i = m_{i1}^{\alpha} \ominus m_{i2}^{\alpha} \ominus m_{ij}^{\alpha} \ominus ... m_{iT}^{\alpha}$.

Finally, we apply the pignistic probability (Eq. 7) to change the last *bba* into a decision which represents the estimated answer.

4   Experimentation

To validate our proposed approach (CIMP-BLA), we performed experiments on a simulated dataset with 50 labellers and 100 questions. We considered three types of workers; the pros, the good and the bad. The pros tend to always give more than 80% exact and precise answers, and the good also have exact answers but more than 50% of them are imprecise and lastly the bad give more than 70% random answers (inexact and imprecise). The experiment consists in showing the robustness of our approach when a high amount of bad labellers are involved. Hence, we vary each time the proportion of low quality workers and observe the accuracy. We compare our approach with the majority voting method (MV) and the (IMP-BLA) proposed in [21]. Figure 2 illustrates results regarding accuracy of the comparison of our approach to majority voting and IMP-BLA when the proportion of bad workers is increased. We can notice that in general the MV drops rapidly compared to IMP-BLA and CIMP-BLA that are similarly accurate at low proportions of bad participants. However, when the proportion reaches 85%, the IMP-BLA accuracy drops from around 0.9 to 0.6. Unlikely, the CIMP-BLA records more than 0.7 accuracy when the proportion hit 90%. In Table 1, the average accuracies relative to the different proportions are noted. We can tell that our approach improves quality results since it records 0.58 average accuracy when the ratio of bad workers is very high.

Table 1. Average accuracies of MV, IMP-BLA and CIMP-BLA when varying % of bad workers

Bad workers	MV	IMP-BLA	CIMP-BLA
[0%–30%]	0.91	1	1
[30%–70%]	0.64	0.98	0.99
[70%–100%]	0.29	0.43	0.58

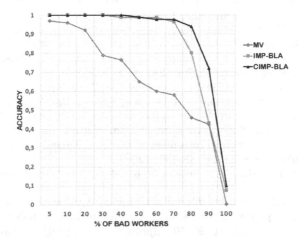

Fig. 2. Accuracy when varying the proportion of bad workers

5 Conclusion and Future Works

In this paper, we proposed a new approach (CIMP-BLA) for the aggregation of imprecise labels within the belief function theory that considers the type of workers after clustering them. The efficiency of our approach is confirmed when compared to baselines through experimentation conducted on simulated dataset. As future work, we intend to address other types of crowd labelling tasks such as ranking tasks and try to improve their results.

References

1. Zheng, Y., Wang, J., Li, G., Feng, J.: QASCA: a quality-aware task assignment system for crowdsourcing applications. In: International Conference on Management of Data, pp. 1031–1046 (2015)
2. Snow, R., O'Connor, B., Jurafsky, D., Ng, A.Y.: Cheap and fast but is it good? Evaluation non-expert annotations for natural language tasks. In: The Conference on Empirical Methods In Natural Languages Processing, pp. 254–263 (2008)
3. Sheng, V.S., Provost, F., Ipeirotis, P.G.: Get another label? Improving data quality and data mining using multiple, noisy labelers. In: International Conference on Knowledge Discovery and Data Mining, pp. 614–622 (2008)

4. Shafer, G.: A Mathematical Theory of Evidence, vol. 1. Princeton University Press, Princeton (1976)
5. Dempster, A.P.: Upper and lower probabilities induced by a multivalued mapping. In: The Annals of Mathematical Statistics, pp. 325–339 (1967)
6. Jousselme, A.-L., Grenier, D., Bossé, É.: A new distance between two bodies of evidence. Inf. Fusion **2**, 91–101 (2001)
7. Lefèvre, E., Elouedi, Z.: How to preserve the confict as an alarm in the combination of belief functions? Decis. Support Syst. **56**, 326–333 (2013)
8. Smets, P.: The combination of evidence in the transferable belief model. IEEE Trans. Pattern Anal. Mach. Intell. **12**(5), 447–458 (1990)
9. Raykar, V.C., Yu, S.: Eliminating spammers and ranking annotators for crowdsourced labeling tasks. J. Mach. Learn. Res. **13**, 491–518 (2012)
10. Dawid, A.P., Skene, A.M.: Maximum likelihood estimation of observer error-rates using the EM algorithm. Appl. Stat. **28**, 20–28 (2010)
11. Khattak, F.K., Salleb, A.: Quality control of crowd labeling through expert evaluation. In: The Neural Information Processing Systems 2nd Workshop on Computational Social Science and the Wisdom of Crowds, pp. 27–29 (2011)
12. Raykar, V.C., et al.: Supervised learning from multiple experts: whom to trust when everyone lies a bit. In: Proceedings of the 26th Annual International Conference on Machine Learning, pp. 889–896 (2009)
13. Smets, P., Kennes, R.: The transferable belief model. Artif. Intell. **66**, 191–234 (1994)
14. Ben Rjab, A., Kharoune, M., Miklos, Z., Martin, A.: Characterization of experts in crowdsourcing platforms. In: Vejnarová, J., Kratochvíl, V. (eds.) BELIEF 2016. LNCS (LNAI), vol. 9861, pp. 97–104. Springer, Cham (2016). https://doi.org/10.1007/978-3-319-45559-4_10
15. Watanabe, M., Yamaguchi, K.: The EM Algorithm and Related Statistical Models, p. 250. CRC Press, Boca Raton (2003)
16. Whitehill, J., Wu, T., Bergsma, J., Movellan, J.R., Ruvolo, P.L.: Whose vote should count more: optimal integration of labels from labelers of unknown expertise. In: Neural Information Processing Systems, pp. 2035–2043 (2009)
17. Abassi, L., Boukhris, I.: Crowd label aggregation under a belief function framework. In: Lehner, F., Fteimi, N. (eds.) KSEM 2016. LNCS (LNAI), vol. 9983, pp. 185–196. Springer, Cham (2016). https://doi.org/10.1007/978-3-319-47650-6_15
18. Abassi, L., Boukhris, I.: A gold standards-based crowd label aggregation within the belief function theory. In: Benferhat, S., Tabia, K., Ali, M. (eds.) IEA/AIE 2017. LNCS (LNAI), vol. 10351, pp. 97–106. Springer, Cham (2017). https://doi.org/10.1007/978-3-319-60045-1_12
19. Abassi, L., Boukhris, I.: Iterative aggregation of crowdsourced tasks within the belief function theory. In: Antonucci, A., Cholvy, L., Papini, O. (eds.) ECSQARU 2017. LNCS (LNAI), vol. 10369, pp. 159–168. Springer, Cham (2017). https://doi.org/10.1007/978-3-319-61581-3_15
20. Abassi, L., Boukhris, I.: A worker clustering-based approach of label aggregation under the belief function theory. Appl. Intell. **49**, 53–62 (2018). ISSN: 1573–7497
21. Abassi, L., Boukhris, I.: Imprecise label aggregation approach under the belief function theory. In: Abraham, A., Cherukuri, A.K., Melin, P., Gandhi, N. (eds.) ISDA 2018 2018. AISC, vol. 941, pp. 607–616. Springer, Cham (2020). https://doi.org/10.1007/978-3-030-16660-1_59
22. Koulougli, D., HadjAli, A., Rassoul, I.: Handling query answering in crowdsourcing systems: a belief function-based approach. In: Annual Conference of the North American Fuzzy Information Processing Society (NAFIPS), pp. 1–6 (2016)

23. Smarandache, K., Martin, A., Osswald, C.: Contradiction measures and specificity degrees of basic belief assignments. In: 14th International Conference on Information Fusion, pp. 1–8 (2011)
24. Kuncheva, L., et al.: Limits on the majority vote accuracy in classifier fusion. Pattern Anal. Appl. **6**, 22–31 (2003)

Combining Machine Learning and Classical Optimization Techniques in Vehicle to Vehicle Communication Network

Mutasem Hamdan[(✉)] and Khairi Hamdi[(✉)]

School of EEE, The University of Manchester, Manchester, UK
{mutasem.hamdan,K.Hamdi}@manchester.ac.uk

Abstract. In this paper a new optimization technique has been proposed to take the advantage of both the Hungarian Algorithm and Deep Q-Learning Neural Network (DQN) to solve the frequency and power resources allocation problem in Vehicle to Vehicle (V2V) future networks. The result shows a better performance for the sum of cellular users throughput, reducing the complexity of the classical optimization methods, overcome the huge State-Action matrix in Q-learning and provides wireless environment features to approximate the Q-values.

Keywords: Deep Q-Learning Network · Vehicle to vehicle network · Reinforcement Learning · Radio resource management · Ultra Reliable Low Latency communications · Maximum weigh matching

1 Introduction

In the era of modern wireless communications, solutions for big challenges such as the optimization of radio resources management (RRM) are extremely important for future massive and heterogeneous wireless networks. The RRM is one of the most sophisticated problems due to the huge diversity of the scenarios and technologies. Adding constraints to the scarce resources of spectrum and transmit power, cause more complexity to maximize the performance of the wireless networks. While Vehicle to everything (V2X) as part of "3GPP" technical standard has a promising services and targeting Ultra Reliable Low Latency communications (URLLC) [1,2], a new level of optimization problem has been created to manage the network quality of service. A critical balance between shared spectrum channels and transmit power levels for the resource blocks used by large number of users may cause degradation instead of upswing, if the allocation of such resources are misused.

As a result, assigning the sub-bands and power levels resources play a critical role in achieving the V2X requirements. The optimization approach in [6] suggested solution for the resources allocation considering the large scale fading effect, then apply a heuristic solution for solving the power allocation, as in [4]

© Springer Nature Switzerland AG 2019
H. Yin et al. (Eds.): IDEAL 2019, LNCS 11871, pp. 350–358, 2019.
https://doi.org/10.1007/978-3-030-33607-3_38

the researchers assumed that Channel Information State (CSI) is a deterministic variable, which is not a realistic assumption in V2V network case.

A new approach in [8] and [9] where Deep Neural Network (DNN) has been used to find optimal values for Q^*, which helps to choose the best action: sub-channel and power level for V2V links. This overcome the huge Q matrix problem.

This paper explores the power of combining classical optimization and Deep Q-Learning Network (DQN) to improve the V2V network for the rapid environment changes in a way that maintains latency and reliability of the V2V network while maximizing the performance of the cellular users (C-UE). This work novelty comes from the fusion of two different techniques no other researchers had investigated before to the best of the authors knowledge. The new approach reduces the complexity of the solution by reducing the DQN size used in [9].

2   V2V Uni-cast Scenario Setup

The resource management problem can be formulated with the help of Fig. 1. The general network consists of C-UEs users $= \{1, 2, \ldots, M\}$ connected to the Bas Station (BS), which are allocated in orthogonal sub-bands and have high capacity links. The pairs of V2V users $= \{1, 2, \ldots, K\}$. The V2V links to share information for traffic safety only. Since up-link resources are less intensively used and the interference at the base station (BS) is more controllable, the considered

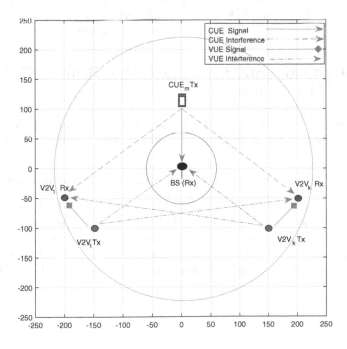

Fig. 1. V2V uni-cast scenario

model will consider the up-link spectrum. So the SINR affecting the CUE and V2V are in Eqs. (2.1) and (2.2) respectively are [8]:

$$\gamma_m^C = \frac{P^C h_m}{\sigma^2 + \sum_{k \in K} \rho_{m,k} P^V \tilde{h}_k} \tag{2.1}$$

$$\gamma_{k,j}^V = \frac{P^V g_k}{\sigma^2 + G_C + G_V} \tag{2.2}$$

$$G_C = \sum_{m \in M} \rho_{m,k} P^C \tilde{g}_{m,k} \tag{2.3}$$

$$G_V = \sum_{m \in M} \sum_{k' \in K, k \neq k'} \rho_{m,k} \rho_{m,k'} P^V \tilde{g}_{k',k,j}^V \tag{2.4}$$

The capacity of mth CUE and k^{th} V2V links are:

$$C_m^C = W \cdot \log(1 + \gamma_m^C) \tag{2.5}$$

$$C_{k,j}^V = W \cdot \log(1 + \gamma_{k,j}^V) \tag{2.6}$$

Where the parameters details: P_m^C is the transmission powers of CUE, P_k^V is the transmission powers of VUE, σ^2 is the noise power, h_m is the channel gain of the channel corresponding to the mth CUE, $\tilde{h}_k$ is the interference power gain of the kth V2V link, and $\rho_{m,k}$ is the spectrum allocation indicator with $\rho_{m,k} = 1$ if the kth V2V link share the spectrum of the mth CUE and $\rho_{m,k} = 0$ if not. g_k is the power gain of kth V2V link, $\tilde{g}_{m,k}$ is the interference power gain of the mth CUE, and $\tilde{g}_{k',k,j}^V$ is the interference power gain of the kth V2V link.

3   Methodologies and Combined Algorithm to Optimize the RRM Problem

Merging classical Hungarian algorithm and the DQN explained in next subsections showing how the mathematical methods used in wireless applications.

3.1   Hungarian Algorithm and Maximum Weight Matching

The solution in [5] and [6] for MWM problem between each CUE and each V2V link to find the best sub-band assignment has been used as an initial step before the DQN. In Fig. 2c the grey, amber and green dashed circles represent three different sub-bands and CUE shares one sub-band with two V2V links to maximize the sum of CUEs capacities. This can be generalized in below optimization problem Eq. (3.1) [3]:

$$\max_{\{\rho_{m,k}\}} \sum_{m \in M} \sum_{k \in K} \rho_{m,k} C_{m,k}^C \tag{3.1}$$

$$\text{s.t.} \quad \sum_{m \in M} \rho_{m,k} \leq 1, \quad \rho_{m,k} \in \{0,1\}, \quad \forall k \in K \tag{3.1a}$$

$$\sum_{k \in K} \rho_{m,k} \leq 1, \quad \rho_{m,k} \in \{0,1\}, \quad \forall m \in M, \tag{3.1b}$$

As we have to assign more than one pairs of V2V (the experiment has two pairs of V2V links assigned per C-UE, not one), an extra sub-step will be introduced. For instant once the first pair for each C-UE has been assigned, the Hungarian algorithm will be conducted again after eliminating the assigned pairs in the previous algorithm run by considering zero value for their $C_{m,k}^C$. At the end of this stage, two pairs of V2V links has been associated to each CUE.

3.2   Deep Q Neural Network

In this sub-section, the DQN used to approximate the Q values of the Reinforcement Learning (RL) policy to achieve the optimal resources allocation for both sub-band and power for each V2V link and this V2V link will be the agent who selects actions.

The V2V link is an agent with a current state. The agent role is to take actions that maximize it's rewards (Depends on the C^C, C^V and latency constraint in Eqs. (2.5) and (2.6) respectively). The environment respond to the agent actions with changing the state of the agent and provide a reward. Both outputs of the environment are used by the agent to update it's policy in taking action [7].

The final result in our wireless case to stay in the optimal state (e.g. the lowest interference, highest power gain, minimum remaining load V2V has to transmit and lowest latency). It is important to define the states and actions to represent the inputs and outputs to use in next sub-section. Figure 2a shows the process of the Q- Learning and Fig. 2b shows the Q values tabulated matrix, this table called Policy. When the Q matrix is huge and updating values process is not practical solution, the natural option to reduce the dimensions of the data processing by DQN [9]. So the main definitions and equations for the Q-Learning function for uni-cast scenario are below.

The state ($s_t \in$ space of S states) definition: $s_t = \{I_{t-1}, H_t, N_{t-1}, G_t, U_t, L_t\}$, where I_{t-1} is the previous time slot interference at BS, H_t is the C-UE channel power gain, N_{t-1} is the neighbours selected sub-channel one time slot before, G_t is the channel power gain for V2V link, U_t is the time value before violating the latency condition and lastly L_t the information bits load that still need transmission to the V2V receiver.

The action ($a_t \in$ set of actions $\mathcal{A}$) definition: one of the sub-channels N_{RB} (the N is the number of available resource blocks) and one of the power level for V2V link agent $P_{level \in [23,10,5]dBm}^V$. The reward is function of Cellular, V2V capacities (C_m^C and C_k^V respectively) and time latency constraints T_0 for how long time to transmit V2V message in time period U_t see below equation. While λ_C, λ_V and λ_p are weights [8]:

$$r_t = \lambda_C \sum_{m \in \mathcal{M}} C_m^C + \lambda_V \sum_{k \in \mathcal{K}} C_k^V - \lambda_p (T_0 - U_t), \tag{3.2}$$

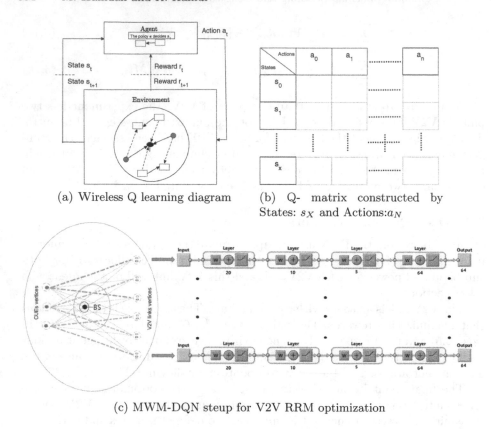

(a) Wireless Q learning diagram

(b) Q- matrix constructed by States: s_X and Actions:a_N

(c) MWM-DQN steup for V2V RRM optimization

Fig. 2. Q Learning Diagram, the Q matrix substituted by DQN (Color figure online)

The goal will be to maximize the expected summation for all rewards discounts:

$$Goal_t = \mathbb{E}[\sum_{n=0}^{\infty} \beta^n r_{t+n}], \tag{3.3}$$

Where $\beta \in [0,1]$ is the discount factor. Next to find the optimal policy π^*, the maximum Q value such that:

$$a_t = \arg\max_{a \in \mathcal{A}} Q(s_t, a). \tag{3.4}$$

Which can be obtained by updating below equation where α is the learning rate:

$$Q_{new}(s_t, a_t) = Q_{old}(s_t, a_t) + \alpha[r_{t+1} + \beta \max_{s \in \mathcal{S}} Q_{old}(s, a_t) - Q_{old}(s_t, a_t)] \tag{3.5}$$

The before mentioned technique can be substituted by DQN that evaluates the Q^*. To overcome the actions-states large space matrix update.

Data: The Inputs of states $s_t = \{I_{t-1}, H_t, N_{t-1}, G_t, U_t, L_t\}$
Result: DQN with weights w
initialization:
a. Initialize the policy π randomly.
b. Initialize the DQN model.
c. Generate V2V links and CUE users.
d. Run MWM to assign each V2V links to each CUE's.

e)**for** *Training the DQN* **do**
> Based on police π iteratively choose sub-channel and power level for each V2V link(agent);
> Calculate and save (State, reward, action,next-state) ;
> Use a mini batch of data saved for training the DQN and update the w;
> Upate the policy π by choosing the action with Max (Q-value).;

end

initialization trained DQN model:
f. Initialize the policy π randomly.
g. Initialize the DQN model;

h)**for** *Testing the DQN* **do**
> Based on police π iteratively choose sub-channel and power level for each V2V link(agent);
> Select the action result in Max (Q-value).;
> Calculate and update (State, reward, action,next-state) ;
> Update Sum Rate of CUEs capacity ;

end

Algorithm 1: Pseudo MWM-DQN combined algorithm

The DQN using the state of the agent as an input for Deep Forward Neural network with Loss function. The deep of DQN in [9] has five hidden layers. The outputs are Q values, each related to one action which is the transmission power of the Agent and the sub-band. See Fig. 2c and the process of representing the Q-learning by DNN with weight w so that the DQN maps the states information (e.g. channel information) to the desired Q values in Eq. (3.8) to tune the NN weights to minimize loss function in Eq. (3.9) [8].

$$y = \lambda_C \sum_{m \in \mathcal{M}} C_m^C + \lambda_V \sum_{k \in \mathcal{K}} C_k^V - \lambda_p(T_0 - U_t) + \max_{a \in \mathcal{A}} Q_{old}(s_t, a, w), \qquad (3.8)$$

and the final form for the loss function taking in consideration D is the training data:

$$Loss(w) = \sum_{(s_t,a_t)\in D} \left(\lambda_C \sum_{m\in\mathcal{M}} C_m^C + \lambda_V \sum_{k\in\mathcal{K}} C_k^V - \lambda_p(T_0 - U_t) \right.$$
$$\left. + \max_{a\in\mathcal{A}} Q_{old}(s_t, a, w) - Q(s_t, a_t, w) \right)^2 \qquad (3.9)$$

It is important to mention that the sum of square errors measures the network performance and once all the training sets or epoch have acceptable square error values, the network considered converged. Algorithm (1) summarise the new technique:

4   The Experiment, Result and Discussion

In this experiment the Matlab 2018a has been used to generate the synthesised data and also to implement MWM-DQN combined Algorithm (1) using parameters in Table 1. The result in Fig. 3 shows that the performance improves when MWM solution and DQN have been combined verses DQN only. It is important to emphasize again that increasing the number of transmit power levels and sub-bands (the more actions the V2V link can take) the network performance improves. The hypothesized explanation for previous outcome, that the DQN are initialized with better organized states that provides at least near local optima solution for some states (e.g. the channel gain of V2V link G_t). Where in the DQN the resources allocation is pure random.

Table 1. Experiment parameters

Parameter name	Parameter value
Carrier frequency	2 GHz
BW	1.5 MHz
CUE power(S^{max})	23 dBm
V2V Tx power(P^{max})	[23,10,5] dBm
No. of sub channels = No. of CUEs	4
BS antenna height	26 m
[$\lambda_C, \lambda_V, \lambda_p$]	[.1, .9, 1]
Latency threshold(T_0)	100 ms
C and V2V-UE antenna height	1.5 m
Noise power σ^2	-117 dBm
DQN structure	5-Layer: [6,20,10,5,64] neurons
DQN activation functions	Relu: $f_r(x) = Max(0, 1)$

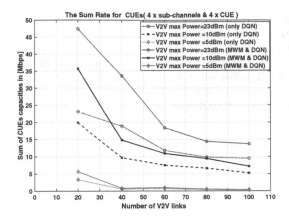

Fig. 3. MWM and DQN result

5 Conclusions and Future Work

In this paper, the result shows that, the increase in the sum of the C-UEs throughput inversely proportional to V2V links. When the Hungarian algorithm has been used before the DQN, the performance has been improved. Future work can include modifying, improving or even extending the experiment, as an example: increasing or decreasing the hidden layers number, the neurons per layer or both to find out what is the impact on the wireless communication throughput performance. Due to the time frame and limitation of number of papers. The result has been provided for the CUEs sum rate for limited sub-channel resources only, more results can be produces as an example: the outage probability vs V2V links and DQN Loss function evaluation vs epochs.

References

1. 3GPP-TR-8.824: Study on physical layer enhancements for NR ultra-reliable and low latency case (URLLC) (2019)
2. 3GPP-TR-TS-122-185: 122 185, LTE; service requirements for V2X services (3GPP TS 22.185 version 15.0.0 release 15) (2018)
3. Liang, L., Li, G.Y., Xu, W.: Resource allocation for D2D-enabled vehicular communications. IEEE Trans. Commun. **65**(7), 3186–3197 (2017)
4. Mei, J., Zheng, K., Zhao, L., Teng, Y., Wang, X.: A latency and reliability guaranteed resource allocation scheme for LTE V2V communication systems. IEEE Trans. Wireless Commun. **17**(6), 3850–3860 (2018)
5. Parsaeefard, S., Sharafat, A.R., Mokari, N.: Robust Resource Allocation in Future Wireless Networks. Springer, Cham (2017). https://doi.org/10.1007/978-3-319-50389-9
6. Sun, W., Ström, E.G., Brännström, F., Sou, K.C., Sui, Y.: Radio resource management for D2D-based V2V communication. IEEE Trans. Veh. Technol. **65**(8), 6636–6650 (2015)

7. Sutton, R.S., Barto, A.G.: Reinforcement Learning: An Introduction. MIT press, Cambridge (2018)
8. Ye, H., Li, G.Y., Juang, B.H.F.: Deep reinforcement learning based resource allocation for V2V communications. IEEE Trans. Veh. Technol. **68**(4), 3163–3173 (2019)
9. Yu, F.R., He, Y.: Reinforcement learning and deep reinforcement learning. Deep Reinforcement Learning for Wireless Networks. SECE, pp. 15–19. Springer, Cham (2019). https://doi.org/10.1007/978-3-030-10546-4_2

Adversarial Edit Attacks for Tree Data

Benjamin Paaßen[⊠][iD]

CITEC, Bielefeld University, Inspiration 1, 33619 Bielefeld, Germany
bpaassen@techfak.uni-bielefeld.de

Abstract. Many machine learning models can be attacked with adversarial examples, i.e. inputs close to correctly classified examples that are classified incorrectly. However, most research on adversarial attacks to date is limited to vectorial data, in particular image data. In this contribution, we extend the field by introducing adversarial edit attacks for tree-structured data with potential applications in medicine and automated program analysis. Our approach solely relies on the tree edit distance and a logarithmic number of black-box queries to the attacked classifier without any need for gradient information.

We evaluate our approach on two programming and two biomedical data sets and show that many established tree classifiers, like tree-kernel-SVMs and recursive neural networks, can be attacked effectively.

Keywords: Adversarial attacks · Tree edit distance · Structured data · Tree kernels · Recursive neural networks · Tree echo state networks

1 Introduction

In recent years, multiple papers have demonstrated that machine learning classifiers can be fooled by *adversarial examples*, i.e. an example x' that is close to a correctly classified data point x, but is classified incorrectly [2,12]. The threat of such attacks is not to be underestimated, especially in security-critical applications such as medicine or autonomous driving, where adversarial examples could lead to misdiagnoses or crashes [8].

Despite this serious threat to *all* classification models, existing research has almost exclusively focused on image data [2,12], with the notable exceptions of a few contributions on audio data [5], text data [7], and graph data [6,19]. In particular, no adversarial attack approach has yet been developed for tree data, such as syntax trees of computer programs or biomedical molecules. Furthermore, all attack approaches for non-image data to date rely on knowledge about the classifier architecture and/or gradient, which may not always be available [12].

In this paper, we address both issues by introducing adversarial edit attacks, a novel black-box attack scheme for tree data. In particular, we propose to select for a point x a neighboring point with a different label y, compute the tree edits

Support by the Bielefeld Young Researchers Fund is gratefully acknowledged.

H. Yin et al. (Eds.): IDEAL 2019, LNCS 11871, pp. 359–366, 2019.
https://doi.org/10.1007/978-3-030-33607-3_39

necessary to change x into y, and applying the minimum number of edits which still change the classifier output.

Our paper is structured as follows. We first introduce background and related work on adversarial examples, then introduce our adversarial attack method, and finally evaluate our method by attacking seven different tree classifiers on four tree data sets, two from the programming domain and two from the biomedical domain.

2   Related Work

Following Szegedy et al. [17], we define an adversarial example for some data point $x \in \mathcal{X}$ and a classifier $f : \mathcal{X} \to \{1, \ldots, L\}$ and a target label $\ell \in \{1, \ldots, L\}$ as the solution z to the following optimization problem

$$\min_{z \in \mathcal{X}, \text{s.t.} f(z) = \ell} \quad d(z, x)^2, \tag{1}$$

where d is a distance on the data space $\mathcal{X}$. In other words, z is the closest data point to x which is still classified as ℓ. For image data, the distance $d(z, x)$ is often so small that z and x look exactly the same to human observers [17].

Note that Problem 1 is hard to solve because $\mathcal{X}$ is typically high-dimensional and the constraint $f(z) = \ell$ is discrete. Accordingly, the problem has been addressed with heuristic approaches, such as the fast gradient sign method [11], which changes x along the sign of the gradient of the classifier loss; or Carlini-Wagner attacks, which incorporate the discrete label constraint as a differentiable term in the objective function [4]. We call these methods *white-box* because they all rely on knowledge of the architecture and/or gradient $\nabla_z f(z)$ of the classifier. In contrast, there also exist *black-box* attack methods, which only need to query f itself, such as one-pixel attacks, which are based on evolutionary optimization instead of gradient-based optimization [2,16].

In the realm of non-image data, prior research has exclusively focused on white-box attacks for specific data types and/or models. In particular, [5] consider audio files, relying on decibels and the CTC loss as measure of distance; [7] attack text data by inferring single character replacements that increase the classification loss; and [6,19] attack graph data by inferring edge deletions or insertions which fool a graph convolutional neural network model.

Our own approach is related to [5], in that we rely on an alignment between two inputs to construct adversarial examples, and to [7], in that we consider discrete node-level changes, i.e. node deletions, replacements, or insertions. However, in contrast to these prior works, our approach is black-box instead of white-box and works in tree data as well as sequence data.

3   Method

To develop an adversarial attack scheme for tree data, we face two challenges. First, Problem 1 requires a distance function d for trees. Second, we need a method to apply small changes to a tree x in order to construct an adversarial tree z. We can address both challenges with the *tree edit distance*, which is defined as the minimum number of node deletions, replacements, or insertions needed to change a tree into another [18] and thus provides both a distance and a change model.

Formally, we define a tree over some finite alphabet $\mathcal{A}$ recursively as an expression $T = x(T_1, \ldots, T_m)$, where $x \in \mathcal{A}$ and where $T_1, \ldots, T_m$ is a (possibly empty) list of trees over $\mathcal{A}$. We denote the set of all trees over $\mathcal{A}$ as $\mathcal{T}(\mathcal{A})$. As an example, a(), a(b), and a(b(a, a), a) are all trees over the alphabet $\mathcal{A} = \{a, b\}$. We define the *size* of a tree $T = x(T_1, \ldots, T_m)$ recursively as $|T| := 1 + \sum_{c=1}^{m} |T_c|$.

Next, we define a *tree edit* δ over alphabet $\mathcal{A}$ as a function $\delta : \mathcal{T}(\mathcal{A}) \to \mathcal{T}(\mathcal{A})$. In more detail, we consider node deletions del_i, replacements $\mathrm{rep}_{i,a}$, and insertions $\mathrm{ins}_{i,c,C,a}$, which respectively delete the ith node in the input tree and move its children up to the parent, relabel the ith node in the input tree with symbol $a \in \mathcal{A}$, and insert a new node with label a as cth child of node i, moving former children $c, \ldots, c+C$ down. Figure 1 displays the effects of each edit type.

We define an *edit script* as a sequence $\bar{\delta} = \delta_1, \ldots, \delta_n$ of tree edits δ_j and we define the application of $\bar{\delta}$ to a tree T recursively as $\bar{\delta}(T) := (\delta_2, \ldots, \delta_n)(\delta_1(T))$. Figure 1 displays an example edit script.

Finally, we define the *tree edit distance* $d(x, y)$ as the length of the shortest script which transforms x into y, i.e. $d(x, y) := \min_{\bar{\delta}:\bar{\delta}(x)=y} |\bar{\delta}|$. This tree edit distance can be computed efficiently via dynamic programming in $\mathcal{O}(|x|^2 \cdot |y|^2)$ [18]. We note that several variations of the tree edit distance with other edit models exist, which are readily compatible with our approach [3,14]. For brevity, we focus on the classic tree edit distance in this paper.

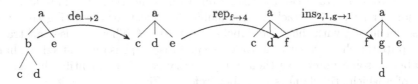

Fig. 1. An illustration of the effect of the tree edit script $\bar{\delta} = \mathrm{del}_2, \mathrm{rep}_{4,f}, \mathrm{ins}_{1,2,1,g}$ on the tree (a(b(c,d), e)). We first delete the second node of the tree, then replace the fourth node with an f, and finally insert a g as second child of the first node, using the former second child as grandchild.

Random Baseline Attack: The concept of tree edits yields a baseline attack approach for trees. Starting from a tree x with label $f(x)$, we apply random tree edits, yielding another tree z, until $f(z) \neq f(x)$. To make this more efficient, we double the number of edits in each iteration until $f(z) \neq f(x)$, yielding an edit script $\bar{\delta} = \delta_1, \ldots, \delta_n$, and then use binary search to identify the shortest prefix $\bar{\delta}_j := \delta_1, \ldots, \delta_j$ such that $f((\delta_1, \ldots, \delta_j)(x)) \neq f(x)$. This reduced the number of queries to $\mathcal{O}(\log(n))$.

Note that this random attack scheme may find solutions z which are far away from x, thus limiting the plausibility as adversarial examples. To account for such cases, we restrict Problem 1 further and impose that z only counts as a solution if z is still closer to x than to any point y which is correctly classified and has a different label than x (refer to Fig. 2).

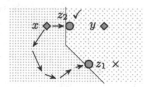

Fig. 2. Two adversarial attack attempts, one random (z_1) and one backtracing attack (z_2). z_1 is constructed by moving randomly in the space of possible trees until the label changes. z_2 is constructed by moving along the connecting line to the closest neighbor with different label y until the label changes. z_1 is *not* counted as successful, because it is closer to y than to x, whereas z_2 is counted as successful. The background pattern indicates the predicted label of the classifier.

Another drawback of our random baseline is that it can not guarantee results after a fixed amount of edits because we may not yet have explored enough trees to have crossed the classification boundary. We address this limitation with our proposed attack method, *backtracing attacks*.

Backtracing Attack: For any two trees x and y, we can compute a co-optimal edit script $\bar{\delta}$ with $\bar{\delta}(x) = y$ and $|\bar{\delta}| = d(x, y)$ in $\mathcal{O}(|x| \cdot |y| \cdot (|x| + |y|))$ via a technique called *backtracing* [13, refer to Algorithm 6 and Theorem 16]. This forms the basis for our proposed attack. In particular, we select for a starting tree x the closest neighbor y with the target label ℓ, i.e. $f(y) = \ell$. Then, we use backtracing to compute the shortest script $\bar{\delta}$ from x to y. This script is guaranteed to change the label at some point. We then apply binary search to identify the shortest prefix of $\bar{\delta}$ which still changes the label (refer to Fig. 2).

Note that we can upper-bound the length of $\bar{\delta}$ by $|x| + |y|$, because at worst we delete x entirely and then insert y entirely. Accordingly, our attack finishes after at most $\mathcal{O}(\log(|x| + |y|))$ steps/queries to f. Finally, because y is the closest tree with label ℓ to x, our attack is guaranteed to yield a successful adversarial example if our prefix is shorter than half of $\bar{\delta}$, because then $d(x, z) = |prefix| < \frac{1}{2}|\bar{\delta}| = \frac{1}{2}d(x, y) = \frac{1}{2}(d(x, z) + d(z, y))$, which implies that $d(x, z) < d(z, y)$. In other words, we are guaranteed to find a solution to problem 1, in the sense that our our label is guaranteed to change to ℓ, and that our solution is closest to x along the shortest script $\bar{\delta}$ towards y.

4   Experiments

In our evaluation, we attack seven different tree classifiers on four data sets. As outcome measures, we consider the success rate, i.e. the fraction of test data points for which the attack could generate a successful adversarial example according to the definition in Fig. 2; and the distance ratio $d(z,x)/d(z,y)$, i.e. how much closer z is to x compared to other points y with the same label as z. To avoid excessive computation times, we abort random adversarial attacks that have not succeeded after 100 tree edits. Accordingly, the distance ratio is not available for random attacks that have been aborted, yielding some n.a. entries in our results (Table 1).

Our experimental hypotheses are that backtracing attacks succeed more often than random attacks due to their targeted nature (H1), but that random attacks have lower distance ratios (H2), because they have a larger search space from which to select close adversarials.

Datasets: We perform our evaluation on four tree classification data sets from [9,14], in particular *MiniPalindrome* and *Sorting* as data sets of Java programs, as well as *Cystic* and *Leukemia* from the biomedical domain. The number of trees in each data set are 48, 64, 160, and 442 respectively. The latter three data sets are (imbalanced) binary classification problems, the first is a six-class problem. We perform all experiments in a crossvalidation with 6, 8, 10, and 10 folds for the respective data sets, following the protocol of [14].

Classifiers: On each data set, we train seven different classifiers, namely five support vector machines (SVM) with different kernels and two recursive neural network types. As the first two kernels, we consider the double centering kernel (*linear*; [10]) based on the tree edit distance, and the radial basis function kernel (*RBF*) $k(x,y) = \exp\left(-\frac{1}{2} \cdot d(x,y)^2/\sigma^2\right)$, for which we optimize the bandwidth parameter $\sigma \in \mathbb{R}^+$ in a nested crossvalidation in the range $\{0.5, 1, 2\} \cdot \bar{d}$, where $\bar{d}$ is the average tree edit distance in the data set. We ensure positive semi-definiteness for these kernels via the clip eigenvalue correction [10]. Further, we consider three tree kernels, namely the subtree kernel (*ST*), which counts the number of shared proper subtrees, the subset tree kernel (*SST*), which counts the number of shared subset trees, and the partial tree kernel (*PT*), which counts the number of shared partial trees [1]. All three kernels have a decay hyper-parameter λ, which regulates the influence of larger subtrees. We optimize this hyper-parameter in a nested crossvalidation for each kernel in the range $\{0.001, 0.01, 0.1\}$. For all SVM instances, we also optimized the regularization hyper-parameter C in the range $\{0.1, 1, 10, 100\}$.

As neural network variations, we first consider recursive neural networks (*Rec*; [15]), which map a tree $x(T_1, \ldots, T_m)$ to a vector by means of the recursive function $G(x(T_1, \ldots, T_m)) := \text{sigm}\left(\boldsymbol{W}^x \cdot \sum_{i=1}^m G(T_i) + \boldsymbol{b}^x\right)$, where $\text{sigm}(a) := 1/(1 + \exp(-a))$ is the logistic function and $\boldsymbol{W}^x \in \mathbb{R}^{n \times n}$ as well as $\boldsymbol{b}^x \in \mathbb{R}^n$ for all $x \in \mathcal{A}$ are the parameters of the model. We classify a tree by means of another linear layer with one output for each of the L classes, i.e. $f(T) := \text{argmax}_\ell [\boldsymbol{V} \cdot G(T) + \boldsymbol{c}]_\ell$, where $\boldsymbol{V} \in \mathbb{R}^{L \times n}$ and $\boldsymbol{c} \in \mathbb{R}^L$ are parameters of the model and $[\boldsymbol{v}]_\ell$ denotes the ℓth entry of vector $\boldsymbol{v}$. We trained the network using the crossentropy loss and Adam as optimizer until the training loss dropped below 0.01. Note that the number of embedding dimensions n is a hyper-parameter of the model,

Table 1. The unattacked classification accuracy (higher is better), attack success rate (higher is better), and distance ratio $d(z,x)/d(z,y)$ between the adversarial example z, the original point x, and the closest point y to z with the same label (lower is better) for all classifiers and all data sets. Classifiers and data sets are listed as rows, attack schemes as columns. All values are averaged across crossvalidation folds and listed $\pm$ standard deviation. The highest success rate and lowest distance ratio in each column is highlighted via bold print. If all attacks failed, results are listed as n.a.

| | No attack | Random | | Backtracing | |
Classifier	Accuracy	Success rate	Dist. ratio	Success rate	Dist. ratio
MiniPalindrome					
Linear	0.96 ± 0.06	0.09 ± 0.09	$\mathbf{0.24 \pm 0.07}$	$\mathbf{0.52 \pm 0.15}$	2.68 ± 3.54
RBF	1.00 ± 0.00	0.06 ± 0.06	$\mathbf{0.27 \pm 0.21}$	$\mathbf{0.52 \pm 0.17}$	1.44 ± 0.51
ST	0.88 ± 0.07	$\mathbf{0.86 \pm 0.08}$	0.29 ± 0.05	0.72 ± 0.10	0.93 ± 0.15
SST	0.96 ± 0.06	$\mathbf{0.78 \pm 0.15}$	0.36 ± 0.08	0.54 ± 0.11	1.91 ± 1.19
PT	0.96 ± 0.06	$\mathbf{0.80 \pm 0.07}$	0.35 ± 0.10	0.54 ± 0.11	1.91 ± 1.19
Rec	0.85 ± 0.13	0.72 ± 0.14	$\mathbf{0.17 \pm 0.05}$	$\mathbf{0.79 \pm 0.08}$	1.26 ± 0.41
TES	0.92 ± 0.06	$\mathbf{0.95 \pm 0.07}$	$\mathbf{0.08 \pm 0.03}$	0.71 ± 0.09	1.57 ± 0.81
Sorting					
Linear	0.94 ± 0.06	0.02 ± 0.04	$\mathbf{0.86 \pm 0.00}$	0.44 ± 0.16	1.55 ± 0.49
RBF	0.94 ± 0.06	0.18 ± 0.14	$\mathbf{0.57 \pm 0.07}$	0.42 ± 0.16	1.64 ± 0.47
ST	0.81 ± 0.16	$\mathbf{0.65 \pm 0.09}$	0.20 ± 0.05	0.61 ± 0.17	3.01 ± 1.91
SST	0.89 ± 0.10	0.42 ± 0.17	0.50 ± 0.14	0.49 ± 0.17	1.67 ± 0.52
PT	0.88 ± 0.12	0.42 ± 0.14	0.52 ± 0.17	0.50 ± 0.14	1.69 ± 0.87
Rec	0.87 ± 0.01	$\mathbf{0.64 \pm 0.20}$	0.44 ± 0.07	0.26 ± 0.17	1.87 ± 0.59
TES	0.70 ± 0.15	$\mathbf{0.84 \pm 0.11}$	0.21 ± 0.08	0.20 ± 0.16	2.40 ± 0.88
Cystic					
Linear	0.72 ± 0.09	0.00 ± 0.00	n.a.	$\mathbf{0.14 \pm 0.07}$	$\mathbf{1.71 \pm 0.65}$
RBF	0.74 ± 0.09	0.00 ± 0.00	n.a.	$\mathbf{0.22 \pm 0.13}$	$\mathbf{1.68 \pm 0.56}$
ST	0.75 ± 0.10	0.00 ± 0.00	n.a.	$\mathbf{0.49 \pm 0.23}$	$\mathbf{0.86 \pm 0.24}$
SST	0.72 ± 0.09	0.00 ± 0.00	n.a.	$\mathbf{0.34 \pm 0.16}$	$\mathbf{1.25 \pm 0.32}$
PT	0.74 ± 0.08	0.00 ± 0.00	n.a.	$\mathbf{0.35 \pm 0.13}$	$\mathbf{1.26 \pm 0.44}$
Rec	0.76 ± 0.11	$\mathbf{0.46 \pm 0.14}$	$\mathbf{0.77 \pm 0.09}$	0.33 ± 0.10	1.45 ± 1.28
TES	0.71 ± 0.11	$\mathbf{0.63 \pm 0.11}$	$\mathbf{0.62 \pm 0.11}$	0.36 ± 0.17	1.23 ± 0.28
Leukemia					
Linear	0.92 ± 0.04	0.00 ± 0.00	n.a.	$\mathbf{0.27 \pm 0.17}$	$\mathbf{3.20 \pm 1.77}$
RBF	0.95 ± 0.03	0.00 ± 0.00	n.a.	$\mathbf{0.20 \pm 0.08}$	$\mathbf{2.88 \pm 1.65}$
ST	0.92 ± 0.03	0.00 ± 0.00	n.a.	$\mathbf{0.21 \pm 0.09}$	$\mathbf{2.64 \pm 0.51}$
SST	0.95 ± 0.03	0.00 ± 0.00	n.a.	$\mathbf{0.19 \pm 0.10}$	$\mathbf{2.57 \pm 0.65}$
PT	0.95 ± 0.02	0.00 ± 0.00	n.a.	$\mathbf{0.20 \pm 0.10}$	$\mathbf{2.54 \pm 0.56}$
Rec	0.93 ± 0.03	$\mathbf{0.41 \pm 0.07}$	$\mathbf{0.73 \pm 0.07}$	0.24 ± 0.10	2.43 ± 0.64
TES	0.88 ± 0.02	$\mathbf{0.69 \pm 0.11}$	$\mathbf{0.53 \pm 0.04}$	0.36 ± 0.16	2.49 ± 1.11

which we fixed here to $n = 10$ as this was sufficient to achieve the desired training loss. Finally, we consider tree echo state networks (TES; [9]), which have the same architecture as recursive neural networks, but where the recursive weight matrices $W^x \in \mathbb{R}^{n \times n}$ and the bias vectors $b^x \in \mathbb{R}^n$ remain untrained after random initialization. Only the output parameters V and c are trained via simple linear regression. The scaling of the recursive weight matrices and n are hyper-parameters of the model, which we optimized in a nested crossvalidation via grid search in the ranges $\{0.7, 0.9, 1, 1.5, 2\}$ and $\{10, 50, 100\}$ respectively.

As implementations, we use the scikit-learn version of SVM, the *edist* package for the tree edit distance and its backtracing[1], the ptk toolbox[2] for the ST, SST, and PT kernels [1], a custom implementation of recursive neural networks using pytorch,and a custom implementation of tree echo state networks [3].We perform all experiments on a consumer grade laptop with an Intel i7 CPU.

Results and Discussion: Table 1 displays the mean classification error $\pm$ standard deviation in crossvalidation, as well as the success rates and the distance ratios for random attacks and backtracing attacks for all data sets and all classifiers.

We evaluate our results statistically by aggregating all crossvalidation folds across data sets and comparing success rates and distance rations between in a one-sided Wilcoxon sign-rank test with Bonferroni correction. We observe that backtracing attacks have higher success rates for the linear and RBF kernel SVM ($p < 10^{-5}$), slightly higher rates for the ST and SST kernels ($p < 0.05$), indistinguishable success for the PT kernel, and lower success rates for the recursive and tree echo state networks ($p < 0.01$). This generally supports our hypothesis that backtracing attacks have higher success rates (H1), except for both neural network models. This is especially pronounced for Cystic and Leukemia data sets, where random attacks against SVM models always failed.

Regarding H2, we observe that random attacks achieve lower distance ratios for the ST, SST, and PT kernels ($p < 0.01$), and much lower ratios for recursive neural nets and tree echo state nets ($p < 10^{-5}$). For the linear and RBF kernel, the distance ratios are statistically indistinguishable. This supports H2.

5 Conclusion

In this contribution, we have introduced a novel adversarial attack strategy for tree data based on tree edits in one random and one backtracing variation. We observe that backtracing attacks achieve more consistent and reliable success across data sets and classifiers compared to the random baseline. Only for recursive neural networks are random attacks more successful. We also observe that the search space for backtracing attacks may be too constrained because random attacks generally find adversarials that are closer to the original sample.

---

[1] https://gitlab.ub.uni-bielefeld.de/bpaassen/python-edit-distances.

[2] http://joedsm.altervista.org/pythontreekernels.htm.

[3] All implementations and experiments are available at https://gitlab.ub.uni-bielefeld. de/bpaassen/adversarial-edit-attacks.

Future research could therefore consider alternative search spaces, e.g. based on semantic considerations. Most importantly, our research highlights the need for defense mechanisms against adversarial attacks for tree classifiers, especially neural network models.

References

1. Aiolli, F., Da San Martino, G., Sperduti, A.: Extending tree kernels with topological information. In: Proceedings of ICANN, pp. 142–149 (2011)
2. Akhtar, N., Mian, A.: Threat of adversarial attacks on deep learning in computer vision: a survey. IEEE Access **6**, 14410–14430 (2018)
3. Bille, P.: A survey on tree edit distance and related problems. Theor. Comput. Sci. **337**(1), 217–239 (2005)
4. Carlini, N., Wagner, D.: Towards evaluating the robustness of neural networks. In: Proceedings of IEEE Security and Privacy, pp. 39–57 (2017)
5. Carlini, N., Wagner, D.: Audio adversarial examples: targeted attacks on speech-to-text. In: Proceedings of SPW, pp. 1–7 (2018)
6. Dai, H., et al.: Adversarial attack on graph structured data. In: Proceedings of ICML, pp. 1115–1124 (2018)
7. Ebrahimi, J., Rao, A., Lowd, D., Dou, D.: HotFlip: white-box adversarial examples for text classification. In: Proceedings of ACL, pp. 31–36 (2018)
8. Eykholt, K., et al.: Robust physical-world attacks on deep learning visual classification. In: Proceedings of CVPR, pp. 1625–1634 (2018)
9. Gallicchio, C., Micheli, A.: Tree echo state networks. Neurocomputing **101**, 319–337 (2013)
10. Gisbrecht, A., Schleif, F.M.: Metric and non-metric proximity transformations at linear costs. Neurocomputing **167**, 643–657 (2015)
11. Goodfellow, I., Shlens, J., Szegedy, C.: Explaining and harnessing adversarial examples. In: Proceedings of ICLR (2015)
12. Madry, A., Makelov, A., Schmidt, L., Tsipras, D., Vladu, A.: Towards deep learning models resistant to adversarial attacks. In: Proceedings of ICLR (2018)
13. Paaßen, B.: Revisiting the tree edit distance and its backtracing: a tutorial. CoRR abs/1805.06869 (2018)
14. Paaßen, B., Gallicchio, C., Micheli, A., Hammer, B.: Tree edit distance learning via adaptive symbol embeddings. In: Proceedings of ICML, pp. 3973–3982 (2018)
15. Sperduti, A., Starita, A.: Supervised neural networks for the classification of structures. IEEE Trans. Neural Networks **8**(3), 714–735 (1997)
16. Su, J., Vargas, D.V., Sakurai, K.: One pixel attack for fooling deep neural networks. CoRR abs/1710.08864 (2017)
17. Szegedy, C., et al.: Intriguing properties of neural networks. In: Proceedings of ICLR (2014)
18. Zhang, K., Shasha, D.: Simple fast algorithms for the editing distance between trees and related problems. SIAM J. Comput. **18**(6), 1245–1262 (1989)
19. Zügner, D., Akbarnejad, A., Günnemann, S.: Adversarial attacks on neural networks for graph data. In: Proceedings of SIGKDD, pp. 2847–2856 (2018)

Non-stationary Noise Cancellation Using Deep Autoencoder Based on Adversarial Learning

Kyung-Hyun Lim, Jin-Young Kim, and Sung-Bae Cho[✉]

Department of Computer Science, Yonsei University, Seoul, South Korea
{lkh1075, seago0828, sbcho}@yonsei.ac.kr

Abstract. Studies have been conducted to get a clean data from non-stationary noisy signal, which is one of the areas in speech enhancement. Since conventional methods rely on first-order statistics, the effort to eliminate noise using deep learning method is intensive. In the real environment, many types of noises are mixed with the target sound, resulting in difficulty to remove only noises. However, most of previous works modeled a small amount of non-stationary noise, which is hard to be applied in real world. To cope with this problem, we propose a novel deep learning model to enhance the auditory signal with adversarial learning of two types of discriminators. One discriminator learns to distinguish a clean signal from the enhanced one by the generator, and the other is trained to recognize the difference between eliminated noise signal and real noise signal. In other words, the second discriminator learns the waveform of noise. Besides, a novel learning method is proposed to stabilize the unstable adversarial learning process. Compared with the previous works, to verify the performance of the propose model, we use 100 kinds of noise. The experimental results show that the proposed model has better performance than other conventional methods including the state-of-the-art model in removing non-stationary noise. To evaluate the performance of our model, the scale-invariant source-to-noise ratio is used as an objective evaluation metric. The proposed model shows a statistically significant performance of 5.91 compared with other methods in t-test.

Keywords: Speech enhancement · Noise cancellation · Non-stationary noise · Generative adversarial networks · Autoencoder

1 Introduction

Speech enhancement can be divided into several specific problems such as noise elimination and quality improvement. Since noise has a very diverse form, it is too difficult to remove it without prior knowledge. In particular, the problem becomes more difficult for non-stationary noises. As shown in Fig. 1, it is difficult to recognize the non-stationary noise in the given signal [1]. That is why modeling it is an essential issue for the noise cancellation.

Several works to eliminate stationary noise have been conducted for a long time. Traditional methods have many limitations because they use filtering based on threshold [1]. If signal data is processed in the frequency domain through Fourier transform (FT) as shown in Fig. 2, the general noise does not overlap much with the

© Springer Nature Switzerland AG 2019
H. Yin et al. (Eds.): IDEAL 2019, LNCS 11871, pp. 367–374, 2019.
https://doi.org/10.1007/978-3-030-33607-3_40

clean signal frequency domain, so filtering can work to some extent. However, it can be confirmed that it is difficult to remove the non-stationary noise by filtering because the frequency domain of the voice signal and the noise overlaps or the area is not constant.

Many researchers have studied to eliminate noise by parameterizing it using deep learning [1–4]. The models based on recurrent neural network (RNN) are used to model signal data which is time series data [5, 6]. Various conversions of sound data are applied into model as input, resulting in high cost. Other models include the generative adversarial network (GAN) method [6]. Pascual et al. proposed a model which had excellent performance in noise removal, but it has a limitation that it is vulnerable to non-stationary noise [6].

To overcome the limitations of the previous works, we propose a model that learns the form of noise itself. The proposed model is trained using two discriminators and takes the adversarial learning structure. In addition, a novel method is proposed to stabilize a process with adversarial learning. The proposed model reduces the pre-processing cost using raw data as input and generalizes various noises through adversarial learning with two discriminators. Besides, non-stationary noise can be removed by additional noise-aware discriminator.

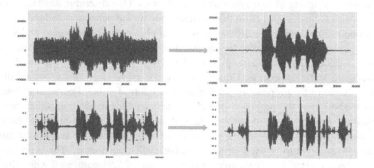

Fig. 1. Example of (top) general noise and (bottom) non-stationary noise cancellation.

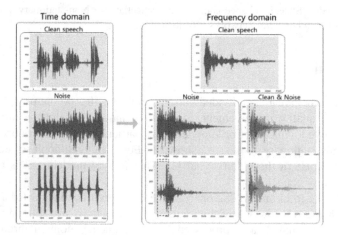

Fig. 2. Comparison of general noise with non-stationary noise in frequency domain.

2 Related Works

There are two main approaches to noise cancellation: filtering and deep learning. The filtering methods previously applied in the noise cancellation are the Wiener filtering and spectral subtraction [7, 8]. Both methods basically eliminate noise by removing a specific signal with high or low frequency based on threshold. However, no matter how well we set the threshold, there can be several types of noise in the real world. It is impossible to set a threshold that can eliminate all these noises. To address the limitation that filtering alone makes it difficult to construct a function to map the complex relationship between noisy and clean signal, a deep learning method is introduced for noise cancellation [1].

Tamura and Waibel used the basic multilayer perceptron [1]. They argued that the neural network can find the relationship between two signals. However, the data used in the experiment were strongly conditioned that only one speaker was used, and it was made up of not many words. Since then, various attempts have been made to create a model that can eliminate independent and diverse noises. Especially, since signal also has the form of time series data, RNN is used as a basic model [9–11]. Parveen and Green succeeded in showing that the noise could be removed independently from the topic with speech data including 55 speakers, but it had the limitation that there were only eleven words included in the signal and experimented with less amount of noise [11]. Another study attempted to remove noise using auto-encoder [12]. This model also showed good performance for general noise but did consider small amount of noise. Since using raw data reduces the information loss and preprocessing costs, Pascual et al. proposed an end-to-end model that first used adversarial learning in the field of noise cancellation [6]. However, it had also limitations that non-stationary noise still became a vulnerability.

In the real environment, since the general noise can also be non-stationary due to various environmental changes, solving this problem is an important issue [13]. The proposed model includes a powerful noise-aware module to eliminate non-stationary noise, and its performance will be evaluated in Sect. 4.

3 The Proposed Method

3.1 Overview

The process of the proposed method is divided into three parts: making powerful discriminators, training proposed generator, and removing the noise with shadow generator. While training the model with adversarial process, there is unstable learning procedure [14, 15]. To prevent this problem, we use two generators having the same structure. Only the first generator is trained in adversarial. In the first phase, original generator and discriminators are learned through adversarial learning, and only the shadow generator is learned using discriminators learned in the second phase. It is intended to create a generator that can learn steadily with a strong teacher. In the third phase, noise is removed using the learned shadow generator.

3.2 Pretraining Discriminators

Figure 3 represents the adversarial learning structure of the original generator and the two discriminators in the first phase. These discriminators are constructed with the same structure. It is composed of the fully-convolution layer to extract the feature from data and learn. One is trained to classify clean and noise as true and classify origin-generated and noisy signal as false. The task of it is to distinguish between the clean signal and the generated signal. The other uses true-pair with the difference of the noisy signal and clean signal and uses false-pair with the difference between the noisy signal and generated signal, which means that it learns the difference between the two signals in the form of noise, residual between two signals. It helps the generator learn to produce a clean signal more closely. The generator is structured like auto-encoder. The encoder and decoder are composed of fully-convolution layer and the U-NET structure is added to prevent the information loss between the encoder and the decoder [16].

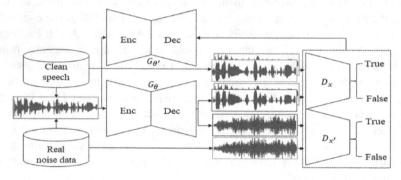

Fig. 3. The overall structure of the proposed method. Real noise data mean noise itself.

After the discriminators are learned, the generator is trained by using discriminators. Since D_x learns true-pair as true and false-pair as false, the objective function can be expressed as Eq. (1). Likewise, $D_{x'}$ has similar learning objectives, so that the loss function can be defined as in Eq. (2).

$$\max_{G_\theta} \min_{D_x} [\{D_x(T(x,\tilde{x})) - 1\}^2 + \{D_x(T(G_\theta(x),\tilde{x}))\}^2] + \|G_\theta(Z,\tilde{x}) - x\|_1 \quad (1)$$

$$\max_{G_\theta} \min_{D_{x'}} [\{D_{x'}(\tilde{x} - x) - 1\}^2 + \{D_{x'}(\tilde{x} - G_\theta(\tilde{x}))\}^2] + \|\{G_\theta(Z,\tilde{x}) - x\} - (\tilde{x} - x)\|_1 \quad (2)$$

where x means clean signal and is the target that the model should make. $\tilde{x}$ means noisy signal, and G_θ means one of the generators. T means concatenation. Noisy signal is used as an input of an encoder, and the output of the encoder and Z, which are prior distributions, are used as inputs of the decoder. The generator adapts its parameters such that discriminators distinguish the outputs of the generator as '*True*'. Besides, additional L1 loss causes the generator to produce a signal closer to the target [17]. Equation (1) is for minimizing the difference between the signal generated and the clean signal. Equation (2) can produce a signal closer to reality by adding the difference between the noise signal and the generated signal difference, which minimizes the noise data [15].

3.3 Training Shadow Generator

Adversarial learning method is difficult to learn in stable [14, 15]. Generators that are in charge of complex tasks compared to discriminators often diverge during learning [13]. To prevent this, we propose a shadow generator $G_{\theta'}$. It is trained using pretrained discriminators that are powerful instructors to produce clean signal. $G_{\theta'}$ just learns with discriminators what is the target distribution, but it is not trained in adversarial. As shown in Fig. 4, G_θ that does adversarial learning is not able to converge easily and its loss oscillates, while $G_{\theta'}$ continues to learn in stable. Loss functions are defined as Eqs. (3) and (4) in a form like G_θ.

$$\min_{G_{\theta'}}\{D_x(G_{\theta'}(Z,\tilde{x})) - 1\}^2 + \|G_{\theta'}(Z,N) - C\|_1 \tag{3}$$

$$\min_{G_{\theta'}}\{D_{x'}(G_{\theta'}(Z,\tilde{x}) - x) - 1\}^2 + \|\{G_{\theta'}(Z,\tilde{x}) - x\} - (\tilde{x} - x)\|_1 \tag{4}$$

4 Experiments

4.1 Dataset and Evaluation Metric

To verify the proposed method, we use noise sounds provided by Hu and Wang [18]. Noise data has 20 categories and has several noises in each category. We use data recorded by 30 native English speakers producing 400 sentences each as clean signal [19]. The dataset for learning is constructed by combining noise data and clean signal data as shown in Fig. 5. Non-stationary noises are selected from the noise data and divided into training data and test data at 8:2 ratio. We use SI-SNR as the evaluation metric to compare the similarity of signal [20]. The SI-SNR value is calculated using Eq. (5).

$$C_{target} = \frac{\langle x,\tilde{x}\rangle x}{\|x\|^2},$$

$$E_{noise} = \tilde{x} - C_{target},$$

$$SI - SNR = 10\log_{10}\frac{\|C_{target}\|^2}{\|E_{noise}\|^2} \tag{5}$$

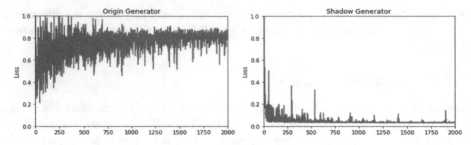

Fig. 4. The change of two generators loss according to epoch.

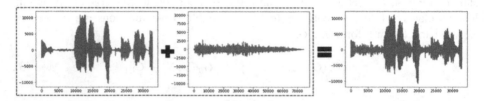

Fig. 5. Process of making noisy signal using noise data and clean signal data.

4.2 Results of Noise Cancellation

The left figures in Figs. 6 and 7 represent the waveforms of the signals in time and frequency domain. Comparing the waveforms of clean signal and noisy signal, it can be confirmed that non-stationary noise is applied to clean signal. The waveform of clean signal and generated signal can be seen to be almost similar. Four comparison groups are used to compare the performance of the proposed model. In the first group, autoencoder is used for learning only the generator without adversarial learning. We conduct the experiment using the traditional low pass filtering method and the model [6] with the second group. The right in Fig. 6 shows the SI-SNR values of each model for the test data in a box plot. Table 1 shows the result of t-test with SI-SNR values.

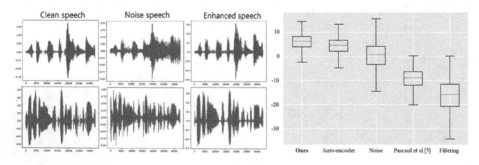

Fig. 6. (Left) Waveform of input and output signals. (Right) Box plot using SI-SNR values of each model.

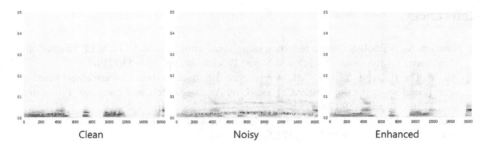

Fig. 7. Spectrogram of signals.

Table 1. Result of t-test on SI-SNR value.

	Ours	Autoencoder	Noisy signal	Pascual et al. [6]	Filtering
Mean	5.971	4.176	1.121	−9.428	−16.982
Variance	11.356	13.174	35.173	11.356	65.659
p-value	–	2.29×10^{-137}	0.	0.	0.

The t-test shows that the experiment result is statistically significant. The experimental results demonstrate that our model outperforms other conventional methods in canceling the non-stationary noise.

5 Conclusions

In this paper, we propose a model that can eliminate more non-stationary noise and unseen noise than other studies. We create a powerful instructor through adversarial learning of the discriminators and the generator. In order to solve the problem that the generator easily diverges during the adversarial learning process, $G_{\theta'}$ with the same structure is provided to facilitate stable learning. $G_{\theta'}$ learns from powerful instructors and generates enhanced signal from a given noisy signal. Several experiments show that the proposed method shows promising results of the best performance with 5.91 in terms of SI-SNR. This means that our model can remove not only non-stationary noise but also various types of noises.

We will add a comparative model to demonstrate the objective performance of the model. Also, several metrics used in the noise reduction domain will be added to evaluate the model. Finally, we will build and release a web demo system to test our model.

Acknowledgment. This work was supported by grant funded by 2019 IT promotion fund (Development of AI based Precision Medicine Emergency System) of the Korean government (Ministry of Science and ICT).

References

1. Tamura, S., Waibel, A.: Noise reduction using connectionist models. In: IEEE International Conference on Acoustics, Speech and Signal Processing, pp. 553–556 (1988)
2. Xu, Y., Du, J., Dai, L.R., Lee, C.H.: A regression approach to speech enhancement based on deep neural networks. In: IEEE/ACM Trans. on Audio, Speech, and Language Processing, vol. 23, no. 1, pp. 7–19 (2014)
3. Kumar, A., Florencio, D.: Speech Enhancement in Multiple-noise Conditions Using Deep Neural Networks. arXiv:1605.02427 (2016)
4. Weninger, F., et al.: Speech enhancement with LSTM recurrent neural networks and its application to noise-robust ASR. In: International Conference on Latent Variable Analysis and Signal Separation, pp. 91–99 (2015)
5. Kim, J.Y., Bu, S.J., Cho, S.B.: Hybrid deep learning based on GAN for classifying BSR noises from invehicle sensors. In: International Conference on Hybrid Artificial Intelligent Systems, pp. 561–572 (2018)
6. Pascual, S., Bonafonte, A., Serra, J.: SEGAN: Speech Enhancement Generative Adversarial Network. arXiv:1703.09452 (2017)
7. Lim, J., Oppenheim, A.: All-pole modeling of degraded speech. In: IEEE Transactions on Acoustics Speech and Signal Processing, vol. 26, no. 3, pp. 197–210 (1978)
8. Berouti, M., Schwartz, R., Makhoul, J.: Enhancement of speech corrupted by acoustic noise. In: International Conference on Acoustics Speech and Signal Processing, vol. 4, pp. 208–211 (1979)
9. Maas, A.L., Le, Q.V., O'Neil, T.M., Vinyals, O., Nguyen, P., Nguyen, A.Y.: Recurrent neural networks for noise reduction in robust ASR. In: InterSpeech, pp. 22–25 (2012)
10. Sun, L., Du, J., Dai, L.R., Lee, C.H.: Multiple-target deep learning for LSTM-RNN based speech enhancement. In: IEEE Hands-free Speech Communications and Microphone Arrays, pp. 136–140 (2017)
11. Parveen, S., Green, P.: Speech enhancement with missing data techniques using recurrent neural networks. In: IEEE International Conference on Acoustics Speech and Signal Processing, pp. 733–736 (2004)
12. Lu, X., Tsao, Y., Matsuda, S., Hori, C.: Speech enhancement based on deep denoising autoencoder. In: InterSpeech, pp. 436–440 (2013)
13. Cohen, I.: Multichannel post-filtering in nonstationary noise environments. In: IEEE Transactions on Signal Processing, vol. 52, no. 5, pp. 1149–1160 (2004)
14. Goodfellow, I., et al.: Generative adversarial nets. In: Neural Information Processing Systems, pp. 2672–2680 (2014)
15. Kim, J.Y., Bu, S.J., Cho, S.B.: Zero-day malware detection using transferred generative adversarial networks based on deep autoencoders. In: Information Sciences, vol. 460–461, pp. 83–102 (2018)
16. Isola, P., Zhu, J.Y., Zhou, T., Efros, A.: Image-to-Image Translation with Conditional Adversarial Networks. arXiv:1611.07004 (2016)
17. He, K., Zhang, X., Ren, S., Sun, J.: Deep residual learning for image recognition. In: IEEE Conference on Computer Vision and Pattern Recognition, pp. 770–778 (2016)
18. Hu, G., Wang, D.L.: A tandem algorithm for pitch estimation and voiced speech segregation. In: IEEE Trans. on Audio Speech and Language Processing, vol. 18, pp. 2067–2079 (2010)
19. Valentini, C.: Noisy speech database for training speech enhancement algorithms and TTS models. In: University of Edinburgh, School of Informatics, Centre for Speech Technology Research (2016)
20. Luo, Y., Mesgarani, N.: TasNet: Surpassing Ideal Time-frequency Masking for Speech Separation. arXiv:1809.07454 (2018)

A Deep Learning-Based Surface Defect Inspection System for Smartphone Glass

Gwang-Myong Go, Seok-Jun Bu, and Sung-Bae Cho[✉]

Department of Computer Science, Yonsei University, Seoul, South Korea
{scooler, sjbuhan, sbcho}@yonsei.ac.kr

Abstract. In recent years, convolutional neural network has become a solution to many image processing problems due to high performance. It is particularly useful for applications in automated optical inspection systems related to industrial applications. This paper proposes a system that combines the defect information, which is meta data, with the defect image by modeling. Our model for classification consists of a separate model for embedding location information in order to utilize the defective locations classified as defective candidates and ensemble with the model for classification to enhance the overall system performance. The proposed system incorporates class activation map for preprocessing and augmentation for image acquisition and classification through optical system, and feedback of classification performance by constructing a system for defect detection. Experiment with real-world dataset shows that the proposed system achieved 97.4% accuracy and through various other experiments, we verified that our system is applicable.

Keywords: Deep learning · Convolutional neural network · Class activation map · Smartphone glass inspection · Defect detection · Augmentation · Image preprocessing

1 Introduction

The surface quality of the smartphone glass is an evaluation factor that interacts with the user, which has a fatal impact on the stability and durability of the final product [1]. Rule-based approaches in the field of smartphone glass surface inspection have been the most common method for solving problems. In this process, there was a problem that the distinction between classes with high similarity was difficult to distinguish by the traditional CV algorithm, and the performance of the whole system was lowered due to the increase of false negative (27.3%, 3/21–4/29). Convolutional neural network (CNN), which is replacing the existing best performance in many fields, is an intuitive idea that can be introduced effectively in the field of industrial manufacturing including surface inspection where a large amount of data can be accumulated [2].

Optical inspection classification systems using deep learning methods have made great achievements in recent decades [3]. In conventional methods, inspection and classification were divided into three separate processes: segmentation, feature extraction and inspection. First, the segmentation process obtains a small patch of suspected defects from a full-size image from line scanner with several traditional

© Springer Nature Switzerland AG 2019
H. Yin et al. (Eds.): IDEAL 2019, LNCS 11871, pp. 375–385, 2019.
https://doi.org/10.1007/978-3-030-33607-3_41

image processing algorithms, such as histogram curves and edge detection [4, 5]. Stark et al. introduced contrast limited adaptive histogram equalization (CLAHE) augmentation method, which normalizes the histogram for each patch to compensate for the traditional flip and shift augmentation for the suspicious patch obtained at the segmentation process [6]. Second, the feature extraction process extracts local and global spatial features from patches to achieve the final classification. G. Fu et al. presented an end-to-end CNN that learns the filters from the data to extract and classify the surface patches [7]. Finally, the inspection process utilizes the constructed model, including localization or model analysis that specifies the defect region from the full-size glass surface. Zhou et al. proposed a class activation map (CAM) visualization method for identifying and analyzing the region used for classification in the input image by using the channels from last convolutional layer of CNN [8]. Although the three processes described above could be considered as a complement of each other, combining them has been troublesome due to the difficulty in implementation and integration. In this paper, we propose a system for smartphone glass surface defect inspection including segmentation, modeling and analysis processes in aspect of integrity and stability. With the combination of augmentation, deep learning-based modeling and the deep model analysis methods, our glass surface inspection system has achieved 0.9754 accuracy for test dataset and verified the statistical significance with 10-fold cross validation. Our system including deep model analysis and augmentation methods shows the results which are statistically significant in chi-square test, t-SNE analysis and CAM analysis, respectively.

The remainder of this paper is organized as follows. In Sect. 2, we review the existing augmentation, modeling and analysis methods based on machine learning and clarify the contributions of this paper by discussing the differences. Section 3 explains how the spatial feature from smartphone glass defect is modeled using CNN. The performance of our system is evaluated in Sect. 4 through various experiments, including visualizations of the decisions the model made, measurements of performance and comparisons with the existing methods.

2   Related Works

In this section, we introduce various works based on deep learning methods for classification with the proposed defect detection system. As the keyword industrial is well known in the field of deep learning in recent research trends, several recent studies have proven to be reliable and valid. The mainstream of glass surface defect inspection field is categorized into segmentation method and modeling method.

Borwankar et al. presented a data-driven approach to the field of surface defect inspection by combining traditional image processing algorithm and machine learning-based classification algorithm [9]. Natarajan et al. proposed the ensemble of multiple CNNs for practical complement in terms of accuracy, and an end-to-end defect classification network combining feature extraction and modeling was constructed [2]. Both studies attempted to benchmark the standard machine learning methodology including CNN in surface defect inspection, resulting in higher performance compared to the existing methods.

Table 1. Related works on defect detection using machine learning algorithm

Authors	Image preprocessing	Spatial feature extraction, modeling
Zhou [8]	–	CNN, Class activation map
Borwankar [9]	Rotated wavelet transform	k nearest neighbor
Natarajan [2]	–	Transfer learning, CNN ensemble
Dong [10]	CNN-based segmentation	U-Net, Random forest
Fu [11]	Non-uniform illumination, Camera noise, Motion blur	Pre-trained SqueezeNet (CNN)
Sun [1]	–	Transfer learning, CNN, ImageNet
Yang [12]	CNN	Convolutional autoencoder
Staar [13]	Flip, Shift augmentation	CNN triplet learning

As an attempt to focus on the aspects of preprocessing and model enhancements including augmentation, Dong et al. and Fu et al. proposed a combination of traditional image augmentation method and deep learning [10, 11]. Experiments showed that augmentation was suitable for deep learning methods that made nonlinear functions from data and verified the need of augmentation component for an inspection system in more practical sense. Finally, focusing on the enhancement of the deep learning model, Sun et al., Yang et al. and Staar et al. used the transfer learning, autoencoder, and triplet learning to obtain higher accuracy for given datasets, respectively.

3 The Proposed System

3.1 Overview

In this section, we propose a standalone system architecture for glass surface inspection. In order to inspect glass of smartphone, a special optical system is constructed to identify defects and to capture the surface. In our system, an optical module of line scan method is constituted to acquire an image. The image obtained through this process is subject to preprocessing. In this preprocessing process, candidates for the defect classification are firstly selected, and the selected images are classified and managed in separate classes through a separate task for learning. The selected image is used as a training/test dataset through the above process and then an assortment process is performed to increase the accuracy of the classification determination. The experiment is conducted to demonstrate the effectiveness of each classification to enhance the performance of the classification. The current system is utilized as a dataset through Flip, Shift and CLAHE.

The images that have been augmented are learned in the model for defect classification in the whole system. In order to utilize the defect meta information, the accuracy is improved by learning the candidate information of the first candidate group.

The proposed system constructs two different learning models for the defect candidate group and the meta data, ensemble them, and finally performs classification. In this process, in order to have high classification accuracy, we proposed an idea to utilize not only the defect image but also the related meta data for learning. After classification is completed, CAM is constructed to decide the characteristics and origin of each feature in latent space, and it is decided whether the classification model is correctly learned. This is used as a feedback to the entire system.

At this time, if the total size of the defect is larger than the size (224 × 224) of the set input, the size of the total defect is adjusted through resizing (112 × 112), and denoising is performed to reduce the number of false negatives in the confusion matrix. In the case of the first class, size, shape, and defect characteristics are used through vision algorithm. The obtained image determines whether the worker has an active defense cognitive class failure and completes the dataset. After that, the normalization module corrects the gaps between different optical systems and equipment, and finally completes the preprocessing for classification. In the early stage of the system configuration, there was no augmentation module. However, the necessity of the augmentation module was justified through the verification and supplementation of the model. We will analyze the results later in the experiment. The model for classification consists of a separate model for embedding location information in order to utilize the defective locations classified as defective candidates and ensemble with the model for classification to enhance the overall system performance.

The CAM of the model constructed based on the classified image is used to decide the consistency of the model configuration. The operator for the classification confirms the result, and finally confirms the feature of the model caused by the failure; when a

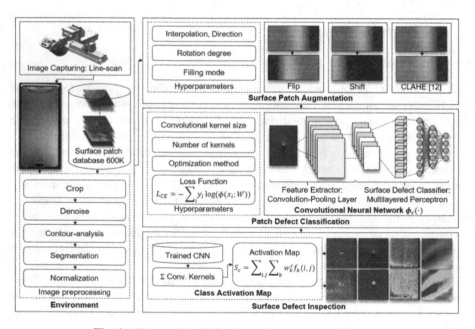

Fig. 1. Glass surface defect inspection system architecture

dataset is created, it is possible to obtain a robust dataset gradually and it is utilized for analysis of the system.

3.2   CNN-Based Modeling

The convolution operation, which preserves the spatial relationships between features by learning filters that extract correlations, is known to reduce the translational variance between features [14]. The hidden correlations between features in images are modeled as a feature-map through emphasis or distortion during the convolution operation:

$$\phi_{c_i}^l = \sum_{k=0}^{f-1} w_i y_{i+k}^{l-1} \tag{1}$$

where the output c_i^l from the ith node of the lth convolutional layer performs the convolution operation on y^{l-1}, which comes from the $(l-1)$ th layer, using a $1 \times f$ sized filter F and a given feature vector $(\overline{x}_i, y_i)$.

Because the dimensionality of the feature-map from the convolutional layer is increased by the number of convolutional filters, it must be carefully controlled [15]. Pooling refers to a dimensionality reduction process used in CNNs in order to impose a capacity bottleneck and facilitate faster computation [16]. The summary statistics of nearby features in the feature-map are derived from the max-pooling operation:

$$\phi_{p_i}^l = \max c_{i \times T}^{l-1} \tag{2}$$

where the output p_i^l from the lth max-pooling layer selects the maximum features from a $k \times k$ area in an $N \times N$ output vector, where T is the pooling stride.

Several convolutional and pooling layers can be stacked to form a deep neural network architecture with more parameters [17]. The proposed inspection system uses two pairs of convolutional and pooling layers to prevent model overfitting or degradation problems.

After the features from the images are encoded through convolution and pooling operations, a shallow multi-layer perceptron (MLP) is used to complete the classification process using the flattened feature vectors $p^l = [p_1, \ldots, p_I]$, where I is the number of units in the last pooling layer:

$$h_i^l = \sum_j \sigma \left(w_{ji}^{l-1} \left(p_i^{l-1} \right) + b_i \right) \tag{3}$$

where w_{ji} is the weight between the ith node from the $(l-1)$ th layer and the jth node from the lth layer, σ is the activation function used in the layer, and b_i is the bias term.

The output vector from the last layer of the MLP represents the probabilities of the roles associated with an input feature vector, where N_R is the total number of roles:

$$p(\hat{y}|\overline{x}_i) = argmax \frac{\exp\left(h^{l-1}w^l + b^l\right)}{\sum_{k=1}^{N_R} \exp(h^{l-1}w_k)} \tag{4}$$

where a softmax activation function is used in the last layer of the MLP so that the output vector is encoded as a probability in the range $[0, 1]$.

The weights of the entire CNN are updated using a backpropagation algorithm based on gradient descent, where $\overline{x_i}$ is mapped to y_i after forward propagation is performed using Eqs. (1)–(4). The details of the CNN architecture are as follows: 32 1×2 convolutional filters and 1×2 max-pooling operations, and 128-64-11 nodes in the MLP, from bottom-to-top, to create a shallow but practical network.

3.3   Data-Preprocessing and Augmentation

Our data set consists of images taken by an optical inspector collected for two weeks in the current manufacturing process. This dataset contains 4 object categories with a total of 600K grayscale images of 224×244 pixels resolution. Each category has at least 21,088 images. For network training, all images are resized to 112×112 pixels resolution and then scaled to values from -1 to 1.

The purpose of surface preprocessing and augmentation module is to derive a candidate defect from images obtained by constructing a line scan camera and an illumination system based on inspection specification for each production process. For the input image obtained through an optical module, resize it so that the defect characteristics are best expressed, and size and normalization are performed to be used as input to the classification model of the proposed system.

Firstly, the decision of the defective candidates is performed through the computer vision algorithm. Through this process, contour analysis is performed on the blobs in the processing image obtained through the calculation process to determine the defective candidates, and the information such as the shape, size, and intensity of the defects are used as a judgment criterion. The process of denoising consists of several modules in order to perform preprocessing on Gaussian noise generated from the optical module or external factors generated in the process. Finally, the results are processed by the inspector to improve the consistency of the generated dataset.

4   Experimental Results

4.1   Inspection System Performance

In Fig. 2, 10-fold cross-validation is performed to verify the classification result of the proposed system, and the performance of the system is verified by using the test data of the constructed dataset. As a result of test by changing the validation set, CNN performance is found to be the best in the case of the basic deep learning model, and accuracy is 10% higher than MLP.

In order to optimize the hyper-parameters of the model used in the proposed system, various experiments are performed on the convolution filter and the depth, and the system tuning is performed on the actual applicable hyper-parameters. Figure 3 shows the results of how the hyper-parameters of our system were experimented and selected.

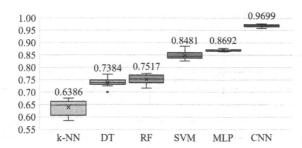

Fig. 2. 10-cross validation result

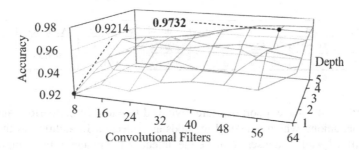

Fig. 3. Hyper-parameter optimization

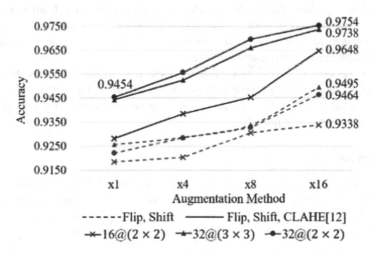

Fig. 4. Augmentation effect for the system

The following experiments are conducted to analyze the accuracy of the system for the effects of augmentation. The results of the experiment are shown in Fig. 4 with two main groups: Flip and Shift, and the database performed with CLAHE, demonstrating the need for assortment in the system. At the same time, the number of channels and the kernel size are changed to 16, 2×2, and 3×3, respectively. In our proposed system,

Table 2. Confusion matrix analysis

		Predicted			
		Dent	Mold	Fold	Control
Actual	Dent	15419	111	61	744
	Mold	1	4916	6	601
	Fold	50	12	3880	234
	Control	260	720	206	87972

Table 3. Precision and recall for defect types

Defect Type	Dent	Mold	Fold	Control	Average	**Accuracy**	F1-score
Precision	0.9802	0.8536	0.9343	0.9824	0.9376	**0.9739**	0.9375
Recall	0.9439	0.8899	0.9291	0.9867	0.9374		

various methods of image assortment are reviewed to enhance the performance of the classification, among which validated methods are selected. In addition to the general flip and shift, CLAHE is chosen to make the defect in the image to be a singular point in the histogram. The performance improvement through augmentation contributed to improve the accuracy of the system as described above.

Tables 1 and 2 show the results obtained through experiments from our system. The model is constructed to perform the classification for four classes. The confusion matrix is obtained to quantitatively measure the performance of each class, and precision and recall are calculated respectively.

Also, the results of Table 3 show that the meta data of the faulty location information used in the system we proposed is used to improve the performance of the system. We select vanilla CNN as a comparative group and find that the ensemble of our model is composed as intended.

4.2 Defect Analysis

Figure 5 shows the dataset of the system by checking the t-SNE (stochastic neighbor embedding) distribution. The characteristics of each class group are schematized through CAM according to the distribution. In our system, the clusters of each class can be confirmed, which is supported by the classification performance.

Table 4. Chi-squared test with vanilla CNNs

Defect type	Dent	Mold	Fold	Control	**Sum**
Chi-square	4848.96	1460.159	3007.727	8280.203	**17597.05**

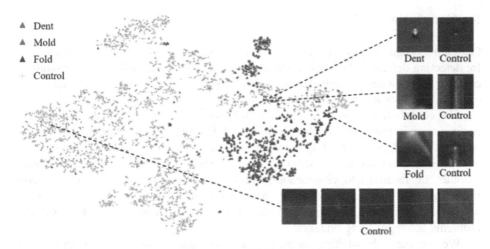

Fig. 5. t-SNE visualization

Figure 6 shows the CAM results for each class. The dent, mold and fold, which are defect classes among the classes in the above diagram, confirm that the learning of the model is well done so that the characteristics of each defect are the result of classification. In the case of the good (OK) class control, the glass surface is activated and shows the conformity of the model.

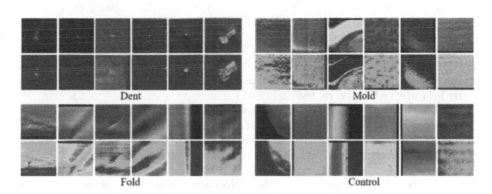

Fig. 6. Class activation map result for each class

As described above, CAM analysis allows the examiner to verify the consistency of the learned features and poor features learned. In our system, CAM images that do not model characteristics well by sampling the output of the model are used as a basis to re-learn the model through analysis or to decide the abnormality of the model.

5 Conclusions

Our entire system proposed for the detection of defects is modeled in a form optimized for detecting defects in the glass of a smartphone. Therefore, the role of each module is clarified so that it can be applied to other surface shapes in the future, and the influence on input and output can be clarified.

The contribution to our research is as follows. First, we show that the proposed performance is practically applicable by using the real-world dataset in the actual process. Second, by constructing the whole system for defect detection, preprocessing and augmentation for image and classification through optical system, and CAM module for feedback to classification are incorporated into a single closed-loop framework for feedback. Finally, we have designed a fusion network to improve detection performance by using meta-data for defect detection, rather than solving problems solely by using images.

We are currently working on an improvement plan to make our systems more capable. More specifically, we are focusing on improvement in the early described ensemble of different meta data that affect the defect detection, and also on optimization of the deep learning model.

Acknowledgements. This research was supported by Samsung Electronics Co., Ltd.

References

1. Sun, J., Wang, P., Luo, Y.K., Li, W.: Surface defects detection based on adaptive multiscale image collection and convolutional neural networks. IEEE Trans. Instrum. Measur. 1–11 (2019)
2. Natarajan, V., Hung, T.Y., Vaikundam, S., Chia, L.T.: Convolutional networks for voting-based anomaly classification in metal surface inspection. In: IEEE International Conference on Industrial Technology, pp. 986–991 (2017)
3. Chen, T., Wang, Y., Xiao, C., Wu, Q.J.: A machine vision apparatus and method for can-end inspection. IEEE Trans. Instrum. Measur. **65**, 2055–2066 (2016)
4. Cao, G., Ruan, S., Peng, Y., Huang, S., Kwok, N.: Large-complex-surface defect detection by hybrid gradient threshold segmentation and image registration. IEEE Access **6**, 36235–36246 (2018)
5. Jian, C., Gao, J., Ao, Y.: Automatic surface defect detection for mobile phone screen glass based on machine vision. Appl. Soft Comput. **52**, 348–358 (2017)
6. Stark, J.A.: Adaptive image contrast enhancement using generalizations of histogram equalization. IEEE Trans. Image Process. **9**, 889–896 (2000)
7. Gupta, E., Kushwah, R.S.: Combination of global and local features using DWT with SVM for CBIR. In: International Conference on Reliability, Infocom Technologies and Optimization, pp. 1–6 (2015)
8. Zhou, B., Khosla, A., Lapedriza, A., Olivia, A., Torralba, A.: Learning deep features for discriminative localization. In: IEEE Conference on Computer Vision and Pattern Recognition, pp. 2921–2929 (2016)

9. Borwankar, R., Ludwig, R.: An optical surface inspection and automatic classification technique using the rotated wavelet transform. IEEE Trans. Instrum. Measur. **67**, 690–697 (2018)
10. Dong, X., Taylor, C.J., Cootes, T.F.: Small defect detection using convolutional neural network features and random forests. In: European Conference on Computer Vision, pp. 1–15 (2018)
11. Fu, G., et al.: A deep-learning-based approach for fast and robust steel surface defects classification. Opt. Lasers Eng. **121**, 397–405 (2019)
12. Yang, H., Chen, Y., Song, K., Yin, Z.: Multiscale feature-clustering-based fully convolutional autoencoder for fast accurate visual inspection of texture surface defects. IEEE Trans. Autom. Sci. Eng. **99**, 1–18 (2019)
13. Staar, B., Lutjen, M., Freitag, M.: Anomaly detection with convolutional neural networks for industrial surface inspection. Procedia CIRP. **79**, 484–489 (2019)
14. Sainath, T., Parada, C.: Convolutional neural networks for small-footprint keyword spotting. In: InterSpeech, pp. 1478–1482 (2015)
15. Donahue, J., et al.: Long-term recurrent convolutional networks for visual recognition and description. In: IEEE Conference on Computer Vision and Pattern Recognition, pp. 2625–2634 (2015)
16. Rippel, O., Snoek, J., Adams, R.P.: Spectral representations for convolutional neural networks. In: Advances in Neural Information Processing Systems, pp. 2449–2457 (2015)
17. Krizhevsky, A., Sutskever, I., Hinton, G.E.: Imagenet classification with deep convolutional neural networks. In: Advances in Neural Information Processing Systems, pp. 1097–1105 (2012)

Superlinear Speedup of Parallel Population-Based Metaheuristics: A Microservices and Container Virtualization Approach

Hatem Khalloof[(✉)], Phil Ostheimer, Wilfried Jakob, Shadi Shahoud, Clemens Duepmeier, and Veit Hagenmeyer

Institute of Automation and Applied Informatics (IAI),
Karlsruhe Institute of Technology (KIT), Karlsruhe, Germany
{hatem.khalloof,wilfried.jakob,shadi.shahoud,clemens.duepmeier,
veit.hagenmeyer}@kit.edu, phil.sidney@gmx.de

Abstract. Population-based metaheuristics such as Evolutionary Algorithms (EAs) can require massive computational power for solving complex and large scale optimization problems. Hence, the parallel execution of EAs attracted the attention of researchers as a feasible solution in order to reduce the computation time. Several distributed frameworks and approaches utilizing different hardware and software technologies have been introduced in the literatures. Among them, the parallelization of EAs in cluster and cloud environments exploiting modern parallel computing techniques seems to be a promising approach. In the present paper, the parallel performance in terms of speedup using microservices, container virtualization and the publish/subscribe messaging paradigm to parallelize EAs based on the Coarse-Grained Model (so-called Island Model) is introduced. Four different communication topologies with scalable number of islands ranges between 1 and 120 are analyzed in order to show that a partial linear/superlinear speedup is achievable for the proposed approach.

Keywords: Parallel EAs · Speedup · Superlinear · Coarse-Grained Model · Microservices · Container · Cluster · Parallel computing · Scalability

1 Introduction

In general, EAs are inspired by the principles of biological evolution [23]. A population of individuals encoding temporal solutions is iteratively altered to generate new solutions. The individuals are assessed by an evaluation function to determine their fitness (suitability as a solution). The evaluation of fitness function is independent of the rest of the population, allowing parallelization for this step [23]. Therefore, different parallelization models have been introduced

© Springer Nature Switzerland AG 2019
H. Yin et al. (Eds.): IDEAL 2019, LNCS 11871, pp. 386–393, 2019.
https://doi.org/10.1007/978-3-030-33607-3_42

to tackle this issue. Cantú-Paz [12] presented and investigated two different approaches to parallelize EAs, namely the parallelizing of the evaluation step which results in the first parallelization model i.e. the so-called Global Model and the structuring of the population to apply the genetic operators in parallel which introduces the second and third parallelization models i.e. the so-called Fine-Grained Model and the so-called Coarse-Grained Model (Island Model).

Over the years, several frameworks and approaches for the execution of parallel EAs according to the above three parallelization models have been introduced. Most of these frameworks, e.g. [3,7,11,14,16], were developed to simplify the implementation of new parallel EAs providing reasonable modularity, reusability and flexibility on the code level. These beneficial traits are usually limited to the code level and do not apply to the algorithm level. For example, they were monolithically developed and do not introduce simple and effective mechanisms to integrate and deploy existing parallel EAs on a cluster or a cloud. Furthermore, they are restricted to one technology stack and do not provide a practical and flexible method to interact with external tools, components and domain-based simulators. In order to overcome the mentioned drawbacks, modern software technologies such as microservice architecture, container virtualization technologies and the publish/subscribe messaging paradigm are utilized to develop distributed, modular, scalable and generic frameworks, e.g. [17–19,22], for parallelizing population-based metaheuristics. An application featuring the microservices architecture is splitted into several independent services, where each service performs a special task [21] and uses its own technology stack. To build parallel and scalable solutions, this is a valuable feature, since each service can be scaled horizontally and use the most suitable technologies. Data is exchanged between the services by calling REST-APIs and using the publish/subscribe messaging paradigm to reduce the coupling. Container technology is used to encapsulate each service to maintain runtime automation and platform independence. These advantages increase the modularity, usability, flexibility, scalability and maintainability of the system. In addition, the interaction with other tools and frameworks can be easily achieved. In [18,19], a microservice and container-based solution to distribute the execution of existing EAs on a cluster based on the Global Model [19] and on the Coarse-Grained Model (Island Model) [18] was developed and evaluated. However, the calculation of speedup of the overall framework presented in [18] has not been introduced and analyzed.

For the Coarse-Grained Model, the works [1,2,4,5,8,20] have theoretically and practically provided a sufficient evidence that the linear and even superlinear speedup is possible. However, they concluded their results based on a monolithic software architecture using conventional distributed technologies such as RPC (Remote Procedure Call) that do not provide simple and flexible mechanisms to exploit the potential of the modern powerful hardware e.g. cloud and cluster (cf. [18,19] for a detailed discussion). All the previous works focus on one communication topology, namely uni-directional ring [1,2,4,5], bi-directional ring [8] or a grid [20]. No results have been concluded with a realistic number of islands which is applicable in a variety of real applications. For example, [1,2,4,5,8] studied a small number of islands ranging between 8 and 32 and [20] studied a large number of islands, i.e. 16384.

Therefore, the present paper extends [1,2,4,5,8,20] in the sense that for the parallel performance, the architecture based on the new technologies introduced in [18] is used and four communication topologies, namely uni-directional ring (Uni-Ring), bi-directional ring (Bi-Ring), ladder (Ladder) and fully connected (Complete) are considered where the number of islands ranges between 1 and 120 (cf. [13] for more description about the mentioned four topologies). The migration among the islands is synchronously occurring and all islands have the same configurations, i.e. they are homogeneous. The performed evaluation is carried out independently of the solution's quality of a concrete optimization problem. We simulate different run-times by applying several delays up to 32 ms per individual to cover several optimization tasks with a more complex fitness evaluation function.

The remainder of the present paper is structured as follows: Sect. 2 shortly explains how modern technologies are used to distribute, execute and deploy EAs on a cluster. In Sect. 3, the parallel performance of using microservices, container virtualization and the publish/subscribe messaging paradigm to parallelize EAs will be analyzed regarding the achievable speedup rates. Section 4 concludes with a summary and future works.

2 Microservices and Container-Based Approach for Distributing EAs

In [18], a framework for the parallelization of EAs using microservices, container technologies and the publish/subscribe messaging paradigm was introduced. The Island Model is integrated and evaluated in terms of Framework Migration Overhead (FMO). The approach is strictly based on a microservice architecture in which each service is developed, deployed and scaled independently. The architecture of the framework is layered into two main tiers, namely the User-Interface Tier and the Cluster Tier. The User-Interface Tier is responsible for the user interaction in the frontend. It delegates data and information to the Cluster Tier (backend) via a set of REST-based service APIs. The Cluster Tier runs the services of the parallelized EAs. It is separated into two layers, namely the Data Layer where the intermediate data and final results are stored. The other layer, namely the Container Layer, contains the eight decoupled microservices that execute the main functionalities of the Island Model where each microservice is deployed and executed in one or more containers (cf. [18,19] for a detailed discussion). Data is exchanged between the services by calling REST-APIs and using the publish/subscribe messaging paradigm to reduce the coupling. The publish/subscribe messaging paradigm is suitable for facilitating the implementation of the communication topology. This can be achieved by assigning a channel for each island. E.g., to create a Uni-Ring topology, the island with ID i publishes the migrants to its channel i and the island with ID $i + 1$ subscribes to this channel.

3   Experimental Setup and Results

Any EA can be executed in the proposed approach, since it is open for importing any EA software implementation. For the evaluation, the EA GLEAM (General Learning Evolutionary Algorithm and Method) [9,10] is chosen as a test case to solve the non-convex and non-linear multimodal Rastrigin function [15]. GLEAM offers the functionality to configure a scalable delay so that the amount of time for the fitness evaluation can be artificially extended. In order to deploy the mentioned services on a four nodes cluster with 128 cores each and a total of 128 GB of RAM, the container orchestration system Kubernetes[1] is used. Kubernetes provides mechanisms to deploy, maintain, manage and scale containerized services. The nodes connected to each other by a LAN with 10 Gbit/s. The Redis in-memory database[2] is utilized to provide the publish/subscribe messaging paradigm and acts as a temporary storage (for more details, see [18]).

The speedup (s_m) is defined as the ratio between the sequential and parallel execution time. In the sequential case, where the algorithm is executed on a single processor, the CPU time to solve the problem will be the base quantity for calculating the computation performance. In the parallel case, the measurement of the parallel runtime should include the parallelism-related overhead. Consequently, the measured time for executing the parallelized algorithm is defined as the difference between starting and finishing the whole process of solving one optimization problem. We denote T_m as the execution time of an algorithm on m processors and T_1 as the execution time of an algorithm on one processor [6]. Due to the non-deterministic nature of EAs, this metric can not be used directly. For non-deterministic algorithms, we have to compute the average for several runs $E[T_1]$ and $E[T_m]$ for both measures:

$$s_m = \frac{E[T_1]}{E[T_m]} \tag{1}$$

Alba [4] introduced a taxonomy depending on the definition of these values. Strong speedup (type I) refers to the speedup achieved compared to the best-so-far sequential algorithm. However, the weak speedup (type II) refers to a comparison between the sequential version and the parallel version of the same algorithm. The two subcategories of weak speedup differ by their respective stopping criterion: For speedup with solution-stop (type II.A), a target solution quality is given. The second subcategory sets a predefined effort as the stopping criterion (type II.B). This is especially useful if the performance of the given EA w.r.t. the quality of solution should not be evaluated, but rather the environment it is executed in. In the present evaluation, we consider (type II.B) speedup as the most suitable measure, since our performance measurement can be taken independently of a concrete optimization problem due to the scalable delay functionality by GLEAM, i.e. the quality of the proposed solution is not considered.

---

[1] kubernetes.io.
[2] redis.io.

In order to measure the speedup of the parallel approach of EAs, the scalable delay feature of the EA GLEAM for the fitness evaluation is varied between 0 ms (just executing the optimization for the Rastrigin function) and 32 ms in order to simulate several optimization tasks with a more complex fitness evaluation function.

The global population size is set to 1024 chromosomes and the migration rate to 4 migrants while the number of islands is varied between 8 and 120. The number of epochs (the interval between two succeeding migrations [13]) is set to 100 with 1 generation per epoch as a termination criterion. For parallel execution, the time between the reception of the configuration from the user and the reception of the final result from all islands is calculated. To measure the execution time of a sequential implementation, the EA GLEAM on one island with an epoch limit of 100 with 1 generation per epoch and the whole population consisting of 1024 individuals is used, resulting in the same effort as executed in parallel. Then, the speedup runs is averaged over 10 runs in order to minimize the possible effect of the delays caused by the OS and the non-deterministic behavior of EAs.

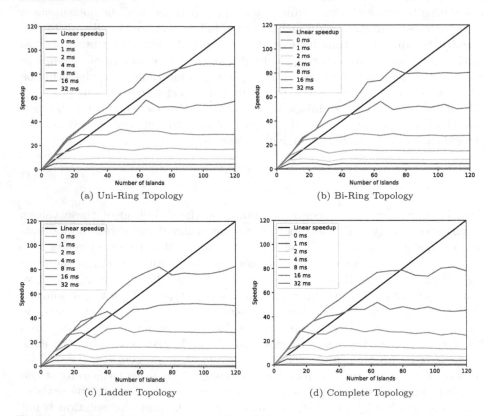

(a) Uni-Ring Topology

(b) Bi-Ring Topology

(c) Ladder Topology

(d) Complete Topology

Fig. 1. Speedup with 4 migrants for four different communication topologies, delay varied between 0 and 32 ms

The results of the speedup measurement for the Uni-Ring, Bi-Ring, Ladder and Complete graph topologies are shown in Fig. 1a, b, c and d respectively. It can be concluded that with an increased delay, the speedup shows a partial linear/superlinear trend for all four topologies, since the computation time of the sequential EA linearly increases from 116 s for executing a fitness function with delay of 0 ms to 15186 s for executing a fitness function with delay of 32 ms. However, the computation time of the parallel implementation increases more slowly, due to the distribution of computation over the islands. Moreover, the migration overhead is independent of the scalable delay therefore, the ratio between the overhead and the parallel computation time decreases. This results in a high speedup for greater delays. In [4], three reasons for the partial super-linear speedup have been introduced. In the context of our measurement of the speedup, the first reason, i.e. the implementation, is the source for the obtained superlinear speedup, since the subpopulations and the computed data structures are smaller compared to the sequential EA GLEAM. Furthermore, the EA GLEAM does not perform a crossover by default, if the parents are too similar, which saves evaluations and therefore the computing time. This effect occurs earlier in small subpopulations than in the larger overall population. Table 1 shows clearly that implementation is the source for the obtained superlinear speedup, since the number of evaluations executed by the EA GLEAM with scalable delay set to 0 ms dramatically decreases with an increasing number of islands.

Table 1. Number of evaluations for each topology with several number of islands to optimize the Rastrigin function with GLEAM scalable delay set to 0 ms

	Uni-Ring	Bi-Ring	Ladder	Complete
1 Island	4833	4833	4833	4833
8 Islands	317	281	285	291
16 Islands	137	142	145	147
24 Islands	102	94	98	98
32 Islands	76	75	70	72
40 Islands	71	70	61	60
48 Islands	58	54	47	48
56 Islands	53	48	44	43
64 Islands	54	43	40	34
72 Islands	41	41	34	30
80 Islands	41	39	31	29
88 Islands	39	32	33	28
96 Islands	45	35	26	24
104 Islands	34	28	30	25
112 Islands	36	25	26	22
120 Islands	34	26	24	23

However, the numerical behavior and the underlying physical hardware are not considerable as reasons for the measured superlinear speedup, since the speedup is calculated according to predefined effort (type II.B). Moreover, the sequential and the Coarse-Grained EA use the same deployed EA and the same underlying hardware. For all considered delays, the speedup starts to decrease at a certain point, since at some points the communication overhead of the increased number of islands outweighs the increased performance of the parallelization. The speedup starts to show a constant trend and then remains constant. Additionally, the FMO for four migrants for the four considered topologies is almost equal as stated in [18]. Hence for the speedup, the respective four evaluated topologies with the migration rate of four migrants show almost the same trend. Regarding the above works mentioned in Sect. 1, it was reported a superlinear speedup for homogeneous, synchronized [1,2,4,5,8,20] and asynchronized [1,2,5,20] island model of EAs according to (type II.A.1) [1,2,4,5], (type II.A.2) [4] and (type II.B) [8].

4   Conclusion and Future Works

In the present paper, the parallel performance in terms of the speedup of using microservices, container virtualization and the publish/subscribe message paradigm to parallelize EAs based on the Coarse-Grained Model is introduced. Four communication topologies, namely Uni-Ring, Bi-Ring, Ladder and Complete are compared. The concluded results show that using modern technologies to execute parallel EAs on a cluster exploiting their enormous computational capabilities is a promising approach not only to increase modularity, usability and maintainability of the system, but also to reach a satisfactory speedup. As future works, the speedup will be measured concerning heterogeneous island configurations and an asynchronized Island Model. Moreover, the quality of the solutions for real-life optimization problems will be considered.

References

1. Abdelhafez, A., Alba, E.: Speed-up of synchronous and asynchronous distributed genetic algorithms: a first common approach on multiprocessors. In: 2017 IEEE Congress on Evolutionary Computation (CEC), pp. 2677–2682. IEEE (2017)
2. Abdelhafez, A., Alba, E., Luque, G.: Performance analysis of synchronous and asynchronous distributed genetic algorithms on multiprocessors. Swarm Evol. Comput. 49, 147–157 (2019)
3. Alba, E., et al.: Efficient parallel LAN/WAN algorithms for optimization. The mallba project. Parallel Comput. 32(5–6), 415–440 (2006)
4. Alba, E.: Parallel evolutionary algorithms can achieve super-linear performance. Inf. Process. Lett. 82(1), 7–13 (2002)
5. Alba, E., Troya, J.M.: Analyzing synchronous and asynchronous parallel distributed genetic algorithms. Future Gener. Comput. Syst. 17(4), 451–465 (2001)
6. Amdahl, G.M.: Validity of the single processor approach to achieving large scale computing capabilities. In: Proceedings of the 18–20 April 1967, Spring Joint Computer Conference, pp. 483–485. ACM (1967)

7. Arenas, M.G., et al.: A framework for distributed evolutionary algorithms. In: Guervós, J.J.M., Adamidis, P., Beyer, H.-G., Schwefel, H.-P., Fernández-Villacañas, J.-L. (eds.) PPSN 2002. LNCS, vol. 2439, pp. 665–675. Springer, Heidelberg (2002). https://doi.org/10.1007/3-540-45712-7_64

8. Belding, T.C.: The distributed genetic algorithm revisited. In: Proceedings of the Sixth International Conference on Genetic Algorithms, pp. 114–121 (1995)

9. Blume, C., Jakob, W.: GLEAM - an evolutionary algorithm for planning and control based on evolution strategy. In: GECCO Late Breaking Papers, pp. 31–38 (2002)

10. Blume, C., Jakob, W.: GLEAM - General Learning Evolutionary Algorithm and Method: Ein evolutionärer Algorithmus und seine Anwendungen, vol. 32. KIT Scientific Publishing (2009)

11. Cahon, S., Melab, N., Talbi, E.G.: ParadisEO: a framework for the reusable design of parallel and distributed metaheuristics. J. Heuristics **10**(3), 357–380 (2004)

12. Cantú-Paz, E.: A survey of parallel genetic algorithms. Calculateurs paralleles, reseaux et systems repartis **10**(2), 141–171 (1998)

13. Cantú-Paz, E.: Topologies, migration rates, and multi-population parallel genetic algorithms. In: Proceedings of the 1st Annual Conference on Genetic and Evolutionary Computation, pp. 91–98 (1999)

14. Fogarty, T.C., Huang, R.: Implementing the genetic algorithm on transputer based parallel processing systems. In: Schwefel, H.-P., Männer, R. (eds.) PPSN 1990. LNCS, vol. 496, pp. 145–149. Springer, Heidelberg (1991). https://doi.org/10.1007/BFb0029745

15. Fogel, G.B., Corne, D.W.: Evolutionary Computation in Bioinformatics. Elsevier (2002)

16. Fortin, F.A., Rainville, F.M.D., Gardner, M.A., Parizeau, M., Gagné, C.: DEAP: evolutionary algorithms made easy. J. Mach. Learn. Res. **13**, 2171–2175 (2012)

17. García-Valdez, M., Merelo, J.J.: evospace-js: asynchronous pool-based execution of heterogeneous metaheuristics. In: Proceedings of the Genetic and Evolutionary Computation Conference Companion, pp. 1202–1208 (2017)

18. Khalloof, H., Ostheimer, P., Jakob, W., Shahoud, S., Duepmeier, C., Hagenmeyer, V.: A distributed modular scalable and generic framework for parallelizing population-based metaheuristics. In: Parallel Processing and Applied Mathematics (PPAM). Springer (2019, accepted)

19. Khalloof, H., et al.: A generic distributed microservices and container based framework for metaheuristic optimization. In: Proceedings of the Genetic and Evolutionary Computation Conference Companion, pp. 1363–1370. ACM (2018)

20. Liu, Y.Y., Wang, S.: A scalable parallel genetic algorithm for the generalized assignment problem. Parallel Comput. **46**, 98–119 (2015)

21. Newman, S.: Building Microservices: Designing Fine-Grained Systems. O'Reilly Media, Inc. (2015)

22. Salza, P., Ferrucci, F.: An Approach for Parallel Genetic Algorithms in the Cloud using Software Containers. Computing Research Repository (CoRR), pp. 1–7 (2016)

23. Whitley, D.: A genetic algorithm tutorial. Stat. Comput. **4**(2), 65–85 (1994)

Active Dataset Generation
for Meta-learning System Quality
Improvement

Alexey Zabashta$^{(\boxtimes)}$ and Andrey Filchenkov

Machine Learning Lab, ITMO University, 49 Kronverksky Pr.,
St. Petersburg 197101, Russia
{azabashta,afilchenkov}@itmo.ru

Abstract. Meta-learning use meta-features to formally describe datasets and find possible dependencies of algorithm performance from them. But there is not enough of various datasets to fill a meta-feature space with acceptable density for future algorithm performance prediction. To solve this problem we can use active learning. But it is required ability to generate nontrivial datasets that can help to improve the quality of the meta-learning system. In this paper we experimentally compare several such approaches based on maximize diversity and Bayesian optimization.

Keywords: Machine learning · Meta-learning · Active learning · Evolutionary Computation · Optimization

1 Introduction

In many areas where researchers are forced to use various heuristic algorithms to solve a certain task, no single optimal algorithm exist for all possible instances of this task due to No Free Lunch Theorems [18]. Classification that is a good example of such task: Labelled datasets are instances of this tasks and it is proved that no instance-agnostic universal classification exist [17]. On contrary, classifiers have algorithmic biases.

Selection of a proper algorithm to solve a data analysis task is always a challenging due to a number of available options. One possible solution to this problem is presented by meta-learning [7]. It is based on the idea that we can extract some prior knowledge about a task and find possible dependencies of algorithm performance on this knowledge. Meta-learning is also used to predict algorithm performance time [3] or select starting points of hyperparameter optimization process [5].

Whatever should be predicted by a meta-learning algorithm, the training process remains the same. Prior knowledge about a task is represented with meta-features of task instances. A set of such instances is collected and their labels are computed.

© Springer Nature Switzerland AG 2019
H. Yin et al. (Eds.): IDEAL 2019, LNCS 11871, pp. 394–401, 2019.
https://doi.org/10.1007/978-3-030-33607-3_43

The first challenge in learning such meta-learning classifiers is data collecting. A dataset should consists of tasks instances, that are labelled dataset in case of classification task. The largest open collection of datasets, OpenML[1] [15] contains 2824 verified datasets. Gathering new datasets is a long process.

A straightforward solution for eliminating this bottleneck is usage of synthetic datasets. Generating synthetic data is a common approach in machine learning [2,11]. Several approaches were also proposed for generating datasets of datasets [19].

However, the dataset synthesis approach meets a bottleneck of high computational costs of labeling such datasets. Whatever should be predicted, label computation involves multiple run of algorithms of interest on each of the datasets, which means that each classifier should be trained on all splits of the datasets. If we begin to synthesize new datasets in uncontrolled manner, their characteristics will be biased towards the generator. Therefore, we need to use methods that synthesize a dataset with specific characteristics.

Active learning [14] is a special case of machine learning in which algorithms work with partially labeled objects. They can receive labels from the external environment for unlabeled objects but the number of such queries is restricted. Therefore, they need to try to select new data for markup so as to maximize information about the data model.

In this paper we use a dataset synthesis method combined with active learning approach in order to create a fast method for training a meta-learning algorithm.

The contribution of this papers includes: improving a known algorithm of classification dataset synthesis and proposing a new active learning based algorithm for training a meta-learning algorithm that guides dataset synthesis.

The rest of the paper is organized as follows. In Sect. 2, we describe algorithms for generating dataset given its meta-feature description. In Sect. 3, we describe strategies for active learning dataset generation. In Sect. 4, we experimentally compare different dataset generation methods and active learning strategies.

2 Dataset Generation

A regular approach for solving the dataset generation problem is its reduction to a minimization problem, in which the objective function is a distance between target and resulting datasets in a meta-feature space. In this Section, we describe approaches for dataset synthesis we consider in our research, including a newer better version of an existing approach.

2.1 DIRECT Method

The **DIRECT** method is based on the idea of direct mapping of a dataset to a vector of numbers and vice versa [13]. The original approach is capable of generating only fixed-size data sets. But we modified it by adding several dimensions

---

[1] https://www.openml.org.

to a numeric search space that determine the number of classes, features and the number of objects of each class. To do it, we calculate maximum number of attributes, classes and objects per class among all used datasets collection. As a result, most processed vectors in search space contain excess information, which is discarded when it is converted to dataset.

2.2   GMM Method

The **GMM** approach [10], is similar to the previous approach, but it does not store the objects of each class. Instead it stores the covariance matrix and the shift vector for the multidimensional Gaussian distribution from which the objects of this class are then sampled. We also modify this approach in the same way as the previous one.

2.3   NDSE Method

The NDSE method (**N**atural **D**ata**S**ets **E**volution) is a modification of similar method from our previous research [19]. Initially, this method could generate only binary classification dataset, but we have generalized it to any number of classes for this paper.

This method is also based on evolutionary algorithms. Its advantage is that it modifies datasets in a more natural way with respect to machine learning operations, such as adding and removing attributes or objects. This method can work with datasets of arbitrary number of attributes and objects. As a result, it can generate datasets without knowing required dimensionality in advice and can use existing datasets as initial population for evolutionary algorithms.

Before describing Mutation and Crossover operators for evolutionary algorithms, first we need to describe basic dataset modifications for changing the numbers of objects, attributes and classes.

Mutation. To perform a mutation, we randomly select one of the following dataset modifications (changing the number of objects, attributes, classes):

- To decrease the number of objects from n to k, we randomly select combination $\binom{n}{k}$ and leave only selected objects in the dataset.
- To increase the number of objects in dataset D, we randomly select the class c, create a new object s with class c and empty attributes, select a random object t with class c and copy from t to s values of random subset of attributes which is still empty in s.
- To decrease the number of attributes, we perform the similar operation as for objects.
- To add a new attribute, we generate random rational function $f : D \to \mathbb{R}$, and apply it to each object in dataset D and use its outputs as values of the corresponding attributes.
- To change the number of classes to k, we generate random categorical function $f : D \to \mathbb{Z}_k$, and apply it to each object in dataset D and use its outputs as values of the corresponding labels.

Crossover. To perform a crossover of two ancestor datasets A_1 and A_2:

1. Randomly select a new number of classes from segment $[\ \#classes(A_1);$ $\#classes(A_2)\ ]$ and apply the modification changing the number of classes to A_1 and A_2 to equalize the number of classes in both datasets. After that rearrange classes in A_1 and A_2 by sorting them by class balance to minimize difference between number of objects in sub datasets of A_1 and A_2 of the same class.
2. For each class c randomly select a new number of objects of this class from segment $[\ \#(A_1|c);\ \#(A_2|c)\ ]$, where $\#(D|c)$ is the number of objects in D with class c. After it assign objects with the same class but different dataset in random pairs.
3. Join assigned pairs from A_1 and A_2 in one big dataset B.
4. Split B to B_1 and B_2 by a random subset of features of B. B_1 and B_2 are the result of crossover of A_1 and A_2.

3 Strategies for Dataset Generation

In this section, we describe strategies we use to generate new synthetic datasets in order to increase performance of the meta-learning system during its training.

3.1 Random Synthesis

The simplest possible strategy for generating new datasets is generating them randomly and independently. We will refer to this approach as **RAND**.

3.2 Diversity Maximization

The **DIV** [1] approach generates new datasets in an attempt to increase the diversity of data in the meta-feature space. Its idea is to obtain a set of datasets distributed uniformly in the meta-feature space. The diversity DIV of datasets collection D is $DIV(D) = \sum_{d_x \in D} \min_{d_y \in D \setminus \{d_x\}} \rho(\mu(d_x), \mu(d_y))/\#D$, where ρ is a distance function in the meta-features space and μ is a mapping from dataset to meta-feature space.

So for this method, the generation strategy is to synthesize each new dataset d_{new} maximizing the distance to the nearest datasets in the current datasets collection D. Formally, it is trying to maximize $\min_{d_x \in D} \rho(\mu(d_x), \mu(d_{new}))$. After d_{new} is synthesized, it is added to the collection D and the process starts again.

So that this algorithm did not generate datasets outside the meta-feature space, we modified the maximizing function as $\min_{d_x \in D} \rho(\mu(d_x), \mu(d_{new})) - \rho(0, \lambda(d_{new}))$, where λ is the out-of-bound penalty $\lambda_i(d_{new}) = \max(0, \mu_i(d_{new}) - \max_{d_x \in D} \mu_i(d_x), \min_{d_x \in D} \mu_i(d_x) - \mu_i(d_{new}))$.

3.3   Maximizing Uncertainty

In a classical active learning setting, a fixed set of object is given, labels of which an algorithm can query. Our situation is different because of an ability to generate (and thus query) any point in a meta-feature space. Interestingly, this situation is very similar to the hyperparameter optimization setting.

This generation strategy **VAR** is based on Bayesian optimization [8]. But since we have direct access to the meta-system we can try to generate datasets in areas where such system most unsure.

We can try to generate datasets in meta-feature space area where our system responds most unsure. We based our meta-classification system on random forest classifier that consists of 30 decision trees (REPTree [12]) that are trained independently. So we can analyze output of several estimators and use variance of it as uncertainty measure.

4   Experiments

Meta-Feature Space. In this paper, we used 10 basic meta-features: average covariance between all pairs of attributes, kurtosis and skewness of attributes, minimum, maximum and average of conditional variance and inner and outer class distance. Also we used 19 meta-features that are determined by the structure of the decision tree, which was built on current dataset. So there is 29 meta-features in total. We took the implementation of these meta-features from [6].

4.1   Dataset Generation

We experimentally compared the described classifications dataset generation approaches: **DIRECT**, **GMM** and **NDSE** on the task of generating datasets with specific meta-features. For this experiment, we used 371 datasets from OpenML [16]. We performed cross-validation over this collection of datasets. For each dataset in a testing set, we tried to generate a dataset with the similar characteristics. We used cross-validation because some approaches can use datasets from training part as initial population for evolutionary algorithms.

To evaluate, how well we generate a new dataset, we used Mahalanobis distance [9] between target and obtained dataset in the meta-feature space. We estimated a covariance matrix for it using the collection of real-world datasets mentioned above.

We use 10 evolutionary minimization algorithms from jMetal library [4].

For each target dataset we used each evolutionary algorithm with each approach twice: without and with real-word datasets as initial population. Each try was limited with 3000 error function evaluations.

Results of Dataset Generation. The results of the datasets generation experiments are shown in Table 1. As you can see, if an algorithm starts with random datasets as initial population, it is better to use **NDSE** approach while **DIRECT** is better from scenario with real-world datasets but **NDSE** is also pretty good in it.

Table 1. The result of datasets generation by different approaches with different evolutionary algorithms. The average Mahalanobis distance between target and obtained datasets in the meta-feature space. The first and the second wide columns represent tries without and with usage real-world datasets respectively.

	DIRECT	GMM	NDSE	DIRECT	GMM	NDSE
CMAES	—	4.3158	—	—	4.4002	—
DE	4.0883	3.7022	—	3.0721	3.7407	—
GDE3	3.8260	3.4788	—	3.3606	3.6214	—
MOCell	3.5173	3.2745	3.6275	**1.9398**	3.5640	2.9385
NSGAII	3.6567	3.2604	2.8400	2.1929	3.5337	2.1634
RS	4.2800	3.9901	3.9514	3.1065	4.2174	3.1673
SMSEMOA	3.7648	3.3472	2.9062	2.3113	3.6596	2.1772
SPEA2	3.7569	3.2677	**2.8160**	2.2488	3.5773	2.1107
SPSO11	3.3951	3.1893	—	1.9530	3.1141	—
gGA	3.8322	3.4008	3.0211	2.6363	3.7488	2.5554

4.2 Active Learning

In this experiment, we tested classifications dataset generation approaches in active learning scenario. According to previous experiment, we selected the best evolutionary algorithms for each generation approach: **MOCell** for **DIRECT**, **SPSO11** for **GMM** and **SPEA2** for **NDSE**. We tested these approaches with the described dataset generation strategies, namely **RAND**, **DIV** and **VAR**.

At each test, we trained a meta-learning system that predicts F-score of kNN classifier for dataset after 10 times repeated 10-folds cross-validation. This task in known to be too complicated. Typically, a learning to rank or classification tasks are used on meta-level.

For each test, we split real-word datasets collection on training and validation parts with ratio 1:2. Then we generated 300 new datasets with current tested generation strategy. For each new dataset generation step, we used training part as initial population for evolutionary algorithms (except for the **RAND** approach, which does not use the concept of initial population). After generating a dataset, we added it to the training part, updated meta-system and test it on validation part. Each test was repeated 10 times and for each new generated dataset evolutionary algorithms were limited by 5000 error function evaluations.

Result. The result of the active learning experiment are shown in Table 2. For each pair of generation approach and strategy is presented minimum average RMSE of meta-system over all generation steps.

We can see from the results that **DIRECT** and **GMM** algorithms shows almost no difference with all active strategies. The only relative improvement is achieved with **VAR** approach that helps to rank the dataset generation methods. As expected, **RAND** is the worst while **NDSE** is the best.

Table 2. The result of the active learning experiment: minimum average RMSE of meta-system over all generation steps.

	DIRECT	GMM	NDSE
RAND	0.1751	0.1790	0.1765
DIV	0.1723	0.1798	0.1633
VAR	0.1777	0.1799	**0.1597**

5   Conclusion

In this paper, we considered 3 classification dataset generation methods and 3 strategies for active dataset generation for meta-learning system quality improvement.

The dataset generation approaches have been modified to generate datasets with specific characteristics without prior knowledge of the number of attributes, objects and classes in the required dataset. We experimentally compared them in the task of generating existing real-world datasets. It was shown that to generate datasets without a datasets collection, it is better to use NDSE approach, while DIRECT approach is better when used with a datasets collection.

We used these dataset generation approaches to build an active learning system for the improvement of meta-system, which predicts classification quality. This active learning system is based on one baseline strategy which randomly oversampled meta-system training set and two strategies that tried to maximize information about meta-feature space. Experiments shown that random oversampling can improve the quality of the meta-system only at the initial interval, but then it decrease quality of it. The remaining two methods were able to improve the quality of the system.

Thus, in this paper, it was shown that by applying dataset generation methods we can use active learning strategies to improve quality of meta-learning system.

Acknowledgments. The work on the dataset generation was supported by the Russian Science Foundation (Grant 17-71-30029). The work on the other results presented in the paper was supported by the RFBR (project number 19-37-90165) and by the Russian Ministry of Science and Higher Education by the State Task 2.8866.2017/8.9.

References

1. Abdrashitova, Y., Zabashta, A., Filchenkov, A.: Spanning of meta-feature space for travelling salesman problem. Procedia Comput. Sci. **136**, 174–182 (2018)
2. Barth, R., Ijsselmuiden, J., Hemming, J., Van Henten, E.J.: Data synthesis methods for semantic segmentation in agriculture: a capsicum annuum dataset. Comput. Electron. Agric. **144**, 284–296 (2018)
3. Brazdil, P.B., Soares, C., Da Costa, J.P.: Ranking learning algorithms: using IBL and meta-learning on accuracy and time results. Mach. Learn. **50**(3), 251–277 (2003)

4. Durillo, J.J., Nebro, A.J., Alba, E.: The jMetal framework for multi-objective optimization: design and architecture. In: IEEE Congress on Evolutionary Computation, pp. 1–8. IEEE (2010)
5. Feurer, M., Springenberg, J.T., Hutter, F.: Initializing Bayesian hyperparameter optimization via meta-learning. In: Twenty-Ninth AAAI Conference on Artificial Intelligence (2015)
6. Filchenkov, A., Pendryak, A.: Datasets meta-feature description for recommending feature selection algorithm. In: 2015 Artificial Intelligence and Natural Language and Information Extraction, Social Media and Web Search FRUCT Conference (AINL-ISMW FRUCT), pp. 11–18 (2015)
7. Giraud-Carrier, C.: Metalearning-a tutorial. In: Tutorial at the 7th international conference on Machine Learning and Applications (ICMLA), San Diego, California, USA (2008)
8. Hutter, F., Hoos, H.H., Leyton-Brown, K.: Sequential model-based optimization for general algorithm configuration. In: Coello, C.A.C. (ed.) LION 2011. LNCS, vol. 6683, pp. 507–523. Springer, Heidelberg (2011). https://doi.org/10.1007/978-3-642-25566-3_40
9. Mahalanobis, P.C.: On the generalized distance in statistics. Proc. Natl. Inst. Sci. (India) **2**(1), 49–55 (1936)
10. Muñoz, M.A., Smith-Miles, K.: Generating custom classification datasets by targeting the instance space. In: Proceedings of the Genetic and Evolutionary Computation Conference Companion, GECCO 2017, pp. 1582–1588. ACM, New York (2017)
11. Myers, G.: A dataset generator for whole genome shotgun sequencing. In: ISMB, pp. 202–210 (1999)
12. Quinlan, J.R.: Simplifying decision trees. Int. J. Man Mach. Stud. **27**(3), 221–234 (1987)
13. Reif, M., Shafait, F., Dengel, A.: Dataset generation for meta-learning. In: Poster and Demo Track of the 35th German Conference on Artificial Intelligence (KI-2012), pp. 69–73 (2012)
14. Settles, B.: Active Learning. Synthesis Lectures on Artificial Intelligence and Machine Learning, vol. 6, no. 1, pp. 1–114 (2012)
15. Van Rijn, J.N., et al.: OpenML: a collaborative science platform. In: Blockeel, H., Kersting, K., Nijssen, S., Železný, F. (eds.) ECML PKDD 2013. LNCS (LNAI), vol. 8190, pp. 645–649. Springer, Heidelberg (2013). https://doi.org/10.1007/978-3-642-40994-3_46
16. Vanschoren, J., Van Rijn, J.N., Bischl, B., Torgo, L.: OpenML: networked science in machine learning. ACM SIGKDD Explor. Newsl. **15**(2), 49–60 (2014)
17. Wolpert, D.H.: The supervised learning no-free-lunch theorems. In: Roy, R., Köppen, M., Ovaska, S., Furuhashi, T., Hoffmann, F. (eds.) Soft Computing and Industry, pp. 25–42. Springer, London (2002). https://doi.org/10.1007/978-1-4471-0123-9_3
18. Wolpert, D.H., Macready, W.G.: No free lunch theorems for optimization. IEEE Trans. Evol. Comput. **1**(1), 67–82 (1997)
19. Zabashta, A., Filchenkov, A.: NDSE: instance generation for classification by given meta-feature description. CEUR Workshop Proc. **1998**, 102–104 (2017)

Do You Really Follow Them? Automatic Detection of Credulous Twitter Users

Alessandro Balestrucci[1(✉)], Rocco De Nicola[2], Marinella Petrocchi[2,3],
and Catia Trubiani[1]

[1] Gran Sasso Science Institute, L'Aquila, Italy
{alessandro.balestrucci,catia.trubiani}@gssi.it
[2] IMT School for Advanced Studies, Lucca, Italy
rocco.denicola@imtlucca.it
[3] Institute of Informatics and Telematics (IIT-CNR), Pisa, Italy
marinella.petrocchi@iit.cnr.it

Abstract. Online Social Media represent a pervasive source of information able to reach a huge audience. Sadly, recent studies show how online
social bots (automated, often malicious accounts, populating social networks and mimicking genuine users) are able to amplify the dissemina-
tion of (fake) information by orders of magnitude. Using Twitter as a
benchmark, in this work we focus on what we define *credulous* users, i.e.,
human-operated accounts with a high percentage of bots among their
followings. Being more exposed to the harmful activities of social bots,
credulous users may run the risk of being more influenced than other
users; even worse, although unknowingly, they could become spreaders of
misleading information (e.g., by retweeting bots). We design and develop
a supervised classifier to automatically recognize *credulous* users. The
best tested configuration achieves an accuracy of 93.27% and AUC-ROC
of 0.93, thus leading to positive and encouraging results.

Keywords: Twitter · Humans-bots interactions · Gullibility ·
Disinformation · Social networks · Data Mining · Supervised learning

1 Introduction

The diffusion of information on Social Media is often supported by automated
accounts, controlled totally or in part by computer algorithms, called bots.
Unfortunately, a dominant and worrisome use of automated accounts is far from
being benign: malicious bots are purposely created to distribute spam, sponsor
public characters and, ultimately, induce a bias within the public opinion [13].
Especially, their malicious activities are of high efficacy when performed on a

Partially supported by the European Union's Horizon 2020 programme (grant agreement No. 830892, SPARTA) and by IMT Scuola Alti Studi Lucca: Integrated Activity
Project TOFFEe 'TOols for Fighting FakEs'. It has also benefited from the computing
resources (ULITE) provided by the IT division of LNGS in L'Aquila.

© Springer Nature Switzerland AG 2019
H. Yin et al. (Eds.): IDEAL 2019, LNCS 11871, pp. 402–410, 2019.
https://doi.org/10.1007/978-3-030-33607-3_44

targeted audience [4, 5] to, e.g., generate misconception or encourage hate campaigns [7]. Recent work in [24, 29] demonstrate that bots are particularly active in spreading low credibility content and amplifying their significance. Moreover, human-operated accounts contribute to the diffusion of disinformation by, e.g., retweeting and/or liking fake content.

In a previous work [3], the authors shed light on so called *credulous* Twitter users assuming, with a harmless abuse of language, that they refer to human-operated accounts with a high percentage of bots as friends. Unlike [3], where the authors performed an analysis involving the friends of a set of human-operated accounts - a highly time consuming task - here we design and develop a classifier to find out credulous Twitter users, by considering a number of features that do not take the friendship with bots into account. Starting by considering a set of features commonly employed in the literature to detect bots [10, 26], we end up with a lightweight classifier, in terms of costs for gathering the data needed for the feature engineering phase. The classification performance achieves very encouraging results – an accuracy of 93.27% and an *AUC* (Area Under the ROC curve) equal to 0.93.

We believe that automatically detecting credulous users is a promising line of research. Such an investigation could help researchers to: 1. better understand the characteristics of those users more polarized and/or more willing to be influenced; 2. unveil low-credibility and/or deceptive content and limite their online diffusion; 3. devise alternative strategies for bot detection by concentrating the analysis on the friends of credulous users; 4. improve the users' awareness about threats to data trustworthiness.

The following section presents the approach for the automatic detection of credulous Twitter users, while Sect. 3 presents the experimental results. Section 4 discusses the outcome and suggests further investigations. Section 5 presents related work in the area, arguing on the differences, the contributions and the novelty w.r.t. our work. Section 6 concludes the paper.

2   The Approach

2.1   Datasets

We consider three publicly available datasets[1]: CR15 [10], CR17 [12], and VR17 [26]. From the merging of these three datasets, we obtain a unique labeled dataset (human-operated/bot) of 12,961 accounts - 7,165 bots and 5,796 humans. We use this dataset to train a bot detector, as described in Sect. 2.2. To this end, we use the Java Twitter API[2], and for each account we collect: tweets (up to 3,200), mentions (up to 100), IDs of friends and followers (up to 5k).

The identification of credulous users follows the approach presented in [3]. To this end, we need to detect the amount of bots which are friends of the 5,796 human-operated accounts. Due to the rate limits of the Twitter APIs and to the

---

[1] Bot Repository Datasets: https://goo.gl/87Kzcr.
[2] Twitter API: https://goo.gl/njcjr1.

huge amount of friends possibly belonging to these human-operated accounts, we consider only those accounts with a list of friends lower than or equal to 400 [3]. This leads to a dataset of 2,838 human-operated accounts, namely Humans2Consider hereafter. By crawling the data related to their friends, we overall acquire information related to 421,121 Twitter accounts.

2.2 Bot Detection

A bot detection phase is required to discriminate bots and genuine accounts in the dataset of selected friends. The literature offers a plethora of successful approaches [13]; however, also due to the capabilities of evolved spambots to evade detection [12], the performances of the diverse techniques degenerate over time [19]. Furthermore, some bot detectors are available online, but not fully usable due to restrictions in their terms of use, see, e.g., [8]. To overcome these issues, we design and develop a supervised approach, which mixes features from popular scientific work and novel features here introduced.

Regarding the features, we consider two sets. The first one derives from Botometer [26], a popular bot detector[3]. In addition to the original Botometer features [26], we also include: the CAP[4] (Complete Automation Probability) score, the Scores[5], the number of tweets and mentions; we call *Botometer+* this augmented set of features. The second feature set is inherited from [10], where a classifier was designed to detect fake Twitter followers. We use almost all their *ClassA* features[6], except the one about duplicated pictures, because it was not possible for us to verify whether the same profile picture was used twice; we call *ClassA−* this reduced set of features. The conjunction of the two sets of features is referred in the following as *ALL_features*.

We use 19 learning algorithms to train our classifier (with a 10-fold cross validation) and we compare their classification capabilities with respect to the three feature sets (*Botometer+*, *ClassA−* and *ALL_features*). The classification performances are evaluated according to: percentage of *accuracy, precision, recall*, F-measure (*F1*), and Area Under the ROC Curve (*AUC*). On the most accurate classifier, Hyper-Parameter tuning is performed. The tuned classifier is then used to label the friends of the Humans2Consider dataset (see Sect. 2.1).

2.3 Identification of Credulous Twitter Users

The identification of credulous users can be performed with multiple strategies, since there are various aspects that may contribute to spot those users more exposed to the malicious activities of bots. In our previous work [3], we introduced a set of rules to discern whether a genuine user is a credulous one. These rules allow to rank users by relying on the ratio of bots over the user's list

---

[3] https://botometer.iuni.iu.edu/.
[4] Complete Automation Probability: https://tinyurl.com/yxp3wqzh.
[5] English/Universal Score: https://tinyurl.com/y2skbmqc.
[6] ClassA features require only information available in the profile of the account [10].

of friends. Here, we inherit these rules to rank the users in our dataset (see Sect. 2.1), but further ranking strategies can be also considered. Our goal is to build a *ground truth* of credulous users to derive an assessed characterization of these accounts. Applying the approach defined in [3], we identified as *credulous* 316 users in Humans2Consider.This constitutes the input data for the next step. We note that the approach in [3] is very expensive in terms of data gathering. For example, for the investigated dataset, it requires 421k users' account information and 833 million of tweets.

2.4 Classification of Credulous Twitter Users

Goal of this phase is to build a decision model to automatically classify a Twitter account as credulous or not. As ground-truth, we consider the 316 accounts identified as credulous according to the process described in Sect. 2.3.

We experiment the same learning algorithms and the same feature sets considered in Sect. 2.2, with 10 cross-fold validation. However, for credulous users classification, the learning algorithms take as input a very unbalanced dataset: we have 2,838 human-operated accounts (see Sect. 2.1) and, among them, 316 have been identified as credulous accounts (see Sect. 2.3). To avoid working with unbalanced datasets, we split the sets of not credulous users into smaller portions, equal to the number of credulous users. We randomly select a number of not credulous users equal to the number of credulous ones; then, we unify these instances in a new dataset (hereinafter referred to as *fold*). Then, we repeat this process on previously un-selected sets, until there are no more not credulous instances. Such procedure has been inspired by the *under-sampling iteration* methodology, for strongly unbalanced datasets [18]. Each learning algorithm is trained on each fold. To evaluate the classification performances on the whole dataset, and not just on individual folds, we compute the average of the single performance values, for each evaluation metric.

3   Experimental Results

All the experiments are performed with Weka [28], a tool providing the implementation of several machine learning algorithms. In the following, we present the main results obtained for bot detection and credulous classification, all the details are publicly available: https://tinyurl.com/y4l632g5.

The first column of Tables 1 and 2 shows the set of features considered for learning (i.e., *ALL_features*, *Botometer+*, *ClassA−*, see Sect. 2.2). The second column reports a subset of the adopted machine learning algorithms whose name is abbreviated according to the Weka's notation and reported in the following:

IBk: K-nearest neighbours [1], NB: Naive Bayes [17], SMO: Sequential Minimal Optimization [22], JRip: RIPPER [9], MLP: Multi-Layer Perceptron [21], RF: Random Forest [6], REP: Reduced-Error Pruning [23], 1R [15]

The remaining columns report the evaluation metrics mentioned above.

Table 1. Results for bot detection

		evaluation metrics				
	alg	accuracy	precision	recall	F1	AUC
	SMO	98.04	0.98	0.98	0.98	0.98
ALL_features	JRip	97.92	0.99	0.98	0.98	0.99
	RF	**98.33**	**0.99**	**0.98**	**0.98**	**1.00**
	SMO	97.64	0.98	0.98	0.98	0.98
Botometer+	JRip	97.61	0.98	0.97	0.98	0.98
	RF	**97.97**	**0.98**	**0.98**	**0.98**	**1.00**
	IBk	91.03	0.91	0.93	0.92	0.91
ClassA-	JRip	94.38	0.96	0.94	0.95	0.96
	RF	**95.84**	**0.98**	**0.95**	**0.96**	**0.99**

Table 2. Results for credulous detection

		evaluation metrics				
	alg	accuracy	precision	recall	F1	AUC
	IBk	89.69	0.74	0.73	0.90	0.96
ALL_features	1R	**93.27**	**0.99**	**0.88**	**0.93**	**0.93**
	REP	93.07	0.99	0.88	0.93	0.94
	IBk	65.03	0.61	0.60	0.63	0.70
Botometer+	JRip	66.42	0.67	0.67	0.66	0.67
	RF	**67.81**	**0.68**	**0.69**	**0.68**	**0.73**
	IBk	92.59	0.74	0.73	0.92	0.97
ClassA-	1R	**93.27**	**0.99**	**0.88**	**0.93**	**0.93**
	REP	93.09	0.98	0.88	0.93	0.95

Regarding bot detection, Table 1 shows that all the machine learning algorithms well behave, regardless of the feature set. Random Forest is the one that performs best. When the set ALL_features is used (see the shaded line in Table 1): accuracy = 98.33%, F1 = 0.98 and AUC = 1.00; and after the tuning phase, we obtain a final accuracy = 98.41%.

Table 2 shows that ALL_features and ClassA− have good and quite similar classification performances, contrary to Botometer+. Both ALL_features and ClassA− demonstrate their efficacy to discriminate credulous users. On the contrary, the Botometer+'s features properly work for bot detection tasks only.

Going into deeper details, in Table 2 we can notice that the 1R algorithm obtains the best accuracy percentage (93.27% with $\sigma = 3.22$) and F-score (0.93), but not the highest AUC (0.93). It is worth noting that the values of the 1R algorithm are exactly the same when considering *ALL_features* and *ClassA−*. This means that the algorithm selects *ClassA−*'s features only, the ones from *Botometer+* are useless in this case. This is a relevant result since we recall that *ClassA−* features refer to the profile of accounts and it is less expensive to collect them.

4   Discussion

The results in Table 2 show the capability of our approach to automatically discriminate those Twitter users with a large number of bots as friends, namely credulous, *without explicitly considering the features of the latter, which would imply a very high cost in terms of data gathering*. To better understand this point, we recall that the approach in [3] for the identification of credulous users needs to crawl a large amount of data, due to the necessity of extending the analysis to the friends of a Twitter account. In the specific case under investigation, this means to retrieve information for more than 400k user accounts, 11 millions of tweet mentions, and more than 820 millions of tweets. As opposite, the credulous detector here proposed requires to gather the profile information of 2,838 accounts only. The classification performances are really promising, with the best accuracy 93.27%, best F1 0.93, best AUC 0.93. We remark that such results have been achieved by relying on so called *ClassA−* features only, i.e., features extracted from the account profile. It is peculiar how the features useful to discriminate credulous genuine accounts are features belonging to the account profile only. This preliminary result calls for three further investigations: 1. to compare the range of values assumed by these features when detecting credulous accounts with the one assumed to detect social bots (as in [10]); 2. to explore the reason why more complex features (such as the ones of Botometer) do not seem to give good results to find credulous users; 3. to perform a deeper analysis on the importance of each specific feature when discriminating credulous users, by means, e.g., of Principal Component Analysis [28]).

Finally, even if the design of a bot detector is not the primary goal of this paper, but only a mean through which we obtain the ground-truth for training the credulous user classifier, we notice that, compared to the performances reported in [12, 29], our bot detector achieves very good classification performances. This strengthens the robustness of the ground truth obtained in Sect. 2.3, since the friends' nature evaluation is assessed by means of a very accurate classifier.

5   Related Work

Our work is related to all those approaches that investigate peculiar features of social networks users. We discuss the ones we find more relevant for our approach, with the caveat that the presented literature review is far from being exhaustive.

A survey on users' behaviour in social networks is proposed in [16]: it is remarked that the recipients of shared information should be chosen, in a more precautionary way, by taking into account more real-life relationships and less virtual links. Our approach works exactly in the direction of enhancing the awareness of users, by classifying the ones more exposed to attacks of social bots.

Information spreading on Twitter is investigated in [20], where the authors demonstrate that the probability of spreading a given piece of information is higher when promoted by multiple sources. This supports our attempt to analyze the percentage of bots within the friends of human-operated Twitter accounts, as a symptom for being more tempted to disinformation.

In [2], human behaviour on Facebook is analyzed by building graphs that capture sequence of activities. Behavioural patterns that do not match any of the known benign models likely signal malicious objectives. Similarly, the realization of a classifier to automatically recognize credulous users is the first step to derive their sequence of activities and, hopefully, peculiar behavioral patterns.

In [11,14], a behavioural analysis of bots and humans on Twitter is performed, to draw fundamental differences between the two groups. Specifically, the former demonstrates how, despite a higher level of synchronization characterizing bot accounts, the human behaviour on Twitter is far from being random. The latter defines a 'credibility score' as a measure of how many tweets by bots are present in the timeline of an account. Our work supports the discrimination of credulous users and it may lead to a deeper characterization of human accounts.

To the best of our knowledge, few research explores ways to automatically recognize those Twitter users susceptible to attacks of social bots or exposed to disinformation. A notable example in [27] builds on interactions (mentions, replies, retweets and friendship) between genuine and bot accounts, to obtain a ground truth of users susceptible to social bots. Then, similar to our approach, different learning algorithms have been adopted to train a classifier. Contrary to their approach, the current work is able to classify users close to social bots with lightweight features, all computed from data available in the user's profile. Another brand new line of research is the detection of users susceptible to fake news. Work in [25] monitors the replies of Twitter users to a priori known fake news, in order to tag the same users as vulnerable to disinformation or not. Then, a supervised classification task is launched, to train a model able to classify gullible users, according to content-, user-, and network-based features.

6   Conclusions

Inspired by recent literature that shows how disinformation is not only promoted by social bots but also emphasized by genuine peers, in this work we proposed a supervised classification engine to discriminate *credulous* users, i.e., human-operated accounts with a high percentage of bots as friends. The classifier achieves very good performances and avoids a heavy feature engineering and extraction phase. Further research efforts will be devoted to investigate the behaviour of credulous users, as well as the posted content, to know more about their peculiarities and the quality of information they contribute to diffuse.

References

1. Aha, D., Kibler, D.: Instance-based learning algorithms. Mach. Learn. **6**, 37–66 (1991)
2. Amato, F., et al.: Recognizing human behaviours in online social networks. Comput. Secur. **74**, 355–370 (2018)
3. Balestrucci, A., et al.: Identification of credulous users on Twitter. In: ACM Symposium of Applied Computing (2019)
4. Bastos, M.T., Mercea, D.: The Brexit botnet and user-generated hyperpartisan news. Soc. Sci. Comput. Rev. **37**(1), 38–54 (2019)
5. Bovet, A., Makse, H.A.: Influence of fake news in Twitter during the 2016 US presidential election. Nat. Commun. **10**(1), 7 (2019)
6. Breiman, L.: Random forests. Mach. Learn. **45**(1), 5–32 (2001)
7. Chatzakou, D., et al.: Mean birds: detecting aggression and bullying on Twitter. In: ACM Web Science Conference, pp. 13–22 (2017)
8. Chavoshi, N., et al.: DeBot: Twitter bot detection via warped correlation. In: Data Mining, pp. 817–822 (2016)
9. Cohen, W.: Fast effective rule induction. In: Machine Learning, pp. 115–123 (1995)
10. Cresci, S., et al.: Fame for sale: efficient detection of fake Twitter followers. Decis. Support Syst. **80**, 56–71 (2015)
11. Cresci, S., et al.: Exploiting digital DNA for the analysis of similarities in Twitter behaviours. In: IEEE Data Science and Advanced Analytics, pp. 686–695 (2017)
12. Cresci, S., et al.: The paradigm-shift of social spambots: evidence, theories, and tools for the arms race. In: 26th World Wide Web, Companion, pp. 963–972 (2017)
13. Ferrara, E., et al.: The rise of social bots. Commun. ACM **59**(7), 96–104 (2016)
14. Gilani, Z., et al.: A large-scale behavioural analysis of bots and humans on Twitter. ACM Trans. Web **13**(1), 7 (2019)
15. Holte, R.: Very simple classification rules perform well on most commonly used datasets. Mach. Learn. **11**, 63–91 (1993)
16. Jin, L., et al.: Understanding user behavior in online social networks: a survey. IEEE Commun. Mag. **51**(9), 144–150 (2013)
17. John, G.H., Langley, P.: Estimating continuous distributions in Bayesian classifiers. In: 11th Uncertainty in Artificial Intelligence, pp. 338–345 (1995)
18. Lee, J., et al.: An iterative undersampling of extremely imbalanced data using CSVM. In: Machine Vision, vol. 9445 (2014)
19. Minnich, A., et al.: BotWalk: efficient adaptive exploration of Twitter bot networks. In: ASONAM, pp. 467–474 (2017)
20. Mønsted, B., et al.: Evidence of complex contagion of information in socialmedia: an experiment using Twitter bots. PLoS ONE **12**(9), e0184148 (2017)
21. Pal, S.K., Mitra, S.: Multilayer perceptron, fuzzy sets, and classification. IEEE Trans. Neural Networks **3**(5), 683–697 (1992)
22. Platt, J.: Fast training of support vector machines using sequential minimal optimization. In: Advances in Kernel Methods - Support Vector Learning (1998)
23. Quinlan, J.R.: Simplifying decision trees. Int. J. Human Comput. Stud. **27**(3), 221–234 (1987)

24. Shao, C., et al.: The spread of low-credibility content by social bots. Nature Commun. **9**(1), 4787 (2018)
25. Shen, T.J., et al.: How gullible are you? Predicting susceptibility to fake news. In: Web Science, pp. 287–288 (2019)
26. Varol, O., et al.: Online human-bot interactions: detection, estimation, and characterization. In: 11th Web and Social Media, pp. 280–289 (2017)
27. Wagner, C., et al.: When social bots attack: modeling susceptibility of users in online social networks. In: Making Sense of Microposts, pp. 41–48 (2012)
28. Witten, I.H., et al.: Data Mining: Practical Machine Learning Tools and Techniques (2016)
29. Yang, K.C., et al.: Arming the public with artificial intelligence to counter social bots. Hum. Behav. Emerg. Technol. **1**(1), 48–61 (2019)

User Localization Based on Call Detail Record

Buddhi Ayesha[1]([✉]), Bhagya Jeewanthi[2], Charith Chitraranjan[1],
Amal Shehan Perera[1], and Amal S. Kumarage[2]

[1] Department of Computer Science and Engineering, University of Moratuwa,
Moratuwa, Sri Lanka
buddhiayesha.13@cse.mrt.ac.lk
[2] Department of Transport and Logistics Management, University of Moratuwa,
Moratuwa, Sri Lanka

Abstract. Understanding human mobility is essential for many fields, including transportation planning. Currently, surveys are the primary source for such analysis. However, in the recent past, many researchers have focused on Call Detail Records (CDR) for identifying travel patterns. CDRs have shown correlation to human mobility behavior. However, one of the main issues in using CDR data is that it is difficult to identify the precise location of the user due to the low spacial resolution of the data and other artifacts such as the load sharing effect. Existing approaches have certain limitations. Previous studies using CDRs do not consider the transmit power of cell towers when localizing the users and use an oversimplified approach to identify load sharing effects. Furthermore, they consider the entire population of users as one group neglecting the differences in mobility patterns of different segments of users. This research introduces a novel methodology to user position localization from CDRs through improved detection of load sharing effects, by taking the transmit power into account, and segmenting the users into distinct groups for the purpose of learning any parameters of the model. Moreover, this research uses several methods to address the existing limitations and validate the generated results using nearly 4 billion CDR data points with travel survey data and voluntarily collected mobile data.

Keywords: Call Detail Records · Localization · Load sharing effect · Mobile Network Big Data · Data fusion · Data mining

1 Introduction

Human mobility patterns are critical sources of information for designing, analyzing and enhancing transport planning activities. These transportation initiatives require mobility rich data to become a continuous process. Currently, manually carried out roadside surveys or household surveys are done once every few years provide input data for such initiatives. Such data collection while being expensive, takes time and is often outdated by the time it is available for analysis. Such

© Springer Nature Switzerland AG 2019
H. Yin et al. (Eds.): IDEAL 2019, LNCS 11871, pp. 411–423, 2019.
https://doi.org/10.1007/978-3-030-33607-3_45

drawbacks of the existing methodologies create the necessity for more advanced data extraction techniques to analyze human motion.

The recent adoption of ubiquitous computing technologies, including mobile devices by very large portions of the world population has enabled the capturing of large-scale spatio-temporal data about human motion [8]. Large-scale penetration of mobile phones serves as a dynamic source to derive insights on human mobility due to the generation of big data. Mobile Network Big Data makes it possible to locate mobile devices in time and space with a certain accuracy using the mobile network infrastructure that includes location information of the mobile device [7]. In the present scenario, Call Detail Records (CDR) data, which is collected by mobile phone carriers for billing purposes is the most common type of MNBD used in a variety of transportation studies. CDRs generate the time-stamped locations of the users. All the phone numbers are replaced by a computer generated unique identifier by the operator to preserve the anonymity and the privacy of the users. A single CDR contains a random ID number of the phone, device and phone number, exact time and date, call duration, location in terms of latitude and longitude of the cell tower that provide the network signal. When a person makes a call, all this information gets recorded as an array of data. These arrays of data provide information on the mobility of the users based on the places they have travelled.

Though CDR has the potential to provide information on human mobility insights at a wider scope, there are certain limitations embedded within CDR data that makes the accuracy of the derived information to be low. One main limitation of the CDR data is that for a given call, there may be frequent changes in the serving tower with no actual displacement of the user, which is referred to as the load sharing effect [2]. This is mainly because the operator often balances call traffic among adjacent towers by allocating a new call to the tower that is handling a lower call volume at that moment. As an example, Fig. 1 indicates the general demonstration of cell towers. User has the possibility of connecting to any tower depicted. But the tower is allocated by the operator based on the total number of calls each one can handle. With that, user might get connected to X1 at one time and next time to X2, even though the A's location does not change.

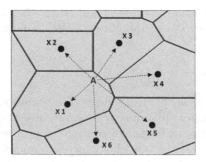

Fig. 1. Load sharing effect

Due to this process, the towers providing the signal varies without actual displacement of the user. This has a large effect on the transportation aspect since the trip counts between origins and destinations may exceed actual values, behavioral structures may incorrectly be identified and etc. Therefore, it is mandatory to fix the existing issues in order to use these real-time data for transportation forecasting processes.

Other than the load sharing effect, one other main issue is the localization error, which means the user's location is identified as anywhere within the area of the connected cell tower. For an example, user A's (Fig. 1) location is identified as anywhere within the coverage area of X1, X2 or X6. Existing studies do not take into account, the load sharing effect or the transmit power of the connected cell tower when localizing the user. The study propose an improved method to localize the user by incorporating the load sharing effect as well as the transmit power levels of the cell towers. Furthermore, when it is required to fit any model parameters to the data, the first attempt is to identify separate groups of users based on demographic attributes and fit the parameters separately to different groups as opposed to the conventional approach of fitting parameters to the entire population of users as one.

This study demonstrates a method to reduce the localization error from CDR data. Second section presents an overview of the previous works done and the drawbacks of the existing studies. Section 3 summarizes the methodology of the research conducted, and Sect. 4 validate the results against the independent data sources from the study area. Based on these findings, the study concludes with a discussion of the limitations and applications of CDR data in future studies.

2   Background and Related Work

Previous studies have used different methodologies to minimize the load sharing effect. Basically there are two techniques that have been used as trajectory smoothing and spatial clustering.

2.1   Trajectory Smoothing

In trajectory smoothing, the sequence of CDR within a certain time threshold is taken into consideration and different smoothing or filtering algorithms were taken into account to reduce the "jumps" in the location sequence. Speed based filtering, time weighted smoothing and assigning a single medoid location to records which are close by were different algorithms used in trajectory smoothing to smooth the location sequence.

In speed based filtering a small trajectory of data points within five minutes are inspected and any points which have irrational travel distance or travel speed (more than 120 kmph) are removed from the database [9]. In the studies which use time weighted smoothing, the preprocessing is initiated by smoothing the positions independently for each user. In time weighted smoothing, the maximum stay region size is set to $d = 300\,m$ to approximate the area that might likely be traversed on foot as part of an urban activity. Then the entire region is divided into rectangular cells of size d/3. Then, all the stay points were mapped and iteratively merge the unlabeled cell with the maximum stay-points and its unlabeled neighbors to a new stay-region and labeled the stay regions [4].

In conclusion, the trajectory smoothing also has its own drawbacks. The user visited locations are removed in the trajectory smoothing method which minimize the possibility to identify exact locations of the users

2.2 Spatial Clustering

Spatial clustering is the other main technique that is used to remove the load sharing effect. There the data points were clustered based on the spatial distribution without considering the temporal distribution of CDR. This process allows to consolidate points that may represent the same location but, visited in different days. Agglomerative [1] and leader clustering [3] are the two main clustering techniques used for this categorization.

Both these methodologies have their own drawbacks. Initially the speed between two consecutive records were calculated using the time difference and the distance parameter. The distance between two consecutive points is the Euclidean distance. Then in speed based filtering of trajectory smoothing, only the data within five minutes were inspected, where the travel speed considered is 120 miles per hour. But the speed should be sensitivity tested rather than using a selected speed.

In trajectory smoothing the positions were smoothed based on a weight. The positions were smoothed for all users based on their caller activities.

In spatial clustering, a stay-point is identified by a sequence of consecutive cell phone records bounded by both temporal and spatial constraints. The spatial constraint is the roaming distance when a user is staying at a location, which should be related to the accuracy of the device collecting location data. The temporal constraint is the minimum duration spent at a location, which is measured as the temporal difference between the first and the last record in a stay. In most of the studies that used spatial clustering techniques for the load sharing removal, only the latitudinal and longitudinal locations were considered without taking the number of caller activities in each location into consideration. Leader clustering is one of the techniques used in spatial clustering, initially the cell towers are sorted and one with the most counts is taken as the centroid of the first cluster. Then, for each subsequent cell tower, it checks whether the tower falls within a threshold radius of the centroid of any existing cluster. If it does not, the tower becomes the centroid of a new cluster. If it does fall within the threshold radius of an existing cluster, the algorithm adds the tower to the cluster and moves the centroid of the cluster to be the weighted average of the locations of all the cell towers in the cluster. Choosing a particular threshold radius around cell towers helps to equalize for the fact that in urban areas towers might be as dense as 200 m apart, while in suburban areas, a spacing of 1–3 miles are more common. These threshold values should be sensitivity tested and validated with another data source, to be used in the study.

Additionally, with respect to spatial clustering only the spatial behavior is taken into account, without considering about the time factor. Specially the sequence of location visits were not considered. Other main limitation in literature is the lack of models to minimize location error using label data.

In summary, There are three main influencing factors for the localization to be stated as load sharing, signal strength of the cell tower and the frequency of connecting to a particular tower. Existing studies has a gap where they have not considered all these three factors together to remove the localization error. Additionally this study combines with user profiling for the localization error reduction in a novel manner.

3   Methodology

This section describes the overall research methodology that was carried out. Each section below, addresses a component in the overall methodology. An overview of the methodology proposed is illustrated in Fig. 2.

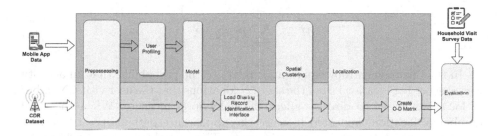

Fig. 2. Flow diagram for the overall methodology

3.1   Data

The study uses Call Detail Records of nearly 13 million SIMs from a mobile operator in Sri Lanka. Data was provided for this research by LIRNEasia - a regional ICT policy and regulation think-tank. Data is completely pseudonymized by the operator, where phone numbers are present.

Study also use voluntarily collected CDR and GPS data in a pseudonymized manner from an unbiased sample of mobile users. Data was collected through a mobile application developed for Android devices and users could download the mobile application from the Google play store. Data was collected from more than 700 users for a period of three months from 21-03-2018 to 21-06-2019. This app collects socioeconomic data, CDR data, signal strength data as well as travel data. For validation purposes, the App also collects signal strength and GPS data every 10 min and it identifies user's travelling movements automatically through GPS data and prompts the user to indicate their travel purpose.

The main data source used for the purpose of validation is obtained from transportation-related data collected from household surveys within the Western Province of Sri Lanka, which was conducted as part of the CoMTrans study [5]. The Household Visit Survey (HVS) data provide transportation information, including social-demographic records, travel time, trip purposes, travel modes, etc. which can be used for validating the results from CDR trip analysis.

3.2   Preprocessing

Preprocessing is done using two filtering techniques. First filter is used for the removal of atypical users. As an example users with low caller activity levels, users with low appearance within the study area (e.g. tourist, visitors). Most of the previous researchers on this was done based on call activity count (CAC). It is required to remove the infrequently visited users who are entering the focused research area, but the removal should be done based on their spatial behaviour but not considering the caller activity counts. Our main approach is to use a filter based on spatial behavior, to filter the users. Shannon Entropy is used for the analysis of user's spatial activities.

$$H = -\sum_{i=1}^{N} P_i \, log_2 \, P_i \tag{1}$$

In the Shannon entropy equation, P refers to the probability that an activity was observed at location i from the set of N locations that the user visits. Number of locations appeared directly affect the behavioral measure. When the total number of locations appeared increase the appearance probability of locations increase by which the entropy value increase. The spatial behavior of 80% of the users are more of a homogeneous behavior compared to others.

Then the Western Province users were filtered since the study area is the Western Province of Sri Lanka.

3.3   User Profiling

One of the major drawbacks of previous research studies is that they consider all users as a single group and use the same model parameters for all users. Especially, when considering mobility patterns at individual level, initially it is essential to profile users based on their socio-economic parameters. Since the travelling behaviour of people highly depend on their socio-economic parameters such as occupation, age, income and etc. As an example, for a public sector worker the general working hours are between 8.30 AM to 4.30 PM and do not work during weekends. But an employee in the private sector may work beyond 4.30 PM and also can appear in work locations during weekends. The behavior of a shift worker or a school worker might be completely different to this. A school employee generally work from 7.30 AM–1.30 PM but a shift worker may be employed in the night. Hence the behavior can be completely different to each other. Considering the age, a student's travelling pattern cannot be expected from a retired adult and income also has a significant influence on the travelling patterns.

Study selected top features for user profiling based on correlation analysis. The identified features are as follows.

- Call frequency for each hour
- Distance for each hour
- Appeared cell frequency
- Appearance in weekday or weekends

The study uses several classification methods such as Nearest Neighbor, Random Forest, Neural Networks and Support Vector Machine for the data collected from mobile app. Since the dataset has some class imbalance problem, study used SVM weighted classes method (SVM_{WC}). Then the same models were used to classify the main CDR dataset, as the next step.

3.4   Load Sharing Record Identification Interface

Main aim of this section is the identification of Load sharing records. Initial attempt was to use a speed based filtering method. In speed based filtering a small trajectory data within the corresponding time are inspected and any points which have irrational travel distance or travel speed are labeled as load shared record CDRs. Then the speed is calculated based on two consecutive CDRs for each user and a speed filter was added. This methodology tries to identify the jumps between location sequence.

The speed limit was selected as 120 kmph based on previous studies. But the results were not as expected. In this method precision is good but recall is very low. In other words, the majority of the records which were categorized as load shared record are actually load shared records. But that method missed a lot of load shared records.

Then the study attempts on a new approach to identify load shared records. Main drawback of the predefined speed based filtering method is that speed of the vehicle in the road is not always equal. For an example in the morning 7 am–9 am average vehicle speed is low compared to 8.30 pm–10 pm. Generally the speed in one of the main corridors to Colombo at peak hours is 6 kmph but during off peak hours the speed can be high up to 100 kmph. Vehicle speed also depends on the geographical area. In urban cities the vehicle speed is slow compared to remote areas. Our approach is to address both scenarios. Speed limit data was taken from previous transportation origin-destination surveys. Since the speed limit is based on the time, time frames were defined for our study as, 07:00–09:00, 09:00–12:00, 12:00–13:00, 13:00–16:30, 16:30–19:00, 19:00–22:00 and 22:00–07:00. Then the study area is divided based on the Divisional Secretariat Divisions (DSD)and the average speeds were taken based on separate time windows. CDR were filtered based on the θ values, which represent in kmph, $\theta = [0, 5, 10, \ldots, 200]$ Then the records were evaluated and labeled as Load shared records using GPS data which represent actual movements. Based on this result, the same process was carried out for the main dataset.

3.5   Reduce Localization Error

Main part of this research is the identification of the user's location. CDR data was clustered based on their spatial behaviour to identify different regions of stay for the user (Stay location clusters). The study use DBSCAN to perform the clustering.

Majority of the previous research considered call frequency related data only. However, that does not provide acceptable results. Therefore, this study propose a new approach to minimize user's localization error. It is noted that the assignment of a cell tower to a user depends, among others, on user's location, signal strength and the call traffic load of the particular cell tower. Following three factors were consider to assign locations for users.

- Load sharing records
- Signal strength of the particular cell tower
- Number of days appeared in the particular cell tower.

Consider the following scenario. Assume that User X is connected with cell towers A, B, C, D with an equal number of caller activities. When the weights are based on the caller activities, then the user's location should be the centroid of the considered 4 locations within the particular time period (Fig. 3).

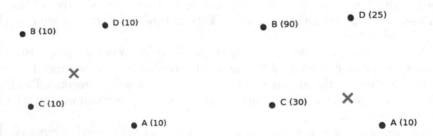

Fig. 3. Use locations based on the number of appearances days

Fig. 4. Use locations based on the signal strength

In order for the above to happen the signal strength of the cells should be equal to each other. But when the signal strength data of cell towers is taken as the weight, the position of the centroid will change from that of the position taken from caller activity levels as shown in the Fig. 4.

Similarly, based on Load shared record frequency, it is possible to assign a location to the particular user.

The study collects signal transmit power for each cell in the study area, and it is used as a measurement to rank the power of connecting to users. P_i demonstrate the signal transmitting power of the ith cell.

Based on those factors, the weight is given by Eq. 2. For a given cell i, Load sharing records, signal transmit power and Number of appearances days are denoted by L_i, P_i, C_i, respectively for particular user.

$$W_i = \alpha L_i + \beta \frac{1}{P_i} + \gamma C_i \qquad (2)$$

Then, the L_i, P_i and C_i values were scaled as in Eq. 3.

$$0 \leq L_i, \ \frac{1}{P_i}, \ C_i \leq 1 \qquad (3)$$

α, β, γ are distinct weights calculated for each user segmentation separately within the range mentioned in the Eq. 4.

$$0 \leq \alpha, \ \beta, \ \gamma \leq 1 \qquad (4)$$

The study used weighted k-means++ algorithm to find the centroid of the cluster using the weights given by Eq. 2. Centroids are assigned as user location for a given time.

Table 1. User profiling results comparison (P = Precision, R = Recall, F1 = F1 Score)

Model	Full-time employees			Part-time employees			Student			Housewife			Retired			Others		
	P	R	F1	P	R	F1	P	R	F1	P	R	F1	P	R	F1	P	R	F1
RF	0.64	0.53	0.58	0.45	0.39	0.42	0.70	0.91	0.79	0.68	0.72	0.70	0.32	0.21	0.26	0.52	0.64	0.57
ANN	0.70	0.44	0.54	0.55	0.40	**0.46**	0.71	0.93	0.81	0.72	0.78	0.75	0.36	0.24	0.28	0.56	0.36	0.44
SVM_{WC}	0.71	0.66	**0.68**	0.50	0.43	**0.46**	0.77	0.95	**0.85**	0.71	0.81	**0.76**	0.31	0.29	**0.30**	0.65	0.53	**0.58**

4   Results

Results of the User Profiling models are shown in Table 1. The SVM_{WC} model showed higher accuracy values for user profiling.

Then Load Sharing Record identification is evaluated. If the last recorded cell tower changes without an actual movement of 100 m of the user, study considered that record as a load shared one. The result is shown in Table 2. Our approach shows significant accuracy improvement, especially in recall.

To evaluate the Localization result, two approaches were used. First, result were evaluated using mobile app data. To calculate home location error, initially the user's home location which is provided by the user was taken and compared with the location predicted by our method. Figure 5 indicates the generated results. The same process was carried out for the work location identification of users and the generated results are shown in Fig. 6. Our proposed methodology outperforms existing results in majority of the population. In the study work location was more accurately identify than the home location. The study reveals that 70% (468) of the users are working in an urban area, but their home locations are in suburban areas. In urban areas cell tower density is high and, in the suburban area cell tower density is low. Therefore, work location error is low compared to the home location.

As the second approach to evaluate the Localization result, Origin-Destination (O-D) matrices [10] were created from the large (13 million subscribers) CDR dataset. Accordingly, the weekends and holidays were removed from the large CDR dataset. Home activities are defined between 8 PM and 5 AM [6] and work activities from 10 am to 12 pm and 1 pm to 4 pm [4]. Then the results were compared with HVS O-D matrix data.

Table 3 indicates the O-D matrices derived from HVS data at district level. This study has been done by covering the Western Province which consists of three districts namely Colombo, Gampaha, Kalutara. As an example cell 1 says that 44% of users have both their home and work location at the Colombo district, cell 2 says that 3% of users have their work location within Gampaha and home location within the Colombo district.

Further Table 4 indicates the user localization based on the number of days the cell tower was connected [3] derived from CDR dataset, which contain nearly 13 million users home-work location distribution. And this table also has the same interpretation as in Table 3.

Table 5 indicates the user localization based on our proposed methodology, which considers the user's Load shared record, Signal strength and Number of appearance days and also has the same interpretation as in Table 3.

In order to statistically compare the extracted O-D matrices with ground truth (HVS data), study used the Pearson's chi-squared test given by Eq. 5.

$$\tilde{\chi}^2 = \sum_{i=1}^{n} \frac{(O_i - E_i)^2}{E_i} \tag{5}$$

χ refers to the Pearson's cumulative test statistic, where the E_i denotes the Expected value, O_i the Observation value and n is the number of cells in the table. Based on Chi-squared test values, p value for Table 4 is 0.64 and for Table 5 is 0.88. Our approach outperforms the result, and it shows similar outcomes based on the HVS dataset according to the statistic.

Table 2. Load sharing record identification results comparison

Method	Precision	Recall	F1
Pre-define speed base filter	0.914	0.166	0.281
Our method	0.864	0.728	**0.790**

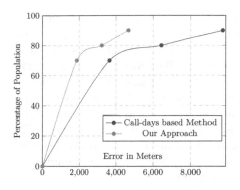

Fig. 5. Home localization error   **Fig. 6.** Work localization error

Table 3. Home work distribution - HVS data

Home/Work		Trip attractors		
		Colombo	Gampaha	Kalutara
Trip generators	Colombo	44%	2%	1%
	Gampaha	10%	27%	1%
	Kalutara	4%	1%	10%

Table 4. Home work distribution - call days based method

Home/Work		Trip attractors		
		Colombo	Gampaha	Kalutara
Trip generators	Colombo	52%	2%	2%
	Gampaha	12%	18%	1%
	Kalutara	5%	1%	7%

Table 5. Home work distribution - proposed methodology

Home/Work		Trip attractors		
		Colombo	Gampaha	Kalutara
Trip generators	Colombo	44%	3%	2%
	Gampaha	8%	30%	1%
	Kalutara	1%	1%	10%

5   Conclusion

As mentioned in the previous studies, real time datasets such as mobile phone traces provide rich information to support transportation planning and operation. Meanwhile, some related limitations should also be addressed when using these datasets in mobility analysis. The most prominent outcome of this research is the formulation of a methodology to reduce the error in localizing the user by considering the load sharing effects of the CDR records and transmit power of the cell towers. This has a significant effect on the travel conclusions obtained from CDR data. The study identifies that there are different optimum speeds for different geographical areas in separate time windows. Findings were validated by comparing with the HVS data. Findings had a significant improvement after reducing the localization error. An additional validation was done using voluntarily collected CDR data. Since the voluntarily collected data also contains GPS data, it provides an accurate way to evaluate localization error. With this, the load sharing effect can be identified easily and hence the accuracy of the findings were also compared with the mobile application data.

Future work is possible through the use of Google traffic data and by setting a speed limit based on the corridors and input load sharing filtering techniques, which are computationally efficient and accurate in result acquisition.

Acknowledgment. The authors wish to thank the DataSEARCH Centre for supporting the project which was partially funded by the Senate Research Council under the grant number SRC/LT/2018/08 of the University of Moratuwa.

References

1. Hariharan, R., Toyama, K.: Project lachesis: parsing and modeling location histories. In: Egenhofer, M.J., Freksa, C., Miller, H.J. (eds.) GIScience 2004. LNCS, vol. 3234, pp. 106–124. Springer, Heidelberg (2004). https://doi.org/10.1007/978-3-540-30231-5_8
2. Iqbal, M.S., Choudhury, C.F., Wang, P., González, M.C.: Development of origin-destination matrices using mobile phone call data. Transp. Res. Part C Emerging Technol. **40**, 63–74 (2014)
3. Isaacman, S., et al.: Identifying important places in people's lives from cellular network data. In: Lyons, K., Hightower, J., Huang, E.M. (eds.) Pervasive 2011. LNCS, vol. 6696, pp. 133–151. Springer, Heidelberg (2011). https://doi.org/10.1007/978-3-642-21726-5_9
4. Jiang, S., Fiore, G.A., Yang, Y., Ferreira Jr, J., Frazzoli, E., González, M.C.: A review of urban computing for mobile phone traces: current methods, challenges and opportunities. In: Proceedings of the 2nd ACM SIGKDD International Workshop on Urban Computing, p. 2. ACM (2013)
5. JICA: Urban transport system development project for Colombo metropolitan region and suburbs (2014)
6. Kung, K.S., Greco, K., Sobolevsky, S., Ratti, C.: Exploring universal patterns in human home-work commuting from mobile phone data. PLoS ONE **9**(6), e96180 (2014)
7. Tiru, M.: Overview of the sources and challenges of mobile positioning data for statistics. In: Proceedings of the International Conference on Big Data for Official Statistics, Beijing, China, pp. 28–30 (2014)
8. Vieira, M.R., Frías-Martínez, E., Bakalov, P., Frías-Martínez, V., Tsotras, V.J.: Querying spatio-temporal patterns in mobile phone-call databases. In: 2010 Eleventh International Conference on Mobile Data Management (MDM), pp. 239–248. IEEE (2010)

9. Wang, M.H., Schrock, S.D., Vander Broek, N., Mulinazzi, T.: Estimating dynamic origin-destination data and travel demand using cell phone network data. Int. J. Intell. Transp. Syst. Res. **11**(2), 76–86 (2013)
10. Zhang, Y., Qin, X., Dong, S., Ran, B.: Daily OD matrix estimation using cellular probe data. In: 89th Annual Meeting Transportation Research Board, vol. 9 (2010)

Automatic Ground Truth Dataset Creation for Fake News Detection in Social Media

Danae Pla Karidi$^{(\boxtimes)}$, Harry Nakos, and Yannis Stavrakas

IMSI Athena RC, Athens, Greece
{danae,xnakos,yannis}@imis.athena-innovation.gr

Abstract. Fake news has become over the last years one of the most crucial issues for social media platforms, users and news organizations. Therefore, research has focused on developing algorithmic methods to detect misleading content on social media. These approaches are data-driven, meaning that the efficiency of the produced models depends on the quality of the training dataset. Although several ground truth datasets have been created, they suffer from serious limitations and rely heavily on human annotators. In this work, we propose a method for automating as far as possible the process of dataset creation. Such datasets can be subsequently used as training and test data in machine learning classification techniques regarding fake news detection in microblogging platforms, such as Twitter.

Keywords: Fake news detection · Automatic dataset creation · Social network

1 Introduction and Motivation

More and more people tend to seek out and consume news from online platforms rather than traditional news organizations, since social media allow millions of users to freely produce, consume and access a vast source of information. Although misinformation and disinformation phenomena exist since the birth of printed press, new online platforms accelerate and boost their diffusion, posing new problems and challenges. Social network openness turns these services into one of the most effective channels for misinformation and disinformation. The extensive spread of digital misinformation and disinformation has a serious negative impact on individuals and society.

To help mitigate the negative effects caused by the extensive spread of fake news it is critical that we develop algorithmic methods to automatically detect misleading content on social media. In this regard, both supervised and unsupervised machine learning classification approaches are data-driven, meaning that the efficiency of the produced models depends on the size and quality of the training dataset. Furthermore, the automatization and standardization of the creation of datasets is essential for the reliable assessment and comparison of fake news detection methods. Although the full

This work is supported by the project MIS 5002437/3, funded by the Action "Reinforcement of the Research and Innovation Infrastructure" co-financed by Greece and the European Union (European Regional Development Fund.

H. Yin et al. (Eds.): IDEAL 2019, LNCS 11871, pp. 424–436, 2019.
https://doi.org/10.1007/978-3-030-33607-3_46

automation of the dataset creation process is not yet possible, we can leverage existing fact-checking websites that provide veracity labels for a large amount of news stories to automate the dataset creation process as far as possible.

In this paper, we describe PHONY, an automatic system for creating fake news datasets. PHONY uses the fake news stories provided by curated fact-checking websites and produces datasets that include social media posts which refer to those fake news stories. Therefore, the resulting datasets are suitable for training and test data in machine learning classification techniques regarding fake news detection in microblogging platforms such as Twitter. Uniform, updatable datasets that can be dynamically constructed are of great importance, since fake news evolve and adapt in order to avoid the current detection methods. We also present the PHONY-EL, a dataset of fake news propagating in the Greek tweetosphere by running the PHONY system in the Greek Twitter stream.

There are certain characteristics of this problem that make it uniquely challenging. First, the nature of fake news is not uniform. Misinformation refers to fake or inaccurate information which is unintentionally spread, while disinformation concerns information intentionally false and deliberately spread. Figure 1 classifies misleading content with respect to facticity, which refers to the truthiness of the facts mentioned, and intention to deceive, which refers to whether the creator of the content intends to mislead. Specifically, misinformation includes parody, mistaken reports and satire, while disinformation includes fabricated news and propaganda, which rank high in their intention to deceive. Propaganda is different from fabricated news in that it contains true facts within an extremely biased context. In this paper, the term fake news is used to denote inaccurate information, both misinformation and disinformation. Several fact-checking websites offer human curated type labelling for their fake news stories; therefore, a dataset creation system should be able to leverage such information.

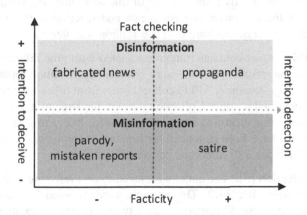

Fig. 1. Types of fake news.

Second, fake news is usually related to newly emerging, time-critical events, which may not have been properly verified by existing knowledge bases and fact-checkers due to the lack of corroborating evidence or claims. Moreover, network-based approaches detect misinformation and disinformation by analyzing content diffusion in the network and require large fractions of content propagations, which results in low detection speed. Since detecting such content early can help prevent further propagation on social media, this work presents a system that collects diffusion data and uses efficient propagation representations, which can be used as features to train machine learning classification models.

Finally, user engagements with fake news in social media produce data that is big, incomplete, unstructured, and noisy. For instance, existing datasets suffer from significant limitations: incomplete news texts, few news sources, partial social network data (due to API limits), small number of news and social posts, and more. Due to this, and to the best of our knowledge, no publicly available dataset exists that provides the necessary information to extract all state-of-the-art features for detection (text, user relationships, network analysis, news propagation). Moreover, the research community lacks datasets which contain spatiotemporal information to understand how fake news propagates over time in different regions. Our method addresses the problems of data quality and completeness by generating broadened tweets comprising the tweet text and metadata as well as web data, for all the URLs contained in each tweet.

2   Existing Datasets and Comparison

Though the research community has produced several datasets for fake news detection, to the best of our knowledge, no automatic system for creating fake news datasets in social media exists. The most significant state-of-the-art datasets are listed below.

BuzzFace [1]: This dataset contains 2282 news stories that were published during September 2016 on nine Facebook pages that correspond to nine news outlets. These articles have been annotated by BuzzFeed, a famous fact-checking organization, into 4 categories: mostly true, mostly false, mixture of true and false, and no factual content. BuzzFace contains the corresponding comments and reactions, article texts, images, links and embedded tweets, reaching over 1.6 million text items.

PHEME [2]: This dataset contains rumors associated with nine different news stories and also Twitter conversations which were initiated by a rumor related tweet. The authors used Twitter's streaming API to collect tweets from breaking news and specific rumors that are identified a priori. Collection through the streaming API was launched straight after a journalist identified a newsworthy event likely to give rise to rumors. The collected tweets have been annotated for support, certainty, and evidentiality. The sampled dataset includes 330 rumorous conversational threads.

USPresidentialElection [3]: The authors retrieved tweets referring to 57 rumors from October 2011 to December 2012. These tweets were gathered by matching specific keywords for each of these 57 rumors. After a tweet preprocessing stage, the authors assembled a set of keywords to identify the rumor-related tweets based on the description of each rumor offered by rumor debunking sites. These keywords were then combined through logical expressions, forming queries that are manually repeatedly tested by two authors. Finally, human coders measured each tweet for two variables: whether the tweet was actually about the rumor and its author's attitude (endorsing, rejecting, or unclear).

FakeNewsNet [4]: This repository contains two datasets with news content, social context, and spatiotemporal information. To collect ground truth labels for fake news, the authors utilized two fact-checking websites to obtain news contents for fake news and true news. User engagements related to news articles were collected using the Twitter's Advanced Search API. The search queries were formed from the headlines of news articles and the obtained tweets were enriched by replies, likes, reposts and user social network information.

CREDBANK [5]: This dataset comprises more than 60 million tweets grouped into 1049 real-world events, each annotated for credibility by 30 human annotators. These tweets were collected from all streaming tweets over a period of three months, computationally summarized into events, and finally routed to crowd workers (Amazon Mechanical Turk) for credibility assessment.

SomeLikeItHoax [6]: This dataset consists of all the public posts and their likes from a list of selected Facebook pages during the second semester of 2016. These Facebook pages were divided into two categories: scientific news sources vs. conspiracy news sources. The authors assumed all posts from scientific pages to be reliable and all posts from conspiracy pages to be "hoaxes". The dataset contains 15,500 posts from 32 pages (14 conspiracy and 18 scientific) with more than 2,300,000 likes.

TraceMiner [7]: The authors retrieved tweets (using Twitter search API) related to fake news by compiling queries with a fact-checking website (Snopes). The dataset contains 3600 messages from which 50% are referring to fake news.

Vosoughi et al. [8]: To construct this dataset the authors, firstly, scraped fact-checking websites (snopes.com, politifact.com, factcheck.org, truthorfiction.com, hoax-slayer. com, and urbanlegends.about.com), collected the archived rumors and parsed the title, body and verdict of each rumor. The rumors were then divided based on their topic (Politics, Urban Legends, Business, Science and Technology, Terrorism and War, Entertainment, and Natural Disasters) using LDA and human annotators. Afterwards, they accessed the full Twitter historical archives to extract all reply tweets containing links to any of these rumors through the fact-checking websites. Then, unrelated tweets were filtered out through a combination of automatic and manual measures. For each reply tweet, they extracted the original tweet that they were replying to and then extracted all the retweets of the original tweet. Each of these retweet cascades is a rumor propagating on Twitter. The dataset contains 126285 rumor cascades corresponding to 2448 rumors.

Existing datasets suffer from various limitations resulting in datasets that do not provide all possible features of interest. For example, some datasets have been produced based on a small number of news outlets (BuzzFace), include a small number of rumors (USPresidentialElection) or have a very restricted definition of real and fake news (SomeLikeItHoax). Moreover, some datasets cover a small period of time (BuzzFace), most of them are outdated or cover a limited topical range (USPresidentialElection). Another disadvantage, within the scope of developing automatic dataset creation systems, is that many of the state-of-the-art datasets rely heavily on human annotators (CREDBANK, USPresidentialElection). In addition, some datasets suffer from a very narrow definition of news propagation (BuzzFace, CREDBANK). For example, BuzzFace uses the Facebook Graph API for the collection of data on "reactions" to the original articles, but independent posts that spread this news are missing as is the temporal information of news propagation. Furthermore, some approaches underestimate or omit the role of social bots in the news propagation process by removing them from their corpus (CREDBANK, Vosoughi et al.). The most important limitation that almost all datasets suffer from are the social network API limits (Facebook, Twitter) that give partial access to social media information. There is one dataset (Vosoughi et al.) that claims to have access to the full Twitter History, but this dataset is not available to the research community. Table 1 depicts a comparison between these datasets and PHONY, the dataset we created for Greek tweets using the process described in this paper.

Table 1. Background work comparison.

Dataset features	Different types of fake news	Automated creation	Full news articles	Social media posts	Network
BuzzFace			✓	✓	
PHEME				✓	
USPresidentialElection				✓	✓
FakeNewsNet			✓	✓	✓
CREDBANK				✓	
SomeLikeItHoax				✓	
TraceMiner			✓	✓	✓
Vosoughi et al.			✓	✓	✓
PHONY	✓	✓	✓	✓	✓

3   The PHONY Dataset Creation Process

In this section, we present in detail PHONY, our automatic fake news dataset generation system, which contains the social media diffusion footprints of fake news stories. PHONY automatic dataset creation is an essential part of our ongoing research for developing models for the automatic detection of fake news content in streaming data. The datasets are created automatically and will be used as training datasets for machine learning classifiers.

Figure 2 depicts the overall architecture of our automatic fake news diffusion dataset generation system. The system consists of two main units: The Tweet Index Creation unit that creates the inverted index of streaming tweets, and the Fake News Collection and Analysis unit that collects and analyzes fake news. The two units are eventually combined through a query process. Specifically, queries are formulated based on the analysis performed in the Fake News Collection and Analysis unit and are then run on the index generated by the Tweet Index Creation unit. These units are described in detail in the following sections.

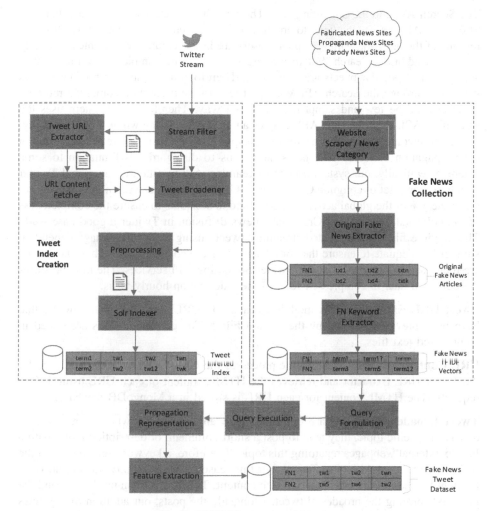

Fig. 2. PHONY system architecture

3.1 Tweet Index Creation

In order to dig out fake news stories scattered throughout the social network we need an efficient way to organize the streaming tweets. Therefore, in this step we describe the creation of an inverted tweet index, a hashmap-like data structure that directs from a word to a tweet. The Tweet Index Creation process is running continuously processing every single streaming tweet regardless of the fake news dataset creation process. As a result, the tweet index is incrementally rebuilt on a regular basis. The basic modules included in this unit are the following.

Stream Filter: The most important limitation that almost all existing datasets suffer from are the social network API limits (Facebook, Twitter) that give partial access to social media information. Specifically, Twitter maintains two publicly accessible APIs:

The Search API and the Streaming API. The Search API returns fewer results than the Streaming API, results that tend to emphasize toward central users and more clustered regions of the network, while peripheral users are less accurately represented or totally absent. In addition, Search API results are not a random sample of overall Twitter activity (most popular tweets are promoted). Therefore, a simple search of fake news keywords through the Search API would not lead to the desired outcome: the returned tweets would be few and sampled in a biased way. Therefore, our system uses the Streaming API. However, this API samples a random 1% of the whole Twitter activity, leading to incomplete and diverse results. Despite that, the fact that we are interested in the propagation of Greek fake news, allows us to disregard this limitation to some extent. Specifically, our system uses the Streaming API by filtering the global Twitter feed based on a set of common Greek words. Given that tweets written in Greek do not reach the 1% of the global activity, this method is adequate to ensure the completeness of the collected data making Greek fake news diffusion in Twitter a good case study This module filters the collected streaming tweets using a set of language-dependent keywords, adequate to ensure the completeness of the collected data. These keywords are then used as parameters of the Twitter Streaming API services. The module stores the collected data in compressed JSON Lines files on an hourly basis.

Tweet URL Extractor: This module extracts the URLs included in the tweets that have been previously stored by the Stream Filter. The extracted URLs are stored in compressed text files.

URL Content Fetcher: This module processes the URLs that have been extracted by the Tweet URL Extractor and retrieves the HTML content of each URL using HTTP requests. The HTML content for each URL is stored in a MongoDB database.

Tweet Broadener: Tweets are of limited size and, therefore, when users share or discuss a specific topic, they tend to post a short comment, or description along with a link to external webpages regarding this topic. Therefore, a keyword search only in the main tweet text would be insufficient. For this reason, Tweet URL Extractor and URL Content Fetcher retrieve this external web content. This text is, then used to expand the post text forming the broadened tweet. Alongside the posts, our addition of the news article content provides a sizeable collection of text that could be used for multiple learning tasks in the future. Specifically, each tweet is enriched with the HTML content extracted in the previous module. From the stored HTML content, the module extracts only the main textual content, in order to exclude advertising and irrelevant content. Finally, the broadened tweets are stored in compressed JSON Lines files.

Preprocessing: In this module the enriched text of each tweet goes through a pre-processing procedure. Specifically, the process includes a text tokenization and a word stemming step. Hence, this module creates a set of terms that are included in each tweet.

Solr Indexer: Our system operates by extending the Twitter Streaming API responses, creating a large set of JSON files. JSON objects offer an intuitive and descriptive way to represent our data and are also used by the Twitter Streaming API as response values. However, directly querying the JSON files would be very slow and, therefore, inefficient. To address this problem, our method creates an inverted index of all the posts that have been previously filtered from the stream and subsequently broadened using the main content of the included URLs. Specifically, this module generates the inverted index using the Apache Solr [9] search engine. The inverted index essentially links each term that exists within the tweet collection to the broadened tweets that contain the term. It is important to note that the tweet collection contains all tweets that have been collected from the Twitter stream, without regard to their possible association with fake news.

3.2   Fake News Collection and Analysis

The Fake News Collection and Analysis process runs on demand whenever we need to create a fake news dataset, collecting ground truth fake news stories from the web and analyzing their text content in order to extract representative keywords. Those keywords are used for querying the tweet index. The basic modules included in this unit are the following.

Website Scraper (Category-Specific): This module retrieves a list of fake news from any fact-checking website the user chooses to use. However, these organizations do not provide API access to their databases or to a coherent fake news dataset. In fact, many fact-checking organizations provide fake news data in the form of a simple web page or a blog. Therefore, in order to create a system that is able to collect fake news data from multiple fact-checking sources, we opted to use the scraping of such websites to obtain a list of fake news content along with the textual information that demonstrates their validation or disproof process. Moreover, our system stores the types of fake news by scraping the corresponding labels, since fact-checkers provide such labels for each fake news item (fabricated news, propaganda, conspiracy theories, unscientific news, clickbait, scam, etc.). Hence, the users of our system are able to choose the fact-checking source and the type of fake news from a customized list and feed the module with relevant websites and blogs. This list is customized, because the scraping process requires a level of customization for each webpage and because fact-checkers do not follow a uniform tagging system for the aforementioned fake news types. Using HTTP requests, the title and text of each article are retrieved and then stored in a MongoDB database.

Original Fake News Extractor: Fact-checkers provide critical analysis and evaluation of the truth of each article. However, this content is not the original fake news story and therefore, it cannot be directly used to trace fake news in social media. In order to find the original fake news, it is necessary to analyze these articles and to retrieve the fake news articles from their original sources. This is achieved through a recursive process of detecting original sources of fake news from the articles of the fact-checkers (extract external links). Then, using HTTP requests, the HTML content of each original URL is retrieved, and the main article content is extracted and stored in the MongoDB. However, news web pages often contain advertisements, pop-up ads, images and external links around the body of an article. Hence, our web content scraping process can be further optimized in order to completely filter irrelevant content out. We plan to improve the filtering of the main news content by using techniques based on DOM (Document Object Model) trees [10].

FN Keyword Extractor: At this point, we could define the diffusion of misinformation in social media as the propagation of those links. However, this approach would miss a large amount of fake news diffusion that is happening without reference to external sources via links, and also would not be able to retrieve fake news from sources that the fact-checkers haven't discovered yet. For this reason, we apply content analysis to the original articles and our system extracts a keyword vector for each original fake news story. In the future, we aim to minimize human effort in the process of creating such datasets by developing fully automated scraping modules that will be able to scrape multiple non-uniform sources. In this module, the system extracts keywords for each fake news article stored in the database. Specifically, keyword vectors are extracted using the TF/IDF method to acquire the most representative keywords for each fake news case.

3.3 Querying Process

This unit forms the queries based on the fake news analysis and runs them against the inverted index. The two modules included in this unit are the following.

Query Formulation: Our system uses the produced keyword vectors for each fake news to form queries that are performed on the inverted index of tweets. However, the query formulation requires a keyword selection strategy. We create queries by combining the top n terms of each fake news story with the AND, OR and NOT operators. We chose number n based on the average number of words per news story (a news story may consist of several articles). Specifically, we evaluated a sample of the collected fake news stories and the results showed that

$$n = \ln \mathrm{avg}(\mathrm{length}(\mathrm{news})) \tag{1}$$

can offer us a balanced representation. This module forms the queries combining the keywords with AND, OR and NOT operators.

Query Execution: In this module, the system runs a query on the inverted index for each fake news story. The module returns a set of tweets from the index that satisfy the query. Hence, a set of tweets is assigned to each fake news story.

3.4 Network Propagation

This module constructs the propagation representation of fake news in Twitter. A news story on Twitter starts with a user posting a claim about a topic or an event using text, multimedia or links to outside sources. Then, people propagate the news by retweeting the post. Users can also comment and discuss the news by replying and by mentioning other users to encourage a larger audience to engage with the news. However, Twitter APIs don't provide the exact retweet path of retweets. Therefore, we have to extract the diffusion network that consists of one or more propagation trees. We assume that a user retweets content that they have first seen on their timeline, hence, content that has been posted by one of their followees. In the case that multiple followees have posted the same tweet, we assume that the user has retweeted the tweet of the followee who most recently posted the tweet, hence closest to the retweeting time. Note that, in order to construct the propagation trees for a news story we don't need the complete follower/followee network of all the users participating in the propagation phenomena. Instead we construct the friendship network, hence, a subgraph of the original follower-followee graph that consists of those users who participated in the propagation process by performing targeted Twitter API calls to check if a participating user is followee/follower of another. The module stores the diffusion graph (propagation trees) and the friendship network of every fake news story in a MongoDB database.

3.5   Feature Extraction

Machine learning models need to be trained based on specific features extracted from the data collected by our system. In the context of another ongoing work, we have recorded, analyzed, and implemented extraction and calculation methods for a large set of various types of existing features: text, user, web content, network structure, geolocation, multimedia and propagation features. All features encountered in the literature can be directly extracted from our dataset by specific scripts and algorithms that will run on the broadened tweets, the propagation trees and the friendship networks that are formed as described in Sect. 3.4.

A detailed presentation of the features and their extraction method falls outside the scope of this work. However, in Table 2, we show a representative feature subset containing network and propagation features (left column). On the right column the corresponding descriptions prove that our system can sufficiently provide all the necessary data to perform these measurements.

4   PHONY-EL Dataset

We used the PHONY method to create a dataset including the fake news diffusion process in Greek tweets regarding the period from January 2016 to April 2018. The streaming tweets were collected as described in Sect. 3.1 and stored in multiple JSON Lines files. The size of the compressed files is 142.2 GB before broadening and 176.7 GB after broadening (text is broadened using main texts from URLs). Specifically, the collection contains 238685450 tweets posted by 1381799 unique users. These tweets contain 57264673 links to web pages. The created index, based on the process described in Sect. 3.1, contains 9237157 terms and takes up 126.6 GB on the hard disk.

Moreover, based on the method presented in Sect. 3.2, the system created a collection of fake news regarding this period. Specifically 677 fake news stories were crawled (by scraping https://www.ellinikahoaxes.gr with the label "fake news"). These stories were hosted on 2835 web pages. Furthermore, the system formed and executed 677 queries against the index as described in Sects. 3.2 and 3.3. The queries returned 381350 tweets, which propagated the fake news stories and were posted by 21223

Table 2. Dataset features

Groups of network and propagation features	Description
Fraction of low-to-high diffusion & fraction of high-to-low diffusion	Propagation from a user with lower influence (number of followers) to a receiver with higher influence. The diffusion of information from users with low influence to users with high influence is a phenomenon which is seen much more frequently when the information is true. The reason for this is perhaps because the user with the high influence would not risk retweeting a less known user's information unless the person had very good reasons to believe the information is true
Fraction of nodes and edges in largest connected component of the diffusion network	Captures the longest diffusion chain of a news story in the propagation tree
Fraction of nodes and edges in largest connected component of friendship network	Captures the longest user engagement tweeting about a news story
Average degree of nodes, average clustering coefficient, density, median in-degree, median out-degree, maximum degree in largest connected component of diffusion network	Structural properties of the diffusion network
Average degree of nodes, average clustering coefficient, density, median in-degree, median out-degree, maximum degree in largest connected component of friendship network	Structural properties of the friendship network
Fraction of nodes without incoming edges in the friendship network	Captures users that do not have any followers tweeting about the news story
Fraction of nodes without outgoing edges in the friendship network	Captures users that do not have any followees tweeting about the news story
Fraction of isolated nodes in the friendship network	Captures users that do not have any followers or followees tweeting about the news story
Fraction of nodes without incoming edges in the diffusion network	Captures users that started a propagation tree about a news story
Fraction of nodes without outgoing edges in the diffusion network	Captures users that participated in the propagation of a news story, but no user continued the propagation under their influence
Fraction of isolated nodes in the diffusion network	Captures users that started a propagation tree about a news story, but no user continued the propagation under their influence
Initial propagation tweets	Captures the number of source-original tweets regarding the news story
Max propagation subtree	Captures the length of the biggest propagation chain of a news story
Avg propagation depth	Captures the average depth of the propagation of a news story

(*continued*)

Table 2. (*continued*)

Groups of network and propagation features	Description
Average depth to breadth ratio	Captures shape of new story diffusion
Ratio of new users	Diversity/novelty of users engaged in the propagation of a news story
Ratio of original tweets	Captures how captivating, engaging and original is the conversation about a news story
Early burst volume	Captures the magnitude of the bursting tweets about a news story. It is the volume of tweets in the first spike of the propagation
Periodicity of external shock	Captures the time periodicity of subsequent bursts of tweet volume regarding a news story
Users avg influence	Captures the influence of users that propagate a news story (user influence = num of followers/num of followees)

unique users. Finally, the system produced 322265 propagation trees, while 68 fake news stories did not propagate in the Twitter network.

As presented in previous sections, the automatic creation of ground truth datasets involves processing and querying the entire set of streaming tweets. In order to efficiently implement our system, we used a cloud (distributed) approach consisting of 12 virtual machines (16 cores and 32 GB RAM each). We used Apache Spark to process tweet data and Apache Solr to create the inverted index and to execute the queries on the index. Specifically, for preprocessing streaming tweets, a Spark application was implemented in Python/PySpark. Moreover, we use NLTK3 and a language-dependent stemmer to preprocess the tweets, which are then given as an input to Apache Solr. Finally, the Solr index querying is performed through HTTP.

5   Conclusion

The efficiency of machine learning methods to detect fake news content on social media depends on the quality of the training datasets. Current ground truth datasets require manual human involvement causing problems to the reliable assessment and comparison of fake news detection methods. In this paper, we describe PHONY, an automatic system for creating fake news datasets, which is based on the fake news stories provided by curated fact-checking websites. Finally, we run PHONY system on the Greek Twitter stream, a dataset of fake news propagating in the Greek tweetosphere.

References

1. Santia, G.C., Williams, J.R.: BuzzFace: a news veracity dataset with Facebook user commentary and egos. In: ICWSM (2018)
2. Zubiaga, A., Liakata, M., Procter, R.: Learning reporting dynamics during breaking news for rumour detection in social media. arXiv:1610.07363 [cs] (2016)

3. Shin, J., Jian, L., Driscoll, K., Bar, F.: Political rumoring on Twitter during the 2012 US presidential election: Rumor diffusion and correction. New Media Soc. **19** (2016). https://doi.org/10.1177/1461444816634054

4. Shu, K., Mahudeswaran, D., Wang, S., Lee, D., Liu, H.: FakeNewsNet: a data repository with news content, social context and dynamic information for studying fake news on social media. arXiv:1809.01286 [cs] (2018)

5. Mitra, T., Gilbert, E.: CREDBANK: a large-scale social media corpus with associated credibility annotations. In: ICWSM (2015)

6. Tacchini, E., Ballarin, G., Vedova, M.L.D., Moret, S., de Alfaro, L.: Some Like it Hoax: Automated Fake News Detection in Social Networks.

7. Wu, L., Liu, H.: Tracing fake-news footprints: characterizing social media messages by how they propagate. In: Proceedings of the Eleventh ACM International Conference on Web Search and Data Mining, pp. 637–645. ACM, New York (2018). https://doi.org/10.1145/3159652.3159677

8. Vosoughi, S., Roy, D., Aral, S.: The spread of true and false news online. Science **359**, 1146–1151 (2018). https://doi.org/10.1126/science.aap9559

9. Apache Solr. https://lucene.apache.org/solr/. Accessed 6 July 2019

10. Gupta, S., Kaiser, G., Neistadt, D., Grimm, P.: DOM-based content extraction of HTML documents. In: WWW (2003)

Artificial Flora Optimization Algorithm for Task Scheduling in Cloud Computing Environment

Nebojsa Bacanin, Eva Tuba, Timea Bezdan, Ivana Strumberger,
and Milan Tuba$^{(\boxtimes)}$

Singidunum University, Danijelova 32, 11000 Belgrade, Serbia
{nbacanin,istrumberger}@singidunum.ac.rs, {etuba,tuba}@ieee.org,
timea.bezdan.17@singimail.rs

Abstract. Cloud computing is a relatively new computing paradigm
that enables provision of storage and computing resources over a network to end-users. Task scheduling represents the allocation of tasks
to be executed to the available resources. In this paper, we propose a
scheduling algorithm, named artificial flora scheduler, with an aim to
improve task scheduling in the cloud computing environments. The artificial flora belongs to the category of swarm intelligence metaheuristics
that have proved to be very effective in solving NP hard problems, such
as task scheduling. Based on the obtained simulation results and comparison with other approaches from literature, a conclusion is that the
proposed scheduler efficiently optimizes execution of the submitted tasks
to the cloud system, by reducing the makespan and the execution costs.

Keywords: Task scheduling · Makespan · Cloud computing ·
Artificial flora · Swarm intelligence · Optimization

1 Introduction

Cloud computing is a relatively novel computing paradigm, that provides shared
resources to the end user over the Internet. According to the National Institute
of Standards and Technology (NIST), cloud computing is defined as a model for
enabling ubiquitous, convenient, on-demand network access to a shared pool of
configurable computing resources (such as networks, servers, storage, applications and services) that can be rapidly provisioned and released with minimal
management effort or service provider interaction [1]. Due to the explosion of
Internet usage and the explosive growth of data, cloud computing technology
had a significant growth in recent years. One of the most important characteristics of the cloud system is that the cloud resources are in the virtual form: the
tasks requested by the users are processed on virtual machines. Users can access
resources on the cloud based on the pay-as-you-go or pay-as-you-use models.
The biggest cloud providers are Amazon Web Services (AWS), Google Cloud
Platform, Microsoft Azure, Oracle Cloud, IBM Cloud.

© Springer Nature Switzerland AG 2019
H. Yin et al. (Eds.): IDEAL 2019, LNCS 11871, pp. 437–445, 2019.
https://doi.org/10.1007/978-3-030-33607-3_47

In the cloud computing environment, resource management and task scheduling play a significant role. The number of tasks to be scheduled increases proportionally with the increase of the cloud users, thus an efficient algorithm is needed to schedule tasks because the efficiency of the cloud depends on the task scheduling. Task scheduling is a process in which the scheduler maps the tasks (requests submitted by the users) to be executed to the suitable and available resources, while taking into account different parameters, such as the calculation of completion time, makespan, cost, resource utilization, etc.

Task scheduling in cloud computing environment represents a typical NP-hard optimization problem. To solve NP-hard optimization problems, meta-heuristics proved to be very effective, thus in this paper we propose adaptations of the artificial flora (AF) swarm intelligence algorithm to address the issue of task scheduling and at the same time reducing the makespan [2]. Swarm intelligence algorithms have many successful applications to different real world NP hard challenges, and some of them include: localization in wireless sensors networks (WSNs) [3], image segmentation [4], portfolio optimization [5], path planning [6], RFID network planning [7] and convolution neural network hyper-parameters' optimization [8]. Swarm intelligence approaches have been also proposed for various cloud computing challenges [9,10].

The remaining of the paper is structured as follows: Sect. 2 gives a description about the proposed model, Sect. 3 provides an overview of the artificial flora optimization algorithm, Sect. 4 is dedicated to the experimental results and discussion, and Sect. 5 summarizes and visualizes the conclusions of the performed simulations.

2 Task Scheduling Model Problem Definition

This section provides a brief description of the problem definition and the system model of task scheduling in the cloud computing environment that was utilized in simulations. The system model is based on the model formulated in [11]. Every task requested by the user has different constraints, such as cost, memory requirement, required budget, deadline.

In the proposed multi-objective model, the cost function of the CPU and the cost function of the memory are calculated, and their sum represents the budget cost function. The task allocation to the virtual machines is based on the fitness value and it is calculated by adding the makespan to the budget cost function.

Assumed is a cloud C, which contains a finite number of resources, p physical machines N_{pm}, each of them containing N_{vm} virtual machines V.

$$C = \{P_1, P_2, ..., P_{N_{pm}}\} \tag{1}$$

$$P_1 = \{V_1, V_2, ..., V_{N_{vm}}\} \tag{2}$$

It is assumed that there are N_{tasks} independent tasks and the set of tasks can be described as follows:

$$Task = \{T_1, T_2, ..., T_{N_{tasks}}\} \tag{3}$$

Cost function of the CPU and memory of the virtual machine V_j can be calculated by using Eqs. (4) and (5).

$$C(x) = \sum_{j=1}^{|VM|} C^{cost}(j) \tag{4}$$

$$M(x) = \sum_{j=1}^{|VM|} M^{cost}(j) \tag{5}$$

where $|VM|$ denotes the number of virtual machines, the CPU cost and memory cost of virtual machine V_j are denoted by $C^{cost}(j)$ and $M^{cost}(j)$, respectively.

The cost of the resource is composed of two components, the cost of the CPU and the cost of the memory. The CPU cost function is calculated according to the Eq. (6) and the memory cost function is calculated according to the Eq. (7).

$$C^{cost}(j) = C_{base} \times C_j \times t_{ij} + C_{Trans} \tag{6}$$

$$M^{cost}(j) = M_{base} \times M_j \times t_{ij} + M_{Trans} \tag{7}$$

In Eqs. (6) and (7), subscript $base$ refers to the base cost, C_j and M_j are CPU and memory of the virtual machine V_j, the execution time of the task T_i in the resource R_j is denoted by t_{ij} and the transmission cost is specified by the subscript $Trans$.

Budget cost function $B(x)$ is calculated as the sum of the CPU cost and memory cost of the virtual machine. Its value should be less than or equal to the budget cost of the user.

$$B(x) = C(x) + M(x) \tag{8}$$

The fitness value $H(x)$ is obtained by the sum of the makespan $F(x)$ and budget cost function $B(x)$. Its value should be less than or equal to the task's deadline value.

$$H(x) = F(x) + B(x) \tag{9}$$

3   Overview of the Artificial Flora Optimization Algorithm

The AF optimization [2] is a recent swarm intelligence algorithm. This algorithm simulates the process of flora migration and reproduction in nature. The authors of the AF used six standard benchmark functions for performance evaluation and the obtained results outperformed other compared algorithms. The algorithm initially generates original plants, in the next step each of them generates new pants within a distance radius. The following steps are the main steps of the AF algorithm:

The first step of the algorithm is the initialization phase, N original plants are created randomly.

In the second step, the radius is calculated for each plant. The value of the radius depends on the radius of the previous two generations. The distance radius is calculated according to the formula:

$$D_j = D_{1j} \times r \times c_1 + D_{2j} \times r \times c_2 \tag{10}$$

where the parent plant's radius and the grandparent plant's radius are D_{2j} and D_{1j}, respectively, the learning coefficients are c_1 and c_2, and r is a random number generated in the range between 0 and 1.

The radius of the new parent plant is calculated as follows:

$$D'_{1j} = D_{2j} \tag{11}$$

The radius of the new grandparent plant is calculated as follows:

$$D'_{2j} = \sqrt{\frac{\sum_{i=1}^{N}(X_{i,j} - X'_{i,j})^2}{N}} \tag{12}$$

The locations of the offspring plants are within the radius of the original plants and it is calculated by the following equation:

$$X'_{i,j \times m} = R_{i,j \times m} + X_{i,j} \tag{13}$$

where the offspring plant location is denoted by $X'_{i,j}$ and m indicates to its number. $R_{i,j \times m}$ is a normally distributed random number with parameters mean 0 and variance D_j.

The surviving probability decides whether an offspring plant will survive or not. The surviving probability is calculated by using the proportionate-based selection:

$$p = \left| \sqrt{\frac{F(X'_{i,j \times m})}{F_{max}}} \right| \times Q^{(j \times m - 1)} \tag{14}$$

where $Q^{(j \times m - 1)}$ represents the selective probability. The maximum fitness of all offsprings is denoted by F_{max}, and the fitness of each individual offspring plant is represented by $F(X'_{i,j \times m})$.

If p is greater than r that means that the offspring is alive, while in the opposite case it is not alive, where r is a randomly generated number in the range between 0 and 1.

In the final step, the selection is made among survived offspring plants. N offspring plants are selected as new original plants for the next run. The steps are repeated until the algorithm reaches the best solution (or other exit criterion).

Algorithm 1. Pseudocode of the AF algorithm

Initialization: Create randomly N original plants; calculate the fitness value; pick the best solution
while $t < MaxIter$ **do**
    for all $N * M$ *solutions* **do**
        Calculate the radius using Eq. (10), Eq. (12), Eq. (11)
        Create offspring plants using Eq. (13)
        if $p > r$ **then**
            Offspring plant survived
        else
            Offspring plant did not survived
        end if
    end for
    Evaluate new solutions
    Choose randomly N new original plants
    Update the solution if the new solution has better value then the old solution
end while
return the best solution

---

4 Simulation Results and Discussion

For simulation purposes, we adapted the AF algorithm for solving tasks scheduling problem in cloud computing environments. We named the proposed approach AF-Scheduler (AF-S).

One of the most important tasks when adapting any metaheuristic for a specific problem is encoding the problem candidate solutions. According to the model formulation given in Sect. 2, if there are N_{tasks} with available N_{vm} VMs in the cloud, the size of the search space and the number of possible combinations (allocations) of tasks to VMs is $(N_{vm})^{N_{tasks}}$. In the AF-S, candidate problem solution is represented as a set of tasks that should be executed, where each tasks is joined (mapped) to the appropriate VM. According to this, every individual in the population is encoded as an array of size N_{tasks}, where each array element has the value from the domain $[1, N_{vm}]$. For example, if there are 8 tasks and 3 VMs, candidate solution can be denoted as $[2, 1, 1, 3, 2, 3, 1, 3]$. We performed experiments with varying number of solutions in the AF-S population N, and the varying number of iterations. On the other side, in all conducted experiments grandparent learning coefficient $c1$ was set to 0.75, while the parent learning coefficient $c2$ was set to 1.25 since it was established that those are the best parameters' values through several trial and error experiments.

As already stated above, in simulations we employed the same model and the same experimental conditions as in [11]. Model parameters were set as follows: $C_{base} = 0.17/hr$, $C_{Trans} = 0.005$, $M_{base} = 0.005\,GB/hr$ and $M_{Trans} = 0.5$.

In order to evaluate the performance of the proposed AF-S approach, we considered the makespan and the cost objectives. Makespan denotes the total amount of time needed to execute all the tasks, while the budget cost represents the total cost needed for scheduling tasks to the VMs.

The formulations of the makespan ($F(x)$) and the total cost ($B(x)$) objectives are given in Eqs. (15) and (16), respectively.

$$F(x) \leq \sum_{i=1}^{|T|} D_i \tag{15}$$

$$B(x) \leq \sum_{i=1}^{|T|} B_i, \tag{16}$$

where D_i and B_i represent the deadline and the budget cost of task T_i, respectively.

Task scheduling simulations were performed in the environment of the standard CloudSim toolkit. In all conducted simulations, each physical machine consisted of 10 VMs. The processing power of each VM was 1860 MIPS or 2660 MIPS, while all VMs had 4 GB of RAM memory. Each task had different requirements in terms of the CPU, memory, deadline and budget costs.

For each test, we executed AF-S in 50 independent runs and took the average of all runs as the final result. In each run, the environment (tasks and VM specifications) were generated randomly within above mentioned constraints. For more details about the experimental conditions, please refer to [11].

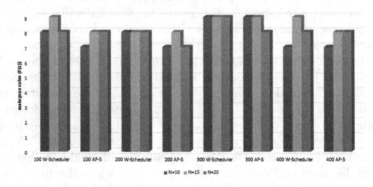

Fig. 1. Comparative analysis - AF-S vs. W-Scheduler for 50 iterations with makespan objective

In the first experiment, we performed comparative analysis with the W-Scheduler [11] for the makespan ($F(x)$) objective. Visualization of makespan results with 50 iterations and 20 physical machines with population sizes (N) of 10, 15 and 20, for number of tasks ranging from 100 to 400, is given in Fig. 1.

In the second experiment, we conducted the comparative analysis with the W-Scheduler [11] for the budget cost ($B(x)$) objective. Visualization of budget cost objective in simulations with 25 iterations and 20 physical machines with population sizes (N) of 10, 15 and 20, and number of tasks ranging from 100 to 400, is given in Fig. 2.

From the presented results, it can be concluded that on average, AF-S accomplishes better results quality than the W-Scheduler algorithm. For example, in simulations with 100 tasks, obtained makespan of the W-Scheduler for 10, 15 and 20 populations sizes is 8, 9 and 8, respectively. For the same indicator, AF-S accomplishes the values of 7, 8 and 8.

In simulations with budget cost, the W-Scheduler only in test instances with 100 tasks and population size of 20, and 400 tasks and population size of 15 obtained better performance than the AF-S, as can bee seen from the Fig. 2. In tests with 200 and 300 tasks, AF-S for every population size accomplished better results and convergence than the W-Scheduler. For example, in test instance with 200 tasks, W-Scheduler obtained objective function results of 2255.188, 2217.960 and 2059.852 for 10, 15 and 20 population sizes, respectively. In the same test instance, AF-S managed to generate budget costs of 2130.305, 2199.845 and 2051.713.

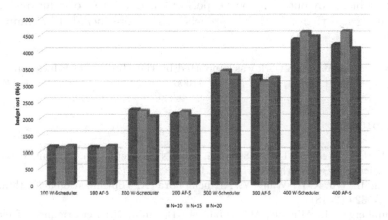

Fig. 2. Comparative analysis - AF-S vs. W-Scheduler for 25 iterations with budget cost objective

Finally, we compare the makespan and the average cost indicator with W-Scheduler, PBACO, SLPSO-SA and SPSO-SA. For more details about this experiment, please refer to [11]. Also, results for all metaheuristics included in comparative analysis were taken from [11]. As can be concluded from results presented in Table 1, the AF-S showed superior performance by obtaining better

Table 1. Comparative analysis for makespan and average cost indicators with other algorithms (best results are marked bold)

	AF-S	W-Scheduler	PBACO	SLPSO-SA	SPSO-SA
Makespan	**6**	7	15	30	15
Average cost	**5.1**	5.8	16	14	13

solutions' quality and the convergence speed than all other algorithms included in comparative analysis.

5   Conclusion

In this paper, we proposed a scheduling algorithm, named artificial flora scheduler (AF-S), with an aim to improve task scheduling in the cloud computing environments. We adapted the AF swarm intelligence approach for tackling cloud scheduling challenges.

For research purposes of this paper we utilized the same model as in [11] and for simulations we employed the CloudSim framework considering the makespan and budget costs objectives. Comparative analysis between the proposed AF-S and other state-of-the-art approaches was performed. According to the obtained empirical results the AF-S proved to be robust metaheuristic for solving this type of problems by obtaining better performance in terms of solutions' quality and convergence than all other approaches that were included in comparative analysis.

Acknowledgment. This work was supported by the Ministry of Education and Science of Republic of Serbia, Grant No. III-44006.

References

1. Mell, P.M., Grance, T.: Sp 800–145. The NIST definition of cloud computing. Technical report, Gaithersburg, MD, United States (2011)
2. Cheng, L., Wu, X.-H., Wang, Y.: Artificial flora (AF) optimization algorithm. Appl. Sci. **8**, 329 (2018)
3. Strumberger, I., Minovic, M., Tuba, M., Bacanin, N.: Performance of elephant herding optimization and tree growth algorithm adapted for node localization in wireless sensor networks. Sensors **19**(11), 2515 (2019)
4. Hrosik, R.C., Tuba, E., Dolicanin, E., Jovanovic, R., Tuba, M.: Brain image segmentation based on firefly algorithm combined with k-means clustering. Stud. Inf. Control **28**(2), 167–176 (2019)
5. Bacanin, N., Tuba, M.: Firefly algorithm for cardinality constrained mean-variance portfolio optimization problem with entropy diversity constraint. Sci. World J. Spec. Issue Comput. Intell. Metaheuristic Algorithms Appl. **2014**, 16 (2014). Article ID 721521
6. Tuba, E., Strumberger, I., Zivkovic, D., Bacanin, N., Tuba, M.: Mobile robot path planning by improved brain storm optimization algorithm. In: 2018 IEEE Congress on Evolutionary Computation (CEC), pp. 1–8, July 2018
7. Strumberger, I., Tuba, E., Bacanin, N., Beko, M., Tuba, M.: Modified monarch butterfly optimization algorithm for RFID network planning. In: 2018 6th International Conference on Multimedia Computing and Systems, pp. 1–6 (2018)
8. Strumberger, I., Tuba, E., Bacanin, N., Zivkovic, M., Beko, M., Tuba, M.: Designing convolutional neural network architecture by the firefly algorithm. In: 2019 International Young Engineers Forum (YEF-ECE), pp. 59–65, May 2019

9. Kalra, M., Singh, S.: A review of metaheuristic scheduling techniques in cloud computing. Egypt. Inf. J. **16**(3), 275–295 (2015)
10. Li, J., et al.: Task scheduling algorithm based on fireworks algorithm. EURASIP J. Wireless Commun. Netw. **2018**, 256 (2018)
11. Sreenu, K., Sreelatha, M.: W-scheduler: whale optimization for task scheduling in cloud computing. Cluster Comput. (2017)

A Significantly Faster Elastic-Ensemble for Time-Series Classification

George Oastler[(⊠)] and Jason Lines[(⊠)]

University of East Anglia, Norwich, UK
{g.oastler,j.lines}@uea.ac.uk

Abstract. The Elastic-Ensemble [7] has one of the longest build times of all constituents of the current state of the art algorithm for time series classification: the Hierarchical Vote Collective of Transformation-based Ensembles (HIVE-COTE) [8]. We investigate two simple and intuitive techniques to reduce the time spent training the Elastic Ensemble to consequently reduce HIVE-COTE train time. Our techniques reduce the effort involved in tuning parameters of each constituent nearest-neighbour classifier of the Elastic Ensemble. Firstly, we decrease the parameter space of each constituent to reduce tuning effort. Secondly, we limit the number of training series in each nearest neighbour classifier to reduce parameter option evaluation times during tuning. Experimentation over 10-folds of the UEA/UCR time-series classification problems show both techniques and give much faster build times and, crucially, the combination of both techniques give even greater speedup, all without significant loss in accuracy.

Keywords: Time series · Classification · Ensembles · Distance measures

1 Introduction

The current state of the art classifier in time series classification (TSC) is the Hierarchical Vote Collective of Transformation-based Ensembles (HIVE-COTE) [8]. The Elastic-Ensemble (EE) [7] is one of five constituent classifiers in the HIVE-COTE meta-ensemble and key in uncovering discriminatory features in the time-domain. The discovery of these features leverages 11 nearest-neighbour classifiers (NN) each coupled with an *elastic* distance-measure. Consequently, EE requires a large amount of time to train and forms a bottle-neck in training HIVE-COTE. Ten of the constituent classifiers in EE use distance measures with $O(m^2)$ run-time complexity (where m is the length of the time-series). Eight of these each require parameter tuning using leave-one-out-cross-validation (LOOCV) over 100 parameter options. Therefore the tuning complexity of EE becomes $O(n^2m^2)$, an often impractically expensive procedure.

Distance-measures have been studied for a long time in TSC research and various distance-measure specific speed-ups have been conceived, such as utilising lower-bounds. However, further speed-ups can be made to the NN and parameter tuning aspects of EE such as:

© Springer Nature Switzerland AG 2019
H. Yin et al. (Eds.): IDEAL 2019, LNCS 11871, pp. 446–453, 2019.
https://doi.org/10.1007/978-3-030-33607-3_48

1. using less parameter options when tuning constituent classifiers. We hypothesise many parameter options perform similarly, therefore a reduced parameter pool still contains a suitable parameter option.
2. using less train cases when estimating the accuracy of different parameter options during tuning. We hypothesise that training a NN on a subset of the train data will perform sufficiently well to evaluate a parameter option and maintain the relative ranks of parameter options during tuning.

We conducted experiments on various configuration of EE to investigate the effectiveness of these techniques. Our experiments use the UEA/UCR TSC problems [2] resampled 10 times at the original train/test distribution. First, we investigate the effectiveness of reduced parameter pools: 10%, 50% and 100% of the original parameter pool, chosen arbitrarily. We use the full training set to evaluate each parameter option. Second, we investigate the effectiveness of reduced training sets for parameter tuning: 10%, 50% and 100% of the original train set size, again chosen arbitrarily. We use the full parameter pool during tuning. Finally, we combine both techniques to investigate both reducing the parameter space and reducing the train set for each parameter option evaluation.

Our results demonstrate that either technique results in substantial reduction of training time without significant loss in test accuracy. Crucially, the subsequent combination of both techniques show further reduction of train time whilst still maintaining no significant loss in test accuracy. We conclude that limiting parameter pool size and train set size during tuning can speed-up EE by nearly two orders of magnitude without any significant loss in test accuracy.

2   Background and Related Work

A time-series, $T < x_1, x_2, x_3, ..., x_l >$, is an ordered sequence of l values with $x_i \in \mathbb{R}$. TSC is the task of prediction a class given a previously unseen time-series for which the class is unknown. A TSC classifier is therefore a function which is learned from the labelled time-series, the train set, to take an input of an unlabelled time-series, a test case, and output the predicted label.

TSC is an active area of research where many diverse algorithms have been proposed. These include, but are not limited to, histogram-based approaches that discriminate cases based on the frequency of reoccurring patterns [10,11]; shapelet algorithms that differentiate class membership through the presence of discriminatory, phase-independent subsequences [4,13]; and forest ensembles built on data transformed into different representations [3,8]. Arguably, most TSC research effort over the last decade has been focused on developing *elastic* distance measures to couple with simple 1-nearest neighbour (1-NN) classifiers. Such *elastic* measures are able to mitigate misalignments and phase-shift within time-series. The most common approach is to use Dynamic Time Warping (DTW) with a warping window set through cross-validation and a 1-NN classifier. Related variants of DTW have also been proposed, such as applying a soft boundaries to warping windows through weighting penalties [5], and warping directly on first-order derivatives [6]. Alternatives exist that are derived from the edit-distance [1], and further hybrid measures have characteristics of both DTW and edit-distance [9,12].

2.1 The Elastic Ensemble

The performance of alternative *elastic* distance measures with 1-NN classifiers were compared in [7] to determine whether one approach outperformed all others. The measures included were: Dynamic Time Warping (DTW), Derivative DTW (DDTW), weighted variants of both DTW and DDTW (WDTW, WDDTW), Edit Distance with Real Penalty (ERP), Longest Common Subsequence (LCSS), Time Warp Edit (TWE) and Move-Split-Merge (MSM). All measures were coupled with 1-NN classifiers and the eight distance-measures with parameters were tuned over 100 parameter options each, respectively. Euclidean distance, full-window DTW and full-window DDTW were used as baselines and all eight subsequent 1-NN classifiers were compared over 85 TSC datasets [2]. It was found that no single measure significantly outperformed all others in test accuracy. However, the diversity in performance of each distance-measure inspired the EE, an ensemble of the 11 1-NN classifiers described above. In training, each constituent is evaluated using a LOOCV to obtain an estimate of train set accuracy and the optimal distance-measure parameter option if required. In testing, each constituent is given a vote weighted by its training accuracy estimate and the ensemble predicts the class with the greatest weighted vote.

3   Proposed Enhancements

3.1   Reduced Parameter Pool Size

Eight of the distance measures in EE have a corresponding pool of 100 parameters. We hypothesise that there is a large amount of redundancy due to the similarity in parameter option performance, therefore the pool of parameters can be reduced whilst still yielding a suitable parameter choice during tuning. In our experiments we arbitrarily use 10%, 50% and 100% of the full parameter pool, sampled randomly.

3.2   Reduced Neighbourhood Size

LOOCV is used to evaluate each parameter option during tuning of the eight distance measures which required parameters. We hypothesise that parameter options can be effectively evaluated in a NN using substantially less train cases, hence reducing the impact of the expensive LOOCV procedure for parameter tuning. Less training cases reduces the neighbourhood of potential nearest neighbours in the NN, decreasing test time and speeding-up LOOCV. We believe a sufficiently large neighbourhood should evaluate a parameter option accurately enough to maintain ranks of parameter options during tuning. In our experiments we arbitrarily use 10%, 50% and 100% of the training set, sampled randomly.

3.3   Combined Strategies

A subsequent technique is to reduce both the parameter pool size Sect. 3.1 and neighbourhood Sect. 3.2 size. The effectiveness of this technique is dependent upon the success of the previous techniques. The two techniques likely impact each other, as introducing less-accurate parameter evaluation through a limited neighbourhood size may not find the optimal parameter option in a reduced parameter option pool. We designed a subsequent experiment to assess all combinations of these techniques over the 10%, 50% and 100% limits of parameter pool size and neighbourhood size respectively.

4   Experimental Design

We ran experiments to assess the impact of techniques from Sect. 3. Our experiments use the UEA/UCR TSC problems over 10 resamples at the original train/test ratio. Only 48 of the smallest datasets were investigated due to infeasible run-time of full EE, demonstrating the importance of speeding up EE for realistic usability. These results are indicative of the performance of the training strategies however, and further results will be added in due course when available to confirm the findings over the complete repository. All source code can be downloaded from the provided link[1,2].

5   Results

The results are organised into three separate experiments and findings: parameter pool size reduction, neighbourhood size reduction, and combining both reduction techniques. For each experiment we report the accuracies and train-times to assess the impact of each technique upon the performance of EE.

The critical difference (CD) diagrams used throughout these experiments visually demonstrate the results of comparing all classifiers using pairwise Wilcoxon signed rank tests, where *cliques* are formed using the Holm correction to represent classifiers where there is no significant difference between them.

5.1   Parameter Pool Size Reduction

The results shown in the CD diagram of Fig. 1 indicate that using 10% of possible parameter options during tuning of each EE constituent is not significantly different to original EE which uses all possible parameter options.

This demonstrates that EE can be trained much faster with no significant loss in test accuracy. It is worth noting there is a significant difference in 10% and 50% parameter pool sizes. This indicates the random sampling of 10% of the parameter pool may be too few and further investigation is required. We provide scatter plots in Fig. 2 comparing the two reduced parameter pool sizes against full EE to demonstrate no significant difference in test accuracy.

---

[1] https://github.com/TonyBagnall/uea-tsc/commit/07408d166072e8fd3057cb1fcbfd
913e603094e3.

[2] https://github.com/alan-turing-institute/sktime.

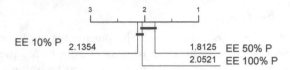

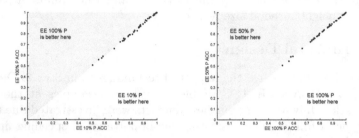

Fig. 1. A CD diagram to compare the test accuracy of EE using reduced parameter pool sizes during tuning of each constituent NN of EE. For clarity, "10% P" uses only 10% of the available parameter pool during tuning for each constituent classifier.

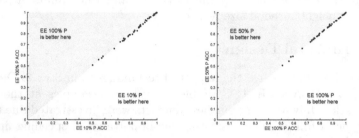

Fig. 2. (a) Scatter plot comparing test accuracy of full EE against EE with 10% parameter pool. (b) Scatter plot comparing test accuracy of full EE against EE with 50% parameter pool.

5.2   Neighbourhood Size Reduction

The results of reducing the neighbourhood size during tuning of parameter options are summarised in the CD diagram in Fig. 3 and scatter plots in Fig. 4.

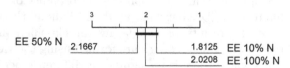

Fig. 3. A CD diagram to compare the test accuracy of EE with reduced neighbourhood sizes for evaluating parameter options during tuning of each constituent NN. For clarity, "10% N" uses 10% of the training data during tuning to evaluate a parameter option.

The results demonstrate that there is no significant loss in test accuracy when using 10% or 50% of training data during tuning of parameter options. This confirms our hypothesis that parameter options can be sufficiently evaluated and ranked using much less training data. Therefore, a substantial amount of time can be saved while training EE by using a smaller neighbourhood. We reinforce the equivalence in test accuracy in the scatter plots in Fig. 4.

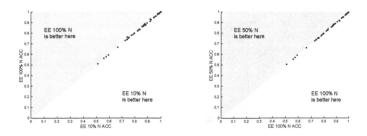

Fig. 4. (a) Scatter plot comparing test accuracy of full EE against EE with 10% neighbourhood size. (b) Scatter plot comparing test accuracy of full EE against EE with 50% neighbourhood size.

5.3 Combined Techniques: Reduced Parameter Pool Size and Reduced Neighbourhood Size

The results presented in Sects. 5.2 and 5.1 can be combined to further speed-up the training of EE. These results show there is no significant difference between full EE against EE with 10% parameter pool size, and no significant difference between full EE against EE with 10% neighbourhood size. These techniques can be combined to investigate further speed-up, again arbitrarily using the values of 10% and 50% to produce four combinations of each (10%/10%, 10%/50%, 50%/10%, and 50%/50%) alongside full EE with 100% neighbourhood size and 100% parameter pool size. These results are summarised in the critical difference diagram in Fig. 5 and scatter plot in Fig. 6.

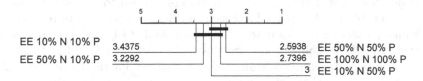

Fig. 5. A CD diagram to compare the test accuracy performance of EE using various neighbourhood sizes (N) and parameter pool sizes (P). For clarity, 50% N and 10% P corresponds to tuning each constituent NN of EE using 10% of the full parameter pool and evaluates each parameter option 50% of the training data.

The results in Fig. 5 confirm that there is no significant difference in the test accuracies of full EE and EE trained using only 10% of parameter options and 10% of training data during tuning (EE-10%). This does not lower the run-time complexity of EE but does decrease the train time to approximately 3% of full EE as outlined in Table 1.

Note that the test accuracy of EE-10%, whilst not significantly different to full EE, is significantly worse than the best performing variants which use 50% of parameter options. Table 1 indicates reducing neighbourhood size provides better speed-up than reducing parameter pool size. Therefore, reducing

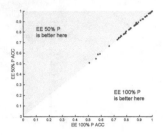

Fig. 6. Scatter plot comparing test accuracy of full EE against recommended EE configuration with 10% neighbourhood size and 50% parameter pool size.

Table 1. Table of accuracies and timings over UEA/UCR TSC datasets comparing EE with different configurations of neighbours and parameters. N corresponds to percentage of neighbours, P corresponds to the percentage of parameters.

Dataset	100% N 100% P	50% N 50% P	10% N 50% P	50% N 10% P	10% N 10% P
Accuracy (%)	0.8478	0.8491	0.8477	0.8466	0.8465
Train time (minutes)	37.1200	17.8101	2.4314	15.1747	0.9948
Train time (% of full EE)	100.0000	47.9800	6.5500	40.8800	2.6800

parameter pool size beyond 50% whilst also reducing neighbourhood size significantly decreases test accuracy. Practitioners are advised to use 50% parameter pool size and 10% neighbourhood size as a sufficient compromise to reduce train time whilst preserving test accuracy. The recommended EE with 50% parameter pool size and 10% neighbourhood size was ranked 3rd overall in Fig. 5 and was not significantly outperformed by any other technique. Furthermore, the timing results in Fig. 1 show that this configuration requires approximately only 6.6% of the time of full EE - nearly two orders of magnitude faster than the original EE. We demonstrate the equivalence in test accuracy between the recommended configuration and the full EE in Fig. 6.

6 Conclusions and Future Work

In this work we have investigated two techniques for reducing the training time required to run EE. First, we proposed a technique to reduce the distance measure parameter pool size (using random sampling) for each constituent NN classifiers of EE during tuning. Second, we proposed a technique to use less neighbours (via random sampling) in the NN constituent classifiers to evaluate a parameter option whilst tuning. We hypothesised that both could lead to substantial speed-ups in train times without significant loss in accuracy as a suitable parameter option is still found. We validated these claims through two independent experiments looking at either technique and conclude that using either 10% neighbourhood size or 10% parameter pool size does not significantly reduce test accuracy.

Inspired by these findings, we combined both techniques to build EE with a reduced parameter pool size and neighbourhood size. We found that EE could be sped-up to approximately 3% of the original train-time of full EE at best. We also conclude that using the recommended, and crucially not significantly worse, configuration of 10% neighbourhood size and 50% parameter pool size takes approximately only 6.6% train-time versus full EE.

References

1. Chen, L., Ng, R.: On the marriage of Lp-norms and edit distance. In: Proceedings 30th International Conference on Very Large Databases (VLDB) (2004)
2. Chen, Y., et al.: The UEA-UCR time series classification archive (2015). http://www.cs.ucr.edu/~eamonn/time_series_data/
3. Deng, H., Runger, G., Tuv, E., Vladimir, M.: A time series forest for classification and feature extraction. Inf. Sci. **239**, 142–153 (2013)
4. Hills, J., Lines, J., Baranauskas, E., Mapp, J., Bagnall, A.: Classification of time series by shapelet transformation. Data Min. Knowl. Disc. **28**(4), 851–881 (2014)
5. Jeong, Y., Jeong, M., Omitaomu, O.: Weighted dynamic time warping for time series classification. Pattern Recogn. **44**, 2231–2240 (2011)
6. Keogh, E., Pazzani, M.: Derivative dynamic time warping. In: Proceedings 1st SIAM International Conference on Data Mining (SDM) (2001)
7. Lines, J., Bagnall, A.: Time series classification with ensembles of elastic distance measures. Data Min. Knowl. Disc. **29**, 565–592 (2015)
8. Lines, J., Taylor, S., Bagnall, A.: Time series classification with HIVE-COTE: the hierarchical vote collective of transformation-based ensembles. ACM Trans. Knowl. Discovery From Data **12**(5) (2018)
9. Marteau, P.: Time warp edit distance with stiffness adjustment for time series matching. IEEE Trans. Pattern Anal. Mach. Intell. **31**(2), 306–318 (2009)
10. Schäfer, P.: The BOSS is concerned with time series classification in the presence of noise. Data Min. Knowl. Disc. **29**(6), 1505–1530 (2015)
11. Schäfer, P., Leser, U.: Fast and accurate time series classification with weasel. In: Proceedings of the 2017 ACM on Conference on Information and Knowledge Management, pp. 637–646. ACM (2017)
12. Stefan, A., Athitsos, V., Das, G.: The Move-Split-Merge metric for time series. IEEE Trans. Knowl. Data Eng. **25**(6), 1425–1438 (2013)
13. Ye, L., Keogh, E.: Time series shapelets: a novel technique that allows accurate, interpretable and fast classification. Data Min. Knowl. Disc. **22**(1–2), 149–182 (2011)

ALIME: Autoencoder Based Approach for Local Interpretability

Sharath M. Shankaranarayana$^{(\boxtimes)}$ and Davor Runje

ZASTI.AI, Chennai, India
sharath@zasti.ai

Abstract. Machine learning and especially deep learning have garnered tremendous popularity in recent years due to their increased performance over other methods. The availability of large amount of data has aided in the progress of deep learning. Nevertheless, deep learning models are opaque and often seen as black boxes. Thus, there is an inherent need to make the models interpretable, especially so in the medical domain. In this work, we propose a locally interpretable method, which is inspired by one of the recent tools that has gained a lot of interest, called local interpretable model-agnostic explanations (LIME). LIME generates single instance level explanation by artificially generating a dataset around the instance (by randomly sampling and using perturbations) and then training a local linear interpretable model. One of the major issues in LIME is the instability in the generated explanation, which is caused due to the randomly generated dataset. Another issue in these kind of local interpretable models is the local fidelity. We propose novel modifications to LIME by employing an autoencoder, which serves as a better weighting function for the local model. We perform extensive comparisons with different datasets and show that our proposed method results in both improved stability, as well as local fidelity.

Keywords: Interpretable machine learning · Deep learning · Autoencoder · Explainable AI (XAI) · Healthcare

1 Introduction and Related Work

Machine learning models with good predictability, such as deep neural networks, are difficult to interpret, opaque, and hence considered to be black box models. Deep neural networks have gained significant prominence in healthcare domain and are increasingly being used in critical tasks such as disease diagnosis, survival analysis, etc. As a result, there is a pressing need to understand these models to ensure that they are correct, fair, unbiased, and/or ethical.

Currently, there is no precise definition of interpretability, and the definitions tend to depend upon the application. Some of the most common types of model explanations are [10]:

© Springer Nature Switzerland AG 2019
H. Yin et al. (Eds.): IDEAL 2019, LNCS 11871, pp. 454–463, 2019.
https://doi.org/10.1007/978-3-030-33607-3_49

1. **Example-based.** In this method, one is interested in knowing the point in the training set that has close resemblance with the test point one is interested in explaining [2,6].
2. **Local.** In these methods, one is interested in deriving explanations by fitting an interpretable model locally around the test instance of interest [12].
3. **Global.** In these methods, one is interested in knowing the overall behaviour of the model. Global explanations attempt to understand underlying patterns in the model behaviour, usually by employing a series of rules for explanations as [7,13].

The interpretable and explainable methods can also be grouped based on other criteria, such as *(i)* Model agnostic or model specific *(ii)* Intrinsic or post-hoc *(iii)* Perturbation or saliency based, etc. Recently, post-hoc explainable methods, such as LIME [12], have gained a lot of interest since *a posteriori* explanations may currently be the only option for explaining already trained models. These methods are model-agnostic and hence do not require understanding of the inner workings of trained model.

Some of the desired characteristics of interpretable models are consistency and stability of explanations, and local fidelity or faithfulness of the model locally. Local post-hoc methods, such as LIME, lack in this regard. In this paper, we propose a modification to the popular framework for generating local explanations LIME [12] that improves both *stability* and *local fidelity* of explanations.

Since our focus is on post-hoc and local interpretable methods, we restrict our literature review to include only those methods. As mentioned above, LIME [12] is one the most popular methods in local interpretability. In LIME, artificial data points are first generated using random perturbation around the instance of interest to be explained, and then a linear model is fit around the instance. The same authors extended LIME to include global explanations in [13]. Most of the modifications to LIME have been in the line of selecting appropriate kind of data for training the local interpretable model. In [5], the authors use K-means clustering to partition the training dataset and use them for local models instead of perturbation-based data generation. In another work called DLIME [15], the authors employ Hierarchical Clustering (HC) for grouping the training data and later use one of the clusters nearest to the instance for training the local interpretable model. These modifications are aimed at addressing the lack of "stability", which is one of the serious issues in the local interpretable methods. The issue of stability occurs in LIME because of the data generation using random perturbation. Suppose we select an instance to be explained, the LIME generates different explanations (local interpretable models with different feature weights) at every run for the same instance. The works such as [15] use clusters from the training data itself to address this problem. Since the samples are always taken from the same cluster of the training set, there is no variability in the feature weights at different runs for a particular instance.

Using the training set would make the black box model (the model for which we seek explanations) overfit by creating a table where each individual example in the training set is assigned to its class (i.e. exact matching of training instances), Therefore, using these results in training the local interpretable model would produce incorrect results, since the explanation instance of interest is from the test set and we do not have information as to how the black box model behaves upon encountering new data points as present in the test set. Thus, we ask the following question:

Can we improve the stability of local interpretable model while sampling randomly generated points?

In other words, we wish to see if we can still improve the stability by following the same sampling paradigm of LIME and therefore not using the training set for local interpretable model as done in [5,15].

Another important issue in local interpretable methods is *locality*, which refers to the neighbourhood around which the local surrogate is trained and in [8], the authors show that it is non-trivial to define the right neighbourhood and how it could impact the *local fidelity* of the surrogate. A straightforward way to improve the stability is to simply generate a large set of points and use it to train the local surrogate. Although doing so would improve stability, it also decreases local fidelity or local accuracy. Thus, we ask another question:

Can we improve the stability while simultaneously maintaining the local fidelity?

In this paper, we mainly focus on answering the questions above by introducing an autoencoder-based local interpretability model ALIME. Our contributions can be summarized as follows:

- we propose a novel weighting function as a modification to LIME to address the issues of stability and local fidelity, and
- we perform extensive experiments on three different healthcare datasets to study the effects and compare with LIME.

2   Methods

Since our model builds upon LIME, we begin by a short introduction to LIME, and then describe our proposed modifications.

2.1   LIME

Local surrogate models use interpretable models (such as ridge regression) in order to explain individual instance predictions of an already trained machine learning model, which could be a black box model. Local interpretable model-agnostic explanations (LIME) is a very popular recent work where, instead of training a global surrogate model, LIME trains a local surrogate model for individual predictions. LIME method generates a new dataset by first sampling from a distribution and later perturbing the samples. The corresponding predictions on this generated dataset given by the black box model are used as ground truth. On these pairs of generated samples and the corresponding black box predictions, an interpretable model is trained around the point of interest by weighting the proximity of the sampled instances to it. This new learned model has the constraint that it should be a good approximation of the black box model predictions locally, but it does not have to be a good global approximation. Formally, a local surrogate model with interpretability constraint is written as follows:

$$\text{explanation}(x) = \arg\min_{g \in G} L(f, g, \pi_x) + \Omega(g) \qquad (1)$$

The explanation model for instance x is the model g (e.g. linear regression model) minimizing loss L (e.g. mean squared error), a measure of how close the explanation is to the prediction of the original model f (e.g. a deep neural network model). The model complexity is denoted by $\Omega(g)$. G is the family of possible explanations which, in our case, is a linear ridge regression model. The proximity measure π_x defines how large the neighborhood around instance x is that we consider for the explanation. In practice, LIME only optimizes the loss part. The algorithm for training local surrogate models is as follows:

- Select the instance of interest for which an explanation is desired for a black box machine learning model.
- Perturb the dataset and use the black box to make predictions for these new points.
- Weight the new samples according to their proximity to the instance of interest by employing some proximity metric, such as euclidean distance.
- Train a weighted, interpretable linear model, such as ridge regression, on the dataset.
- Explain the prediction by interpreting the local linear model by analyzing the coefficients of the local linear model.

Algorithm 1: LIME

Input   : Dataset D_{train} with K features, model M, instance x, number of points sampled m

Output: feature importance at x

begin

    # Sample new dataset from Gaussian distribution

    $D_{sample} \xleftarrow{\text{m}} \textbf{Gaussian}(K)$

    # For all sampled points calculate:

    foreach $y \in D_{sample}$ **do**

        # euclidean distance from x

        $d(y) \longleftarrow |y - x|$

        # weight

        $W(y) \longleftarrow e^{-d(y)}$

    end

    # Fit linear model

    $L \longleftarrow \textbf{LinearModel.fit}(D, W)$

    # Return weights of the linear model

    return L_w

end

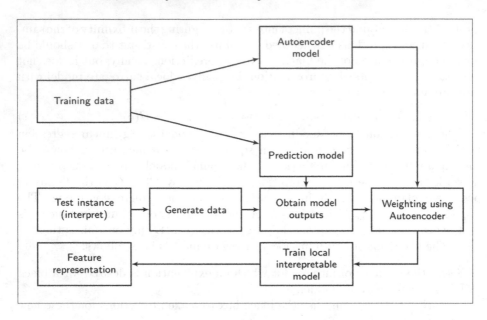

Fig. 1. Block diagram depicting our overall approach in ALIME

2.2   ALIME

The high-level block diagram of our approach is shown in Fig. 1 and also described in Algorithm 2. Once the black box model is trained, we need to train a local interpretable model. Our first focus is on improving stability of the local interpretable model. For this, instead of generating data by perturbation every time for explaining an instance (as done in LIME), we generate a large number of data points beforehand by sampling from a Gaussian distribution. This has an added advantage, as we reduce the time-complexity by reducing the sampling operations. However, since we need to train a local model, we must ensure that for a particular instance, only the generated data around the instance is used for training the interpretable model. For this we use an autoencoder [1,14] and thus, the most important change comes from the introduction of autoencoder.

Autoencoder [1] is a neural network used to compress high dimensional data into latent representations. It consists of two parts: an encoder and a decoder. The encoder learns to map the high dimensional input space to a latent vector space, and the decoder maps the latent vector space to the original uncompressed input space. We use a variant of autoencoder called denoising autoencoder [14], where the input is corrupted by adding a small amount of noise and then trained to reconstruct the uncorrupted input. Looked in another way, denoising is used as a proxy task to learn latent representations. We train an autoencoder with the help of the training data to be used for building the black box model. We first standardize the training data and then corrupt the training data by adding a small amount of additive white Gaussian noise. Then, the autoencoder is made to reconstruct the uncorrupted version of the input using the standard L_2 loss. Once trained, we employ the autoencoder as a weighting function, i.e., instead of computing the euclidean distances for the generated data and the instance to be explained on the original input space, we compute the distance on the latent vector space. For this, we compute the latent embeddings for all the generated points and the instance to be explained, and compute the distance on the embedded space. We discard the points with a distance larger than a predefined threshold and then, for the selected data points, we weight the points by using an exponential kernel as a function of distance. This way, we ensure locality and, since the autoencoders have been shown to better learn the manifold of data, it also improves local fidelity.

Algorithm 2: ALIME

Input : Dataset D_{train} with K features, model M, instance x, number of
points used n, number of points sampled m

Output: feature importance at x

\# Precompute embeddings using autoencoder

begin

    \# Train autoencoder model

    $AE \longleftarrow$ **AutoEncoder.fit**(D_{train})

    \# Sample new dataset from Gaussian distribution

    $D_{sample} \xleftarrow{m}$ **Gaussian**(K)

    \# Calculate embeddings for D_{sample}

    $E \longleftarrow [AE(y) \mid y \in D_{sample}]$

end

\# Given x, calculate feature importance

begin

    \# Calculate distances from x in embedded space

    $d \longleftarrow |E - E(x)|$

    \# Find n-th minimum distance in d

    $d_{min} \longleftarrow$ **min**(n, d)

    \# Collect n closest points into a local dataset

    $D \longleftarrow [y \in D_{train}$ such that $|d_x - d_y| < d_{min}]$

    \# Calculate weights

    $W \longleftarrow [e^{-y} \mid y \in D]$

    \# Fit linear model

    $L \longleftarrow$ **LinearModel.fit**(D, W)

    \# Return weights of the linear model

    return L_w

end

---

3   Experiments and Results

For the sake of experiments, we use three datasets belonging to healthcare domain from the UCI repository [4]:

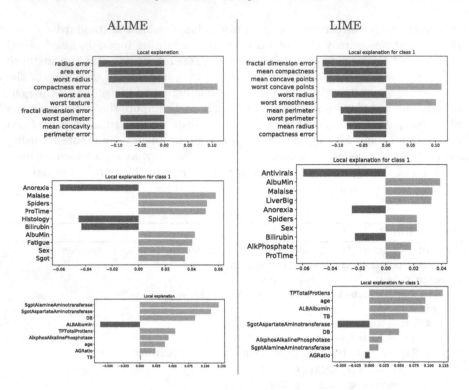

Fig. 2. Comparisons between our method and LIME [12]. The first row corresponds to the Breast Cancer dataset, the second row corresponds to the Hepatitis Patients dataset and the third row corresponds to the Liver Patients dataset

1. Breast Cancer dataset [9]. A widely used dataset that consists of 699 patient observations and 11 features used to study breast cancer.
2. Hepatitis Patients dataset [3]. Dataset consisting of 20 features and 155 patient observations.
3. Liver Patients dataset [11]. Indian liver patient dataset consisting of 583 patient observations and 11 features used to study liver disease.

As a black box model, we train a simple feed forward neural network with a single hidden layer having 30 neurons and 2 neurons in the output layer for the two classes, and train the network using binary cross entropy loss. For all the three datasets, we use 70 − 30 split for training and testing. We obtained accuracies of 0.95, 0.87 and 0.83 on the above mentioned three datasets respectively. The sample results for instances from the three datasets are shown in Fig. 2. The red bars in the figure show the negative coefficients, and green bars show the positive coefficients of the linear regression model. The positive coefficients show the positive correlation among the dependent and independent attributes. On the other hand, negative coefficients show the negative correlation among the dependent and independent attributes.

Currently, there exists no suitable metric for proper comparison of the two different interpretable models. Since our focus is on local fidelity and stability, we define and employ suitable metrics for the two issues of focus. For local fidelity, the local surrogate model should fit the global black box model locally. To test this, we compute the R^2 score of the local surrogate model using the results from the black box model as the ground truth. This tells us how good the model has fit on the generated data points. We compute the mean R^2 scores considering all the points in the test set. We also test the local model for fidelity by computing the mean squared error (MSE) between the local model prediction and the black box model prediction for the instance of interest that is to be explained. We again compute mean MSE considering all the points in the test set for the three respective datasets. Additionally, to study the effects of the dataset size used for the local surrogate model, we vary the number of generated data points used for training the local surrogate model. The results for the local fidelity experiments are shown in Fig. 3. It can be seen that in terms of both metrics, ALIME clearly outperforms LIME by providing a better local fit. The results seem consistent across the three datasets.

It is even more difficult to define a suitable metric for the interpretable model stability. Since the explanations are based on the surrogate model's coefficients, we can compare the change in the values of the coefficients for multiple iterations. Randomly selecting a particular instance from the test set, we run both LIME and ALIME for 10 iterations. Because of their nature, both methods sample different set of points at every iteration. Because of the different dataset used

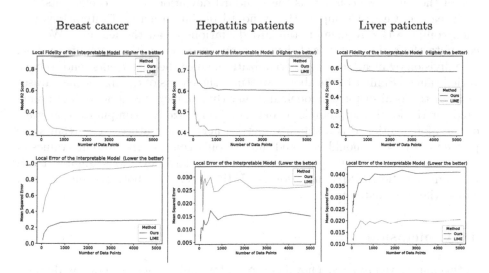

Fig. 3. Comparisons between our method and LIME [12] for local fidelity. The first row corresponds to the R2 score of the local surrogate model, the second row corresponds to the mean of MSE for all the points in the test set. In all cases, the number of data points for the local surrogate model is varied in log scale.

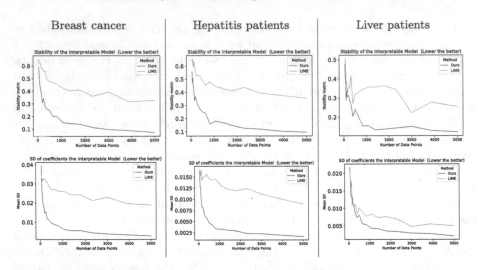

Fig. 4. Comparisons between our method and LIME [12] for stability. The first row corresponds to the mean stability metric of 10 iterations for the same instance and for all the features, the second row corresponds to the mean standard deviations of 10 iterations for the same instance and for all the features. The number of data points for the local surrogate model is varied in log scale.

at every iteration, the coefficients' values change. As a measure for stability, one of the things we compare is the standard deviations of the coefficients. For each feature, we first compute the standard deviation across 10 iterations and then compute the average of standard deviations of all the features. We also compute the ratio of standard deviation to mean as another stability metric. The division by mean serves as normalization, since the coefficients tend to have varied ranges. Similar to the above, we study the effects of the size of the dataset used for the local surrogate model, and vary the number of generated data points used for the local interpretable model. For every size, we compute the average of the aforementioned two stability metrics across all features. The results are plotted in Fig. 4. It should be noted that we only consider the absolute values of coefficients while computing the means and standard deviations. Again, it can be seen that the ALIME outperforms LIME in terms of both the metrics and across the three datasets.

4   Conclusion

In this paper, we proposed a novel approach for explaining the model predictions for tabular data. We built upon the LIME [12] framework and proposed modifications by employing an autoencoder as the weighting function to improve both stability and local fidelity. With the help of extensive experiments, we showed that our method yields in better stability as well as local fidelity. Although we have shown the results empirically, a more thorough analysis is needed to substantiate the improvements. In future, we would work on performing a theoretical analysis and also exhaustive empirical analysis spanning different types of data.

References

1. Bengio, Y., Lamblin, P., Popovici, D., Larochelle, H.: Greedy layer-wise training of deep networks. In: Advances in Neural Information Processing Systems, pp. 153–160 (2007)
2. Bien, J., Tibshirani, R., et al.: Prototype selection for interpretable classification. Ann. Appl. Stat. **5**(4), 2403–2424 (2011)
3. Diaconis, P., Efron, B.: Computer-intensive methods in statistics. Sci. Am. **248**(5), 116–131 (1983)
4. Dua, D., Graff, C.: UCI machine learning repository (2017). http://archive.ics.uci.edu/ml
5. Hall, P., Gill, N., Kurka, M., Phan, W.: Machine learning interpretability with H2O driverless AI, February 2019. http://docs.h2o.ai
6. Koh, P.W., Liang, P.: Understanding black-box predictions via influence functions. In: Proceedings of the 34th International Conference on Machine Learning-Volume 70, pp. 1885–1894 (2017). JMLR.org
7. Lakkaraju, H., Bach, S.H., Leskovec, J.: Interpretable decision sets: a joint framework for description and prediction. In: Proceedings of the 22nd ACM SIGKDD International Conference on Knowledge Discovery and Data Mining (2016)
8. Laugel, T., Renard, X., Lesot, M.J., Marsala, C., Detyniecki, M.: Defining locality for surrogates in post-hoc interpretablity. arXiv preprint arXiv:1806.07498 (2018)
9. Mangasarian, O.L., Street, W.N., Wolberg, W.H.: Breast cancer diagnosis and prognosis via linear programming. Oper. Res. **43**(4), 570–577 (1995)
10. Molnar, C.: Interpretable machine learning (2019). https://christophm.github.io/interpretable-ml-book/
11. Ramana, B.V., Babu, M.S.P., Venkateswarlu, N., et al.: A critical study of selected classification algorithms for liver disease diagnosis. Int. J. Database Manag. Syst. **3**(2), 101–114 (2011)
12. Ribeiro, M.T., Singh, S., Guestrin, C.: Why should i trust you?: Explaining the predictions of any classifier. In: Proceedings of the 22nd ACM SIGKDD International Conference on Knowledge Discovery and Data Mining, pp. 1135–1144. ACM (2016)
13. Ribeiro, M.T., Singh, S., Guestrin, C.: Anchors: high-precision model-agnostic explanations. In: Thirty-Second AAAI Conference on Artificial Intelligence (2018)
14. Vincent, P., Larochelle, H., Bengio, Y., Manzagol, P.A.: Extracting and composing robust features with denoising autoencoders. In: Proceedings of the 25th International Conference on Machine Learning, pp. 1096–1103. ACM (2008)
15. Zafar, M.R., Khan, N.M.: DLIME: a deterministic local interpretable model-agnostic explanations approach for computer-aided diagnosis systems. In: In Proceeding of ACM SIGKDD Workshop on Explainable AI/ML (XAI) for Accountability, Fairness, and Transparency. ACM, Anchorage (2019)

Detection of Abnormal Load Consumption in the Power Grid Using Clustering and Statistical Analysis

Matúš Cuper, Marek Lóderer$^{(\boxtimes)}$, and Viera Rozinajová

Faculty of Informatics and Information Technologies, Slovak University of
Technology, Ilkovičova 2, 841 04 Bratislava, Slovakia
{xcuper,marek_loderer,viera.rozinajova}@stuba.sk

Abstract. Nowadays, the electricity load profiles of customers (consumers and prosumers) are changing as new technologies are being developed, and therefore it is necessary to correctly identify new trends, changes and anomalies in data. Anomalies in load consumption can be caused by abnormal behavior of customers or a failure of smart meters in the grid. Accurate identification of such anomalies is crucial for maintaining stability in the grid and reduce electricity loss of distribution companies. Smart meters produce huge amounts of load consumption measurements every day and analyzing all the measurements is computationally expensive and very inefficient. Therefore, the aim of this work is to propose an anomaly detection method, that addresses this issue. Our proposed method firstly narrows down potential anomalous customers in large datasets by clustering discretized time series, and then analyses selected profiles using statistical method S-H-ESD to calculate final anomaly score. We evaluated and compared our method to four state-of-the-art anomaly detection methods on created synthetic dataset of load consumption time series containing collective anomalies. Our method outperformed other evaluated methods in terms of accuracy.

Keywords: Anomaly detection · Time series · Clustering · Statistical analysis

1 Introduction

One of the main issues, distribution companies are facing nowadays, is to find a reliable approach for anomaly detection of customers' behavior. The primary goal of detection is to identify customers' profiles, that look non-standard due to unauthorized manipulation with measured values or as a result of malfunctioning of smart meter device. Another factor that has to be considered in the analysis is unpredictable customer behavior.

Early detection of unusual behavior can enhance stability of distribution network, improve distributors' predictions accuracy and also increase the capacity

© Springer Nature Switzerland AG 2019
H. Yin et al. (Eds.): IDEAL 2019, LNCS 11871, pp. 464–475, 2019.
https://doi.org/10.1007/978-3-030-33607-3_50

of the network. Based on the data that are produced by smart meter devices, it is possible to model a behavior of individual customers or groups of similar customers.

The aim of this article is to propose a method which can identify such intervals in electricity consumption time series, that by its characteristics do not match with the behavior of other consumers in dataset. A manual identification of such intervals is time consuming and it requires a lot of labor, therefore we decided to base our proposed method on unsupervised learning. The proposed method performs the best if a large dataset of customers' load consumption measurements is available. The dataset is used to calculate differences between customers' profiles and consequently determine the degree of anomaly in analyzed data. The processing and transformation of big datasets are accomplished by statistical methods and machine learning.

2   Anomaly Detection in Load Consumption

In general, anomalous load consumption of consumer can be defined as behavior that does not correspond to standard behavior of neighboring group of consumers. It is a challenging task to define accurate borders between normal and anomalous behavior. Furthermore, consumption profiles of consumers can evolve during the observation period or they may change according to unobserved exploratory variables.

Currently, we distinguish point, contextual and collective anomalies [2]. The point anomaly is defined as an individual measurement which significantly differs from normal measurements. A measurement is considered as the contextual anomaly if its value significantly differs from neighboring measurements. The collective anomaly is defined as sequence of multiple measurements marked as anomalies when they appear together. All three types of anomalies are present in load consumption measurements.

2.1   Anomaly Detection Methods

Over the last decades several anomaly detection methods have been proposed. According to work of Hu-Sheng Wu [18], the anomaly detection methods can be divided into five categories based on used approaches such as statistics, clustering, deviation, distance and density. There are also other authors who provide more or less complex categorization of anomaly detection methods [4,13].

Table 1 provides several anomaly detection methods for time series that have been published recently.

In our work, we focus mainly on unsupervised techniques of clustering and statistical based anomaly detection methods.

Clustering is an unsupervised or semi-supervised technique used to organize similar data instances into groups called clusters [16].

Table 1. Comparison of various anomaly detection methods for time series data.

Method name	Type	Properties
PEWMA [1]	Statistical-based	Robust to anomalous fluctuations Adapts to changes in the model of data Works in stationary environments
SD-EWMA [12]	Statistical-based	Suitable for real-time run Reduction of false-alarms
KNN-LDCD [7]	Distance-based Clustering-based	Suitable for quasi-periodic time series Adapts to non-stationarity in the data streams Uses sliding windows to keep relevant observations Based on the conformal prediction paradigm High computational complexity
S-H-ESD [17]	Statistical-based	Detects long term trends Suitable for seasonal time series Designed for datasets with up to 50% anomalies Works in non-stationary environments
Proposed method	Clustering-based Statistical-based	Applies clustering to reduces large datasets Uses properties of S-H-ESD statistical analysis Anomalies are determined by multistep scoring Uses sliding windows to keep relevant observations Estimation of threshold values is needed

The idea behind clustering used as anomaly detection method is that it can be assumed that normal instances belong to the big and consistent clusters, while anomalous ones to the small and sparse clusters, or the normal instances are closer to centroids, while the anomalous ones are further.

A number of clustering algorithms that have been used to detect anomalies are presented in the literature. For example k-means or k-medoids clustering, DBSCAN, Cluster-Based Local Outlier Factor or SNN clustering [2,9,14,15].

Statistical methods are based on mean, median, standard deviation, moving averages [5], various distributions [18], ARMA and ARIMA models [11].

A commonly used statistical method for anomaly detection is Extreme Studentized Deviate (ESD) [6]. The method is based on Grubb's test and has several modifications, such as seasonal ESD (S-ESD) [17], seasonal hybrid ESD (S-H-ESD) and more [6].

In S-H-ESD, the modified time series decomposition is applied, and ESD algorithm is modified by replacing mean and standard deviation for median absolute deviation (MAD). The MAD is more robust and capable of identifying up to 50% anomalies [6]. It is calculated as a median of absolute differences between given observation and median, also described by equation:

$$MAD = median_i(|X_i - median_j(X_j)|) \tag{1}$$

where $median_j(X_j)$ represents median of all observed measurements and X_i is observation i.

3 Proposed Method

Our proposed anomaly detection method for time series based on clustering and statistical analysis can be divided into multiple steps, where each step transforms input data and calculates new data for subsequent steps (see Fig. 1).

The method consists of two main phases. The first phase focuses on data preparation, time series clustering and selecting abnormal looking customers' time series profiles based on defined scoring. In the second phase, selected time series are analyzed, and final anomaly scoring is calculated. The aim is to identify and locate intervals with non-standard behavior in the time series as accurately as possible.

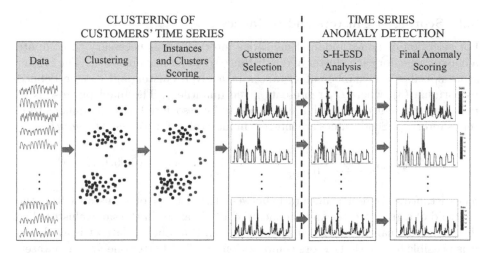

Fig. 1. Visual representation of the proposed method depicting step-by-step data transformation from measured customers' time series of electricity load consumption, through cluster scoring into final anomaly detection.

Based on the score, we are able to determine the degree of confidence that there is an anomaly within the given time series interval. The advantage of this solution is its universality, low time complexity and also comprehensible visualization of anomalies. While doing that, we must consider the cycle of days and weeks, but also understand the habits of customers individually and consider them when marking anomalies.

This method is repeated incrementally when new data are obtained which helps to cope with trend, seasonalities and changes in the data. For this purpose, a sliding window of fixed size is used. It means that the oldest data from the window are removed and new data are added.

3.1   Clustering

Before clustering consumers' data, some preprocessing of raw data has to be done. Data are split into workdays and non-workdays, due to different curve of electricity consumption. Also, z-score normalization is applied, since we mainly focus on clustering by similarity and not by absolute distance between data points. The preprocessed data are clustered step by step using sliding window. The length of sliding window, overlapping and step length should be determined experimentally. Each window frame is firstly aggregated by equation:

$$\hat{x}_i = \frac{1}{n} \sum_{j=0}^{n} x_{j*m+i} \tag{2}$$

where m represents number of measurements per day, n is number of aggregated weeks and $\hat{x}_i$ is aggregated value. That creates representative profiles of consumers, which are finally clustered by k-medoid clustering algorithm.

3.2   Scoring of Clusters and Instances

For selection of non-standard time series intervals, it is necessary to compute a cluster score for each cluster and a distance score for each instance. The cluster score specifies the number of instances in a given cluster and can be defined by formula (3), where Q_j represents given quantile j. The main purpose is to penalize small clusters, which are assumed to be associated with anomalous time series intervals. Score values are integers in interval $< 1, number\_of\_quantiles >$.

$$score_{cluster_i} = \sum_{j=1}^{n} \begin{cases} 1 \text{ if } P(size_i \leq Q_j) \\ 0 \text{ else} \end{cases} \text{, for j } \in (0.05, 0.1, ..., 1) \tag{3}$$

The distance score represents a ratio between the distance of an instance from the centroid of a given cluster and the mean of all distances in the same cluster. The distance score can be described by formula (4). If a labeled dataset is available, it is possible to adjust the score computation and find better one for the dataset.

$$score_{distance_i} = \frac{distance_i}{\frac{1}{m} \sum_{j=1}^{m} distance_j}, \text{ for} \tag{4}$$

$$instance_i \in cluster_k \wedge instance_j \in cluster_k \wedge m = |cluster_k|$$

Both scores are computed for each instance and cluster on every single frame in the sliding window. Subsequently, both scores are combined and create consumer score, which can be described by formula 5.

$$score_{consumer_i} = score_{distance_i} * score_{cluster_j}, \text{ for } instance_i \in cluster_j \tag{5}$$

The score computation is shown in the Fig. 2. Clusters are divided into minor and major based on the cluster score. If the number of instances in a cluster in run j is greater than quantile Q_j, then it is a major cluster, otherwise it is labeled as minor.

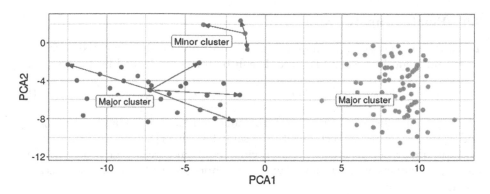

Fig. 2. Visual representation of instance and cluster scoring. The grey arrow represents a distance between given instance and centroid of a cluster.

3.3 Consumer Selection

The next step after computation of consumer's score, is selection of consumers with the highest and lowest score. Only instances with the score that fall out of interval (6) are selected for further analysis.

$$< Q_1 - 1.5 \cdot IQR,\ Q_3 + 1.5 \cdot IQR > \tag{6}$$

$$IQR = Q_3 - Q_1 \tag{7}$$

The defined interval (6) is based on the interquartile range IQR between upper quartile Q_3 and lower quartile Q_1.

There are also other algorithms that can be applied. For example dimension reduction methods PCA and t-SNE [10], or sequence and features extracting method FeaClip proposed by authors in article [8].

3.4 S-H-ESD Analysis

Selected consumers are further analyzed by statistical method S-H-ESD. Due to the fact that analyzed raw data are highly volatile, a smoothing process is needed to eliminate numerous local insignificant anomalies. As mentioned earlier, our proposed method focuses mainly on global anomalies that can considerably influence distribution of total electricity in given region. The degree of smoothing can be adjusted by an input parameter called radius.

There are two smoothings in the process. Firstly, the raw data are smoothed with a lower value of radius in order to eliminate local anomalies. The second smoothing with higher radius is applied on the output of S-H-ESD analysis. The S-H-ESD output consists of binary flags indicating anomalousness of given measurements. Some time series analyzed by S-H-ESD may contain numerous crossings between these two flags. Therefore, multiple positive and negative flags are grouped together by smoothing, which reduces local insignificant anomalies.

The smoothing is performed by Locally Estimated Scatterplot Smoothing (LOESS) [3].

3.5   Final Anomaly Scoring

The final anomaly score is computed as a sum of previously calculated consumer score and smoothed S-H-ESD score. It can be formally defined by Eq. (8)

$$score_{anomaly_t} = score_{consumer_i} + score_{S-H-ESD_t} \qquad (8)$$

where t is an index of observation in the analyzed time series, and $score_{anomaly}$ and $score_{S-H-ESD}$ are vectors with the same length as the analyzed time series.

It is important to realize that previous consumer scores are computed on individual moving windows and day types, while S-H-ESD method analyzes all consumer data at once.

The main disadvantage of using just the cluster score (without S-H-ESD score) is that the cluster score has low granularity. The cluster score of a consumer does not change over several days and the value of the score is the same for every hour in these days. This disadvantage can be overcome by combining S-H-ESD score together with the cluster score, which creates more refined result with higher granularity.

4   Evaluation

The data used to evaluate our method come from the Irish CER Smart Metering Project and are publicly accessible at the Irish Social Science Data Archive (ISSDA)[1]. The project was designed to measure the performance of smart meters, their influence on consumers' energy consumption and various impact on national economy.

The data contain half-hourly measurements of electricity consumption of more than 4 621 customers collected in years 2009 and 2010. Each measurement contains timestamp, load consumption and bank holiday flag. The majority of customers were households. The rest consists of small and medium enterprises.

Basic statistical characteristics of the data are shown in the Table 2.

Table 2. Characteristics of used dataset.

Mean	Standard deviation	Minimum	Q1	Median	Q3	Maximum
0.673	1.373	0	0.121	0.269	0.666	66.815

Consumers' profiles differ during workdays and non-workdays, which needs be taken into consideration during data preprocessing. Also, the load consumptions of customers differ from each other.

---

[1] http://www.ucd.ie/issda/data/commissionforenergyregulationcer/.

4.1   Experimental Setup

The quality of clustering methods depends on proper hyperparameter setup. In our proposed method, there are input parameters to k-medoid algorithm like number of clusters (k) and distance metric, and also parameters to windowing method, such as window size, step size or day type. Based on our experiments, the best values of Silhouette's and Davies-Bouldin's indexes were achieved by the setup, where number of clusters $k = 25$, windows size $l = 2$, step size was one week and distance metric was GAK. Results of Silhouette's and Davies-Bouldin's indexes are very similar, even though there are some small differences in final order and time complexity of given solution. It is possible to use daily step size in order to apply incremental analysis of streaming data.

Based on literature and experimental evaluation, we set hyperparameters for the other anomaly detection methods which are used in the final evaluation. We used parameters $\alpha_0 = 0.8$ and $\beta = 0.1$ for PEWMA and a threshold $t = 0.000075$ in SD-EWMA. The limit multiplier $l = 3$ was set for both methods. KNN-LDCD method achieved best results with threshold $t = 1$, window length $l = 48$ and neighborhood size $k = 27$. Method S-H-ESD and our proposed method used identical parameters α set to 0.001 and maximal number of anomalies with value of 50%.

4.2   Synthetic Dataset

Due to the lack of labeled anomaly datasets, we decided to create a synthetic one to evaluate precision of the proposed method.

The synthetic dataset was created by swapping selected days profiles of given consumers with each other. An example of such profile can be found in the Fig. 3. The dataset consists of twelve weeks, ten were used for model training and two weeks for testing. We swapped several workdays selected from testing set.

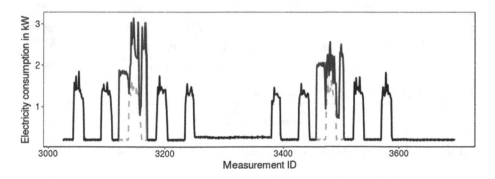

Fig. 3. Example of the synthetic dataset created from profile of consumer no. 2812. The black line represents original data, the blue line represents data that have been swapped from other dataset, and the grey dashed line shows original data that have been replaced. (Color figure online)

Each swapped measurement is considered as anomalous. The dataset contains 960 anomalous and 5 760 non-anomalous measurements. Since the proposed method is not deterministic, each experiment was repeated multiple times and final results were calculated by mean.

4.3   Results

In our experiments, we evaluated the proposed method and compared its results with results of four other anomaly detection methods. The tests were performed on previously created synthetic datasets. Added anomaly labels in the synthetic datasets are used only for evaluation purposes. They are not used in any step of the anomaly detection process or the hyperparameter estimation. The hyperparameter configuration used in the methods are listed in Sect. 4.1.

As mentioned earlier in the paper, our proposed method builds on the output of the S-H-ESD method. Figures 4 and 5 provide a detailed view on output of both methods. The parts of the customer's consumption detected as anomalies are illustrated in red.

According to results visualized in Figs. 4 and 5, both methods detected anomalous measurements correctly on tested synthetic dataset. Both methods also marked falsely a high number of measurements as anomalies, but the proposed method assigned lower relative score to these anomalous measurements. The method computes scores with higher granularity comparing to S-H-ESD method. These higher granularity scores allow to define a certain threshold value that can be applied to exclude insignificant or potentially falsely detected anomalies and mark them as normal measurements. This approach is able to increase the overall accuracy of anomaly detection as shown in Table 3.

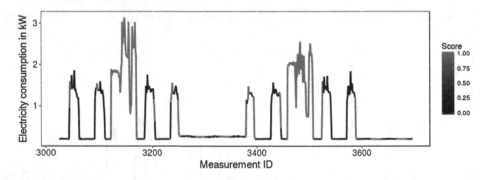

Fig. 4. Example of synthetic dataset scored by S-H-ESD. (Color figure online)

Since the experiments were performed on labeled synthetic datasets, the confusion matrix and F1-score could have been calculated. Results displayed in Table 3 show that the tested methods detected anomalies with various accuracy.

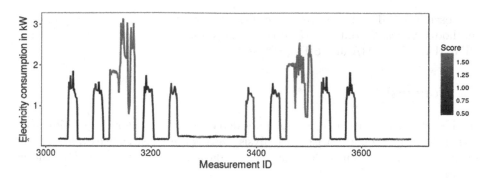

Fig. 5. Example of synthetic dataset scored by the proposed method. (Color figure online)

Table 3. Results from evaluation on synthetic dataset containing confusion matrices (TP - true positive, TN - true negative, FP - false positive and FN - false negative), Precision, Recall, F1 score and p-value computed by Wilcoxon two-sample paired signed rank test. Values in brackets represents number of observations.

Method name	TP	TN	FP	FN	Precision	Recall	F1 score	p-value
PEWMA	0.6% (47)	82.5% (5 541)	3.3% (219)	13.6% (913)	17.67%	4.90%	7.67%	5.4074e-06
SD-EWMA	0.7% (45)	83.1% (5 584)	2.6% (176)	13.6% (915)	20.36%	4.69%	7.62%	1.0378e-03
KNN-LDCD	0.04% (3)	85.6% (5 758)	0.03% (2)	14.2% (957)	60.00%	0.31%	0.62%	0.9652840
S-H-ESD	6.4% (431)	58.1% (3 901)	27.6% (1 859)	7.9% (529)	18.82%	44.90%	26.52%	2.8829e-217
Proposed method	4.4% (297)	80.6% (5 418)	5.1% (342)	9.9% (663)	42.84%	39.27%	40.98%	–

PEWMA and SD-EWMA achieved similar results in terms of F1 score. They were the fastest ones, but the overall accuracy measured by F1 score was low, only around 7.6%. The methods lacked to identify real anomalies.

The KNN-LDCD method had high precision but very low recall. It was caused by the fact that the method labeled almost all measurements (99.93%) as normal data.

Both S-H-ESD and our proposed method achieved higher recall compared to other methods. They also marked falsely a lot of measurements as anomalies. Although, the proposed method identified less true positives compared to S-H-ESD, it managed to radically reduce number of false alarms (false positives) from 27.6% to 5.1%, which helped the method increase its precision and overall accuracy. The reduction of false alarms was achieved by applying a certain threshold to calculated anomaly score. Only measurements with score greater or equal to 1.15 are consider as anomalies.

We also performed Wilcoxon signed rank test to determine the significance of the results. The p-value in Table 3 represents significance calculated between our proposed method and a given anomaly detection method showing if the proposed method achieves better results. According to results, our proposed method significantly outperformed three of the four compared methods; namely PEWMA, SD-EWMA and S-H-ESD method.

5   Conclusion

In this paper, we presented an anomaly detection method for time series data based on clustering and statistical analysis. During our research, we evaluated the proposed method on unlabeled smart meter data as well as on created synthetic datasets.

We also compared our proposed method with four other anomaly detection methods. According to the results, our solution achieves more accurate outcome measured by F-score. This is achieved by reduction of false alarms (false positives) and S-H-ESD flag smoothing. Moreover, the final score is computed 4 times faster than original S-H-ESD analysis. It is caused by pre-selection of consumers by clustering and application of S-H-ESD analysis only to selected profiles.

The possible direction of future research is to increase the method's accuracy by developing an approach for dynamic selection of the anomaly score threshold, or to modify the current method for online processing of stream time series data.

Acknowledgment. This work was partially supported by the Slovak Research and Development Agency under the contract APVV-16-0213, and by the Scientific Grant Agency of the Slovak Republic VEGA, grant No. VG 1/0759/19. The authors would also like to thank for financial assistance from the STU Grant scheme for Support of Excellent Teams of Young Researchers (Grant No. 1391).

References

1. Carter, K.M., Streilein, W.W.: Probabilistic reasoning for streaming anomaly detection. In: 2012 IEEE Statistical Signal Processing Workshop (SSP). IEEE, August 2012. https://doi.org/10.1109/ssp.2012.6319708
2. Chandola, V., Banerjee, A., Kumar, V.: Anomaly detection: a survey. ACM Comput. Surv. **41**(3), 15:1–15:58 (2009). https://doi.org/10.1145/1541880.1541882
3. Cleveland, W.S.: Robust locally weighted regression and smoothing scatterplots. J. Am. Stat. Assoc. **74**(368), 829–836 (1979)
4. Fernandes Jr., G., Rodrigues, J.J., Carvalho, L.F., Al-Muhtadi, J.F., Proença Jr., M.L.: A comprehensive survey on network anomaly detection. Telecommun. Syst. **70**(3), 447–489 (2019). https://doi.org/10.1007/s11235-018-0475-8
5. Hasani, Z.: Robust anomaly detection algorithms for real-time big data: comparison of algorithms. In: 2017 6th Mediterranean Conference on Embedded Computing (MECO), pp. 1–6, June 2017. https://doi.org/10.1109/MECO.2017.7977130
6. Hochenbaum, J., Vallis, O.S., Kejariwal, A.: Automatic anomaly detection in the cloud via statistical learning. CoRR abs/1704.07706 (2017)
7. Ishimtsev, V., Nazarov, I., Bernstein, A., Burnaev, E.: Conformal k-NN anomaly detector for univariate data streams (2017)

8. Laurinec, P., Lucká, M.: Interpretable multiple data streams clustering with clipped streams representation for the improvement of electricity consumption forecasting. Data Min. Knowl. Disc. **33**(2), 413–445 (2019)
9. Lima, M.F., Zarpelão, B.B., Sampaio, L.D.H., Rodrigues, J.J.P.C., Abrão, T., Proença, M.L.: Anomaly detection using baseline and k-means clustering. In: Soft-COM 2010, 18th International Conference on Software, Telecommunications and Computer Networks, pp. 305–309, September 2010
10. Maaten, V.D.L., Hinton, G.: Visualizing data using t-SNE. J. Mach. Learn. Res. **9**, 2579–2605 (2008)
11. Qi, J., Chu, Y., He, L.: Iterative anomaly detection algorithm based on time series analysis. In: 2018 IEEE 15th International Conference on Mobile Ad Hoc and Sensor Systems (MASS), pp. 548–552, October 2018
12. Raza, H., Prasad, G., Li, Y.: EWMA model based shift-detection methods for detecting covariate shifts in non-stationary environments. Pattern Recogn. **48**(3), 659–669 (2015). https://doi.org/10.1016/j.patcog.2014.07.028
13. Salehi, M., Rashidi, L.: A survey on anomaly detection in evolving data: [with application to forest fire risk prediction]. SIGKDD Explor. Newsl. **20**(1), 13–23 (2018). https://doi.org/10.1145/3229329.3229332
14. Salvador, S., Chan, P.: Learning states and rules for detecting anomalies in time series. Appl. Intell. **23**(3), 241–255 (2005)
15. Shao, X., Zhang, M., Meng, J.: Data stream clustering and outlier detection algorithm based on shared nearest neighbor density. In: 2018 International Conference on Intelligent Transportation, Big Data Smart City (ICITBS), pp. 279–282 (2018)
16. Tan, P.N., Steinbach, M., Kumar, V.: Introduction to Data Mining. Addison Wesley, Boston (2005)
17. Vallis, O., Hochenbaum, J., Kejariwal, A.: A novel technique for long term anomaly detection in the cloud. In: Proceedings of the 6th USENIX Conference on Hot Topics in Cloud Computing, p. 15. USENIX Association, Berkeley (2014)
18. Wu, H.: A survey of research on anomaly detection for time series. In: 2016 13th International Computer Conference on Wavelet Active Media Technology and Information Processing (ICCWAMTIP), pp. 426–431, December 2016

Deep Convolutional Neural Networks Based on Image Data Augmentation for Visual Object Recognition

Khaoula Jayech[(✉)]

Université de Sousse, Ecole Nationale d'Ingénieurs de Sousse,
LATIS- Laboratory of Advanced Technology and Intelligent Systems,
4023 Sousse, Tunisie
jayech_k@yahoo.fr

Abstract. Deep Neural Networks (DNNs) have achieved a great success in machine learning. Among a lot of DNN structures, Deep Convolutional Neural Networks (DCNNs) are currently the main tool in the state-of-the-art variety of classification tasks like visual object recognition and handwriting and speech recognition. Despite wide perspectives, DCNNs have still some challenges to deal with. In previous work, we demonstrated the effectiveness of using some regularization techniques such as the dropout to enhance the performance of DCNNs. However, DCNNs need enough training data or even a class balance within datasets to conduct better results. To resolve this problem, some researchers have evoked different data augmentation approaches. This paper presents an extension of a later study. In this work, we conducted and compared the results of many experiments on CIFAR-10, STL-10 and SVHN using variant techniques of data augmentation combined with regularization techniques. The analysis results show that with the right use of data augmentation approaches, it is possible to achieve good results and outperform the state-of-the-art in this field.

Keywords: Deep learning · DCNN · Image data augmentation · Object recognition

1 Introduction

Visual object recognition is a fundamental yet still challenging problem in machine learning and computer vision community. Over the past years, Artificial Neural Networks (ANN) is common machine learning algorithms with a great training ability and good generalization. They have different structures. Layered ANN is one of the most practical classifiers with numerous neurons in each layer which should be defined by the user [1].

In recent years, the deep networks have made tremendous progress, achieving impressive results in a range of domains with a vast number of real-world applications such as handwritten digits, face recognition, sentiment analysis, object detection and recognition, image classification and game development [2–6]. In fact, deep networks are capable of representing more complex functions than shallow networks. Indeed,

© Springer Nature Switzerland AG 2019
H. Yin et al. (Eds.): IDEAL 2019, LNCS 11871, pp. 476–485, 2019.
https://doi.org/10.1007/978-3-030-33607-3_51

they have a great potential to extract important features and can remove unimportant features.

Over the last few years, advances in both machine learning algorithms and computer hardware have led to more efficient methods for training deep neural networks. The main proper types of deep networks are the Convolutional Neural Networks (CNNs). They have achieved a great progress and have attracted huge attention among computer vision research communities thanks to its high performance in classification tasks [7, 8]. It has produced extremely promising results for various tasks of pattern recognition issues. Those models have some pros and cons. In fact, the benefit of using DCNN is supposed to be good at developing an internal representation of a two-dimensional image. This allows the model to learn position and scale in variant structures in the data, which is important when working with images. Indeed, the DCNNs let the use of shared weight in convolutional layers, which means that the same filters (weights bank) are used for each pixel in the layer; this both reduce memory footprint and improve performance [3, 9–16]. Nevertheless, these models include a large number of layers, a huge number of units, and connections. So, to train these models, we need a lot of training data. This paper presents an extension of a previous work.

In this study, we benefit from the previous results and we search to enhance it by experiment some data augmentation techniques with and without regularization method. The CIFAR-10, STL-10 and SVHN datasets are used to evaluate the performance of proposed models. The contributions of our work are summarized as follows: (1) A new architecture of deep learning models called DCNN is proposed. (2) Some data augmentation techniques are studied with and without regularization method. (3) In the experiments, we implement and evaluate the model to classify images into 10 categories.

The remainder of this paper has been organized according to the following: some related works are reviewed in Sect. 2. Section 3 presents the background of DCNNs and image data augmentation techniques. In Sect. 4, we describe the overall architecture of the proposed system. Section 5 presents experimental results with details on databases. Finally, conclusions and future directions of this work are given in Sect. 6.

2   Related Work

Several approaches were introduced to face the main challenges in visual object recognition. Recently, most of the proposed works were based on deep learning algorithms, such as the CNNs. Good results have been achieved thanks to the powerful learning capabilities offered by DCNNs. However, the task of object recognition is still a very active and challenging problem.

In this context, a CNN was developed by Liang et al. in [9] for visual object recognition. In fact, they proposed to incorporate recurrent connections into each convolutional layer. Despite the input is static, the activities of CNN units evolve over time so that the activity of each unit is modulated by the activities of its neighboring units. This property enhances the ability of the model to integrate the context information, which is important for object recognition. The proposed CNN was tested on several benchmark object recognition datasets. With fewer parameters, CNN achieved better results than the state-of-the-art.

Zheng et al. in [5] proposed to optimize the feature boundary of deep CNN via a two-stage training method. The pre-training process elaborated in order to train a network model to extract the image representation for anomaly detection. However, the implicit regularization training process used to re-train the network based on the anomaly detection results to regularize the feature boundary and make it converge in the proper position. Experimental results on five image classification benchmarks show that the two-stage training method achieves a state-of-the-art performance.

Devries et al. in [12] suggested the use of new approach named cutout to improve the robustness and overall performance of CNNs. This technique is easy to implement and can be combined with the different existing forms of data augmentation and other regularizers to further improve model performance. In fact, this approach is a simple regularization technique of randomly masking out square regions of input during training. This approach was tested on several popular image recognition datasets and achieved a new state-of-the- art results. In addition, this approach proved to be very effective on low data and higher resolution problems.

A combination of DCNN and random forests was proposed by Bai et al. in [13] for scene categorization. In fact, they used DCNNs pre-trained on the large-scale image database to extract features from scene images. Then, they concatenated multiple layers of features of the deep neural networks as image representations. After that, random forests are used for scene classification. Finally, they proved by the obtained results, the effectiveness of the proposed method.

An incremental deep learning network for on-line unsupervised feature extraction was elaborated by Liang et al. in [3]. The architecture of the deep network is composed of a cascaded incremental orthogonal component analysis network, a binary hashing step elaborated to enhance the nonlinearity of IOCANet and reduce the quantity of the data and block wise histograms. The developed network has potential results for on-line unsupervised feature extraction.

Almodfer et al. in [6] proposed VGG_No for Handwriting digit recognition. This model has minimized the overall complexity of VGGNet. VGG_No constructed by thirteen convolutional layers, two max-pooling layers, and three fully connected layers. They proved that the classification performance of VGG_No is superior to existing techniques using multi-classifiers since it has achieved better results using very simple and homogeneous architecture.

This brief state-of-the-art proves the effectiveness of DCNN in very challenging tasks. These DCNNs have a powerful learning capability and the ability to obtain a high multi-classification accuracy on very complex problem by extracting deep hidden features. DCNNs need a sufficient amount of the training data or even class balance within the datasets to conduct better results. To resolve this problem, some researchers evoke different data augmentation approaches.

3   Background

Convolutional Neural Networks have been extensively applied for image classification and recognition problems and outperform the state-of-the-art in this field. However, to learn these models, we need a lot of training data in the training step. Thus, image data augmentation is used to conduct better results. In this section, we present these networks and we detail image data augmentation techniques.

3.1   Deep Convolutional Neural Networks

In recent years, the CNNs has achieved great success in many computer vision tasks including face recognition [17, 18] image classification and recognition [19] and object detection [20] because it is one of the most efficient methods for extracting critical features for non-trivial tasks. CNN is a feed forward DNN that consists of a pipeline of alternative several different layers. Unlike neural network, CNN has four distinct types of layers. The first layer named convolution layer is the core building block of a CNN. Its parameters consist of a set of learnable filters. During the forward pass, each filter is convolved with the image i.e. slide over the image spatially and calculate a dot product between the input and the filter which will produce a feature map with two dimensions. The second layer is the pooling layer. This layer is inserted between successive convolution layers. It down-samples the input and reduces the number of parameters in the network by reducing the spatial size of the representation. Besides, it controls the problem of over fitting. Usually several pairs of the convolution and pooling layers are repeated and then a fully connected layer and a classification layer are followed. Add to these layers, there are some intermediate layers between those main layers. In the recent years, to improve the recognition accuracy further, deeper and complex network architectures have been developed. Indeed, several image data augmentation techniques and regularization methods have been proposed to enhance or speed up the training of the deep CNNs.

3.2   Image Data Augmentation Techniques

The performance of deep learning neural networks often enhances with the amount of data available. Training DCNNs models on more data can be an effective method where there is little training data available and result powerful models.

In fact, the data augmentation involves applying a small mutation to training images to create diversity, avoid over fitting, increase amount of data, improve generalization ability and overcome different kinds of variances.

This technique creates new and different training examples from an existing one as shown in Fig. 1. It involves creating modified versions of images in the training dataset that belongs to the same class as the original image. Many ways are followed to augment the images grouped into two strategies:

- *Geometric strategies*: These techniques alter the geometry of the image by mapping the individual pixel values to new destinations. It includes but not limited to:
  (1) Horizontal and Vertical Shift Augmentation: A shift to an image means moving

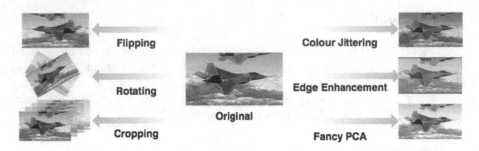

Fig. 1. Some image data augmentation strategies [21]

all pixels of the image in one direction, such as horizontally or vertically, while keeping the image dimensions the same. (2) Horizontal and Vertical Flip Augmentation: An image flip means reversing the rows or columns of pixels in the case of a vertical or horizontal flip respectively. (3) Random Rotation Augmentation: A rotation augmentation randomly rotates the image clockwise by a given number of degrees from 0 to 360. The rotation will likely rotate pixels out of the image frame and leave areas of the frame with no pixel data that must be filled in.

- *Photometric strategies:* These techniques alter the RGB channels by shifting each pixel value to new pixel values according to pre-defined heuristics. It includes color jittering, edge enhancement, etc.

4   Architecture of Proposed DCNN

The proposed system is presented in Fig. 2. The overall architecture of the DCNN is composed of two main parts such as feature extraction and classification.

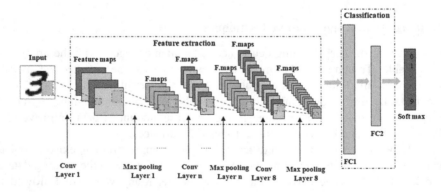

Fig. 2. DCNN architecture

In the feature extraction layers, each layer of the network receives the output from its immediate previous layer as its input, and passes current the output as input to the next layer.

The CNN architecture consists with the combination of four types of layers: convolution, max-pooling, fully connected and classification.

In fact, the system contains eight convolution layers with 16, 32, 64, 64, 128, 128, 512, and 512 filters, where each filter has a size of 5 * 5 and each convolution layer is followed by a max pooling layer of a size of 2 * 2. Each node of the convolution layer extracts the features from the input images by convolution operation on the input nodes. The max-pooling layer abstracts the feature through average or propagating the operation on input nodes. The higher level features have been extracted from the propagated feature of the lower level layers. As the feature propagate to the highest layer or level the dimension of the feature is reduced depending on the size of the convolution and max-pooling masks respectively.

Two fully connected layers with 600 and 300 units performed after the convolution layers, and the softmax layer, which is the final layer of the CNN model classifying the output of the last layer of a CNN into 10 class labels.

A dropout layer is applied to all hidden layers with a probability of 0.5. With this fixed architecture, we then proceed to test the effects of the different data augmentation strategies on the set classification task.

5 Experimentation

The performances of the different described image data augmentation methods combined with regularization techniques are studied. The approaches are conducted on the benchmark datasets: CIFAR-10, STL-10 and Street View House Numbers (SVHN). In this section, we present our experimental results by giving a description of the datasets and analyzing the performance of the DCNN model.

5.1 Data Base Description

- SVHN dataset

The SVHN is a real-world image dataset for developing machine learning and object recognition algorithms with minimal requirement on data pre-processing and formatting. It can be seen as similar in flavor to MNIST but incorporates an order of magnitude more labeled data (over 600,000 digit images) and comes from a significantly harder, unsolved, real world problem. SVHN is obtained from house numbers in Google Street View images.

- CIFAR-10 dataset

The CIFAR-10 is an established computer-vision dataset used for object recognition. The CIFAR-10 dataset consists of 60,000 32 × 32 color images in 10 classes, with 6,000 images per class. There are 50,000 training images and 10,000 test images. The CIFAR-10 classification is a common benchmark problem in machine learning.

The problem is to classify RGB 32 × 32 pixel images across 10 categories: airplane, automobile, bird, cat, deer, dog, frog, horse, ship, and truck.

- STL-10 dataset

We use also the STL-10 dataset that contains 96 × 96 RGB images in 10 categories This dataset has 500 images per class for training and 800 for testing. Additionally, it includes 100,000 unlabeled images for unsupervised learning algorithms, which are extracted from a similar but broader distribution of images.

5.2   Experiments, Results and Discussion

This study evaluates the effects of various data augmentation techniques such as cropping, rotating, shearing, shifting, flipping, Color jittering and edge enhancement on the CIFAR-10, STL-10 and SVHN datasets with and without regularization techniques. Recognition rate is defined as to the number of correctly recognized samples divided by the total number of test samples. The objective is to classify the input image into 10 class labels. Table 1 shows the recognition rate obtained by the different data augmentation with SVHN, CIFAR-10 and STL-10.

Table 1. Results conducted by DCNNs for each data augmentation techniques and without regularization

	Recognition rate (%)		
	SVHN	CIFAR-10	STL-10
Baseline	89.5	85.3	86.6
Flipping	90.5	86.4	87.6
Cropping	94.5	93.7	92.4
Shearing	90.4	87.6	90.3
Rotating	91.5	90.2	90.7
Shifting	91.3	89.6	89.6
Color jittering	88.3	86.4	88.4
Edge enhancement	90.2	88.2	87.6

Table 1 show that cropping as data augmentation technique gives us the greatest improvement in classification accuracy. We theorize that the *cropping* scheme outperforms the other methods as it generates more sample images than the other augmentation schemes. This increase in training data reduces the likelihood of over-fitting, improving generalization and thus increasing overall classification task performance.

The regularized DCNNs used are based on max pooling dropout, fully connected dropout and Rectified Linear Units (ReLUs). The results of each data augmentation technique with regularization method are shown in Table 2.

As shown in Table 2, regularized DCNNs with cropping method achieves higher recognition rate of 95.5% using SVHN datasets, 95.7% with CIFAR-10 and 96.5% with STL-10.

Table 2. Results conducted by DCNNs for each data augmentation techniques and with regularization

	Recognition rate (%)		
	SVHN	CIFAR-10	STL-10
Baseline	92.1	89.3	90.6
Flipping	94.5	90.4	92.6
Cropping	95.5	95.7	96.5
Shearing	92.4	91.6	93.3
Rotating	93.5	93.2	94.7
Shifting	93.3	92.6	92.7
Color jittering	90.3	91.4	91.7
Edge enhancement	92.2	91.8	90.1

Experimental results conducted using 4-fold cross-validation proves that **cropping** in **geometric augmentation** significantly increases DCNN task performance.

6   Conclusion

In this article, we proposed a deep convolutional neural network architecture which, in combination with a set of image data augmentations, produces state-of-the-art results for object recognition. We showed that the improved performance stems from the combination of a deep, high-capacity and regularized model and an augmented training set. Also, we show that cropping in geometric augmentation significantly increases DCNN.

References

1. Sermanet, P., Eigen, D., Zhang, X., Mathieu, M., Fergus, R., LeCun, Y.: OverFeat: integrated recognition, localization and detection using convolutional networks. arXiv preprint arXiv:1312.6229 (2013)
2. Mellouli, D., Hamdani, T.M., Ayed, M.B., Alimi, A.M.: Morph-CNN: a morphological convolutional neural network for image classification. In: Liu, D., Xie, S., Li, Y., Zhao, D., El-Alfy, E.S. (eds.) Neural Information Processing, ICONIP 2017. LNCS, vol. 10635, pp. 110–117. Springer, Cham (2017). https://doi.org/10.1007/978-3-319-70096-0_12
3. Liang, Y., Yang, Y., Shen, F., Zhao, J., Zhu, T.: An incremental deep learning network for on-line unsupervised feature extraction. In: Liu, D., Xie, S., Li, Y., Zhao, D., El-Alfy, E.S. (eds.) International Conference on Neural Information Processing, vol. 10635, pp. 383–392. Springer, Cham (2017). https://doi.org/10.1007/978-3-319-70096-0_40
4. Liang, X., Wang, X., Lei, Z., Liao, S., Li, S.Z.: Soft-margin softmax for deep classification. In: Liu, D., Xie, S., Li, Y., Zhao, D., El-Alfy, E.S. (eds.) Neural Information Processing, ICONIP 2017. Lecture Notes in Computer Science, vol. 10635, pp. 413–421. Springer, Cham (2017). https://doi.org/10.1007/978-3-319-70096-0_43

5. Zheng, Q., Yang, M., Yang, J., Zhang, Q., Zhang, X.: Improvement of generalization ability of deep CNN via implicit regularization in two-stage training process. IEEE Access **6**, 15844–15869 (2018)
6. Almodfer, R., Xiong, S., Mudhsh, M., Duan, P.: Very deep neural networks for hindi/arabic offline handwritten digit recognition. In: Liu, D., Xie, S., Li, Y., Zhao, D., El-Alfy, E.S. (eds.) International Conference on Neural Information Processing, ICONIP 2017. Lecture Notes in Computer Science, vol. 10635, pp. 450–459. Springer, Cham (2017). https://doi.org/10.1007/978-3-319-70096-0_47
7. LeCun, Y., Bengio, Y., Hinton, G.: Deep learning. Nature **521**(7553), 436 (2015)
8. Chen, W., Wilson, J., Tyree, S., Weinberger, K., Chen, Y.: Compressing neural networks with the hashing trick. In: International Conference on Machine Learning, pp. 2285–2294, June 2015
9. Liang, M., Hu, X.: Recurrent convolutional neural network for object recognition. In: Proceedings of the IEEE Conference on Computer Vision and Pattern Recognition, pp. 3367–3375 (2015)
10. Tobías, L., Ducournau, A., Rousseau, F., Mercier, G., Fablet, R.: Convolutional neural networks for object recognition on mobile devices: a case study. In: 2016 23rd International Conference on Pattern Recognition (ICPR), pp. 3530–3535. IEEE, December 2016
11. Li, H., Xu, B., Wang, N., Liu, J.: Deep convolutional neural networks for classifying body constitution. In: Villa, A., Masulli, P., Pons Rivero, A. (eds.) International Conference on Artificial Neural Networks and Machine Learning – ICANN 2016. Lecture Notes in Computer Science, vol. 9887, pp. 128–135. Springer, Cham (2016). https://doi.org/10.1007/978-3-319-44781-0_16
12. Devries, T., Taylor, G.W.: Improved regularization of convolutional neural networks with cutout. arXiv preprint arXiv:1708.04552 (2017)
13. Bai, S.: Growing random forest on deep convolutional neural networks for scene categorization. Expert Syst. Appl. **71**, 279–287 (2017)
14. Szegedy, C., et al.: Going deeper with convolutions. In: Proceedings of the IEEE Conference on Computer Vision and Pattern Recognition, pp. 1–9 (2015)
15. He, K., Zhang, X., Ren, S., Sun, J.: Delving deep into rectifiers: surpassing human-level performance on imagenet classification. In: Proceedings of the IEEE International Conference on Computer Vision, pp. 1026–1034 (2015)
16. Srivastava, N., Hinton, G., Krizhevsky, A., Sutskever, I., Salakhutdinov, R.: Dropout: a simple way to prevent neural networks from overfitting. J. Mach. Learn. Res. **15**(1), 1929–1958 (2014)
17. Sun, Y., Chen, Y., Wang, X., Tang, X.: Deep learning face representation by joint identification-verification. In: Advances in Neural Information Processing Systems, pp. 1988–1996 (2014)
18. Taigman, Y., Yang, M., Ranzato, M.A., Wolf, L.: DeepFace: closing the gap to human-level performance in face verification. In: Proceedings of the IEEE Conference on Computer Vision and Pattern Recognition, pp. 1701–1708 (2014)
19. Krizhevsky, A., Sutskever, I., Hinton, G.E.: ImageNet classification with deep convolutional neural networks. In: Advances in Neural Information Processing Systems, pp. 1097–1105 (2012)
20. He, K., Zhang, X., Ren, S., Sun, J.: Deep residual learning for image recognition. In: Proceedings of the IEEE Conference on Computer Vision and Pattern Recognition, pp. 770–778 (2016)

21. Taylor, L., Nitschke, G.: Improving deep learning using generic data augmentation. arXiv preprint arXiv:1708.06020 (2017)
22. Jayech, K.: Regularized deep convolutional neural networks for feature extraction and classification. In: Liu, D., Xie, S., Li, Y., Zhao, D., El-Alfy, E.S. (eds.) Neural Information Processing. ICONIP 2017. Lecture Notes in Computer Science, vol. 10635, pp. 431–439. Springer, Cham (2017). https://doi.org/10.1007/978-3-319-70096-0_45

An Efficient Scheme for Prototyping kNN in the Context of Real-Time Human Activity Recognition

Paulo J. S. Ferreira[1]($\boxtimes$)(iD), Ricardo M. C. Magalhães[1]($\boxtimes$)(iD),
Kemilly Dearo Garcia[2]($\boxtimes$)(iD), João M. P. Cardoso[1]($\boxtimes$)(iD),
and João Mendes-Moreira[1]($\boxtimes$)(iD)

[1] INESC TEC, Faculty of Engineering, University of Porto, Porto, Portugal
{up201305617,up201502862,jmpc,jmoreira}@fe.up.pt
[2] University of Twente, Enschede, The Netherlands
k.dearogarcia@utwente.nl

Abstract. The Classifier kNN is largely used in Human Activity Recognition systems. Research efforts have proposed methods to decrease the high computational costs of the original kNN by focusing, e.g., on approximate kNN solutions such as the ones relying on Locality-sensitive Hashing (LSH). However, embedded kNN implementations need to address the target device memory constraints and power/energy consumption savings. One of the important aspects is the constraint regarding the maximum number of instances stored in the kNN learning process (being it offline or online and incremental). This paper presents simple, energy/computationally efficient and real-time feasible schemes to maintain a maximum number of learning instances stored by kNN. Experiments in the context of HAR show the efficiency of our best approaches, and their capability to avoid the kNN storage runs out of training instances for a given activity, a situation not prevented by typical default schemes.

Keywords: k-Nearest Neighbor · Classification · kNN prototyping · LSH · Human Activity Recognition (HAR)

1 Introduction

Human Activity Recognition (HAR) (see, e.g., [1]) aims to recognize the activities performed by humans through the analyses of a series of observations using sensors, e.g., carried by users.

k-Nearest Neighbor (kNN) [2] is one of the most popular classifiers used in HAR systems. kNN is based on lazy learning, which means that it does not have an explicit learning phase. Instead, it memorizes the training objects, keeping them in a buffer memory. The kNN popularity in classification problems comes from its simplicity and straightforward implementation. The main disadvantage

This work has been partially funded by FCT project POCI-01-0145-FEDER-016883.

of kNN is the necessity of high storage requirements in order to retain the set of training instances and the number of distances to be calculated [5]. This problem becomes more evident when incremental/online learning is used in devices with limited memory and computing capacity. In such devices, it is necessary to limit the number of instances kNN can store and provide efficient real-time schemes to substitute/update the training instances.

In this paper, we propose substitution schemes that, when it is necessary to store a new instance during the incremental learning phase, allow to keep the number of instances limited and at the same time maintain the distribution of the number of instances per activity, thus preventing, during the incremental phase, that certain activities run out of instances. In addition, the paper evaluates the substitution schemes and compares their results with the results of using the default scheme which substitutes the oldest instance stored.

This paper is organized as follows. Section 2 introduces some related work. Section 3 describes the implication of kNN and LSH [11] + kNN limitation as well as descriptions of the substitution schemes proposed. Section 4 describes the setup used in the experiments performed. Then, Sect. 5 shows experimental results and, finally, Sect. 6 draws some conclusions and summarizes future work planned.

2   Related Work

This section presents previous work using kNN for classification. The presented approaches use different methods to select and remove instances from the kNN buffer memory.

The method ADWIN (ADaptive sliding WINdowing) [9] monitors statistical changes in a sliding window. The window stores all the instances since the last detected change. The window is partitioned into two sub-windows of various size until the difference of their average error exceeds a threshold, depending on the size of the sub windows and a confidence parameter, a change is detected and the older window is dropped.

The method PAW (Probabilistic Approximate Window) [8], stores in a sliding window only the most recent and relevant instances according to a probabilistic measure. To decide which instance to maintain or discard, the algorithm randomly select the instances that is kept in the window, which makes a mix of recent and older instances.

In [8], ADWIN is coupled with kNN with PAW. In this case, the ADWIN is used to keep only the data related to the most recent concept of the stream and the rest of instances are discarded.

The method SWC (Sliding Window Clustering) [10], instead of storing all instances that fit in a sliding window (for representing both old and current concepts), stores compressed information about concepts and instances close to uncertainty border of each class. The clusters are compressed stable concepts and the instances are possible drifts of these concepts.

The approaches proposed in this paper are focused on simple schemes in order to avoid execution time and energy consumption overhead and bearing in mind the implementation of kNN in memory and computing power constrained wearable devices.

3 kNN Substitution Schemes

In our HAR system, features are extracted from the raw data from sensors, are normalized, and then input to a machine learning (ML) classifier, such as kNN. The classifiers were implemented using the MOA library [6], since it allows dealing with data streams and offers a collection of incremental ML algorithms.

The kNN used already includes its own substitution scheme. In this scheme (herein mentioned as "default"), whenever a new training instance arrives and the maximum limit of instances that kNN can store is reached, the new instance substitutes the oldest instance.

When the number of instances that can be stored by the kNN is limited and with the continued substitution of the oldest instance, whenever a new training instance arrives, it may happen that, due to this scheme, activities run out of training instances stored in kNN. This may cause classification errors when kNN tries to classify an instance for which it has none or insufficient training instances. Because of this, it is important to propose efficient substitution schemes able to prevent a given activity from running out of training instances stored by kNN.

The following eight simple, energy/computationally efficient, and real-time feasible substitution schemes are proposed and implemented:

- **Default:** In the kNN used (present in the MOA library [6]) when the instance limit was reached, whenever a new training instance arrives and needs to be stored, it replaces the oldest instance stored in kNN (regardless of the activity);
- **SS1:** randomly selects an instance and replaces it with the new instance of the same class;
- **SS2:** selects the oldest instance of a certain class and replaces it with the new instance of the same class;
- **SS3:** classifies the new instance. If the new instance is incorrectly classified, SS1 is applied. Otherwise the new instance is discarded;
- **SS4:** classifies the new instance. If the new instance is incorrectly classified, SS2 is applied. Otherwise the new instance is discarded;
- **SS5:** classifies the new instance. If the new instance is correctly classified, the scheme randomly selects to apply SS1 or to discard the new instance. If the instance is classified incorrectly it applies SS1;
- **SS6:** classifies the new instance. If the new instance is correctly classified, the scheme randomly selects to apply SS2 or to discard the new instance. If the instance is classified incorrectly it applies SS2;
- **SS7:** classifies the new instance. If the new instance is incorrectly classified, the new instance replaces its nearest neighbor;
- **SS8:** classifies the new instance. If the new instance is correctly classified, the scheme randomly selects to replace the nearest neighbor of the new instance by the new instance, or to discard the new instance. If the new instance is incorrectly classified, the new instance replaces its nearest neighbor.

kNN is a lazy learner, which means that it does not have an explicit learning phase, but instead stores the training instances. For each learning instance, a vector of features is stored in conjunction with its label (class). Imposing a maximum limit of N means that the kNN can only store N instances. For LSH [11] + kNN, a maximum limit implies that a maximum number of instances are stored in the hash tables of the method. In our experiments, we use the LSH parameters empirically selected in our preliminary experiments, i.e., 20 hash tables, 1 random projection per hash value, and 10 as the width of the projection.

4   Experimental Setup

4.1   Dataset, Feature Extraction and Normalization

On our experiments, we used PAMAP2 (Physical Activity Monitoring for Aging People) [7], a dataset where 18 different activities were recorded for 9 different individuals, each wearing 3 IMUs (Inertial Measurement Units) and a heart-rate monitor. Each IMU presents 3D-data from accelerometer, gyroscope and magnetometer.

For each sliding window with sensor data, and according to the number of axis of each sensor, several features are extracted. In the case of our prototypes the features are associated to 3D and 1D sensors. Each 1D and 3D sensor has, respectively, 2 (mean and standard deviation) and 10 features (x mean, y mean, z mean, mean of the sum of the 3 axes, x standard deviation, y standard deviation, z standard deviation, xy Pearson Correlation, xz Pearson Correlation, and yz Pearson Correlation) extracted. Additionally, all features were subsequently normalized by dividing each value in a vector with the vector's magnitude.

4.2   Setup

In the experiments presented in this paper, our HAR system begins with a model built offline using the $N-1$ users. Then, we emulate the use of the HAR system and the real-time behavior using data from a different user representing the actual user of the system. After preliminary studies, user 5 was selected to represent the user of the system. This selection was because user 5 is the one of the users with more activities collected and also being the user with the highest number of instances. A sliding window of 300 sensor readings and a 10% overlap were used. With these settings and the data of the remaining 8 users, 6186 training instances are considered for creating the offline model.

In these experiments, the maximum number of instances the kNN (k = 3) could store was limited to 100, 200, 500, and from 1000 to 8000 with increments of 1000. The limit of 8000 allows storing the entire original training set plus the instances for incremental training. In order to reduce the initial training set with 6186 instances, a random-based filter reduces the original training set for the various limits tested, maintaining the global distribution of instances per activity of the original set.

5   Experimental Results

5.1   Substitution Schemes Comparison

In these experiments, all substitution schemes presented above were evaluated varying the kNN and LSH kNN storage maximums. Figure 1 shows the results obtained.

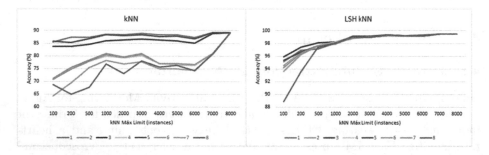

Fig. 1. Impact on accuracy of substitution schemes for the different limits considered: (a) kNN; (b) LSH kNN

According to Fig. 1(a), the best substitution schemes are 1, 3 and 5, with the scheme 1 showing slightly better results than the other two. The substitution scheme 8 is the one that presents the worst results of all. The results show a significant difference between the schemes with better results (1, 3 and 5) and those with worst results (2, 4, 6, 7 and 8). On average, there is a decrease of about 9% in accuracy between the best and worst schemes.

The convergence of all the lines of the chart to the same point is due to the fact that for 8000 all the learning instances are kept stored in the kNN. So, for the 8000 limit, substitution schemes are not applied.

According to Fig. 1(b), all substitution schemes behave similarly and differences in accuracy are minimal. However, scheme 3 achieves the best average accuracy ($\approx 98.56\%$), while scheme 7 presents the worst result ($\approx 98.17\%$). Globally, there is a difference of $\approx 0.39\%$ on average.

Since all schemes achieve an already high accuracy, the substitutions done incrementally are not capable of improving or decreasing the accuracy of the classifier in a considerable amount on average.

The best scheme, 1, is compared with the methods kNN (default scheme), kNN with PAW (kNN$_P$) and kNN with PAW with ADWIN (kNN$_{PA}$). The results are present in Table 1.

The results presented in Table 1 show that all methods obtained very close accuracies, with kNN$_{PA}$ being the only one that got lower accuracy than our best scheme. However, none of the other methods guarantee that no class runs out of instances in the kNN feature space.

Table 1. Average accuracy for each method.

	kNN (SS 1)	kNN (default)	kNN$_P$	kNN$_{PA}$
Average accuracy	88.54	88.90	88.75	88.51

5.2 Proposed Substitution Schemes vs Default Scheme

In order to study the impact of the substitution schemes, an experiment was carried out where in one case the substitution scheme which obtained better results in the previous experiment was used and in the other case, no substitution scheme was used. In this experiment, the accuracy was measured throughout the incremental training, that is, the accuracy is calculated whenever a new instance is classified. Figure 2 shows the results obtained.

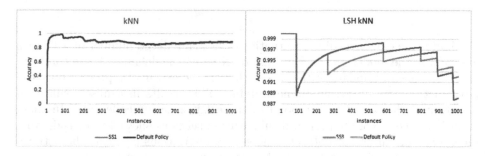

Fig. 2. Accuracy of kNN throughout the incremental training, using a maximum limit of 5000 instances, the best substitution scheme over the default scheme.

Analyzing the results of the left chart of Fig. 2, it can be concluded that the best proposed substitution scheme can match the results of the default scheme of kNN, failing to improve on the default scheme. Focusing on the right chart, we can conclude the proposed best scheme for LSH kNN is not capable of improving the default scheme for the most part, only achieving higher accuracy on the last classifications performed.

The limits used so far, were high enough to keep instances for all activities. In order to test the behavior in one extreme case of the default scheme as compared to the proposed substitute schemes, an experiment was performed where all 1008 instances of User 5 were used for incremental training and then used again for testing the accuracy of kNN. For this experiment, the maximum limit of instances that kNN can store has been set to 100, 200, and 500. PAMAP users data is organized by activity. Using the default scheme and such small limits, it turns out that in incremental training the kNN feature space will only save instances of the latest User 5 activities.

The results presented in Table 2 show that the default substitution scheme achieves very low accuracy. This is due to the fact that the default scheme always

Table 2. Impact on accuracy for the different limits and when considering online learning with longer time periods followed by classification of activities.

Substitution scheme	kNN limits			LSH kNN limits		
	100	200	500	100	200	500
1	68.25	86.81	91.27	86.21	86.41	85.12
2	57.04	64.68	73.91	77.78	73.91	74.11
3	70.83	85.81	89.78	97.42	99.50	99.80
4	57.54	64.38	73.91	91.17	98.41	96.53
5	67.16	85.81	90.67	88.00	87.40	90.18
6	57.44	64.19	74.01	84.42	85.91	87.10
7	58.23	68.25	74.11	94.84	97.92	98.81
8	54.96	63.10	73.71	80.46	83.63	87.00
Default	11.71	20.54	50.40	11.81	20.73	51.88

eliminates the oldest, leading to situations where a reduced number of activities is stored in the model. In this case, the kNN using the default scheme conducts to an incremental update of the model that in the end may represent only the latest activities used in the online training. This results in a misclassification of activities that do not correspond to those latest activities learned. In these situations, the use of the proposed substitution schemes allowed to considerably increase the accuracy of both classifiers, as substitutions occur at the activity level, thus preventing certain activities from running out of training instances stored and keeping the distribution of instances by activity.

As expected, the results show the importance of updating the model with instances of the user using the HAR system. This is the reason why the substitution scheme of the oldest instance in the model provides good results whenever the limit is enough or the sequence of activities do not override completely the model and make it not representative. This is what happens in the scenario forced by the learning of the instances of user 5 and then testing with the same instances.

6   Conclusion

Although kNN is one of the most used classifiers in Human Activity Recognition (HAR) systems, its storage requirements increase with the number of learning instances. This paper proposed schemes to substitute/update the set of learning instances for kNN classifiers in order to maintain a maximum number of training instances. The proposed schemes are evaluated in the context of a Human Activity Recognition (HAR) system. Eight different substitution schemes were proposed and tested to allow incremental online training when there is a maximum limit of instances that the kNN can store. This approach is useful in the case of implementing a HAR system in devices with limited memory (e.g., wearable devices). The experimental results show the efficiency of some of the eight schemes evaluated.

Ongoing work is focused on additional experiments with datasets from other domains. As future work, we plan to continue researching substitution/update schemes and compare with other more computational intensive schemes and verify the overheads regarding response time and energy consumption.

References

1. Zhang, S., Wei, Z., Nie, J., Huang, L., Wang, S., Li, Z.: A review on human activity recognition using vision-based method. J. Healthc. Eng. **2017**, 31 (2017). Article ID 3090343
2. Cover, T.M., Hart, P.E.: Nearest neighbor pattern classification. IEEE Trans. Inf. Theory **13**(1), 21–27 (1967)
3. Su, X., Tong, H., Ji, P.: Activity recognition with smartphone sensors. Tsinghua Sci. Technol. **19**(3), 235–249 (2014)
4. Calvo-Zaragoza, J., Valero-Mas, J.J., Rico-Juan, R.J.: Improving kNN multilabel classification in Prototype Selection scenarios using class proposals. Pattern Recogn. **48**(5), 1608–1622 (2015)
5. Garcia, S., Derrac, J., Cano, J., Herrera, F.: Prototype selection for nearest neighbor classification: taxonomy and empirical study. IEEE Trans. Pattern Anal. Mach. Intell. **34**(3), 417–435 (2012)
6. Bifet, A., Holmes, G., Kirkby, R., Pfahringer, B.: MOA: massive online analysis. J. Mach. Learn. Res. **11**, 1601–1604 (2010)
7. Reiss, A., Stricker, D.: Introducing a new benchmarked dataset for activity monitoring. In: The 16th IEEE International Symposium on Wearable Computers (ISWC) (2012)
8. Bifet, A., Pfahringer, B., Read, J., Holmes, G.: Efficient data stream classification via probabilistic adaptive windows. In: Proceedings of the 28th Annual ACM Symposium on Applied Computing, pp. 801–806. ACM, March 2013
9. Bifet, A., Gavalda, R.: Learning from time-changing data with adaptive windowing. In: Proceedings of the 2007 SIAM International Conference on Data Mining, pp. 443–448. Society for Industrial and Applied Mathematics, April 2007
10. Garcia, K.D., de Carvalho, A.C.P.L.F., Mendes-Moreira, J.: A cluster-based prototype reduction for online classification. In: Yin, H., Camacho, D., Novais, P., Tallón-Ballesteros, A. (eds.) IDEAL 2018. Lecture Notes in Computer Science, vol. 11314, pp. 603–610. Springer, Cham (2018). https://doi.org/10.1007/978-3-030-03493-1_63
11. Indyk, P., Motwani, R.: Approximate nearest neighbors: towards removing the curse of dimensionality. In: Proceedings of the Thirtieth Annual ACM Symposium on Theory of Computing, pp. 604–613. ACM (1998)

A Novel Recommendation System for Next Feature in Software

Victor R. Prata[1(✉)], Ronaldo S. Moreira[2], Luan S. Cordeiro[1], Átilla N. Maia[1],
Alan R. Martins[1], Davi A. Leão[1], C. H. L. Cavalcante[1],
Amauri H. Souza Júnior[1], and Ajalmar R. Rocha Neto[1]

[1] Federal Institute of Ceará, Fortaleza, Brazil
{victor.prata,luan.sousa,atilla.negreiros}@ppgcc.ifce.edu.br,
{henriqueleitao,amauriholanda,ajalmar}@ifce.edu.br,
alanrabelo13@gmail.com, davileao2014@gmail.com
[2] Fortes Tecnologia, Rua António Fortes 330, Fortaleza 60813-460, Brazil
ronaldox2@gmail.com
https://www.fortestecnologia.com.br

Abstract. Software that needs to fulfill many tasks requires a large number of components. Users of these software need a lot of time to find the desired functionality or follow a particular workflow. Recommendation systems can optimize a user's working time by recommending the next features he/she needs. Given that, we evaluate the use of three algorithms (Markov Chain, IndRNN, and LSTM) commonly applied in sequence recommendation/classification in a dataset that reflects the use of the accounting software from Fortes Tecnologia. We analyze the results under two aspects: accuracy for top-5 recommendations and training time. The results show that the IndRNN achieved the highest accuracy, while the Markov Chain reached the lowest training time.

Keywords: Recommendation system · Independently Recurrent Neural Network · Sequential recommendation · Deep learning · Commercial applications

1 Introduction

Recommendation systems are real-world applications to help users quickly select items that they could choose or might be interested in. Moreover, recommendation systems can reduce work time by recommending the next features to users based on what they are working on. The problem of feature recommendation diverges from classic recommendation systems mainly by the temporal/sequential component.

In this context, [10] presents a review of proposed models and architectures for sequence-aware recommendation systems. Among the most common approaches, we highlight probabilistic models based on the Markov chain, such as MDP (Markov Decision Processes) [11]; Context Trees [4], which correspond

© Springer Nature Switzerland AG 2019
H. Yin et al. (Eds.): IDEAL 2019, LNCS 11871, pp. 494–501, 2019.
https://doi.org/10.1007/978-3-030-33607-3_53

to Markov Chains of variable order; and Hidden Markov Models [7], where latent variables are introduced to capture longer-lived dependencies.

One may also employ recurrent neural networks (RNN) in the problem of interest. Recently, [5] and [8] use RNN and nearest neighbors for session-based recommendations, as well as [14] for prediction of clicks. Moreover, [13] shows how RNN models can work with time series classification. [15] used RNN to remove the dynamic preference of the user's listening histories. The work proposed by [12] uses deep RNNs for improved recommend user interfaces and Gated Recurrent Unit (GRU) to deal with long-term dependencies.

In this work, we investigate the usage data of accounting software developed by Fortes Tecnologia[1], to help their users by recommending the next features. The recommendation is based on the user's usage history in a particular software or the usual routine of all users for that software within the same company. Therefore, we analyze the use of probabilistic algorithms such as Markov Chains and recurrent neural networks, from which the Independently Recurrent Neural Network (IndRNN) obtained the best results in recommending the next features.

2 Feature Sequence Recommender Systems

In simple words, we are interested in recommending the next features to users of software systems from historical user data. Here, features represent 'clickable' items of the software system. One could also refer to them as frames, components, services, or webpages.

We employ an auto-regressive approach in which previous features are used to predict the current feature state. Clearly, the problem can be formulated as a sequence modeling or time-series prediction task, where each prediction consists of one or more recommendations of the system features.

2.1 Problem Formulation

Formally, we wish to predict the feature $f_t \in \{1, \ldots, N_f\}$ at time t from previously observed features $f_{t-1}, f_{t-2}, \ldots, f_1$. In particular, we are interested in predictions at (i) user-level, and (ii) system-level. The first aims to predict features for specific-users of specific-systems (in fact, users only participate in one system), whereas system-level predictions take into account data from all users of a system to provide general recommendations.

In our formulation, features, users, and systems are modeled as discrete random variables. Let F_t be a discrete random variable denoting the feature at discrete-time t taking values in $\mathcal{F} = \{1, \ldots, N_f\}$. Likewise, let U and S denote the user and system variables, with support $\mathcal{U} = \{1, \ldots, N_u\}$ and

---

[1] https://www.fortestecnologia.com.br.

$\mathcal{S} = \{1, \ldots, N_s\}$, respectively. In the user-level, we are interested in top-K recommendations given by:

$$\hat{f}_t^{(1)} = \underset{f_t \in \mathcal{F}}{\arg\max} \quad P(F_t = f_t \mid F_{t-1} = f_{t-1}, \ldots, F_1 = f_1, U = u, S = s),$$

$$\hat{f}_t^{(2)} = \underset{f_t \in \mathcal{F} \setminus \{\hat{f}_t^{(1)}\}}{\arg\max} \quad P(F_t = f_t \mid F_{t-1} = f_{t-1}, \ldots, F_1 = f_1, U = u, S = s),$$

$$\vdots$$

$$\hat{f}_t^{(K)} = \underset{f_t \in \mathcal{F} \setminus \{\hat{f}_t^{(1)}, \ldots, \hat{f}_t^{(K-1)}\}}{\arg\max} \quad P(F_t = f_t \mid F_{t-1} = f_{t-1}, \ldots, F_1 = f_1, U = u, S = s).$$

$$(1)$$

From now on, we simplify notation by writing the probability mass function $P(F_t = f_t \mid F_{t-1} = f_{t-1}, \ldots, F_1 = f_1, U = u, S = s)$ as $P(f_t \mid f_{t-1}, \ldots, f_1, u, s)$. Note that K is a free parameter that represents the number of the recommended features. The value $\hat{f}_t^{(1)}$ denotes the (first) most likely choice for the feature at time t; thus, $\hat{f}_t^{(k)}$ denotes the k-th most likely choice.

The term $P(f_t \mid f_{t-1}, \ldots, f_1, u, s)$ expresses the distribution of the feature at time t, given the features in all previous instants, the user u, and the system s. System-level predictions are those based on $P(f_t \mid f_{t-1}, \ldots, f_1, s)$.

3   Materials and Methods

3.1   Dataset

Fortes Tecnologia provided the dataset we use in this work. The dataset is composed of the interactions of their accounting software users. In this scenario, a feature means a functionality present in each window of the software. A sequence represents a user's usage history of an application throughout a session. In summary, we use data from 6 software applications described in 81211 sequences.

3.2   Markov Chain

One of the sequence modeling approaches considered in this work is the first-order, discrete-time, discrete-state Markov Chain method. A sequence of random variables $X_0, X_1, \ldots$ assuming values in state-space $\{1, 2, \ldots, M\}$ is called a Markov chain if for all $n \geq 0$,

$$P(X_{n+1} = j | X_n = i, X_{n-1} = i_{n-1}, \ldots, X_0 = i_0) = P(X_{n+1} = j | X_n = i). \quad (2)$$

The term $P(X_{n+1} = j | X_n = i)$ denotes the transition probability from state i to state j. In addition, we assume that the chain is homogeneous, i.e., $P(X_{n+1} = j | X_n = i)$ is the same for all n. In this case, we can represent the Markov Chain using a transition matrix Q, with $q_{ij} = P(X_{n+1} = j | X_n = i)$. Q is a non-negative matrix whose row-wise sum is equal to 1.

The application of Markov Chains to the problem of feature recommendation is straightforward. The Markovian assumption states that the feature at time t only depends on its value at time $t-1$. For instance, the user-level predictive distribution $P(f_t \mid f_{t-1}, \ldots, f_1, u, s)$ reduces to $P(f_t \mid f_{t-1}, u, s)$. This consists of different Markov chains for each pair user-system. We can represent these Markov chains by the transition matrices $Q(u, s)$, each of them operating on the state-space of features $\mathcal{F} = \{1, \ldots, N_f\}$.

The probabilities $Q(u, s)$ can be estimated from data using, for instance, the maximum likelihood principle, i.e., the estimated values $\hat{q}_{ij}(u, s)$ are

$$\hat{q}_{ij}(u, s) = P(F_t = j | F_{t-1} = i, U = u, S = s) = \frac{\#(i \to j, u, s)}{\#(i \to *, u, s)}, \tag{3}$$

where $\#(i \to j, u, s)$ denotes the number of occurrences of feature transitions $i \to j$ in the sequences associated with the user u and system s. By the same token, $i \to *$ denotes transitions from i to any state. For system-level recommendations $Q(s)$, we have

$$\hat{q}_{ij}(s) = P(F_t = j | F_{t-1} = i, S = s) = \sum_u P(F_t = j, U = u | F_{t-1} = i, S = s)$$

$$= \sum_u P(F_t = j | F_{t-1} = i, U = u, S = s) P(U = u | F_{t-1} = i, S = s)$$

$$= \sum_u \left(\frac{\#(i \to j, u, s)}{\#(i \to *, u, s)} \frac{\#(i \to *, u, s)}{\sum_u \#(i \to *, u, s)} \right) = \frac{\sum_u \#(i \to j, u, s)}{\sum_u \#(i \to *, u, s)}. \tag{4}$$

3.3 Long Short-Term Memory (LSTM)

LSTMs [6] were proposed to try to solve the vanishing gradient problem in RNNs. The LSTM architecture consists of a chain of recurrently connected sub-nets, known as memory blocks. Each block contains one or more self-connected memory cells to store important information at each input instance. The information flows through the memory cells and is moderated by three multiplicative units, usually called gates. The *input gate* controls the input from the previous state. The *output gate* calculates the output from the current memory cell to make predictions at the corresponding input instance. Finally, the *forget gate* controls information from the previous state that will be ignored. This multiplicative units provide continuous analogs of write, read and reset operations for the cells [3].

Considering the problem as a time series $X = x^{(1)}, x^{(2)}, \ldots, x^{(n)}$, where each point $x^{(t)} \in R^m$ in the time series is an m-dimensional vector $\{x_1^{(t)}, x_2^{(t)}, \ldots, x_m^{(t)}\}$, whose elements correspond to the input variable, a prediction model learns how to predict the next feature.

3.4 Independently Recurrent Neural Network (IndRNN)

According to [9], RNNs are commonly tricky to train because of the gradient vanishing phenomena, exploding problems, and hard to learn long-term patterns.

LSTM and GRU were developed to address these problems, but the use of hyperbolic tangent and the sigmoid action functions results in gradient decay over layers.

To address these problems, [9] proposed a network where neurons in the same layer are independent of each other and connected across layers. [9] had shown that an IndRNN could be easily regulated to prevent the gradient exploding and vanishing problems while allowing the network to learn long-term dependencies.

Each neuron in IndRNN deals with one type of spatial-temporal pattern independently, only receiving information from the input and each hidden neuron state at the previous time step. Different from the conventional RNNs, [9] proposed the IndRNN providing a new perspective of recurrent neural networks as independently aggregating spatial patterns over time.

4   Experiments

In this section, we present and discuss the results achieved with the Markov Chain, LSTM and IndRNN on the sequence recommendation problem. We evaluate the algorithms' performances for the *top-5* recommendations with the following metrics: accuracy when predicting the next feature, standard deviation of the accuracy, training and prediction time. We emphasize that a hit occurs when the true next feature is in the *top-5* recommendations. Considering the potential commercial use, the training and prediction time are important metrics, because large training time requires more computational resources.

4.1   Computational Environment

We carried out the experiments using Python 3.6 with the following open source libraries: keras 2.2.2 [2] with tensorflow 1.1.0 [1] as backend, numpy 1.14.4, and sklearn 0.20.1. Also, we highlight that all experiments were performed using a 4th Generation Intel Core i3 3.4 GHz, 10 GB of DDR3 RAM.

4.2   Experimental Setup

In each run, we train the algorithms in two contexts: (i) for all sequences of a given user, we randomly select 80% from the data to train the models; (ii) for all sequences of a given system, we also randomly select 80% to train the respective models. Then, we use the remaining 20% of both contexts to assess the models' generalization performances. Both analyses aim to examine if a user's experiences can help other users that use the same system.

In sequence-aware recommendation systems, the learning algorithms usually consider a limited number of the user's past actions to predict the next feature. This concept is named "window". The window length may vary according to the algorithms. For the Markov Chain, we use a window length of 1. For the LSTM and IndRNN, we use a window length of 2. We punctuate that we chose those values by trial-end-error.

4.3   Results and Discussion

Table 1 presents the average accuracy, training and prediction time of all models for both user and system recommendations. One can note that the IndRNN outperforms all the other algorithms for the *top-5* recommendations, with an average accuracy of 71.23% for the users and 69.97% for the system.

Table 1. Average accuracy and training time for the *top-5* recommendations. We highlight the best results in bold.

	Algorithms	Accuracy (%)	Time (sec)	
			Training	Prediction
User	Markov Chain	63.68 ± 15.21	**0.0006**	**0.0074**
	LSTM	69.70 ± 17.37	22.35	0.18
	IndRNN	**71.23 ± 16.91**	14.23	0.1209
System	Markov Chain	62.71 ± 07.23	**0.2747**	**36.30**
	LSTM	68.80 ± 07.47	13093.48	37.48
	IndRNN	**69.97 ± 08.24**	10617.14	40.57

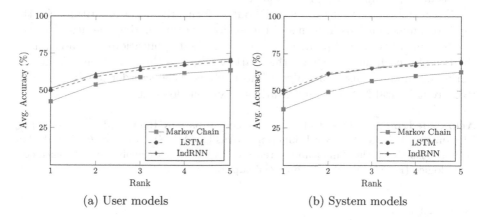

(a) User models   (b) System models

Fig. 1. Average accuracy and rank of all algorithms for each context.

Despite having the lowest average accuracy, the Markov Chain is the fastest algorithm, spending an average of 0.0006 s per user and 0.2747 s per system. The justification relies on its simple mathematical formulation, which performs fast computations over the transition matrix. For the recurrent networks, IndRNN has a lower average training time. This outcome can be evidence that architectures with several independent neurons may accelerate the training, without jeopardizing the network performance. The Markov Chain is also the fastest algorithm in the prediction phase, spending an average of 0.0074 s per user and 36.30 s per system.

Figure 1 shows a comparison between Markov Chains, LSTM, and IndRNN for *top-1*, ..., *top-5* recommendations. One can observe that for the user models, IndRNN outperforms the other algorithms in every rank. For the system models, LSTM presents the highest average accuracy for *top-1*, *top-2*, and *top-3* ranks. The Markov Chain achieves the lowest average accuracy for all the ranks, in both user and system models.

The results indicate that using LSTM and IndRNN incur better accuracy when compared to Markov Chains. Nonetheless, choosing such algorithms culminates in a considerable increase in training time.

5   Conclusions

In this paper, we propose a new recommendation system for the next feature in software using Markov Chains, LSTM, and IndRNN. We use the users' action sequences, and the system's user sequences to verify if the user's experiences in the same software help the feature recommendation for other users. The sequences were collected and provided by Fortes Tecnologia from their accounting software. We evaluate the algorithms' performances for the *top-5* recommendations with accuracy when predicting the next feature, standard deviation of the accuracy, training and prediction time.

The results show that the IndRNN outperforms the other algorithms for the top-5 recommendations in terms of the accuracy of predicting the next feature. As the time of train and prediction matters in real applications, the analysis of the time for each method shows the Markov Chain is the fastest one. As future works, we intend to use other algorithms such as variable-order Markov models, Recurrent Neural Network, and Markov Decision Processes.

Acknowledgments. This study was financed in part by the Coordenação de Aperfeiçoamento de Pessoal de Nível Superior – Brasil (CAPES) – Finance Code 001. The authors also thank the Fundação Cearense de Apoio ao Desenvolvimento Científico e Tecnológico (FUNCAP), for the financial support.

References

1. Abadi, M., et al.: TensorFlow: large-scale machine learning on heterogeneous systems (2015). https://www.tensorflow.org/
2. Chollet, F., et al.: Keras (2015). https://keras.io
3. Graves, A.: Supervised Sequence Labelling with Recurrent Neural Networks. Studies in Computational Intelligence. Springer, Heidelberg (2008). https://doi.org/10.1007/978-3-642-24797-2
4. He, Q., et al.: Web query recommendation via sequential query prediction. In: IEEE 25th International Conference on Data Engineering, pp. 1443–1454 (2009)
5. Hidasi, B., Karatzoglou, A., Baltrunas, L., Tikk, D.: Session-based recommendations with recurrent neural networks. CoRR 1511.06939 (2015)

6. Hochreiter, S., Schmidhuber, J.: Long short-term memory. Neural Comput. **9**(8), 1735–1780 (1997)
7. Hosseinzadeh Aghdam, M., Hariri, N., Mobasher, B., Burke, R.: Adapting recommendations to contextual changes using hierarchical hidden Markov models. In: Proceedings of the 9th ACM Conference on Recommender Systems, RecSys 2015, pp. 241–244 (2015)
8. Jannach, D., Ludewig, M.: When recurrent neural networks meet the neighborhood for session-based recommendation. In: Proceedings of the Eleventh ACM Conference on Recommender Systems, RecSys 2017, pp. 306–310 (2017)
9. Li, S., Li, W., Cook, C., Zhu, C., Gao, Y.: Independently Recurrent Neural Network (IndRNN): Building A Longer and Deeper RNN. arXiv e-prints arXiv:1803.04831, March 2018
10. Quadrana, M., Cremonesi, P., Jannach, D.: Sequence-aware recommender systems. CoRR **1802.08452** (2018)
11. Shani, G., Heckerman, D., Brafman, R.I.: An MDP-based recommender system. J. Mach. Learn. Res. **6**, 1265–1295 (2005)
12. Soh, H., Sanner, S., White, M., Jamieson, G.: Deep sequential recommendation for personalized adaptive user interfaces. In: Proceedings of the 22nd International Conference on Intelligent User Interfaces, pp. 13–16. ACM IUI (2017)
13. Usken, H.M., Stagge, P.: Recurrent neural networks for time series classification. Neurocomputing **50**, 223–235 (2003)
14. Zhang, Y., et al.: Sequential click prediction for sponsored search with recurrent neural networks. In: Proceedings of the Twenty-Eighth AAAI Conference on Artificial Intelligence, AAAI 2014, pp. 1369–1375 (2014)
15. Zheng, H.T., Chen, J.Y., Liang, N., Sangaiah, K.A., Jiang, Y., Zhao, C.Z.: A deep temporal neural music recommendation model utilizing music and user metadata. Appl. Sci. **9**, 703 (2019)

Meta-learning Based Evolutionary Clustering Algorithm

Dmitry Tomp[1,2], Sergey Muravyov[1,2(✉)], Andrey Filchenkov[1,2],
and Vladimir Parfenov[2]

[1] Machine Learning Lab, ITMO University, 49 Kronverksky Pr.,
St. Petersburg 197101, Russia
dmitrytomp@gmail.com, {smuravyov,afilchenkov}itmo.ru
[2] Information Technologies and Programming Faculty, ITMO University,
49 Kronverksky Pr., St. Petersburg 197101, Russia
parfenov@mail.ifmo.ru

Abstract. In this work, we address the hard clustering problem. We present a new clustering algorithm based on evolutionary computation searching a best partition with respect to a given quality measure. We present 32 partition transformation that are used as mutation operators. The algorithm is a $(1 + 1)$ evolutionary strategy that selects a random mutation on each step from a subset of preselected mutation operators. Such selection is performed with a classifier trained to predict usefulness of each mutation for a given dataset. Comparison with state-of-the-art approach for automated clustering algorithm and hyperparameter selection shows the superiority of the proposed algorithm.

Keywords: Clustering · Evolutionary clustering · Meta-learning · Evolutionary computation

1 Introduction

Clustering is an unsupervised learning problem that has real-world data mining applications in fields such as bioinformatics [1], market segmentation [2], sociological studies [3] and many others.

The hard clustering task is formulated as splitting a dataset into partitions of similar objects; however, a specific measure of partition *quality* depends on a particular subject field and cannot be given in a general case due to the Kleinberg's theorem [4].

There is a wide variety of state-of-the-art clustering algorithms, including K-Means, Affinity Propagation, Agglomerative Clustering and Mean Shift, implemented in popular libraries like Scikit-Learn [5] and WEKA [6]. Most of them are designed based on some expectation on how a good partition should look like, making them indirectly tied to one or some particular partition quality criteria.

An alternative approach to the problem involves determining a partition quality criterion *separately* from selecting a clustering algorithm to seek a partition that satisfies the chosen criterion. For instance, one may choose to optimize

H. Yin et al. (Eds.): IDEAL 2019, LNCS 11871, pp. 502–513, 2019.
https://doi.org/10.1007/978-3-030-33607-3_54

an internal scalar partition quality measures called cluster validity indices (CVI) such as Silhouette [7], Davies-Bouldin [8], Calinski-Harabasz [9], DBCV [10] or many others. A sound review of them can be found in [11]. In this case, any optimization algorithm (possibly dedicated to solving a clustering problem in particular) can be used to optimize such a measure.

Evolutionary algorithms have been used as a stochastic method of searching optimal solutions for a wide variety of problems. For clustering, an evolutionary algorithm may consider a partition as an individual, apply mutation and/or crossover operators to modify partitions and compare them using a specified cluster validity index. A nice survey [12] describing many attempts to solve a clustering problem using evolutionary algorithms.

Meta-learning is a popular approach to machine learning automation that is a development of algorithms that helps to choose machine learning models and tune their hyperparameters to perform better on specific kinds of input data. In meta-learning framework, the choice of a machine learning algorithm and its parameters is made by applying a supervised learning model that treats a dataset as a single vector of its scalar and/or categorical *meta-features*. The meta-learning approach has been applied to clustering problem previously in [13, 14], but these algorithms choose between state-of-the-art clustering methods without any regard for alternatives like evolutionary algorithms.

In this work, we focus on developing new clustering algorithm based on evolutionary computation. In this case, mutation operators are partition transformation. We propose 32 mutation operators inspired by ideas described in [12]. Experiments showed that only some of these mutations are useful in finding the best partitions of a dataset. What is even more important, sets of useful mutations vary for different datasets. Based on this, we propose an evolutionary-based clustering algorithm that preselects useful mutations for a particular dataset using meta-learning, adopting an approach presented in [15].

Section 2 contains description of our evolutionary algorithm and mutation operators. In Sect. 3, we explain how we preselect mutations via meta-learning. In Sect. 4, our we describe experimental setup and provide measurement results. Finally, in Sect. 5, we discuss efficiency of the proposed algorithm based on experimental results and point out some directions of further work that can be done in the field of clustering automation.

2   Proposed Evolutionary Clustering Algorithm

2.1   Scheme of the Evolutionary Algorithm

We use the $(1+1)$ as a basic evolution strategy, which is a best among the strategies we tested (see Sect. 2.2). Given a subset of mutations, it chooses a random mutation from the set, applies it to the current individual and get another partition. In a general case, the random choice can be performed according to an arbitrary (and possibly dynamically adjusted) probability vector. We initialize the partition for a given dataset by a random number of clusters build along correspondent number of axes that go through the centroid of the dataset. The success of the mutation is defined by the value of the chosen fitness function, if it is better that its ancestor then mutation is treated successful. Pseudocode of our algorithm is provided in listing Algorithm 1. We have chosen a stopping criterion of 250 consequent unsuccessful iterations based on early experimental runs. The algorithm also stops if either our mutation operator appears to be inapplicable to the current individual, or current number of clusters exceeds a certain threshold (for our experiments, we chose 70 as a fixed upper bound).

The latter part of the stopping condition is explained by two reasons. First, some CVIs considered in this work appear to perform very slowly on partitions with large number of clusters, second, it has been observed experimentally that the proposed algorithm sometimes converges to an "optimal" partition with most clusters containing from 1 to 5 objects.[1]

Algorithm 1. Algorithm $(1 + 1)$ with an externally selected mutation subset

---

   function ONEPLUSONE(cvi, dataset, mutationSubset)
      currentIndividual ← INITIALIZE(dataset)
      currentIndex ← EVALUATEPARTITION(cvi, currentInidividual)
      unsuccessfulSteps ← 0
      while unsuccessfulSteps < 250 **do**
         newIndividual ← MUTATE(individual, mutationSubset)   ▷ see listing 2
         if (newIndividual = ∅) and (NUMCLUSTERS(newIndividual) ≥ 70) **then**
            return currentIndividual
         end if
         newIndex ← EVALUATEPARTITION(cvi, newIndividual)
         if newIndex ≻ currentIndex **then**
            currentIndividual ← newIndividual
            currentIndex ← newIndex
            unsuccessfulSteps ← 0
         else
            unsuccessfulSteps ← unsuccessfulSteps + 1
         end if
      end while
      return currentIndividual
   end function

A pseudocode for mutation application routine is shown in listing Algorithm 2. In certain cases, a particular mutation cannot be applied to a partition, for example mutation, if applied to a current partition, would produce an inconsistent partition with a single cluster in it. The mutations used by our algorithm are described in Sect. 2.3.

---

[1] This phenomenon is most likely related to properties of a specific CVI and can be further mitigated, e.g. by applying different initialization method or using a more complex mutation/evolutionary scheme.

Sometimes it is also possible that *all* available mutations cannot be applied to the partition; upon such an occurrence, the algorithm terminates.

Algorithm 2. Mutation application

---

function MUTATE(individual, usefulMutations)
    unusedMutations ← COPY(usefulMutations)
    mutation ← ∅
    repeat
        mutation ← RANDOMCHOICE(usefulMutations)
        if MUTATIONNOTAPPLICABLE(mutation, currentIndividual) **then**
            if HASREPLACEMENT(mutation) **then**
                mutation ← GETREPLACEMENT(mutation)
            else
                unusedMutations ← unusedMutations \{mutation}
                mutation ← ∅
            end if
        end if
    until (mutation ≠ ∅) and (usefulMutations = ∅)
    if mutation ≠ ∅ **then**
        return APPLYMUTATION(mutation, individual)
    else
        return ∅
    end if
end function

2.2 Choosing Scheme of the Evolutionary Algorithm

The choice of the $(1+1)$ scheme was based on early experimental results, which will be described in this Section.

In this work, we considered evolutionary schemes from the $(1 + \lambda)$ family, namely:

- $(1+1)$ with random mutation from listing Algorithm 1,
- $(1+\lambda = 10)$ that applies 10 random mutations on each step, possibly choosing a particular stochastic mutation several times;
- $(1 + \lambda = 32)$ that applies each of the implemented 32 mutations once.

We observed that all tested evolutionary schemes demonstrate approximately the same partition quality, while the $(1 + 1)$ algorithm outperforms two other schemes in execution time. Some relevant results are illustrated in Fig. 1.

2.3 Mutations for Clustering

We used a total number of 32 mutations, all of them inspired by a survey in [12] and works [1,16,17] in particular. In this section, each of them will be described in detail. For lucidity, each mutation operator is given its unique number index **M*** to clarify how a total number of 32 mutations is obtained.

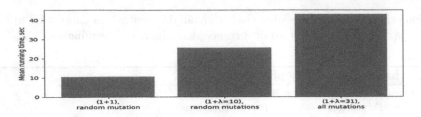

Fig. 1. Average running times of different evolutionary schemes

Cluster-Oriented Mutation. A cluster-oriented mutation chooses a cluster (or, in a certain case, a pair of clusters) and modifies its content by moving its objects to or from other clusters.

Note that every mutation in this section is presented in either 5 or 3 variations depending on the number of interestingness measures it is used with – the logic of choosing a cluster according to interestingness values is described in the next subsection. For instance, (**M1–M5**) notation means that the mutation is used with trivial separation, diameter-based separation, mean centroid distance separation, density sparseness and validity index, respectively. *Hyperplane split-gene mutation.* A cluster is split into a pair new clusters by a hyperplane. The hyperplane is chosen in either of two ways:

- with random norm and position (**M1–M5**);
- directly between centroid and the point, which is the farthest from centroid in the cluster, perpendicular to line between them (**M6**).

Spanning Tree Split-Gene Mutation. A minimum spanning tree is built based on objects in the cluster and distances between them. The mutation chooses an edge in the tree probabilistically based on edge weights; the chosen edge is removed and the tree is split into a pair of connected components, each of them corresponding to a new cluster (**M7–M10**). *Merge-gene mutation.* Two chosen clusters are merged into one (**M11–M13**). *Cluster elimination mutation.* A cluster is removed; all its objects are reclassified into nearest clusters based on their centroids (**M14–M18**). *Remove-and-reclassify mutation.* A small set of l random objects from the cluster is attached to other nearest clusters based on distances to their centroids. The number $l \sim \mathrm{B}\left(|c_i|, \frac{1}{|c_i|}\right)$ is chosen according to the binomial distribution, where $|c_i|$ is number of objects in the cluster (**M19–M23**). *Cluster expansion mutation.* For a chosen cluster, a small set of l random objects from other clusters are attached to it. The objects are chosen selectively based on their distances to centroid of the target cluster – an object that is closer to the target cluster is more likely to be reclassified into it by the mutation (**M24**).

Choosing Clusters Selectively. A cluster-oriented mutation from previous subsection chooses a set of clusters (or pairs of clusters) and operates on them sequentially. More specifically: a number $l \sim \mathrm{B}\left(k, \frac{1}{k}\right)$ of clusters undergoing mutation is chosen out of k total clusters or l clusters are chosen according to probabilities proportional to *interestingness* values of the clusters.

In this work, we use the interestingness measures based on [10,18]:

- Mean centroid distance separation, which is used as a separation for generalized Dunn indices gD_{x3}, where x stands for any cohesion of a generalized Dunn measure.
- Diameter separation, which is used as a separation for generalized Dunn indices gD_{x1}.
- Density sparseness separation $DSC(i)$ from the DBCV index formula.
- Validity index separation $V_C(i)$ from the DBCV index formula.
- Centroid distance cohesion, which is used as a cohesion for generalized Dunn indices gD_{4x}.
- Density cohesion, which is originally called density separation of pair of clusters $DSPC(i, j)$ in description of the DBCV index.
- Trivial interestingness measure, in which clusters or pairs of clusters are chosen pseudo-randomly.

It is worth noting that all cohesion measures and V_C separation are *descending*, while others are ascending. The less is value of the descending measure, the greater should be probability of a cluster (pair of clusters) with this measure to be chosen. For this condition to hold, descending measure values are flipped symmetrically based on their maximum and minimum values – after that transformation, the clusters are chosen using the method described above.

Prototype-Based Mutation. A partition will be referred to as *centered* if it is encoded with a set of synthesized objects called *centres*, with a direct correspondence between clusters and centres. In a centered partition, an object belongs to a cluster if and only if center of this cluster is closest to the object among others.

Similarly, a partition will be referred to as *prototype-based* if it is encoded with a set of real objects called *prototypes* and an object belongs to a cluster if and only if prototype of this cluster is closest to the object.

Some of the mutations listed below operate on centered or prototype-based partitions only and cannot be applied to integer-encoded partitions. See Sect. 2.1 for logic that is performed when a mutation cannot be performed with a partition.

K-Means Step. Centroids of old clusters are computed; a new centered partition is composed with these centroids as centres (**M25**). *Centroid hill climbing.* A single randomly chosen coordinate of a centered partition is either multiplied by $1+2\delta$ if it is non-zero or set to 2δ, otherwise. δ is chosen from uniform distribution $\mathrm{U}([-1, 1])$ (**M26**). *Mean Shift step.* Centroids of old clusters are computed; for each of them, an object closest to this centroid is found. A new prototype-based partition is composed with these real objects as prototypes (**M27**). *Prototype hill climbing.* A random set of l^+ prototypes is added to the prototype-based partition; another random set of l^- prototypes is removed from it. Values of

l^+ and l^- are chosen based on binomial distributions: $l^+ \sim \mathrm{B}\left(n - k, \frac{1}{n-k}\right)$, $l^- \sim \mathrm{B}\left(k, \frac{1}{k}\right)$ (**M28**).

Minimum Spanning Tree Mutation. An individual can be represented as a set of edges in the minimum spanning tree built for the whole dataset. More specifically, k clusters can be represented as connectivity components of this spanning tree after $k - 1$ edges are removed from it. We use two mutations treating partition as a set of edges in the spanning tree. The first one can be performed with any partition, while the last one operates on partitions which are already represented using this format. Both of them are represented as (**M29**), because choice between is determined by partition encoding. *Creating a tree-based partition.* For an arbitrary source partition, the mutation finds all weights in the spanning tree connecting objects from different clusters $L = \{(p, q) \in E | \exists i, j : i \neq j, p \in c_i, q \in c_j\}$, where E are edges of the spanning tree, c_i is the i-th cluster in source partition, and p, q are objects in the dataset connected with the edge. Number l of edges to be removed from the spanning tree to form clusters of the new partition is chosen according to binomial distribution $l \sim \mathrm{B}\left(|L| - 1, \frac{k-1}{|L|-1}\right)$. After that, l edges are selected with probability of an edge to be chosen proportional to weight of that edge: $P_{p,q} \sim d(p, q)$. With the selected edges removed from the tree, a new tree-based partition is obtained. *Modifying a tree-based partition.* For a tree-based partition, the mutation adds and removes some edges from the spanning tree similarly to prototype hill-climbing mutation described above. The numbers of edges to be added and removed are determined according to similar binomial distributions, while the choice of edges is performed in a guided fashion with probabilities proportional to their weights. The more is distance between objects connected by the edge, the greater is probability for this edge to appear in the resulting partition.

Other Mutations. We also utilize two non-cluster-oriented mutations operating on a partition with integer encoding—i.e. a cluster is explicitly assigned for each object in the dataset. The first one reclassifies an object with its k nearest neighbors to another cluster [19] (**M30**). The second one is a special case of this mutation with $k = 1$ (**M31**). Another mutation (**M32**) used just transfer some random points from one cluster to another. The main aspect is that the CVI is computed in incremental way on each iteration [20].

3   Prediction of Useful Mutations via Meta-learning

As it has been mentioned in Sect. 2.1, the algorithm requires a subset of mutations to be preselected for particular dataset and CVI. This subset is determined at the beginning of the algorithm according to prediction performed by a trained meta-classifier. Idea behind applying meta-learning here is simple. We run the evolutionary algorithm many times with all mutations and evaluated how many times each mutation improves an individual. The results indicate that datasets are associated with various subsets of useful mutations. This means that each mutation may be useful or not useful depending on a dataset the evolutionary algorithm is applied to. Predicting mutations in meta-learning framework can be considered as hyper-heuristic [21,22] and was applied in a similar manner for selecting mutations of an evolutionary algorithm applied to a routing problem [15].

3.1 Meta-features

The meta-features a meta-classifier include statistical and distance-based meta-features from [13] and landmarks which are composed as follows: four state-of-the-art clustering algorithms are run on the dataset. If an algorithm works with a predetermined number of clusters, it is initialized with a value $\sqrt[3]{n}$, where n is length of the dataset, then for each of the resulting four partitions, values of different CVIs are computed. In our setup, we used 10 CVIs mentioned in Sect. 4.

3.2 Determining Target Value

The proposed setup also requires *usefulness* of each mutation to be determined for each pair of a test dataset and CVI.

We perform this evaluation by running $(1 \mid \lambda)$ algorithm on datasets with the following modification: on every step, each mutation described in Sect. 2.3 is applied to current partition once. This implies that $\lambda = 32$, which is the exact number of mutations we used in our setup.

We consider a mutation useful if it has led to improvement of the CVI value at least once during execution of $(1 + \lambda)$. For partial noise reduction, we run the $(1 + \lambda)$ algorithm twice for each pair of CVI and dataset and consider a union of useful mutations from both runs as a target set for the classification task. We have collected statistics for our runs of $(1 + \lambda)$, it can be found at https://bit.ly/30Kb7kZ.

4 Experimental Setup

We used 93 datasets from the OpenML repository [23] and 4 synthetic datasets generated as unions of normal distributions, similarly to [11]. The algorithm was tested with 10 CVIs, namely: Silhouette [7], Davies-Bouldin [8], Generalized Dunn [18] (Dunn, gD_{13}, gD_{41}, gD_{43}, gD_{51}, gD_{53}), Calinski-Harabasz [9], DBCV [10].

For quality and time measurement of our approach in comparison with [24], we performed 10 runs of automated $(1+1)$ algorithm, 10 runs of non-automated version of our $(1 + 1)$ algorithm (which considers all mutations useful) and 10 runs of automated clustering algorithm from [24]. This procedure was performed for every pair of CVI and dataset, resulting in a total number of 29100 runs.

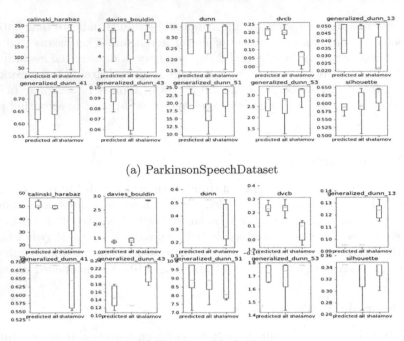

(a) ParkinsonSpeechDataset

(b) HappinessRank_2015

Fig. 2. Boxplots comparing CVI values reached by clustering algorithms for 3 example datasets. "predicted" stands for $(1 + 1)$ with predicted mutations, "all" means non-automated version of $(1+1)$ and "shalamov" is partition quality of reinforcement-based algorithm proposed by Shalamov et al. [24]

Resulting quality value is computed as mean value and standard deviation of a CVI for a particular dataset – Fig. 2 illustrates this comparison method for some datasets[2]. Time is computed as mean execution time over all 10 runs for each pair of CVI and dataset.

4.1 Effectiveness of Proposed Automation

We compared performance of the $(1 + 1)$ algorithm with two configurations: the first one uses mutations predicted by the meta-classifier, while the last one considers all mutations applicable to any CVI and dataset. In other words, we analyze effectiveness of our automation by comparing performance of automated $(1 + 1)$ algorithm with its non-automated version.

The comparison results are presented in Table 1.

---

[2] Full collection of comparison boxplots can be found at https://bit.ly/2Zr3WwG.

Table 1. Number of $(1 + 1)$ runs, grouped by CVI. "worse" stands for automated version reaching worse quality than non-automated; "not worse", conversely, means that partition quality was not degraded by applied automation. t_p, t_a are running times of automated and non-automated versions of the algorithm, respectively.

	Dunn		Davies-Bouldin		Calinski-Harabasz	
	Worse	Not worse	Worse	Not worse	Worse	Not worse
$t_p > t_a$	2	48	0	45	1	6
$t_p < t_a$	18	29	3	49	18	72
	Generalized Dunn$_{41}$		Generalized Dunn$_{13}$		DBCV	
	Worse	Not worse	Worse	Not worse	Worse	Not worse
$t_p > t_a$	2	32	0	27	0	48
$t_p < t_a$	16	47	17	53	0	49
	Generalized Dunn$_{51}$		Generalized Dunn$_{43}$		Silhouette	
	Worse	Not worse	Worse	Not worse	Worse	Not worse
$t_p > t_a$	1	17	2	33	4	15
$t_p < t_a$	10	69	7	55	20	58
	Generalized Dunn$_{53}$		**All CVIs**			
	Worse	Not worse	Worse		Not worse	
$t_p > t_a$	3	14	15		285	
$t_p < t_a$	8	72	117		553	

4.2 Comparison with Reinforcement-Based Automated Clustering Method

In a recent work [24], an optimal clustering partition is obtained by performing SMAC (sequential model-based algorithm configuration [25]) iterations to optimize hyperparameters of state-of-the-art algorithms. To choose an algorithm

Table 2. Number of $(1 + 1)$ runs compared to reinforcement-based approach [24], grouped by CVIs. "Worse" stands for $(1 + 1)$ reaching worse partition quality, "same" stands for quality values being equal with accuracy within standard deviation, "greater" means that the evolutionary algorithm outperformed the reinforcement-based approach.

	Worse	Same	Better		Worse	Same	Better
Calinski-Harabasz	11	50	36	Davies-Bouldin	5	54	38
Dunn	5	57	35	DBCV	4	22	71
Generalized Dunn$_{13}$	25	63	9	Silhouette	12	72	13
Generalized Dunn$_{43}$	28	64	5	Generalized Dunn$_{51}$	4	65	28
Generalized Dunn$_{53}$	22	64	11	Generalized Dunn$_{41}$	4	76	17
				All CVIs	121	587	263

that will be optimized in the next iteration, the multi-armed bandit problem is solved based on previous performance improvements of the trained models. In this Section, partition quality obtained by our automated approach is compared to this method.

For each CVI and dataset, we launch the reinforcement-based method with time budget equal to the average running time of our algorithm with these configuration and input. Table 2 shows the comparison results.

5   Conclusion

In this paper, we introduced 32 mutations for transforming partitions. Based on this, we propose an evolutionary clustering algorithm, which uses not all, but preselect the most useful for a given dataset using meta-learning. The proposed approach is based on predicting the set of useful mutations for each clustering task instead of choosing all the mutations available inside the evolutionary clustering algorithm. It appears that our approach achieves better results in most of the considered cases in terms of partition quality measured by different CVIs and working time than the evolutionary clustering approach without gathering a set of useful mutations. Also we provide the comparison with recently introduced reinforcement-based automation algorithm for clustering; we find that our approach gains similar results in terms of partition quality with the same working time budgets. The implementation of all the methods in the paper can be found at https://github.com/dimatomp/MultiClustering.

The suggested method is planed to be tested on more datasets, also it can be improved by applying different automation techniques, for example reinforcement learning approach.

Acknowledgments. The authors would like to thank Maxim Buzdalov for useful comments. The research was financially supported by The Russian Science Foundation, Agreement 17-71-30029.

References

1. Ma, P.C., Chan, K.C., Yao, X., Chiu, D.K.: An evolutionary clustering algorithm for gene expression microarray data analysis. Trans. Evol. Comp. **10**, 296–314 (2006)
2. Punj, G., Stewart, D.W.: Cluster analysis in marketing research: review and suggestions for application. J. Mark. Res. **20**(2), 134–148 (1983)
3. Farseev, A., Samborskii, I., Filchenkov, A., Chua, T.-S.: Cross-domain recommendation via clustering on multi-layer graphs. In: Proceedings of the 40th International ACM SIGIR Conference on Research and Development in Information Retrieval, pp. 195–204. ACM (2017)
4. Kleinberg, J.: An impossibility theorem for clustering. In: Proceedings of the 15th International Conference on Neural Information Processing Systems, NIPS 2002, pp. 463–470. MIT Press, Cambridge (2002)

5. Pedregosa, F., et al.: Scikit-learn: machine learning in Python. J. Mach. Learn. Res. **12**, 2825–2830 (2011)
6. Hall, M., Frank, E., Holmes, G., Pfahringer, B., Reutemann, P., Witten, I.H.: The weka data mining software: an update. ACM SIGKDD Explor. Newsl. **11**(1), 10–18 (2009)
7. Rousseeuw, P.J.: Silhouettes: a graphical aid to the interpretation and validation of cluster analysis. J. Comput. Appl. Math. **20**, 53–65 (1987)
8. Davies, D.L., Bouldin, D.W.: A cluster separation measure. IEEE Trans. Pattern Anal. Mach. Intell. **2**, 224–227 (1979)
9. Caliński, T., Harabasz, J.: A dendrite method for cluster analysis. Commun. Stat.-Theory Methods **3**(1), 1–27 (1974)
10. Moulavi, D., Jaskowiak, P.A., Campello, R.J.G.B., Zimek, A., Sander, J.: Density-based clustering validation, April 2014
11. Arbelaitz, O., Gurrutxaga, I., Muguerza, J., PéRez, J.M., Perona, I.N.: An extensive comparative study of cluster validity indices. Pattern Recogn. **46**, 243–256 (2013)
12. Hruschka, E.R., Campello, R.J.G.B., Freitas, A.A., De Carvalho, A.C.P.L.F.: A survey of evolutionary algorithms for clustering. Trans. Syst. Man Cyber. Part C **39**, 133–155 (2009)
13. Ferrari, D.G., de Castro, L.N.: Clustering algorithm selection by meta-learning systems. Inf. Sci. **301**, 181–194 (2015)
14. Muravyov, S., Filchenkov, S.: Meta-learning system for automated clustering. In: AutoML@ PKDD/ECML, pp. 99–101 (2017)
15. Shalamov, V., Filchenkov, A., Shalyto, A.: Heuristic and metaheuristic solutions of pickup and delivery problem for self-driving taxi routing. Evol. Syst. **10**, 11 (2017)
16. Cole, R.: Clustering with genetic algorithms. Ph D. thesis (1998)
17. Hruschka, E.R., Ebecken, N.F.F.: A genetic algorithm for cluster analysis. Intell. Data Anal. **7**, 15–25 (2003)
18. Bezdek, J.C., Pal, N.R.: Some new indexes of cluster validity. Trans. Sys. Man Cyber. Part B **28**, 301–315 (1998)
19. Handl, J., Knowles, J.: An evolutionary approach to multiobjective clustering. Trans. Evol. Comp **11**, 56–76 (2007)
20. Muravyov, S., Antipov, D., Buzdalova, A., Filchenkov, A.: Efficient computation of fitness function for evolutionary clustering. MENDEL **25**, 87–94 (2019)
21. Pillay, N., Qu, R.: Hyper-Heuristics: Theory and Applications. Springer, Switzerland (2018). https://doi.org/10.1007/978-3-319-96514-7
22. Woodward, J.R., Swan, J.: The automatic generation of mutation operators for genetic algorithms. In: Proceedings of the 14th Annual Conference Companion on Genetic and Evolutionary Computation, pp. 67–74. ACM (2012)
23. Vanschoren, J., van Rijn, J.N., Bischl, B., Torgo, L.: Openml: networked science in machine learning. SIGKDD Explor. **15**(2), 49–60 (2013)
24. Shalamov, V., Efimova, V., Muravyov, S., Filchenkov, A.: Reinforcement-based method for simultaneous clustering algorithm selection and its hyperparameters optimization. Procedia Comput. Sci. **136**, 144–153 (2018)
25. Hutter, F., Hoos, H., Leyton-Brown, H.: An evaluation of sequential model-based optimization for expensive blackbox functions. In: Proceedings of the 15th Annual Conference Companion on Genetic and Evolutionary Computation, pp. 1209–1216. ACM (2013)

Fast Tree-Based Classification
via Homogeneous Clustering

George Pardis[1], Konstantinos I. Diamantaras[1], Stefanos Ougiaroglou[1,2(✉)],
and Georgios Evangelidis[2]

[1] Department of Information and Electronic Engineering, International Hellenic
University, 57400 Sindos, Thessaloniki, Greece
george.pardis@gmail.com, kdiamant@it.teithe.gr
[2] Department of Applied Informatics, School of Information Sciences,
University of Macedonia, 54636 Thessaloniki, Greece
{stoug,gevan}@uom.edu.gr

Abstract. Data reduction, achieved by collecting a small subset of representative prototypes from the original patterns, aims at alleviating the
computational burden of training a classifier without sacrificing performance. We propose an extension of the Reduction by finding Homogeneous Clusters algorithm, which utilizes the k-means method to propose
a set of homogeneous cluster centers as representative prototypes. We
propose two new classifiers, which recursively produce homogeneous clusters and achieve higher performance than current homogeneous clustering methods with significant speed up. The key idea is the development
of a tree data structure that holds the constructed clusters. Internal tree
nodes consist of clustering models, while leaves correspond to homogeneous clusters where the corresponding class label is stored. Classification
is performed by simply traversing the tree. The two algorithms differ on
the clustering method used to build tree nodes: the first uses k-means
while the second applies EM clustering. The proposed algorithms are
evaluated on a variety datasets and compared with well-known methods.
The results demonstrate very good classification performance combined
with large computational savings.

Keywords: Classification · k-means · EM · Prototype generation

1 Introduction

Many Data Reduction Techniques (DRTs) [5,8] have been proposed in an
attempt to reduce the classification cost of the k-Nearest Neighbour classifier
(k-NN) [2]. The goal of DRTs is to build a small representative set of the training
set. This set is called condensing set and has the advantage of low computational
cost without sacrificing accuracy. A DRT can either select representative prototypes form the original training set or generate new prototypes by summarizing
similar items. Most DRTs are based on a simple idea: The instances that lie

© Springer Nature Switzerland AG 2019
H. Yin et al. (Eds.): IDEAL 2019, LNCS 11871, pp. 514–524, 2019.
https://doi.org/10.1007/978-3-030-33607-3_55

far from the class decision boundaries are redundant and increase the computational cost. Hence, DRTs try to select or generate many prototypes for the close borders areas. It is worth mentioning that editing algorithms are a special type of DRTs that aim to remove noise.

Reduction by finding Homogeneous Clusters (RHC) [7] is a simple and fast DRT that is based on a recursive procedure that builds homogeneous clusters, i.e. clusters that contain items of a specific class. For each non-homogeneous cluster, RHC applies k-means clustering using as initial seeds the means of the classes that are present in the cluster. This procedure is repeated on all non-homogeneous clusters and terminates when all become homogeneous. The means of the homogeneous clusters constitute the condensing set. Certainly, k-means can be replaced by EM clustering.

RHC discards the means of the non-homogeneous clusters. However, they can be used in order to speed-up the classification processes. This simple idea constitutes the motive behind the present work. More specifically, the non-homogeneous clusters can be used to build a tree data structure, where the non-leaf nodes store the means of non-homogeneous clusters while the leaf-nodes store the means of homogeneous clusters. Thus, instead of applying k-NN over the condensing set (i.e., the set of leaf nodes), the classification could be performed by traversing the tree. The contribution of the paper is the development of two classifiers that utilize a tree-based model in order to achieve significantly faster classification than that of k-NN classifier that runs over the condensing set generated by RHC. When an unclassified instance is to be classified, the prediction process of the proposed algorithms traverses the tree and the instance gets assigned to the unique class stored in the leaf node. The two algorithms differ to each other on the clustering algorithm that they use, i.e., k-means and EM, respectively. We find that the proposed methods have similar or slightly better performance compared to classical data reduction techniques and k-NN but they offer faster classification by many orders of magnitude.

The rest of the paper is organized as follows: Sect. 2 reviews RHC and ERHC. Section 3 presents in detail the proposed tree-based classifiers. Section 4 presents the results of the experimental study. Section 5 concludes the paper.

2   The RHC and ERHC Algorithms

The RHC algorithm recursively applies k-means and keeps on constructing clusters until all of them are homogeneous. Initially, RHC considers the whole training set as a non-homogeneous cluster. The algorithm begins by computing the mean for each class by averaging the attribute values of the corresponding instances. Therefore, for a dataset with n classes, RHC computes n class means. Then, RHC executes k-means using the n aforementioned class-means as initial seeds and builds n clusters. For each homogeneous cluster, its mean is placed in the condensing set as prototype. For each non-homogeneous cluster, the above procedure is applied recursively. RHC stops when all clusters are homogeneous. In the end, the condensing set contains all the mean items of the homogeneous clusters. Using the class means as initial means for k-means clustering, the number of clusters is determined automatically.

RHC generates more prototypes for the close-class-border data areas and less for the central class data areas. Therefore, datasets with many classes and noisy items create more prototypes and thus lower reduction rate is achieved.

Editing and Reduction through homogeneous clusters (ERHC) [6] is a simple variation of RHC. It is based on a following idea: a cluster that contains only one item is probably noise or outlier. ERHC removes that type of clusters. Therefore, ERHC is more noise tolerant than RHC and achieves higher accuracy and reduction rates as well as lower computational cost, especially on datasets with noise.

3   The Proposed Algorithms

We propose two classifiers. Both recursively apply clustering methods to create a tree data structure whose leaves represent homogeneous clusters. The first algorithm applies k-means while the second one employs EM clustering [3]. The nodes store cluster models trained on items belonging to more than one classes. The idea behind this is that, during classification stage, instead of comparing the distance between each unclassified item with all prototypes, the algorithm will only compare the distances to the closest ones. Also, the homogeneous clustering technique comes up with a weakness. Since more prototypes are generated to close-class-borders, datasets with noisy items create more prototypes and thus lower reduction rate is achieved. To overcome this weakness, the proposed algorithms remove the noise with a smart technique that leads to a higher rate of reduction without negatively affecting the processing cost.

The two algorithms operate in the same way and the only difference is the applied clustering technique. Initially, they consider the whole training set to be a non-homogeneous cluster. If the cluster contains items from c classes, they compute c clusters and then apply k-means or EM on each cluster recursively. The model is saved in the tree as a root node and the children of this node are the c different clusters. For each child node, homogeneity is checked. If the cluster is homogeneous or is "almost" homogeneous (i.e., a large percentage of its items share the same class label) this node becomes an appropriately labeled leaf. In case of a large percentage of homogeneity, the small remaining percentage of items probably lies in a region of a different class (noise) or close to a decision-boundaries region, and thus, it is removed and the node is considered to be homogeneous. If the cluster node is not homogeneous this procedure is applied recursively. Each unclassified item is classified only by descending the tree. The pseudo-code in Algorithm 1 describes the process.

3.1   Training HC_EM

The first algorithm (HC_EM) uses EM. In the beginning, the method checks if the dataset consists of items of only one class and if it does, this set is homogeneous so

---

Algorithm 1. Proposed algorithms

Input: TS
Output: $HomogeneousTree$
 1: $HomogeneousTree \leftarrow \varnothing$
 2: $queue \leftarrow$ enqueue(clusters obtained by applying EM or $k - means$ on TS)
 3: **repeat**
 4:     $C \leftarrow$ dequeue($queue$)
 5:     $m \leftarrow$ trained model of C
 6:     **if** C is homogeneous **then**
 7:         Insert leaf m in $HomogeneousTree$
 8:     **else**
 9:         Insert internal node m in $HomogeneousTree$
10:         $queue \leftarrow$ enqueue(clusters obtained by applying EM or $k - means$ on C)
11:     **end if**
12: **until** IsEmpty($queue$)
13: **return** $HomogeneousTree$

---

a leaf is created that carries the label of this class. In case it is not homogeneous, the method finds the most popular label and compares its items with all the rest. If the percentage of the most popular label is more than ninety-five, the set is considered as homogeneous and a leaf is created carrying this label. In case there are less than five items in total, a leaf that carries the label of the most popular class is created. Therefore, the noisy items are being removed and they do not have any negative impact on accuracy. The numbers five and ninety-five were chosen after a trial and error procedure to find the combination that gives the best results on average. The initial dataset is not homogeneous so the algorithm will try to find the best parameters to create an effective model. The first parameter that the training method uses, is the weight of each class. The other parameters are the number of classes, the means of each class and the initial precisions [4]. Now that the model is ready, we train it using the dataset and it produces some predictions about the class each item belongs to. In this point, we create different clusters consisting of the items of each cluster label that the model predicted. Then, we produce a node that carries the trained model and has as many children as the different clusters created before. For each different child cluster we repeat the whole training procedure from the beginning given as input the set of items with their real class label and not the predicted one.

3.2   Training HC_KMEANS

HC_KMEANS uses the k-means as a training method. The reason is that EM is slow, although it may lead to better results. By using the k-means, we achieve faster execution. The first difference between HC_KMEANS and HC_EM is that the number of the minimum items capable to constitute a cluster is three and the percentage of superiority of a class that can make it a homogeneous cluster is ninety-eight. Again, these numbers were carefully chosen after a trial and error procedure in order to effectively remove noise. The other difference is that the model created takes as a parameter only the number of the different classes and the mean (representative) of each one. The rest of the procedure remains the same.

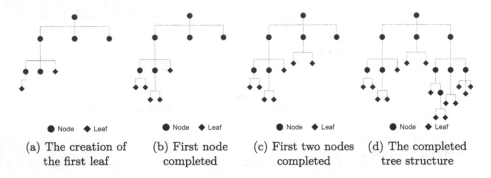

(a) The creation of   (b) First node   (c) First two nodes   (d) The completed
the first leaf   completed   completed   tree structure

Fig. 1. General training procedure

Figure 1 visualizes the training procedure. Once the training set is loaded in the root of the tree, EM or k-means algorithm trains a model to predict as many clusters as the different classes in the set. This model is stored in the node. For every cluster we test if there are any misclassified items and the procedure repeats until all clusters are homogeneous. In Fig. 2, one can visualize a complete example of training when using k-means. The initial dataset contains three class labels. Eventually, a tree model is constructed using recursive application of k-means. The nodes of the tree contain the cluster models, in our case cluster centroids and a class label. The leaves of the tree correspond to the homogeneous clusters whose items form the initial dataset.

Figure 3 illustrates the prediction procedure. The predicted label matched the cluster labels of internal nodes 1 and 5 and finally of the leaf node 15. During the traversal the item gets assigned to different clusters until it adopts the final predicted class label of a leaf. The trained model that is stored in each node predicts a cluster label for each item. The way each model makes that prediction is related to the algorithm we use. In case of k-means, the only criterion is the Euclidean distance which determines how similar an item is to a group of items. The model predicts the closest cluster mean each unclassified item belongs to. On the other hand, for each cluster, EM uses the weight, the mean and the precision to create a likelihood logarithm value of the most probable mixture component for the given sample.

Each cluster of a parent node, will be dominated by data items belonging to a certain class label. This means that the model created a cluster that consists of items with similar properties. It is normal for some items to be mislabelled at first, because their properties are more alike those of items belonging to a different label than theirs. These items may constitute noise and in most cases

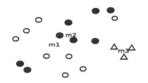

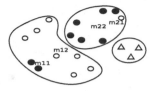

(a) Initial dataset with m_i indicating the class representatives

(b) Resulting clusters after applying k-means using m_i's as initial seeds. c3 is homogeneous.

(c) Computing class representatives m_{1i} and m_{2i} for clusters c1 and c2, respectively.

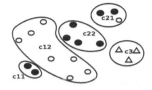

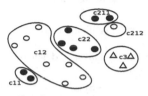

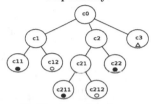

(d) Resulting clusters after applying k-means on c1 and c2. c11, c12 and c22 are homogeneous.

(e) Resulting clusters after applying k-means on c21. c211 and c212 are homogeneous.

(f) The completed tree structure that replaces the initial dataset. Notice the labeled leaves.

Fig. 2. HC_KMEANS Training procedure. In the final model, leaves correspond to the centroids of the homogeneous clusters and internal nodes to the centroids of the non-homogenous clusters

they lay in the border of two or more different classes. In some cases, these items may be a very small cluster within the boundaries of a bigger one. In our tree structure, during the prediction procedure, each model will predict a new cluster label that may or may not be the same with the previously predicted one. Each item needs at most as many predictions as the depth of the tree to be classified.

4   Performance Evaluation

4.1   Experimental Setup

The proposed algorithms were evaluated using fourteen datasets distributed by the KEEL repository [1] and summarized in Table 1. For comparison purposes, we used 1-NN classification on the original datasets, 1-NN classification after applying reduction with RHC and ERHC, and the Decision Tree C4.5 algorithm on the original datasets. We selected these methods because: like RHC and ERHC, our algorithms are based on the concept of homogeneity. Also, the Decision Tree (DT) C4.5 constructs a tree structure similar to ours. It is also, one of the most well-known algorithms [9].

The thirteen datasets (except the KDD dataset) were used without data normalization. Also, we wanted to evaluate our algorithms on datasets with noise. We built two additional datasets by adding 10% random noise in the LS and

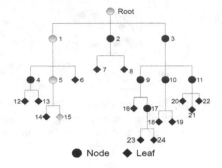

Fig. 3. Classification procedure

Table 1. Dataset description

Dataset	Size	Attributes	Classes
Letter Recognition (LR)	20000	16	26
Magic G. Telescope (MGT)	19020	10	2
Pen-Digits (PD)	10992	16	10
Landsat Satellite (LS)	6435	36	6
Shuttle (SH)	58000	9	7
Texture (TXR)	5500	40	11
Phoneme (PH)	5404	5	2
KddCup (KDD)	494020/141481	36	23
Balance (BL)	625	4	3
Banana (BN)	5300	2	2
Ecoli (ECL)	336	7	8
Yeast (YS)	1484	8	10
Twonorm (TN)	7400	20	2
MONK 2 (MN2)	432	6	2

PD datasets. We refer to these datasets as LS10 and PD10 respectively. The noise was added by setting the class label of the 10% of the training items to a randomly chosen different class label. KDD has 41 attributes. However, for simplifying the experimentation process, we removed the three nominal and the two fixed value attributes. In addition, we removed all the duplicates. Furthermore, the attribute value ranges of the KDD dataset vary extremely. Therefore, we normalized them in the range [0, 1]. We then randomized the transformed KDD dataset and divided it into the appropriate folds.

For each dataset and algorithm, we report three average measurements obtained via five-fold cross-validation: (i) Accuracy, (ii) Training time in seconds and (iii) Testing cost in total number of distance computations. Algorithm implementations were written in Python and all the experiments were performed on an 8 GB laptop with an AMD A8-6410 CPU and Windows 10 operating system.

In order not to favor our algorithms against DT C4.5, after a trial and error procedure, we determined the maximum depth of the DT C4.5 that gives the best average results for all datasets. Maximum depth can be used as a parameter to set a limit on how deep a tree can be. The depth that gave the best results is nineteen (19), although the results for different values do not have significant difference in accuracy. Our measurements show that both k-means and EM implementations of the proposed algorithms are superior to the best tuned decision trees.

Table 2. Experimental results: accuracy (%)

Dataset	1-NN	DT C4.5	RHC	ERHC	HC_KMEANS	HC_EM
LR	**95.83**	87.09	93.61	92.89	90.33	91.27
MGT	78.14	83.06	72.45	76.94	76.70	**85.38**
PD	**99.35**	96.21	98.30	98.63	97.78	97.71
LS	**90.60**	85.04	89.08	89.16	88.02	81.26
SH	99.82	**99.98**	99.05	98.69	98.48	99.73
TXR	99.02	92.68	96.96	97.22	95.98	**99.84**
PH	**90.10**	87.02	85.70	86.72	85.29	85.98
KDD	99.71	**99.77**	99.39	99.39	99.44	99.39
BL	78.40	76.94	69.52	75.97	78.06	**91.61**
BN	86.91	87.08	84.32	84.32	**88.93**	86.99
ECL	**79.78**	74.62	69.19	79.19	74.64	73.43
YS	52.02	52.74	48.48	**53.00**	50.77	52.00
TN	94.88	85.42	89.41	91.55	**97.82**	97.66
MN2	90.51	**100**	94.62	95.08	98.35	**100**
PD10	89.50	87.29	77.15	97.25	85.57	**97.45**
LS10	82.02	80.25	79.81	**87.08**	81.98	80.92
Avg	87.91	85.95	84.19	87.69	86.76	**88.79**

4.2 Results and Discussion

Table 2 reports the accuracy results. Table 3 reports the time needed to train the classifiers. K-NN is not included there, since it does not build any model. Table 4 reports the total number of distances computed during testing. Notice that for HC_EM, we report the number of likelihood functions computed, which are only slightly more expensive than plain euclidean distance computations. Also, we do not report the testing cost for C4.5 since it outperforms all the other algorithms by traversing its trees using plain comparisons and not distance computations. The tests confirm that HC_EM algorithm achieves the highest accuracy, with 1-NN and ERHC being close.

We observe that HC_EM has better results in accuracy than HC_KMEANS. However, the total training process time is about three times as much for HC_EM, which we attribute to the higher computational cost of EM. The lower testing times are due to the higher rates of reduction HC_EM achieves. That means that the depth of the produced tree is smaller and so the computations made are less. Regarding the performance of the algorithms on the noisy datasets, ERHC achieves the best overall accuracy, with HC_EM achieving the highest accuracy in PD10. We attribute this to the pruning mechanism used by HC_EM.

It must be noted that our tree-based approach resembles the well-known decision tree (DT) models. However, there is a major difference on the discrimination function applied in each node of the tree. In a typical classification tree, applied on numerical data, each node corresponds to an area in the input space which is divided by a linear separating surface. This surface must be perpendicular to one of the data dimensions. Clearly, imposing such a restriction limits the separating capabilities of DTs. Nonlinear discriminating surfaces as well as linear surfaces forming some random angle with respect to the axes are not allowed. In our case each tree node is divided using the much more versatile Gaussian split, in the case of the HC_EM, or the Voronoi tessellation in the case of HC_KMEANS.

Table 3. Experimental results: training time (seconds)

Dataset	DT C4.5	RHC	ERHC	HC_KMEANS	HC_EM
LR	<1	32.85	30.94	20.46	261.10
MGT	1.11	71.25	82.94	42.35	55.88
PD	<1	5.22	5.20	3.06	30.79
LS	<1	7.80	8.64	5.39	21.41
SH	<1	8.94	9.23	4.60	44.86
TXR	<1	3.77	3.97	3.00	5.22
PH	<1	12.80	13.34	7.76	20.97
KDD	4.38	26.41	25.71	7.79	20.45
BL	<1	1.19	1.16	<1	1.49
BN	<1	11.11	11.40	6.05	4.08
ECL	<1	1.50	<1	<1	1.83
YS	<1	8.65	5.72	2.73	13.81
TN	<1	7.86	6.51	<1	<1
MN2	<1	<1	<1	<1	<1
PD10	<1	30.32	23.67	26.57	39.83
LS10	<1	16.22	12.00	14.10	42.05

As Table 2 shows, the results of the DT models are inferior to our HC_EM method in most of the cases we have studied. In some cases the difference is significant, such as, for example, the case of the TN (HC_EM accuracy = 97.66%, vs. DT accuracy = 85.42%, a difference of 12.24%) and BL datasets (HC_EM accuracy = 91.61%, vs. DT accuracy = 76.94%, a difference of 14.67%). The highest accuracy difference in favor of the DT models is observed for the LS dataset where HC_EM accuracy = 81.26%, and DT accuracy = 85.04%, a difference of 3.78%). We can claim that our approach is a decision tree method with nonlinear discriminating surfaces per node.

Table 4. Experimental results: Distance computations in thousands. For HC_EM, likelihood functions computed in thousands.

Dataset	1-NN	RHC	ERHC	HC_KMEANS	HC_EM
LR	64000	7573	5057	405	254
MGT	57881	15219	8984	440	252
PD	19335	673	492	96	44
LS	6625	669	466	72	33
SH	538228	2465	1679	501	135
TXR	4840	257	196	40	30
PH	4673	899	557	78	57
KDD	3202767	6581	5234	78	56
BL	62	13	8	5	0.7
BN	4494	217	196	70	40
ECL	18	5	2	2	2
YS	352	176	72	20	23
TN	8761	286	212	33	11
MN2	30	1	1	1.6	1.4
PD10	19335	5205	2567	228	139
LS10	6625	1605	792	122	70
Avg	246127	2615	1657	137	72

5   Conclusions

The paper proposes two data reduction based classifiers that achieve similar or slightly superior classification performance compared to most state-of-the-art data reduction schemes but with significant classification speed. Compared to the k-NN classifier the classification speed is larger by many orders of magnitude. Our methods improve the quality of the training data through fast preprocessing that removes noisy items. Both are based on RHC. The main advantage is that the two algorithms use a tree structure of trained models that lead to improved accuracy and significantly fewer computations than RHC.

The algorithms were experimentally evaluated on known datasets and compared to RHC and ERHC, the k-NN classifier and the C4.5 decision tree. We measured the accuracy, the training time and the testing cost. Through experimental studies, we have demonstrated that they have met the goals for which they were developed and led to improved performance.

References

1. Alcalá-Fdez, J., Fernández, A., Luengo, J., Derrac, J., García, S.: Keel data-mining software tool: data set repository, integration of algorithms and experimental analysis framework. Multiple-Valued Logic Soft Comput. **17**(2–3), 255–287 (2011)
2. Cover, T., Hart, P.: Nearest neighbor pattern classification. IEEE Trans. Inf. Theor. **13**(1), 21–27 (2006). https://doi.org/10.1109/TIT.1967.1053964
3. Dempster, A.P., Laird, N.M., Rubin, D.B.: Maximum likelihood from incomplete data via the EM algorithm. J. Roy. Stat. Soc. Ser. B (Methodol.) **39**(1), 1–22 (1977)
4. Scikit-learn developers: scikit-learn user guide, March 2019. https://Scikit-learn.org
5. Garcia, S., Derrac, J., Cano, J., Herrera, F.: Prototype selection for nearest neighbor classification: taxonomy and empirical study. IEEE Trans. Pattern Anal. Mach. Intell. **34**(3), 417–435 (2012). https://doi.org/10.1109/TPAMI.2011.142
6. Ougiaroglou, S., Evangelidis, G.: Efficient editing and data abstraction by finding homogeneous clusters. Ann. Math. Artif. Intell. **76**(3), 327–349 (2015). https://doi.org/10.1007/s10472-015-9472-8
7. Ougiaroglou, S., Evangelidis, G.: RHC: non-parametric cluster-based data reduction for efficient k-NN classification. Pattern Anal. Appl. **19**(1), 93–109 (2016)
8. Triguero, I., Derrac, J., Garcia, S., Herrera, F.: A taxonomy and experimental study on prototype generation for nearest neighbor classification. Trans. Sys. Man Cybern. Part C **42**(1), 86–100 (2012). https://doi.org/10.1109/TSMCC.2010.2103939
9. Wu, X., et al.: Top 10 algorithms in data mining. Knowl. Inf. Syst. **14**(1), 1–37 (2008). https://doi.org/10.1007/s10115-007-0114-2

Ordinal Equivalence Classes for Parallel Coordinates

Alexey Myachin[1,2] and Boris Mirkin[1(✉)]

[1] National Research University Higher School of Economics,
20 Myasnitskaya Street, Moscow 101000, Russia
bmirkin@hse.ru
[2] Institute of Control Science of Russian Academy of Science,
65 Profsoyuznay Street, Moscow 117997, Russia

Abstract. We give a mathematical treatment to the concept of ordinal equivalence defined relative to all m! possible permutations of parallel axes. We prove that the ordinal equivalence is determined by the pair-wise co-monotonicity equivalence relations, thus leading to simple algorithmic procedures for finding the corresponding partition. Each ordinal equivalence class can be visualized as a profile of co-monotone polylines, in this way preventing any clutter at the image. We illustrate our approach with two datasets taken from the literature.

Keywords: Parallel coordinates · Co-monotonicity · Ordinal equivalence · Clutter

1 Introduction

Parallel coordinates is a well-known framework for data visualization (see [1, 2, 4, 7–9, 13, 14, 17, 19, 20] for reviews). Given m features are represented by m vertical axes which are parallel to each other. Each m-dimensional data point is represented by a polyline combining m-1 lines between m positions of the point's coordinates at neighboring axes (see Fig. 1 further on). Of course, an ordering of the features is prespecified, usually implicitly, so that the ordering of the axes from the left to the right reflects the pre-specified ordering of features. Unfortunately, at larger data sets the polylines tend to form clutters – incomprehensible patterns of lines crossing each other in various directions and thus preventing the user from meaningful observations over the data.

There have been various attempts at overcoming the issue such as edge bundling [22], or coordinate reordering [11, 17], or (nested) data clustering [3, 18, 21], etc. A workable taxonomy of approaches for reducing clutter at parallel coordinates is given in [5] a dozen years ago. Although some newer approaches have been published since (see, for example, [4, 10, 13, 15, 20], most of them may be covered by categories of the taxonomy in [5]).

One specific category of our attention is filtering. In [5], filtering is defined quite broadly: "filtering is the selection of a subset of data that satisfies given criteria" [5], p. 1217. However, out of an extensive collection of papers, under the category "filtering" the authors of [5] cited only those dealing with a rather narrow type of operation

© Springer Nature Switzerland AG 2019
H. Yin et al. (Eds.): IDEAL 2019, LNCS 11871, pp. 525–533, 2019.
https://doi.org/10.1007/978-3-030-33607-3_56

- by filtering out elements from a certain local area of the visual plot (see references in Table 1 in [5]). In fact, some other techniques should be classified as filtering too, even those referred to in different segments of the taxonomy in [5]. This relates to, first of all, the idea of separating outliers from the rest [16], see also [6]. More recent approaches include a proposition for using only those elements that fall into an association over a subset of features [20] or stacking out inconvenient parts [4].

The approach being developed in this paper may be considered a far-reaching extension of the approach of identifying outliers in [16]. This defines outliers as those entities whose polylines deviate from a common pattern of lines between a pair of neighboring feature axes. Also, [16] heavily relies on subdividing feature ranges in a number of intervals, so that only distributions of entities over the intervals are considered, which provides for scalability of the approach. This latter line is barely touched here. The idea of the pair-wise co-monotonicity of polylines is elaborated on here instead.

Specifically, given a pair of features, k and l, we define the (k, l)-co-monotonicity as an equivalence relation on the set of objects. The two classes of the (k, l)-comonotonicity are: (1) those objects at which lines grow from k-th axis to l-th axis and (2) those at which the lines decrease. We are interested at ordinally equivalent polyline patterns. Given an ordering $p = (f_1, f_2, ..., f_m)$ of data features corresponding to the parallel axis arrangement from the left to the right, we refer to objects as p–equivalent if the parallel coordinate polylines for them are co-monotone for any neighboring axes, f_u and f_{u+1} ($u = 1, 2, ..., m - 1$). We refer to the two objects as ordinally equivalent if they are p-equivalent at any order p of parallel coordinates. Obviously, the ordinal equivalence is an equivalence relation, thus, corresponding to a partition of the observations in classes of ordinally equivalent objects. Of course, no clutter may emerge at the parallel coordinate graphs of classes of the ordinal equivalence. Therefore, we filter out rare patterns of ordinally equivalent polylines and present visual plots for those most frequent ordinally equivalent patterns.

On the first glance, finding ordinal equivalence classes is a non-polynomially complex combinatorial problem involving an enumeration of all or almost all the m! orderings. We prove however, that the ordinal equivalent objects are those (k, l)-co-monotone for any pair of features k, l. This shows that the problem is quadratic over m, that is, relatively simple. We give two different algorithms for finding ordinal equivalence classes, and provide real-world data examples of using them for visual parallel coordinate plotting them.

The remainder of the paper is structured accordingly. Section 2 presents our theoretical results and algorithms. Frequent ordinal equivalence classes in two datasets, Pollen and Athletes, are visualized in Sect. 3. Section 4 concludes the paper.

2 Ordinal Equivalence: Definitions, Properties, Algorithms

Let I be a set of N objects characterized by m features forming columns $f_1, f_2, ..., f_m$ of $N \times m$ data matrix $F = (f_1, f_2, ..., f_m)$. This dataset can be presented by $m! = 1*2*...*m$ ways in parallel coordinates depending on the ordering of the features and respective ordering of the axes. Take, for example, dataset F in Table 1.

Table 1. An illustrative 4 × 3 data set.

Objects	Feature 2	Feature 3	Feature 3
1	1.5	1.7	1.6
2	1.0	1.7	0.7
3	1.6	1.2	0.9
4	0.4	0.8	0.7

Figure 1 presents two views of the data in Table 1: that on the left corresponds to the feature order ABC, whereas that on the right, to the order ACB.

Consider a serial ordering, or permutation, of features $p = (f_1, f_2, ..., f_m)$. Let us refer to objects i and j as p-equivalent if, for any $u = 1, ..., m - 1$, if $f_u(i) > f_{u+1}(i)$, then $f_u(j) > f_{u+1}(j)$; if $f_u(i) = f_{u+1}(i)$, then $f_u(j) = f_{u+1}(j)$; and if $f_u(i) < f_{u+1}(i)$, then $f_u(j) < f_{u+1}(j)$. It is not difficult to see that the relation π_p of p-equivalence is reflexive so that $(i, i) \in \pi_p$ for any $i \in I$, and symmetric so that if $(i, j) \in \pi_p$ then $(j, i) \in \pi_p$ too, and transitive so that if $(i, j) \in \pi_p$ and $(j, k) \in \pi_p$ then $(i, k) \in \pi_p$. That means that π_p is an equivalence relation and it divides the set I in non-overlapping non-empty classes of mutually p-equivalent objects.

Unfortunately, the equivalence relation π_p and, thus, the corresponding partition $S = \{S_1, S_2, ..., S_K\}$, where S_k are equivalence classes of π_p ($k = 1, 2, ..., K$), may differ at different permutations p, as the example in Fig. 1 clearly shows.

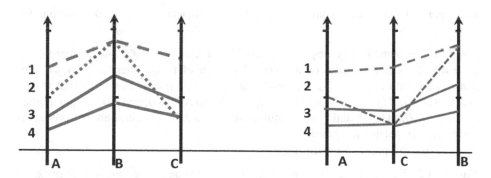

Fig. 1. Three features and four objects from dataset in Table 1 presented in parallel coordinates in the order ABC, on the left, and in the order ACB, on the right. All four polylines are mutually co-monotone at the order ABC, whereas object 2 falls out at the order ACB on the right, to be classified as an outlier.

Therefore, it is natural to define a permutation-independent equivalence relation as the intersection of p-equivalence relations for all possible permutations p of the set of objects I. While doing this we remember that the intersection of equivalence relations is an equivalence relation itself. We will refer to an equivalence relation ε as the ordinal equivalence relation if

$$\varepsilon = \cap_p \pi_p \tag{1}$$

where the intersection is taken over all the permutations of I.

On the computational side, the definition (1) looks rather disappointingly since the number of all permutations, $m!$, is extremely large and manifests an extra-exponential growth at growing numbers of features m. This, however, is a superficial impression. In fact, the number of really differing permutation equivalence relations is much smaller. We are going to show that, to find the overall ordinal equivalence relation ε, there is no need to compute p-equivalence relations at all: it suffices to use only co-monotonicity relations corresponding to just pairs of features. The number of pairs is of the order of m^2.

Consider a pair of features, f_k and f_l (k, $l = 1, 2,..., m$). We say that objects i and j are (f_k, f_l)-co-monotone, if $f_k(i) > f_k(j)$ if and only if $f_l(i) > f_l(j)$; $f_k(i) = f_k(j)$ if and only if $f_l(i) = f_l(j)$; and $f_k(i) < f_k(j)$ implies $f_l(i) < f_l(j)$. On Fig. 1, one can see that objects 1 and 2 are (A, B)-co-monotone and not (A, C)-co-monotone. We denote the relation of (k, l)-co-monotonicity as ε_{kl}. Clearly, ε_{kl} is an equivalence relation and, moreover, $\varepsilon_{kl} = \varepsilon_{lk}$.

Consider a permutation of features $p = (f_1, f_2, ..., f_m)$ and take the pairs of neighboring features in p, (f_u, f_{u+1}), $u = 1, 2,..., m - 1$. It is not difficult to prove the following statement.

Theorem 1. Objects are p-equivalent if and only if they are (f_u, f_{u+1})-co-monotone for all $u = 1,..., m - 1$.

Proof. Assume that i and j are p-equivalent. Then i and j are (f_u, f_{u+1})-co-monotone for all $u = 1,..., m - 1$ because of the very definition of p-equivalence. Now assume, conversely, that i and j are (f_u, f_{u+1})-co-monotone for all $u = 1,..., m - 1$. That means that $f_u(i) > f_u(j)$ if $f_{u+1}(i) > f_{u+1}(j)$; $f_u(i) = f_u(j)$ if $f_{u+1}(i) = f_{u+1}(j)$; and $f_u(i) < f_u(j)$ implies $f_{u+1}(i) < f_{u+1}(j)$. Then i and j are p-equivalent because of the definition of p-equivalence. This completes the proof.

Now we can prove the main result.

Theorem 2. The ordinal equivalence relation ε in (1) equals the intersection of (k, l)-co-monotonicity relations ε_{kl} for all pairs of variables k, $l = 1, 2,..., m$.

Proof. Consider any permutation $p = (f_1, f_2, ..., f_m)$ of the variables. Then, the relation of p-equivalence is the intersection of relations $\varepsilon_{u,u+1}$ for all $u = 1, 2,..., m - 1$. The proof now follows from the fact that, for any k, $l = 1, 2,..., m$ there exists a permutation p at which k and l are neighbors.

The proven theorem 2 suggests a set of natural algorithms, referred to as ORDEQ, for deriving the overall co-monotonicity relation ε in (1).

Below, we give formulations of two ORDEQ algorithms. Both use the following general subroutine referred to as COMON.

COMON(k, l) Algorithm: "(k, l)-co-monotonicity". This algorithm takes in a subset $S \subset I$ with features f_k and f_l defined over the set and divides S in three parts, one part consisting of those i in S for which $f_k(i) > f_l(i)$, the other of those j in S for which $f_k(j) < f_l(j)$, and the third part consisting of those $i \in S$ for which $f_k(i) = f_l(i)$.

The output is the partition of S corresponding to the equivalence relation ε_{kl}. Of course, some of the three defined classes may be empty and thus omitted from the partition.

Here are outlines of two algorithms for obtaining the ordinal equivalence partition.

ORDEQ-a Algorithm: "Adding features one-by-one". This algorithm begins with a set of features G consisting of any two features f_1 and f_2 over I and applies COMON(1, 2) to obtain a partition of I in three parts according to ε_{12}. On general step, the algorithm adds to G any feature g, which is not in G yet, and, in a loop over features $f \in G$, applies COMON(f, g) to every part of the partition obtained to the current step. Algorithm stops when all the features have been added to G.

ORDEQ-o Algorithm: "Testing an ordering". This algorithm begins with an ordering $p = (f_1, f_2, ..., f_m)$ of the variables and applies COMON(f_u, f_{u+1}), at each $u = 1$, 2,..,$m - 1$, to every part of the partition obtained to the current step. Then, in a double loop over $u = 1, 2,..., m - 2$, $v = u+1$, $u + 2,..., m - 1$, test whether the pair (f_u, f_v) has been applied already; if not, apply COMON(f_u, f_v) to every part of the partition obtained to the current step.

Obviously, both ORDEQ-a and ORDEQ-o produce the ordinal equivalence relation in (1).

3   Application to Real-World Data

Consider Pollen dataset that was under investigation in [3] with respect to much clutter in its parallel coordinate visualization. The dataset refers to 3848 entities characterized by 5 features; its first four entities are presented in Table 2, both in the original form (on the left), and after max-min normalization (on the right).

Table 2. A fragment of Pollen data, both normalized (on the right) and not (on the left).

p1	p2	p3	p4	p5	p1	p2	p3	p4	p5
−2.348	3.631	5.029	10.872	−1.385	0.47	0.60	0.59	0.64	0.47
−1.152	1.480	3.237	−0.594	2.123	0.50	0.53	0.56	0.48	0.62
−2.524	−6.863	−2.804	8.463	−3.413	0.46	0.28	0.46	0.61	0.38
5.752	−6.509	−5.151	4.348	−10.326	0.65	0.29	0.43	0.55	0.07

We consider this set with features normalized: unfortunately, the character of feature co-monotonicity may change when the operation of normalization is applied.

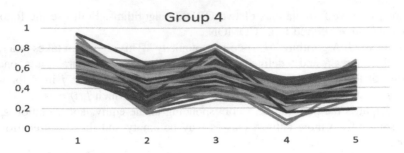

Fig. 2. The set of polylines of the maximal *p*-equivalence class *S* to embrace 559 objects.

The claim that this dataset is complex is supported by the high total number of ordinal equivalence classes, 358, of which 90 are singletons and obvious outliers. The number of ordinal equivalence classes with 50 or more elements is 17 embracing, in total, 1249 entities, which is less than one third of the total. To increase the number of non-outliers, let us specify the natural feature ordering $p = (p1, p2, p3, p4, p5)$ and take the p-equivalence only rather than all orderings possible. Classes of this equivalence are much larger: there are 12 equivalence classes containing 100 or more elements each. These embrace 3343 entities, which is about 87% of the total. The largest *p*-equivalent class *S* is visualized in parallel coordinates in Fig. 2. Two of its ordinal equivalence subclasses are visualized in Fig. 3. Since they all are p-equivalent, to show the difference, we use a different ordering of the axes, $q = (p5, p2, p4, p1, p3)$, in Fig. 3.

Table 3. A fragment of the Athlete data from [1]. Out of 15 features, the last six are taken to expose the main clutter according to Fig. 4 in [1]. All the rows here are ordinally equivalent.

Athlete	Med Ball(ft)	40yd (sec)	Sit-ups/min,	% Body Fat	Sit-n-Reach(in,)	Wt
1	35.6	4.75	61	18	20.25	167.4
3	34	4.85	50	13.8	21.5	164.4
5	25	5.58	50	11.7	19.75	130.4
7	30.6	4.3	57	11.9	18.25	162

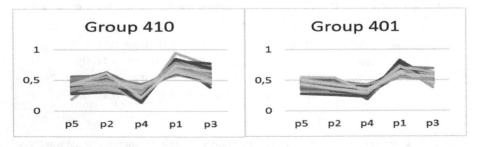

Fig. 3. Parallel coordinate profiles of two ordinal equivalence classes, Groups 410 and 401, both being subclasses of Group 4 from Fig. 2. The difference between Groups 410 and 401 is in the character of (p5, p2)-co-monotonicity: at Group 410, all feature values grow from p5 to p2, whereas they decrease from p5 to p2 at Group 401.

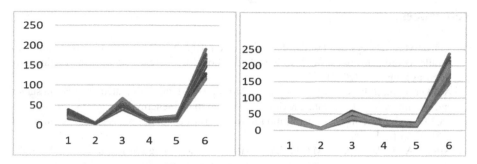

Fig. 4. Two ordinal equivalence classes in Athlete data from [1]; they differ in the direction of the lines between feature axes 4 and 5 – increasing on the left, decreasing on the right.

Consider one more example of a cluttered visualization – the data of 62 athletes from [1]; a fragment of the data is presented in Table 3. This data form two ordinal equivalence classes of 22 and 26 elements presented in Fig. 4. The second ordinal class is part of a p-equivalence class embracing 37 elements, where p is the order of features in Table 3 and Fig. 4. Therefore, only 3 athletes are outliers here, each forming a singleton of the ordinal equivalence.

4   Conclusion

This paper gives a mathematical treatment to the concept of ordinal equivalence introduced in [13, 14]. It is proven that the ordinal equivalence, defined relative to all $m!$ possible permutations of the parallel axes, is in fact determined by the pair-wise co-monotonicity equivalence relations, thus leading to simple algorithmic procedures for finding the corresponding partition. Each ordinal equivalence class can be visualized as a profile of co-monotone polylines, in this way preventing any clutter at the image. As the number of "outliers", that are small ordinal equivalence classes, can be rather large, we propose visualize the data by using classes of p-equivalence, where p is a single order of parallel axes. These classes are much larger, still with no clutter whatsoever.

As it happens, the concept of ordinal equivalence has something to do with the level of convenience in using parallel coordinates for data visualization: the greater the number of ordinal equivalence classes, the greater the clutter. The celebrated Iris dataset (see, for example, [12]), for example, has only two ordinal equivalence classes; one consisting of the first Iris taxon, the second, of the rest – which once more illustrates its attractiveness for visualization.

Acknowledgment. The authors acknowledge continuing support by the Academic Fund Program at the National Research University Higher School of Economics (grant №19-04-019 in 2018-2019) and by the International Decision Choice and Analysis Laboratory (DECAN) NRU HSE, in the framework of a subsidy granted to the HSE by the Government of the Russian Federation for the implementation of the Russian Academic Excellence Project "5-100".

References

1. Akbar, M.S., Gabrys, B.: Data analytics enhanced data visualization and interrogation with parallel coordinates plots. In: 2018 26th International Conference on Systems Engineering (ICSEng), pp. 1–7 (2018)
2. Aleskerov, F., Belousova, V., Egorova, L., Mirkin, B.: Methods of pattern analysis in statics and dynamics, part 1: Literature review and definition of the term. Bus. Inform.3 (25), 3–18 (2013). Part 2: Examples of application to real-world economics processes. Bus. Inform. 4 (26), 3–20 (in Russian)
3. Artero, A.O., de Oliveira, M.C.F., Levkowitz, H.: Uncovering clusters in crowded parallel coordinates visualizations. In: IEEE Symposium on Information Visualization, pp. 81–88 (2004)
4. Dang, T.N., Wilkinson, L., Anand, A.: Stacking graphic elements to avoid over-plotting. IEEE Trans. Vis. Comput. Graph. 16(6), 1044–1052 (2010)
5. Ellis, G., Dix, A.: A taxonomy of clutter reduction for information visualisation. IEEE Trans. Vis. Comput. Graph. 13(6), 1216–1223 (2007)
6. Ferdosi, B.J., Tarek, M.M.: Visual verification and analysis of outliers using optimal outlier detection result by choosing proper algorithm and parameter. In: Abraham, A., Dutta, P., Mandal, J., Bhattacharya, A., Dutta, S. (eds.) Emerging Technologies in Data Mining and Information Security. AISC, vol. 813, pp. 507–517. Springer, Singapore (2019). https://doi.org/10.1007/978-981-13-1498-8_45
7. Heinrich, J., Weiskopf, D.: State of the art of parallel coordinates. In: Eurographics (STARs), pp. 95–116 (2013)
8. Hismutova, A., Yatskiv, I.: Clutter reduction in parallel coordinates using aesthetic criteria. Front. Artif. Intell. Appl. 312, 81–100 (2019)
9. Johansson, J., Forsell, C.: Evaluation of parallel coordinates: overview, categorization and guidelines for future research. IEEE Trans. Vis. Comput. Graph. 22(1), 579–588 (2015)
10. Liu, S., Maljovec, D., Wang, B., Bremer, P.T., Pascucci, V.: Visualizing high-dimensional data: advances in the past decade. IEEE Trans. Vis. Comput. Graph. 23(3), 1249–1268 (2016)
11. Lu, L.F., Huang, M.L., Zhang, J.: Two axes re-ordering methods in parallel coordinates plots. J. Vis. Lang. Comput. 33, 3–12 (2016)
12. Mirkin, B.: Clustering: A Data Recovery Approach, Chapman and Hall/CRC (2012)
13. Myachin, A.: Analysis of global data education and patent activity using new methods of pattern analysis. Procedia Comput. Sci. 31, 468–473 (2014)
14. Myachin, A.L.: Pattern analysis: ordinal-invariant pattern clustering. Upravlenie Bol'shimi Systemami 61, 41–59 (2016). (in Russian)
15. Nguyen, H.: Rosen, P: DSPCP: a data scalable approach for identifying relationships in parallel coordinates. IEEE Trans. Vis. Comput. Graph. 24(3), 1301–1315 (2017)
16. Novotny, M., Hauser, H.: Outlier-preserving focus+ context visualization in parallel coordinates. IEEE Trans. Vis. Comput. Graph. 12(5), 893–900 (2006)
17. Van Long, T., Linsen, L.: Efficient reordering of parallel coordinates and its application to multidimensional biological data visualization. In: Linsen, L., Hamann, B., Hege, H.-C. (eds.) Visualization in Medicine and Life Sciences III. MV, pp. 309–328. Springer, Cham (2016). https://doi.org/10.1007/978-3-319-24523-2_14
18. Wang, J., Liu, X., Shen, H.W., Lin, G.: Multi-resolution climate ensemble parameter analysis with nested parallel coordinates plots. IEEE Trans. Vis. Comput. Graph. 23(1), 81–90 (2017)

19. Weissenbock, J., et al.: Fiberscout: an interactive tool for exploring and analyzing fiber reinforced polymers. In: Visualization Symposium (PacificVis), 2014 IEEE Pacific, pp. 153–160 (2014)
20. Zhang, C., Chen, Y., Yang, J., Yin, Z.: An association rule-based approach to reducing visual clutter in parallel sets. Vis. Inf. **3**(1), 48–57 (2019)
21. Zhou, H., Yuan, X., Qu, H., Cui, W., Chen, B.: Visual clustering in parallel coordinates. Comput. Graph. Forum **27**(3), 1047–1054 (2008)
22. Zhou, H., Xu, P., Yuan, X., Qu, H.: Edge bundling in information visualization. Tsinghua Sci. Technol. **18**(2), 145–156 (2013)

New Internal Clustering Evaluation Index Based on Line Segments

Juan Carlos Rojas Thomas[1] and Matilde Santos Peñas[2(✉)]

[1] Departamento de Informática y Automática, UNED, Madrid, Spain
correorojas@gmail.com
[2] Facultad de Informática, Universidad Complutense de Madrid, Madrid, Spain
msantos@ucm.es

Abstract. This work proposes a new internal clustering evaluation index, based on line segments as central elements of the clusters. The data dispersion is calculated as the average of the distances of the cluster to the respective line segment. It also defines a new measure of distance based on a line segment that connects the centroids of the clusters, from which an approximation of the edges of their geometries is obtained. The proposed index is validated with a series of experiments on 10 artificial data sets that are generated with different cluster characteristics, such as size, shape, noise and dimensionality, and on 8 real data sets. In these experiments, the performance of the new index is compared with 12 representative indices of the literature, surpassing all of them. These results allow to conclude the effectiveness of the proposal and shows the appropriateness of including geometric properties in the definition of internal indexes.

Keywords: Clustering evaluation · Internal indexes · Geometry · Line segment

1 Introduction

The process of clustering consists of classifying in an unsupervised manner a set of samples or data in sets [1]. The goal of a clustering algorithm is to perform a partition where objects within a cluster are similar and objects grouped in other clusters are dissimilar [2]. An important issue in the cluster analysis is to find the partition that best fits the underlying structure of the data [3]. One of the most used techniques for this purpose is the internal evaluation indexes. These indexes require no extra information about the data, as they are only based on the properties of the clusters themselves, such as compaction, roundness, and separation [4]. Most indices combine these concepts of compaction and separation, either as a ratio or as a weighted sum. Both measures are usually calculated by the distance between pairs of data or between representative points (centroids). The first makes the indexes very sensitive to noise, and the second makes these indexes only suitable for spherical clusters.

To overcome these disadvantages, this work proposes the use of geometric concepts to define an internal index. In particular, we draw a line segment that captures the length of the cluster in its maximum variance dimension. Then, the cluster dispersion is calculated as the average of the distances of its data to the segment. To measure the distances between two clusters, the new index calculates a representative point of the

© Springer Nature Switzerland AG 2019
H. Yin et al. (Eds.): IDEAL 2019, LNCS 11871, pp. 534–541, 2019.
https://doi.org/10.1007/978-3-030-33607-3_57

edge of each cluster in the direction of the line segment that connects both centroids. This is done through a procedure that selects the closest data using a hypersphere centered at the midpoint of that segment.

The rest of the paper is organized as follows. Section 2 summarize some related works. In Sect. 3 the new proposal is explained. Section 4 presents the experimental methodology and discusses the results. The paper ends with the conclusions and future work.

2   Related Works

As a measure of compaction, the indices CH [5], I [6], Davies-Bouldin [7] and XB [8] use the distance from the data to the centroids of each cluster. Index I also includes a measure of global compaction with respect to the centroid of the entire data set. Another widely used concept is the diameter of the cluster, defined as the maximum distance between two pairs of data belonging to the same cluster. The Dunn index [9], in its original version, uses this measure, and the CS index [10] calculates the average of the relative diameters of each cluster data. Another approach, used by the C [11] and Silhouette [12] indexes, considers the distances between all the data pairs in a cluster. Finally, the BHG index [13] counts the number of times a pair of data in the same cluster has a distance greater than a pair of data belonging to different clusters.

Regarding separation, the CH, I, Davies-Bouldin, SD [14], XB and CS indexes use the distances between centroids, while the Dunn index uses the closest pair of data between two clusters. The Silhouette index calculates all distances from the data to those of the nearest cluster. The C index makes a selection of the closest and furthest data pairs in the set, regardless of the membership cluster. Finally, the BHG index counts the number of times that a pair of data in the same cluster are at a distance smaller than that of a pair of data belonging to different clusters.

3   New Proposed Internal Clustering Validation Index

The new index is based on the calculation of the cluster dispersion and the distance between the clusters, regarding the line segment we define in the maximum variance dimension of the cluster.

3.1   Cluster Dispersion

The cluster dispersion is calculated as the average distance of all data to a central line segment which extends over the maximum variance vector of the cluster, inspired by [15]. To do so, first the central line segment is obtained and then the average distance of the cluster data to this segment is calculated. Let $C_k = \{p_1, p_2,..,p_n\}$ be the k-th cluster with n data points, $c_k = \{x_1^c, x_2^c,..,x_n^c\}$ its centroid, and v^{max} its maximum variance vector. The distance vectors of the cluster data with respect to the centroid are calculated. Let p_i be a data point belonging to the cluster C_k. Its associated distance vector d_i with respect to the centroid c_k is:

$$\vec{d_i} = \vec{p_i} - \vec{c_k} \tag{1}$$

The C_k cluster is then divided by a hyperplane that passes through the centroid c_k perpendicular to the v_{max} vector, generating two subsets, C_k^+ and C_k. To determine which data belong to each of these subsets, the dot product is used:

$$C_k^+ = \left\{ p_i, p_i \in C_k \wedge \vec{d_i} \bullet \vec{v}^{max} > 0 \right\} \quad C_k^- = \left\{ p_i, p_i \in C_k \wedge \vec{d_i} \bullet \vec{v}^{max} < 0 \right\} \tag{2}$$

The lengths of the central line segment on each side of the hyperplane, L^+ and L^-, are calculated based on the projection averages:

$$L^+ = \frac{\sum_{i=1}^{|C_k^+|} \left| \vec{d_i} \bullet \vec{v}^{max} \right|}{|C_k^+|}, \vec{d_i} \in C_k^+ \quad L^- = \frac{\sum_{i=1}^{|C_k^-|} \left| \vec{d_i} \bullet \vec{v}^{max} \right|}{|C_k^-|}, \vec{d_i} \in C_k^- \tag{3}$$

Finally, the coordinates of the ends of the central line segment, $e^+ = \{x_1^+, x_2^+, ..., x_n^+\}$ and $e^- = \{x_1^-, x_2^-, ..., x_n^-\}$, are calculated as:

$$x_i^+ = L^+ \cdot x_i^{max} + x_i^c \qquad x_i^- = L^- \cdot x_i^{max} + x_i^c \tag{4}$$

Where x_i^{max} corresponds to the *i-th* coordinate of the maximum variance vector v^{max}. A general representation of the process is shown in Fig. 1.

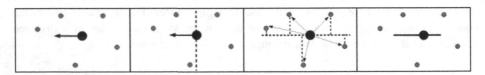

Fig. 1. From left to right, the process of obtaining the central line segment: centroid and maximum variance vector of a cluster, division of the cluster by a hyper-plane, projections of the distance vectors, and finally the length and coordinates of the line segment.

A triangle is built, where the data and the extreme points of the central line segment are its vertices. Its sides are defined as:

$$a = (p_i, e^+), b = (p_i, e^-), c = (e^-, e^+) \tag{5}$$

Let α, β and γ be the internal angles of the triangle formed by the intersections of the sides b-c, c-a and a-b respectively (Fig. 2, left). If the cosine theorem is applied, the distance between a point p_i and the center line segment can be estimated as shown in Fig. 2 (right). If the angles of the base are both smaller than 90°, the distance corresponds to the height of the triangle (middle). Otherwise the distance is the length of the side adjacent to the angle (right).

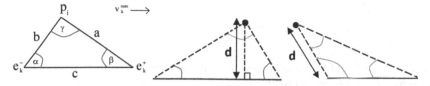

Fig. 2. Triangle between a data and the central line segment (left) and distance between a data (black circle) and the central line segment (triangle base), (middle and right).

$$d(p_i, c) = \begin{cases} b \cdot \sqrt{1 - \cos(\alpha)^2}, & \cos(\alpha) > 0 \wedge \cos(\beta) > 0 \\ b, & \cos(\alpha) \le 0 \\ a, & \cos(\beta) \le 0 \end{cases} \tag{6}$$

Finally, the dispersion of the C_k cluster is defined as:

$$dispersion(C_k) = \frac{\sum\limits_{i=1}^{|C_k|} d(p_i, sg_k)}{|C_k|} \tag{7}$$

Where sg_k corresponds to the central line segment of the C_k cluster.

3.2 Distance Between Clusters

The "spherical distance" is defined to obtain the distance between two pairs of clusters, using a line segment that connects the centroids and a hypersphere built around the midpoint of this segment. The goal is to obtain representative points of the edges of both clusters and then compute the Euclidean distance between both points. The whole process is shown in Fig. 3. Let c_i and c_j be the centroids of clusters C_i and C_j, respectively. Let pm_{ij} be the midpoint between the two, defined as:

$$pm_{ij} = \frac{c_i + c_j}{2} \tag{8}$$

The radius of the hypersphere centered on pm_{ij} is:

$$r_{ij} = \frac{d(c_i, c_j)}{2} \tag{9}$$

The data of both clusters that are within the hypersphere are selected, grouping them into two subsets, S_i and S_j, according to the following expressions:

$$S_i = \{p_i, p_i \in C_i \wedge d(p_i, pm_{ij}) \le r_{ij}\} \quad S_j = \{p_i, p_i \in C_j \wedge d(p_i, pm_{ij}) \le r_{ij}\} \tag{10}$$

The unit vector v_{ij} is calculated from centroid c_i to the centroid c_j and vice versa (v_{ji}):

$$\vec{v}_{ij} = \frac{\vec{c}_j - \vec{c}_i}{\|\vec{c}_j - \vec{c}_i\|} \qquad \vec{v}_{ji} = -\vec{v}_{ij} \qquad (11)$$

For each subset S_i and S_j, the averages of the projections of the distance vectors on the unit vectors v_{ij} and v_{ji}, respectively, L_i and L_j, are obtained:

$$L_i = \frac{\sum_{i=1}^{|S_i|} |\vec{d}_i \bullet \vec{v}_{ij}|}{|S_i|}, \ p_i \in S_i \qquad L_j = \frac{\sum_{i=1}^{|S_j|} |\vec{d}_i \bullet \vec{v}_{ji}|}{|S_j|}, \ p_i \in S_j \qquad (12)$$

Let the coordinates of the edges of the clusters C_i and C_j, b^i and b^j, be defined as:

$$b^i = \left\{ x_1^{bi}, x_2^{bi}, \ldots, x_n^{bi} \right\} \qquad b^j = \left\{ x_1^{bj}, x_2^{bj}, \ldots, x_n^{bj} \right\} \qquad (13)$$

The unit vectors v_{ij} and v_{ji} are defined as:

$$\vec{v}_{ij} = \left\{ x_1^{vij}, x_2^{vij}, \ldots, x_n^{vij} \right\} \qquad \vec{v}_{ji} = \left\{ x_1^{vji}, x_2^{vji}, \ldots, x_n^{vji} \right\} \qquad (14)$$

Then, the i-th coordinates of the representative points b_i and b_j are calculated as:

$$x_i^{bi} = L_i \cdot x_i^{vij} + x_i^{ci} \qquad x_i^{bj} = L_j \cdot x_i^{vji} + x_i^{cj} \qquad (15)$$

Where x_i^{ci} corresponds to the i-th coordinate of the centroid c_i, and x_i^{cj} is the i-th coordinate of centroid c_j. Finally, the distance between clusters C_i and C_j is obtained as the Euclidean distance between b_i and b_j, i.e.:

$$separation(C_i, C_j) = d(b_i, b_j) \qquad (16)$$

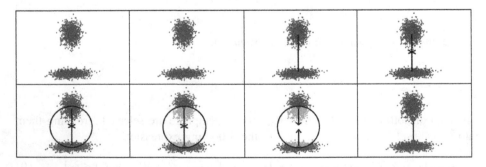

Fig. 3. From top to bottom and from left to right, the process of calculating the distance between two clusters: initial clusters, centroids of the clusters, line segment between both centroids, midpoint of the segment, mid-centered hypersphere, selection of data contained in the hypersphere, unit vectors over the segment and finally, the distance between the representative points.

3.3   Definition of the New Index

Let $D = \{p_1, p_2,.., p_n\}$ be a set that contains n data points. Let $P = \{C_1, C_2,.., C_k\}$ be a partition of set D in k clusters. The following expression is calculated for each i-th cluster:

$$F(C_i) = \frac{dispersion(C_i)}{MIN\left(separation\left(C_i, C_j\right)\right)}, \forall C_j \in P, i \neq j \qquad (17)$$

The new index, called SG (segment), is defined as:

$$SG = \frac{\sum_{i=1}^{k} F(C_i)}{k} \qquad (18)$$

The partition that minimizes this evaluation index will be considered the best one.

4   Methodology, Experiments and Results

The *k-means* clustering algorithm is used to obtain the partitions. For each dataset up to 14 partitions were generated, starting with a minimum of 2 clusters up to a maximum of 15 clusters. The target partition is directly included in the set of partitions to be evaluated by the indexes, instead of generating this partition with the clustering algorithm. For each data set, the number of clusters selected by each internal index was recorded.

The average error (difference, in absolute terms, between the number of clusters found by the index (*prediction*) and the target number of clusters (*target*) for the n data sets), was used to compare the performance of the indices,

$$avg\_error = \frac{\sum_{i=1}^{n} |target - prediction|}{n} \qquad (19)$$

4.1   Materials: Artificial and Real Datasets

Artificial datasets were generated using a pseudo-random generator with a standard distribution to create the basic clusters, that were later stretched, rotated and shifted. The following 10 artificial datasets were generated (Fig. 4): *compact* (9 clusters, 2 features and 450 objects), noisy (9 clusters, 2 features and 765 objects), *parallel* (8 clusters, 2 features and 400 objects), *growing* (8 clusters, 2 features and 300 objects), *intercalated* (7 clusters, 2 features and 380 objects), *satellites* (7 clusters, 2 features and 400 objects), *subsets* (6 clusters, 2 features and 600 objects), *random* (8 clusters, 2 features and 700 objects), *3-dimensional* (4 clusters, 3 features and 330 objects) and *4-dimensional* (4 clusters, 4 features and 330 objects).

We worked with 8 public real datasets [16]: *Iris* (3 classes, 4 features and 150 objects), *Wine* (3 classes, 13 features and 178 objects), *Ecoli* (8 classes, 7 features and 336 objects), *Breast Tissue* (6 classes, 9 features and 106 objects), *Seeds* (3 classes, 7 features and 210 objects), *Statlog Vehicle* (4 classes, 18 features, 846 objects), *Diabetic* (2 classes, 20 features, 1151 objects), and *User Knowledge* (4 classes, 5 features, 403 objects). Previously, each feature of the data was normalized to the range 0–100.

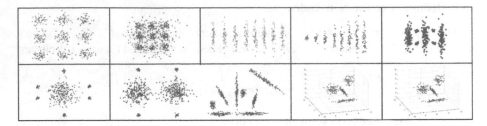

Fig. 4. From top to bottom and from left to right: compact, noisy, parallel, growing, intercalated, satellites, subsets, random, 3-dimentional and 4-dimentional (projection).

4.2 Results

The well-known internal indices used in this study for comparison purposes are: Dunn, G+, CS, C, BHG, SF, CH, Davies-Bouldin, I, XB, Silohuette and SD. The experiments results on both series of experiments (artificial and real datasets) are shown in Fig. 5.

As it can be observed, the performance of the proposed index (SG) is always the best, both on artificial and real data sets. Indeed, in 9 out of the 10 artificial sets this index obtained the smallest possible error (0), much lower than the average error rate of the rest of the internal indices (Fig. 6). Only on the "noisy" datasets did it give a positive error rate (3), slightly higher than the average of the other indices. As for the real sets, the error rate was always much lower than the other indexes, though quite near for the "Ecoli" set, but always lower (Fig. 6, right).

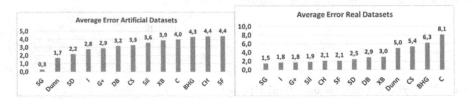

Fig. 5. Average error of the internal indices from lowest to highest, on artificial data sets (left) and real data sets (right).

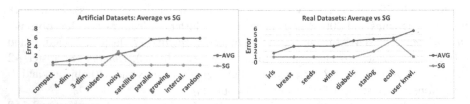

Fig. 6. Error rate of the proposed index (SG) vs. the average of the other internal indices (AVG), artificial (left) and real (right) datasets, sorted in ascending order according to the latter value.

These good results of the SG index may be due to the fact that this index is able to capture and take into account the geometry of the clusters, and it works with better measures of the dispersions and distances. The indices of the literature, on the other hand, do not consider the geometric interpretation of the data, but just statistical measures from single data.

5 Conclusions and Future Works

A new internal clustering evaluation index has been proposed. It considers the geometry of the clusters to enhance the dispersion and separation measures of clusters. It has been validated on 10 artificial datasets and on 8 real datasets, showing a better performance that 12 well-known internal indexes.

As future work, what kind of clustering approaches would be more likely to have favourable index values could be analysed and also other geometric figures [17].

References

1. Jain, A.K., Murty, M.N., Flynn, P.J.: Data clustering: a review. ACM Comput. Surv. **31**(3), 264–323 (1999)
2. Arbelaitz, O., Gurrutxaga, I., Muguerza, J., Pérez, J.M., Perona, I.: An extensive comparative study of cluster validity indices. Pattern Recogn **46**(1), 243–256 (2013)
3. Rojas-Thomas, J.C., Santos, M., Mora, M.: New internal index for clustering validation based on graphs. Expert Syst. Appl. **86**, 334–349 (2017)
4. Brun, M., et al.: Model-based evaluation of clustering validation measures. Pattern Recogn. **40**(3), 807–824 (2007)
5. Caliński, T., Harabasz, J.: A dendrite method for cluster analysis. Commun. Stat. Theory Methods **3**(1), 1–27 (1974)
6. Maulik, U., Bandyopadhyay, S.: Performance evaluation of some clustering algorithms and validity indices. IEEE Trans. Pattern Anal. Mach. Intell. **24**(12), 1650–1654 (2002)
7. Davies, D., Bouldin, D.: A cluster separation measure. IEEE PAMI **1**(2), 224–227 (1979)
8. Xie, S.L., Beni, G.: A validity measure for fuzzy clustering. IEEE Trans. Pattern Anal. Mach. Intell. **8**, 841–847 (1991)
9. Dunn, J.: Well separated clusters and optimal fuzzy partitions. J. Cybern **4**(1), 95–104 (1974)
10. Chou, C-H., Mu-Chun S., Lai, E.: A new cluster validity measure for clusters with different densities. In: IASTED International Conference on Intelligent Systems and Control (2003)
11. Hubert, L.J., Levin, J.R.: A general statistical framework for assessing categorical clustering in free recall. Psychol. Bull. **83**(6), 1072 (1976)
12. Rousseeuw, P.J.: Silhouettes: a graphical aid to the interpretation and validation of cluster analysis. J. Comput. Appl. Math. **20**, 53–65 (1987)
13. Baker, F.B., Hubert, L.J.: Measuring the power of hierarchical cluster analysis. J. Am. Stat. Assoc. **70**, 31–38 (1975)
14. Halkidi, M., Batistakis, Y., Vazirgiannis, M.: Clustering validity checking methods: part II. ACM Sigmod Rec. **31**(3), 19–27 (2002)
15. Thomas, J.C.R.: A new clustering algorithm based on k-means using a line segment as prototype. In: San Martin, C., Kim, S.-W. (eds.) CIARP 2011. LNCS, vol. 7042, pp. 638–645. Springer, Heidelberg (2011). https://doi.org/10.1007/978-3-642-25085-9_76
16. Dua, D. Graff, C.: UCI Machine Learning Repository. University of California, School of Information and Computer Science, Irvine, CA (2019). http://archive.ics.uci.edu/ml
17. Rojas-Thomas, J.C., Santos M., Mora, M., Duro, N.: Performance analysis of clustering internal validation indexes with asymmetric clusters. IEEE Lat. Am. Trans. (5) (2019, in press)

Threat Identification in Humanitarian Demining Using Machine Learning and Spectroscopic Metal Detection

Wouter van Verre[✉], Toykan Özdeğer, Ananya Gupta, Frank J. W. Podd, and Anthony J. Peyton

University of Manchester, Oxford Road, Manchester M13 9PL, UK
wouter.vanverre@manchester.ac.uk

Abstract. The detection of buried minimum-metal anti-personnel landmines is a time-consuming problem, due to the high false alarm rate (FAR) arising from metallic clutter typically found in minefields. Magnetic induction spectroscopy (MIS) offers a potential way to reduce the FAR by classifying the metallic objects into threat and non-threat categories, based on their spectroscopic signatures. A new algorithm for threat identification for MIS sensors, based on a fully-connected artificial neural network (ANN), is proposed in this paper, and compared against a classifier based on Support Vector Machines (SVM). The results demonstrate that MIS is a potentially viable option for the reduction of false alarms in humanitarian demining. It is also shown that the ANN outperforms the SVM-based approach for threat objects containing minimal amounts of metal.

Keywords: Magnetic induction spectroscopy · Machine learning · Landmine detection

1 Introduction

The detection, and subsequent removal, of buried minimum-metal anti-personnel landmines is a slow and expensive process, in large part due to the high false-alarm rate (FAR) [5]. These false alarms are often caused by the large number of metallic clutter items which are found minefields. Typically the number of metallic clutter items can exceed one-hundred for every landmine found, and potentially be as high as three-hundred per landmine [19].

To alleviate the FAR problem, dual-modality detection systems, combining metal detection (MD) and ground-penetrating radar (GPR), such as MINEHOUND [4], HSTAMIDS [2], and ALIS [16], have been developed and evaluated in field trials in areas such as Afganistan, Cambodia and Angola. During these field trials, rejection rates of up to 95% have been reported by MINEHOUND [4] and HSTAMIDS [18], and up to 77% for ALIS [16].

Supported by the Sir Bobby Charlton Foundation and EPSRC.

The GEM-3 sensor [22], developed by Geophex, aims to deal with the clutter problem through the use of magnetic induction spectroscopy (MIS). The GEM-3 detector has typically been used for the detection of unexploded ordnance (UXO) and other forms of explosive remnants-of-war (ERW) [9], rather than minimum-metal AP landmines.

However, subsequent work demonstrated the viability of the spectroscopic approach to the landmine problem [8,21]. In a further development, Geophex have explored the concept of combining MIS and GPR into a dual-modality detector, named the GEM-3M [6]. This idea has also been investigated by Knox, Rundel and Collins [10].

Simultaneously, work has been ongoing to develop a prototype dual-modality landmine detector employing both MIS and GPR sensors [13,14]. Previous work has been published on the development of an algorithm for the detection of metallic objects using MIS [20]. This algorithm is limited to the detection of buried metallic objects, and does not perform any classification. It combines traditional digital signal processing and a modified linear regression algorithm.

Neural network based methods have shown promise in a number of fields, from medical analysis and computer vision to robotic control [7,12]. In this paper, an analysis of the performance of classical machine learning based methods is presented, and the use of an artificial neural network (ANN) is proposed for the challenging task of identifying buried landmines using MIS.

Neural networks have previously been used in landmine detection, for example in the data-fusion between a metal detector and a ground-penetrating radar [17]. They have also been used in target recognition in data from GPR-only sensors [11].

It is important to note that UN International Mine Action Standards (IMAS), from which the National Mine Action Standards (NMAS) are derived, require the complete removal of all metallic objects, including clutter, from the ground, down to a depth of 13–15 cm. As such it is necessary for landmine detector systems to separately identify both metallic clutter and threat objects such as landmines. The metallic clutter can then be excavated quickly, while extreme care can be taken with the suspected threat objects.

In this paper a new method has been proposed to aid in the discrimination between innocuous metallic clutter items and threat objects such as landmines, using spectroscopic metal detection. This builds on our previous work where we first introduced the sensor hardware [14] and subsequently developed an algorithm for object detection, without threat discrimination [20].

2   Experimental Methodology

2.1   Data Collection

A large dataset was collected using the dual-modality detector, containing measurements of 13 buried objects or empty ground, in three different soil types, at three different depths. The soil type varied from non-mineralised to heavily mineralised. The sensor head and processing electronics have been described in previous work [14], and the experimental setup has also been described previously [20].

The handheld detector was passed over the object and the measurements from both the MIS and GPR sensor were recorded on a laptop. At least twelve sweeps were recorded for each combination of object, depth of cover and soil type.

The aim of the algorithm presented in this paper is to correctly classify each object into one of three classes: no metal (NM), innocuous metal (M) or threat (T). Table 1 lists all the buried objects, as well as the total number of sweeps recorded for that object type, for all soils and depths.

As described in previous work [14], the MIS sensor generates a primary time-varying magnetic field using one coil, and a second coil is used to record the response signal. The excitation signal contains five frequency components, and the system measures the in-phase and quadrature phase responses at each of these frequencies. The output frequency of the demodulated data is approximately 120 Hz.

Figure 1 shows an example of the data collected by the MIS sensor at each timepoint. The magnitudes of the measured in-phase (I) and quadrature phase (Q) are shown for four example objects; the two landmines (a PMA2 and a PMA3) and two metallic clutter objects (a 1 Euro coin and a ring pull). The different objects have different spectral shapes, which can be used to classify them [8].

Due to the large difference in magnitude of the response between the small threat objects and the typical clutter objects, the data acquisition system in the MIS sensor needs to have a large dynamic range. A further challenge exists due to the relatively large magnitude of the response to heavily mineralised soils.

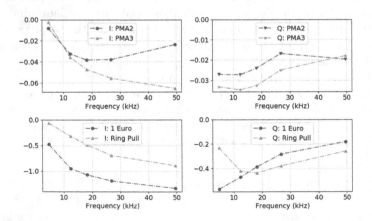

Fig. 1. Example of typical MIS in-phase (I) and quadrature phase (Q) measurements for four objects, taken in air at a distance of 5 cm.

The input to the identification algorithm will be a block of data containing 60 consecutive MIS samples (approximately 0.5 s). For those scans over metallic and threat objects the blocks are extracted from the sweep when the detector was close to the objects. For the remaining scans the blocks are extracted throughout the scan. In total 27,619 blocks were extracted, and Table 1 shows how many blocks were extracted for every type of object.

Table 1. Data about the buried objects

Object	Class	Sweeps	Blocks	Total
Empty	NM	174	3,306	
Wood S	NM	122	2,318	
Wood M	NM	126	2,394	
Rock 1	NM	106	2,014	
Rock 2	NM	118	2,242	12,274
10 Cent	M	111	1,665	
1 Euro	M	114	1,710	
Ball Bearing	M	108	1,620	
Bullet M	M	105	1,575	
Bullet L	M	110	1,650	
Ring Pull	M	110	1,650	
Mortar Tail	M	110	1,650	11,520
PMA2	T	122	1,830	
PMA3	T	133	1,995	3,825
Total		1,669	27,619	

2.2 Processing Algorithm

Digital signal processing is applied to the recorded data to improve the signal quality. First a low-pass filter is applied, with a cutoff frequency of 0.75 Hz, to reduce high-frequency noise. Then the complex-valued response at each frequency is rotated such that the response from the ground lies along the in-phase axis. Finally a high-pass filter is applied, with a cutoff frequency of 0.005 Hz, to remove any potential drift from the sensor response.

In this paper the performance of a fully-connected ANN will be compared to a baseline level of performance given by an SVM. A further point of study is the effect of dimensionality reduction of the input to the classifier. For this, the dimension of the input data is reduced to 100 points using Scalar Value Decomposition (SVD), using the TruncatedSVD algorithm in *scikit-learn* [15].

The SVM (from *scikit-learn*) uses a linear kernel with a penalty parameter, C, of 1.0 and the class weights are chosen to balance the three classes (NM, M, T). A linear SVM was chosen because it offers a good balance between classification performance and computational complexity. The fully-connected neural network contains an input layer with 50 nodes, 3 hidden layers and output layer with three output nodes. The first hidden layer contains 25 nodes, the second hidden layer 20 nodes and the final hidden layer contains 10 nodes. The input layer is followed by a dropout layer with a dropout rate of 0.80, the first hidden layer is followed by a dropout layer with a rate of 0.60 and the second hidden layer by a dropout layer with a rate of 0.40. There is no dropout layer following the final hidden layer.

The Adam optimizer was used during training, with the categorical cross-entropy loss function. The network was trained for 200 epochs. The ANN was developed with Keras [3] using the TensorFlow backend [1]. The class weights were set to balance out the uneven sizes of the classes in the dataset, using functionality built into Keras.

3   Results

The performance of the classifiers is evaluated using confusion matrices and one-vs-rest receiver operating characteristic (ROC) curves. In these ROC curves the multi-class classification problem is broken up into multiple binary classification tasks. This type of ROC curve demonstrates how well a single class can be classified against all remaining classes.

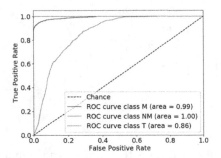

Fig. 2. ROC curve using SVM

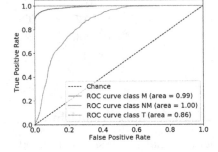

Fig. 3. ROC curve using SVD & SVM

Figure 2 shows the baseline threat identification performance which can be achieved with an MIS sensor using the traditional SVM algorithm. The classifier works well for the detection of metal objects, but does not identify threat objects as well. From Fig. 3 it can be seen that the SVD does not change the performance of the SVM classifier. The confusion matrices for these classifiers are shown in Tables 2 and 3.

Table 2. Confusion matrix using SVM

	M	NM	T
M	5671	0	106
NM	56	6047	34
T	654	13	1253

Table 3. Confusion matrix using SVD & SVM

	M	NM	T
M	5669	0	106
NM	55	6048	34
T	649	13	1258

Figure 4 shows the performance of the ANN using a one-vs-rest ROC curve. The ANN has an improved ability to identify threat objects, but this comes at the cost of slightly reduced performance in pure metal detection. The confusion matrix for this classifier is shown in Table 4.

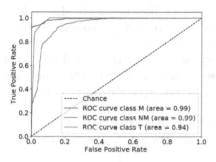

Fig. 4. ROC curve using ANN

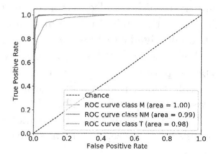

Fig. 5. ROC curve using SVD & ANN

Finally Fig. 5 and Table 5 show the performance of the ANN after SVD has been applied to the input data. This shows a clear improvement over the previous approach using the ANN only. The network's ability to correctly identify objects of all classes has improved, and the identification of threat objects is significantly better than the baseline performance recorded with the SVM.

Table 4. Confusion matrix using ANN

	M	NM	T
M	5537	68	170
NM	0	5956	181
T	619	579	722

Table 5. Confusion matrix using SVD & ANN

	M	NM	T
M	5749	0	26
NM	1	6072	64
T	436	181	1303

4 Discussion and Future Work

It has been shown that MIS can be useful beyond pure object detection, and can be used to identify buried threat objects based on their spectroscopic signatures. The results from this work demonstrate that the fully-connected neural network, containing 3 hidden layers, has an improved ability to identify threats than an SVM classifier. Furthermore, using SVD as a method of dimensionality reduction improves the accuracy in the case of the neural network.

The algorithms were tested on a large dataset containing 1,669 separate passes over 14 objects buried in different soil types and at different depths. This demonstrates that the algorithm is robust to changes in the environment. The final classification algorithm was able to correctly identify threat objects with a true positive rate of 95% at a false positive rate of 12.0%. Even at a true positive rate of 100% the false positive rate was below 45%, this means that up to 55% of false alarm due to innocuous metallic clutter could be reject. Currently the typical false alarm rate is around 100 to 300 per landmine, hence this could significantly speed up landmine clearance.

These results are based on just a single measurement of the buried object. In typical operations the detector would be passed over a suspected buried landmine multiple times, potentially allowing for even better rejection rates as multiple measurements can be combined.

Future work in this area could consider how to extend this algorithm to classify whole sweeps, with a sensitivity that can be chosen based on the ROC curves.

References

1. Abadi, M., et al.: TensorFlow: large-scale machine learning on heterogeneous systems (2015). https://www.tensorflow.org/
2. Amazeen, C., Locke, M.: Developmental status of the U.S. Army's new Handheld STAndoff MIne Detection System (HSTAMIDS). In: Second International Conference on Detection of Abandoned Land Mines, vol. 1998, pp. 193–197 (1998)
3. Chollet, F., et al.: Keras (2015). https://keras.io
4. Daniels, D., Braustein, J., Nevard, M.: Using MINEHOUND in Cambodia and Afghanistan. J. ERW Mine Action **18**(2), 14 (2014)
5. Daniels, D.J.: A review of GPR for landmine detection. Sens. Imaging: Int. J. **7**(3), 90–123 (2006)
6. Geophex: GEM-3M: A Ground Imager with a Local Navigator. Technical report (2012)
7. Goodfellow, I., Bengio, Y., Courville, A.: Deep Learning. MIT Press (2016). http://www.deeplearningbook.org
8. Huang, H., Won, I.J.: Automated identification of buried landmines using normalized electromagnetic induction spectroscopy. IEEE Trans. Geosci. Remote Sens. **41**(3), 640–651 (2003)
9. Huang, H., Won, I.J.: Characterization of UXO-like targets using broadband electromagnetic induction sensors. IEEE Trans. Geosci. Remote Sens. **41**(3), 652–663 (2003)

10. Knox, M., Rundel, C., Collins, L.: Sensor fusion for buried explosive threat detection for handheld data. In: Detection and Sensing of Mines, Explosive Objects, and Obscured Targets XXII 10182, May 2017. 101820D (2017). https://doi.org/10.1117/12.2263013

11. Lameri, S., Lombardi, F., Bestagini, P., Lualdi, M., Tubaro, S.: Landmine detection from GPR data using convolutional neural networks. In: 25th European Signal Processing Conference, EUSIPCO, January 2017, pp. 508–512 (2017). https://doi.org/10.23919/EUSIPCO.2017.8081259

12. LeCun, Y., Bengio, Y., Hinton, G.: Deep learning. Nature **521**(7553), 436 (2015)

13. Marsh, L.A., et al.: Spectroscopic identification of anti-personnel mine surrogates from planar sensor measurements. In: Proceedings of IEEE Sensors, pp. 1–3 (2016)

14. Marsh, L.A., et al.: Combining electromagnetic spectroscopy and ground-penetrating radar for the detection of anti-personnel landmines. Sensors **19**(15), 3390 (2019)

15. Pedregosa, F., et al.: Scikit-learn: machine learning in Python. J. Mach. Learn. Res. **12**, 2825–2830 (2011)

16. Sato, M., Kikuta, K., Chernyak, I.: Dual sensor "ALIS" for humanitarian demining. In: 2018 17th International Conference on Ground Penetrating Radar (GPR), pp. 1–4 (2018)

17. Stanley, R.J., Gader, P.D., Ho, K.C.: Feature and decision level sensor fusion of electromagnetic induction and ground penetrating radar sensors for landmine detection with hand-held units (2002)

18. The Halo Trust: HALO Utilises Dual-sensor Detector — The HALO Trust (2011). https://www.halotrust.org/media-centre/news/halo-utilises-dual-sensor-detector/

19. UN Secretary General: Assistance in mine clearance: Report of the Secretary-General A/49/357. Technical report, United Nations, September 1994

20. van Verre, W., Marsh, L.A., Davidson, J.L., Cheadle, E., Podd, F.J.W., Peyton, A.J.: Detection of Metallic Objects in Mineralised Soil Using Magnetic Induction Spectroscopy (2019, submitted)

21. Won, I.J., Keiswetter, D.A., Bell, T.H., Miller, J., Barrow, B.: Electromagnetic induction spectroscopy for landmine identification. IEEE Trans. Geosci. Remote Sens. **39**(4), 801–809 (2001)

22. Won, I.J., Keiswetter, D.A., Hanson, D.R., Novikova, E., Hall, T.M.: GEM-3: Monostatic Broadband Electromagnetic Induction Sensor (1997)

Author Index

Printed in the United States
By Bookmasters